VSC-FACTS-HVDC

VSC-FACTS-HVDC

Analysis, Modelling and Simulation in Power Grids

Professor Dr Enrique Acha
Laboratory of Electrical Energy Engineering
Tampere University
Tampere, Finland

Dr Pedro Roncero-Sánchez
Department of Electronics
Electrical Engineering and Control Systems
University of Castilla-La Mancha, Spain

Dr Antonio de la Villa Jaén
Department of Electrical Engineering
University of Seville, Spain

Dr Luis M. Castro
Faculty of Engineering
National University of Mexico (UNAM)
Mexico City, Mexico

Dr Behzad Kazemtabrizi
School of Engineering
Durham University, UK

WILEY

This edition first published 2019

Registered Offices
John Wiley & Sons, Inc., 111 River Street, Hoboken, NJ 07030, USA
John Wiley & Sons Ltd, The Atrium, Southern Gate, Chichester, West Sussex, PO19 8SQ, UK

Editorial Office
The Atrium, Southern Gate, Chichester, West Sussex, PO19 8SQ, UK

For details of our global editorial offices, customer services, and more information about Wiley products visit us at www.wiley.com.

Wiley also publishes its books in a variety of electronic formats and by print-on-demand. Some content that appears in standard print versions of this book may not be available in other formats.

Library of Congress Cataloging-in-Publication Data

Names: Acha, Enrique, author.
Title: VSC-FACTS-HVDC : analysis, modelling and simulation in power grids / Professor Dr Enrique Acha, Tampere University, Tampere, Finland, Dr Pedro Roncero-Sanchez, Universidad de Castilla-La Mancha, Ciudad Real, Espana Dr Antonio de la Villa Jan, Universidad de Sevilla, Sevilla, Espana, Dr Luis M Castro, Universidad Nacional Autonoma de Mexico (UNAM), Mexico City, Mexico, Dr Behzad Kazemtabrizi, Durham University, Durham, England.
Description: First edition. | Hoboken, NJ : John Wiley & Sons Ltd, 2019. | Includes bibliographical references and index. |
Identifiers: LCCN 2018051883 (print) | LCCN 2018055480 (ebook) | ISBN 9781118965801 (Adobe PDF) | ISBN 9781118965849 (ePub) | ISBN 9781119973980 (hardcover)
Subjects: LCSH: Smart power grids. | Flexible AC transmission systems. | Electric power transmission–Direct current.
Classification: LCC TK3105 (ebook) | LCC TK3105 .A25 2019 (print) | DDC 621.319–dc23
LC record available at https://lccn.loc.gov/2018051883

Cover Design: Wiley
Cover Images: Background: © Teka77/iStock.com, Diagram: Courtesy of Enrique Acha

Set in 10/12pt WarnockPro by SPi Global, Chennai, India

Printed and bound by CPI Group (UK) Ltd, Croydon, CR0 4YY

10 9 8 7 6 5 4 3 2 1

To the memory of Jos Arrillaga, the one who wrote the most and best about HVDC transmission.

Contents

Preface

Electrical power transmission using high voltage direct current (HVDC) is a well-established practice. There is common agreement that the world's first commercial HVDC link was the Gotland link, built in 1954, designed to carry undersea power from the east coast of Sweden to the Island of Gotland, some 90 km away. The original design was rated at 20 MW, 100 kV and used mercury-arc valve converters. Its power and voltage ratings were increased in 1970 to 30 MW and 150 kV, respectively. Solid-state electronic valves were used for the first time in the upgrade, with the new type of valve, termed silicon-controlled rectifier (SCR) or thyristor, being connected in series with the mercury-arc valves. This kind of HVDC link, and ancillary technology, has been magnificently described in earlier treatises by Adamson and Hingorani, Kimbark, Uhlmann and Arrillaga.

By the turn of the second millennium, there had been 56 HVDC links of various topologies and capacities built around the world: 22 in North and South America, 14 in Europe, 2 in Africa and 18 in Australasia. They ranged from the small, 25 MW, Corsica tapping of the Sardinia–Italy HVDC link to the large, 6300 MW HVDC link, a part of the awesome Itaipu hydro-electric development on the Brazil–Paraguay border. At the time, three other large capacity HVDC links were at the planning/construction stage in China, to transport hydro-electric power from the Three Gorges to the east and southeast of the country, each spanning distances of around 1000 km, rated at 3000 MW – the Three Gorges is a gigantic hydro resource in central China, with an estimated power capacity of 22 MW. Heretofore all the HVDC links in the world had employed either mercury-arc rectifier or thyristor bridges and phase control to enable the rectification/inversion process. These converters are said to be line commutated and when applied to HVDC transmission are termed LCC-HVDC converters. The LCC-HVDC technology has continued its upward trend and five other high-power, high-voltage, long-distance DC links have been built in China since 2010. The most recent LCC-HVDC in operation was commissioned in 2012; it is the Jinping-Sunan link in East China, rated at 7200 MW and ±800 kV, spanning a distance of 2100 km.

In LCC-HVDC systems the current is unidirectional, flowing from the rectifier to the inverter stations. Such a fundamental physical constraint in thyristor-based converters limits applicability to the following HVDC system topologies: point-to-point, back-to-back and radial, multi-terminal links. In this context, the conventional, or classical, HVDC transmission technology is not a meshed grid maker; rather, its role has been to interconnect AC systems where an AC interconnection is deemed too expensive or technically infeasible.

However, one has to bear in mind that nowadays, in many situations, robust AC interconnections may be achieved more economically using one or more of the options afforded by the Flexible Alternating Current Transmission Systems (FACTS) technology, an array of power electronics-based equipment and control methods which became commercially available in around 1990. It is widely acknowledged that N.G. Hingorani and L. Gyugyi stand out prominently as the intellectual driving force behind the development of the FACTS technology.

The main aim of the FACTS technology is to enable almost instantaneous control of the nodal voltages and power flows in the vicinity of where the FACTS equipment has been installed. We should not forget that power flows over an AC line can be manipulated very effectively by controlling the line impedance, or the phase angles, or the voltages, or a combination of these parameters up to the thermal rating of the equipment. A key element of the FACTS technology is the so-called static compensator (STATCOM), which, in the parlance of a power electronics engineer, is a voltage source converter (VSC) and serves the purpose of injecting/absorbing reactive power to enable tight voltage magnitude regulation at its point of connection with the AC power grid. The advent of the STATCOM in the mid-1990s was made possible by the development of power semiconductor valves with forced turn-off capabilities, like the gate turn-off (GTO) first and the insulated gate bi-polar transistor (IGBT) soon afterwards. GTOs are like thyristors, which can be turned on by a positive gate pulse when the anode–cathode voltage is positive, and, unlike thyristors, can be turned off by a negative gate pulse. This turn-off feature led to new circuit concepts and methods such as self-commutated, pulse-width-modulated, soft-switching, voltage-driven and multi-level converters. These circuits may be made to operate at higher internal switching frequencies than the fundamental level, at several hundreds of hertz, which, in turn, reduces low-order harmonics and allows operation at unity and leading power factors. This contrasts sharply with what can be achieved with the normal thyristors.

Advances in the design of the power GTO and its applications in Japan and the USA continued apace by virtue of strategic collaborative R&D projects funded by utilities, manufacturers and governments. In Japan there was a target to develop 300 MW GTO converters for back-to-back HVDC interconnections, while in the USA a 100 MVAR GTO-STATCOM was commissioned in 1996 for the Tennessee Valley Authority. Meanwhile, similar efforts were conducted in Europe in the design of the power IGBT. It is reported that on 10 March 1997, power was first transmitted between Hellsjön and Grängesberg in central Sweden using an HVDC link employing IGBT converters driven by pulse-width-modulation (PWM) control. The link is 10 km long, rated at 3 MW, 10 kV and is used to test new components for HVDC.

In spite of the great many technical advantages and operational flexibility of the VSC compared with the thyristor bridge, the GTO-based converters did not make inroads into HVDC applications because of the much higher power losses and cost of GTOs compared with thyristors. A further reason is that the ratings of GTOs are low compared with those of thyristors. All this conspired to make VSC-HVDC installations expensive. The impasse was broken with the use of IGBT valves, which exhibit lower switching losses than GTO valves, and decreasing manufacturing costs. Three years after the commissioning of the Hellsjön-Grängesberg, four other VSC-HVDC links had been commissioned in very distant parts of the world: a 50 MVA DC link in the emblematic Island of Gotland to evacuate wind power, an 8 MVA DC link in West Denmark to link an offshore wind farm, the 180 MVA Directlink or Terranora project in Australia for power export from New South Wales into Southern Queensland, and a 36 MVA DC link for system interconnection on the Mexican–Texan border. The undersea Estlink 1, linking the Estonian and Finnish power grids, was commissioned in 2006, rated at 350 MW and using VSC stations. Intriguingly, the Estlink 2, rated at 650 MW and commissioned in 2014, uses the classical thyristor-converter technology.

It should be noted that all the VSC stations used in HVDC projects until 2010 had been of the so-called two- and three-level power converters. In around 2008, a new breed of VSCs was introduced into the market, the modular multilevel converters (MMCs), which switch at low frequencies, yield minimum harmonic production and have power losses just above those of the classical thyristor-based HVDC converters. Equally important is the fact that it has been possible to increase the capacity of VSC-HVDC links using MMC, by a very considerable margin, say 1000 MW per circuit, such as in the INELFE DC link between Baixas, France, and Santa Llogaia, Spain. Two identical circuits make up for a transmission capacity of 2000 MW. The link was commissioned at the end of 2013. Note that this application comes into the realm of bulk power transmission and is already eating into the niche area of classical thyristor-based HVDC technology, namely asynchronous bulk power transmission, an area until recently thought to be unassailable. The Trans Bay Cable link was the first MMC VSC-HVDC, commissioned in 2010, transmitting up to 400 MW of power from Pittsburg in the East Bay to Potrero Hill in the centre of San Francisco, California.

Furthermore, there are new application areas in which VSC-HVDC transmission does not seem to have a competitor in sight – the connection of wind sites lying more than 70 km away from the shore is one of the most obvious applications, but there are a few others. For instance, the connection of microgrids with insufficient local generation and little or no inertia (inertia-less power grids), the electricity supply of oil and gas rigs in deep waters, the infeed of densely populated urban centres with power grids already experiencing high short-circuit ratios. Moreover, the unassailable characteristic of the HVDC transmission using the VSC technology is that it is a natural enabler of meshed DC power grids, with such a high level of operational flexibility, reliability and efficiency that one day may surpass that of the meshed AC power grids. To get to this point, though, further technological breakthroughs are still awaited in the ancillary areas of DC circuit breaker technology and high-temperature superconductor cables and circuit breakers, as well as more affordable VSCs.

About the Book

The purpose of this book is to facilitate the study of technology that has emerged over the past 15 years in the area of flexible alternating current transmission systems (FACTS) and its technological convergence with the long-standing application of high voltage direct current (HVDC) but now using voltage source converters (VSC). This includes the back-to-back, point-to-point and multi-terminal VSC-HVDC applications. The subject is addressed from a modern perspective, including the latest development in the power systems industry that will extend the applicability of the VSC-FACTS-HVDC technology.

Contrary to *FACTS Modelling and Simulation of Power Networks*, published by the leading author and colleagues in 2004, which was limited to material on FACTS power flows and optimal power flows (OPFs), this book will address new FACTS power system application areas which have received much attention from industry over the past 12 years. These areas include FACTS state estimation, FACTS-constrained OPF, studies of FACTS dynamic performance and control, and the all-important topic of electromagnetic transients. These applications areas coincide with research areas, which the authors have developed over the past 15 years and have published widely in the top journals and presented in their research work at international forums.

The book is aimed at a very wide sector of the power engineering community, encompassing utility and equipment manufacturing engineers, researchers, university professors and PhD and MSc students, and undergraduate students in their final year. The reader is expected to have a sound knowledge of electrical and electronic circuits, algebra and numerical methods, and a working knowledge of electrical power and control engineering – an undergraduate student embarking on their final year in an electrical and electronics degree course should be well qualified to read the book. It goes without saying that utility engineers and managers with a background in electrical power would also take to the book like a 'duck to water'. Students conducting research in any of the topics covered in the book will find useful the modelling approach adopted by the authors which has resulted in flexible and comprehensive FACTS and HVDC-VSC models with which to carry out a wide range of power network-wide simulation studies, ranging from steady-state to dynamic and transient studies.

Chapter 1 gives an overview of the role that the VSC plays in the area of power systems VAR compensation. To qualify its prowess in this arena, a qualitative comparison is carried out against the long-enduring static var compensator. The VSC is a rather flexible piece of equipment which may be connected either in shunt or in series with the AC system, according to requirement. Two or more of them may be made to combine to give rise to compound equipment or systems, such as the UPFC and the various flavours of VSC-HVDC systems. Moreover, it is shown that the VSC combines well with the DC-DC converters and plays a pivotal role in enabling the grid connection of the renewable sources of electricity and energy storage systems. Equally important is the fact that the VSC technology is a builder of multi-terminal HVDC systems, with a DC network which may be a single node, a radial system or a meshed system. This is in stark contrast to the classical CSC-HVDC technology, whose flexibility is very limited as far as multi-terminal schemes is concerned. This chapter illustrates that a strategic feature of the VSC technology is to enable the conversion of AC transmission systems into DC transmission systems with an unassailable power transfer capacity and having, at least, an equal level of operational flexibility. Future developments in distribution systems seem to lie squarely in the incorporation of VSCs to enable greater operational flexibility, greater power throughputs and the incorporation of renewable electricity sources and storage.

Chapter 2 presents the theory of power electronics, which is essential to a good understanding of modern power converters topologies. The most popular semiconductor valves are presented first, followed by the classical two-level, single-phase and three-phase power converters. This forms the preamble of the study of multi-level converters, which is the technology used today by industrial vendors. The chapter presents a comparison between the HVDC systems based on voltage source converters and those that employ current source converters. A comprehensive list of current VSC-HVDC installations around the world is given at the end of the chapter.

Chapter 3 addresses the theory of power flows. The chapter may be seen as consisting of two parts: (i) the conventional power flow theory, including a Newton-Raphson power flow method in rectangular coordinates used to solve the set of nodal power equations describing the power grid during steady-state operation; and (ii) the power flow equations of the VSC model, which are derived from first principles and then extended to establish the power flow models of the STATCOM, VSC-HVDC links and generalized multi-terminal VSC-HVDC systems. The algebraic, non-linear equations describing the steady-state performance of the VSC, the STATCOM and VSC-HVDC systems are solved using the Newton-Raphson method in rectangular coordinates. The ensuing solutions fulfil the quadratic convergence characteristic which is the hallmark of the Newton-Raphson method. The VSC is the basic building block with which all the VSC-FACTS and VSC-HVDC equipment is assembled; hence, a Newton-Raphson power flow computer program written in Matlab, with the model of the VSC included, is available at www.wiley.com/go/acha_vsc_facts for the user to gain hands-on experience.

Chapter 4 introduces the topic of optimal power flow (OPF) used by transmission system operators for optimal economic and security assessments of their power grids. The chapter is divided into three main sections: (i) a general overview of the OPF problem and its applications in power systems operational planning; (ii) the introduction of the OPF problem as a non-linear optimization problem and possible solution methods; and (iii) an extension to the OPF formulation to incorporate hybrid AC-DC networks using VSC-HVDC systems. The OPF methodology introduced in this chapter uses the VSC model developed in Chapter 3 to formulate versatile models of hybrid AC-DC networks suitable for minimum-cost assessments of power systems subject to realistic operational constraints in both the AC and the DC grids. The overall solution algorithm adopted for solving the non-linear system of equations is the de facto industry's standard Newton's method. Concerning the introduction of models of VSC-based equipment, the chapter follows a similar line of development as in Chapter 3, starting with the VSC and progressing to develop models of the various kinds of VSC-HVDC systems. However, in this chapter the complex nodal voltages are represented in polar form as opposed to rectangular form because the former is widely employed in the OPF literature. The models are formulated and solved using a general-purpose mathematical solver package called AIMMS (Advanced Interactive Multidimensional Modelling System), employing its nonlinear augmented Lagrangian solver. However, the reader may implement these models using any equivalent general-purpose simulation platform. Alternatively, the more advanced readers may wish to write their own OPF computer program using MATLAB scripting. In any case, a free academic licence of AIMMS can be obtained for academic research purposes.

Chapter 5 presents the theory of power systems state estimation. The chapter is divided into two main parts, addressing the following issues: (i) the classical power systems state estimation theory using the weighted least squares method as the solution algorithm; and (ii) the state estimation models of the VSC, the STATCOM, UPFC, VSC-HVDC links and generalized multi-terminal VSC-HVDC systems. The timely topic of PMUs in power systems state estimation is addressed in this chapter. Each major topic in this chapter is accompanied by a set of well-designed numerical exercises using a MATLAB environment (WLS-SE) for the user to gain hands-on experience.

Chapter 6 is dedicated to the study of power systems dynamics in time domain. It uses a similar outline to the previous four chapters. In the first part, it introduces the theory of conventional power systems dynamics, where the synchronous generators and their controls are the only equipment that exhibit a dynamic behaviour following a disturbance in the power grid. The transmission lines, transformers and loads are taken to exhibit a static behaviour, although provisions are made for the models of these equipment to have a voltage and frequency dependency. In this chapter the interest is the study of power system dynamic phenomena which exhibit a relatively low variation in time. Hence, the dynamics of the power grid is described well by a set of algebraic-differential equations which are discretized and linearized in order to carry out the solution by iteration, which is valid for a single point in time. The solution algorithm used is an implicit simultaneous method employing the Newton-Raphson method. The resulting mathematical model is coded in software and applied to assess the dynamic behaviour of a test system, in order to illustrate the usefulness of the overall dynamic model. In the second part of the chapter, the dynamic model of the VSC-STATCOM, receives a similar treatment to the synchronous generator but having a detailed representation of the dynamics of its DC bus. This dynamic model is then suitably extended to encompass the dynamic models of the back-to-back, point-to-point and multi-terminal VSC-HVDC system. The VSC-HVDC model is applied to study the timely issues of frequency support in power grids with near-zero inertia and supplied by a VSC-HVDC link.

Chapter 7 is devoted to the simulation of the transient responses of various FACTS and HVDC systems using PSCAD/EMTDC, a commercial software package for electromagnetic transient analysis, which is widely used in industry and academia. Four different systems are simulated: (i) a STATCOM based on a conventional two-level voltage source converter; (ii) an extension of the STATCOM using a three-level flying capacitor converter as an example of a multilevel converter; (iii) a two-terminal

HVDC system based on a multilevel voltage source converter topology; and (iv) a multi-terminal HVDC system which also employs multilevel VSCs. Furthermore, the control schemes of the different power systems are comprehensively explained and control design specifications are provided.

Acknowledgements

Bringing this book project to a close has been an endeavour made possible only with the support of colleagues and institutions from across the world, having started in the research laboratories of the University of Glasgow, Scotland, in 2008 and completing today, when the authors work in the following universities: Tampere University, Finland, Universidad de Castilla-La Mancha, Spain, Universidad de Sevilla, Spain, Universidad Nacional Autónoma de México, Mexico, and Durham University, England. Our appreciation goes foremost to the University of Glasgow and our respective home universities for the time that allowed us to bring this project to fruition. We would like to thank Dr Rodrigo Garcia Valle from Ørsted, Denmark, and Dr Luigi Vanfreti from Rensselaer Polytechnic Institute, USA, for their early contribution to the book project. We would like to thank our respective families for the time that we were lovingly spared throughout the project.

Enrique Acha would like to thank Antonio Gómez Expósito, Jose Maria Maza-Ortega and Sigridt Garcia for having written the following award-winning paper: J.M. Maza-Ortega, E. Acha, S. Garcia, A. Gomez-Exposito, 'Overview of power electronics technology and applications in power generation, transmission and distribution', *J. Mod. Power Syst. Clean Energy* – Springer (2017) 5(4):499–514, which provided the inspiration for Chapter 1.

Luis Miguel Castro and Enrique Acha would like to acknowledge the financial assistance of Consejo Nacional de Ciencia y Technología (CONACYT), México, and Professor Pertti Järventausta from the Tampere University, Finland, through the SGEM project, to conduct fundamental research on the modelling and simulation of multi-terminal Voltage Source Converter High-Voltage Direct Current (VSC-HVDC) systems. This research forms the basis of Chapters 3 and 6.

Behzad Kazemtabrizi would like to thank Ahmad Asrul Bin-Ibrahim of Durham University, England, for his help in producing and verifying the results for the AC/DC optimal power flow (OPF) test case used in Chapter 4.

Antonio de la Villa would like to thank the Spanish Ministry of Economy and Competitiveness (MINECO) under grants ENE 2010-18867, which provided the facilities for the work of Chapter 5. Thanks are also expressed to the following faculty staff of Universidad de Sevilla: Antonio Gómez Expósito, Esther Romero Ramos and Pedro Cruz Romero, for their useful suggestions during the preparation of this chapter.

Pedro Roncero would like to thank MINECO, whose financial support, at various stages in the preparation of the book, proved instrumental in seeing its completion.

The large number of simulations in the book were enabled by the use of a wide range of open source and commercial software (educational versions): Matlab, Simulink, MATPOWER, the Advanced Interactive Multidimensional Modelling System (AIMMS) and Power System Computer Aided Design/Electromagnetic Transient Direct Current (PSCAD/EMTDC). We would like to extend our most ample gratitude to all the owners and developers of such powerful simulation platforms.

We are grateful to the staff of John Wiley & Sons for their utmost patience and continuous encouragement throughout the preparation of the manuscript.

About the Companion Website

This book is accompanied by a companion website:

www.wiley.com/go/acha_vsc_facts

The website includes software files associated to Chapters 3, 5 and 7:

- Matlab files corresponding to Chapters 3 and 5.
- Two PSCAD files of Cases 1 and 4 corresponding to Chapter 7.

Scan this QR code to visit the companion website.

1

Flexible Electrical Energy Systems

1.1 Introduction

Following a sustained programme of expansion of high-voltage power grids in the 1960s and their widespread interconnection in the 1970s, by the end of that decade the expansion programmes of many utilities had become thwarted by a variety of well-founded, environmental, land-use and regulatory pressures, preventing the licensing and building of new transmission lines and electricity generating plants. This was in the face of sustained global demand for electricity.

An in-depth analysis of the options available for increasing power throughputs with high levels of reliability and stability pointed towards the use of modern power electronics equipment, control techniques and methods [1]. Such a far-reaching work was carried out first at the Electric Power Research Institute (EPRI) in Palo Alto, CA, under the leadership of N.G. Hingorani. The result was an integrated philosophy for AC network reinforcement using electronics principles, endowing AC transmissions lines with a degree of operational flexibility and power-carrying capacity that had not been possible before. Flexible alternating current transmission systems (FACTS) was the name given to the family of power electronic-based equipment, control techniques and methods emanating from this initiative [2].

In the same time span, electricity distribution companies were experiencing a marked increase in the deployment of end-user equipment which was highly sensitive to poor-quality electricity supply. Several large industrial users reported experiencing significant financial losses as a result of even minor lapses in the quality of electricity supply. A great many efforts were made to remedy the situation, with solutions based on the use of the latest power electronic technology of the time [3].

A range of custom-made equipment and solution techniques was put to the fore, with the key ideas emanating from EPRI. This initiative, aimed at ameliorating adverse power quality phenomena at the interface between the low-voltage distribution power grid and the industrial user, was given the name 'custom power' by its creator, N.G. Hingorani. Indeed, custom power technology was announced as the low-voltage counterpart of the FACTS technology, aimed at high-voltage power transmission applications and emerging as a credible solution to many of the problems relating to continuity of supply at the end-user level [2].

It is fair to say that many of the ideas upon which the foundation of FACTS rests evolved over a period of several decades, building on the experience gained in the areas

VSC-FACTS-HVDC: Analysis, Modelling and Simulation in Power Grids, First Edition.
Enrique Acha, Pedro Roncero-Sánchez, Antonio de la Villa Jaén, Luis M. Castro and Behzad Kazemtabrizi.

Companion website: www.wiley.com/go/acha_vsc_facts

of high-voltage direct current (HVDC) transmission and reactive power compensation equipment, methods and operational experiences [4, 5]. Nevertheless, there is widespread agreement that FACTS, as an integrated philosophy, was a novel concept brought to fruition at EPRI in the 1980s. Since those early days of the FACTS technology, a great many breakthroughs have taken place in the area of power electronics, encompassing new valves, control methods and converter topologies [6]. To a greater or lesser extent, the recent technological developments have all been incorporated into the fields of FACTS and HVDC, giving rise to a new generation of power transmission equipment in either AC or DC, with unrivalled operational flexibility [7].

The original boundaries between HVDC and FACTS were drawn along the type of solid-state converters employed and their control [1], but these boundaries became blurred with the arrival of newer technology. For instance, the static compensator (STATCOM), which is essentially a voltage source converter (VSC), is a product of the FACTS technology used to provide reactive power support [8]. Two such devices connected in series on their DC sides results in the modern expression of an HVDC transmission system. This has been designated VSC-HVDC to distinguish it from the classical HVDC transmission using thyristor-based bridges and phase control [9]. The largest vendors of power electronics equipment, ABB, Siemens and Alstom, have proprietary equipment termed HVDC *Light*, HVDC *Plus* and *MaxSine* HVDC, respectively. It is documented that the use of a VSC in a utility-level application was in the form of a STATCOM, which falls squarely within the realm of the FACTS technology. The VSC application in HVDC transmission came next, which, it may be argued, is an application comprising two STATCOMs connected back-to-back or through a DC cable. Of course, such an argument is more difficult to sustain when we progress into the realm of multi-terminal VSC-HVDC [10].

From a traditional perspective, artificial lines have been drawn between the FACTS and the HVDC technologies. It is argued here that these lines be removed and that, instead, the focus should be on flexible transmission systems (FTS), a unifying concept bridging the FACTS and HVDC technologies – the aim being to enable the best-of-breed solutions underpinning the new power-carrying structures that the smart grids demand [11].

The breakthroughs in power electronics impacted not only the transmission and distribution sectors of the electrical energy industry but also the generation sector, particularly the renewable generation and energy storage technologies [12]. The use of advanced power electronic converters enabled the wind power equipment manufacturers to transit from the first generation of fixed-speed wind turbines to the second generation of variable-speed wind turbines, which are larger, more efficient and fully compliant with modern grid codes [13]. The use of advanced power electronic converters also led to the proliferation, on a global scale, of grid-connected photo-voltaic generators, with full compliance to modern grid codes [14]. More recently, with the widespread availability of affordable lithium-ion batteries suitable for power applications, battery energy storage systems (BESS) are becoming off-the-shelf products [15]. It is very likely that, once BESS prices decrease further, this equipment will become ubiquitous in the power grid since it has a potentially major role to play in electrical energy retailing.

In a more ample technological sense than FTS or flexible power generation, a wide range of enabling technologies has become cost-effective, such as extruded cables, smart meters, phasor measurement unit (PMU), advanced protection systems, accessible satellite communications and distributed energy resources such as EV charging stations [16]. Equally important is the fact that information and communication technologies (ICTs), the internet, the web and distributed computing, have become even more powerful and popular than, say, only one decade ago.

The widespread availability of all these technological developments has been seized on by the proponents of the smart grid philosophy, who argue that the coming together, in an all-encompassing manner, of these technologies should provide a solid foundation on which to build new, smarter energy grids, now that a large portion of the existing infrastructure in many countries is ageing and up for renewal [17, 18]. Key drivers of the smart grid technology are enhanced security of supply and self-healing properties, its market-oriented philosophy, progressive demand-side response (DSR) and demand-side management (DSM) policies [19].

The available technologies and drivers of the smart grid concept are voltage-level independent and network structure independent. Hence, it is argued here that it should be feasible to talk about smart grids at either the low-voltage distribution system level or the high-voltage transmission system level. Admittedly, each has its own peculiarities but with a great many common objectives and interrelated technology issues yet to be resolved.

The power network is expected to incorporate increasing amounts of wind and solar power, leading to new challenges in its operation owing to the intermittent nature of these two new forms of renewable power generation. Frequency oscillations, resulting from temporary power imbalances, may become more common. Also, voltage control may become a significant problem if no suitable FTS equipment is in place, such as static var compensators (SVCs) or VSCs in the form of STATCOMs, BESSs or as part of VSC-HVDC systems [7]. Moreover, the power-carrying structures of AC low-voltage smart grids are likely to use mainly underground cables – which is already happening in Denmark – and inductive reactive power compensation equipment may be required [20]. Alternatively, smart grids using AC and DC microgrids may reach the commercial stage, following on from the experience gained with current prototypes deployed by some of the distribution companies in Finland [21].

Indeed, structures based on multi-terminal VSC-HVDC systems are just the kind of transmission structures that are likely to be used by the next generation of smart grids, both at the transmission level and at the distribution level [7]. This may be a system within a larger AC system, as exemplified in the Figure 1.1. In this one-line diagram, an AC power system incorporating a fair amount of FACTS controllers and a multi-terminal VSC-HVDC system is also equipped with synchronized measurement systems at key points of the power grid using PMU and satellite communications. The deployment of such equipment would go a long way towards establishing a high-voltage smart grid, particularly if the FTS equipment is fitted with processor agents, sensors and a fibre optic network – or some other means of fast communications – linking all the FTS equipment, so that a coordinated action takes place at the system level as opposed to the local, individual level [22]. Such arguments would also apply to low-voltage, low-power microgrids where the core system could be a multi-terminal VSC-HVDC system interconnecting an arbitrary number of AC systems – this point is elaborated further in Section 1.3.

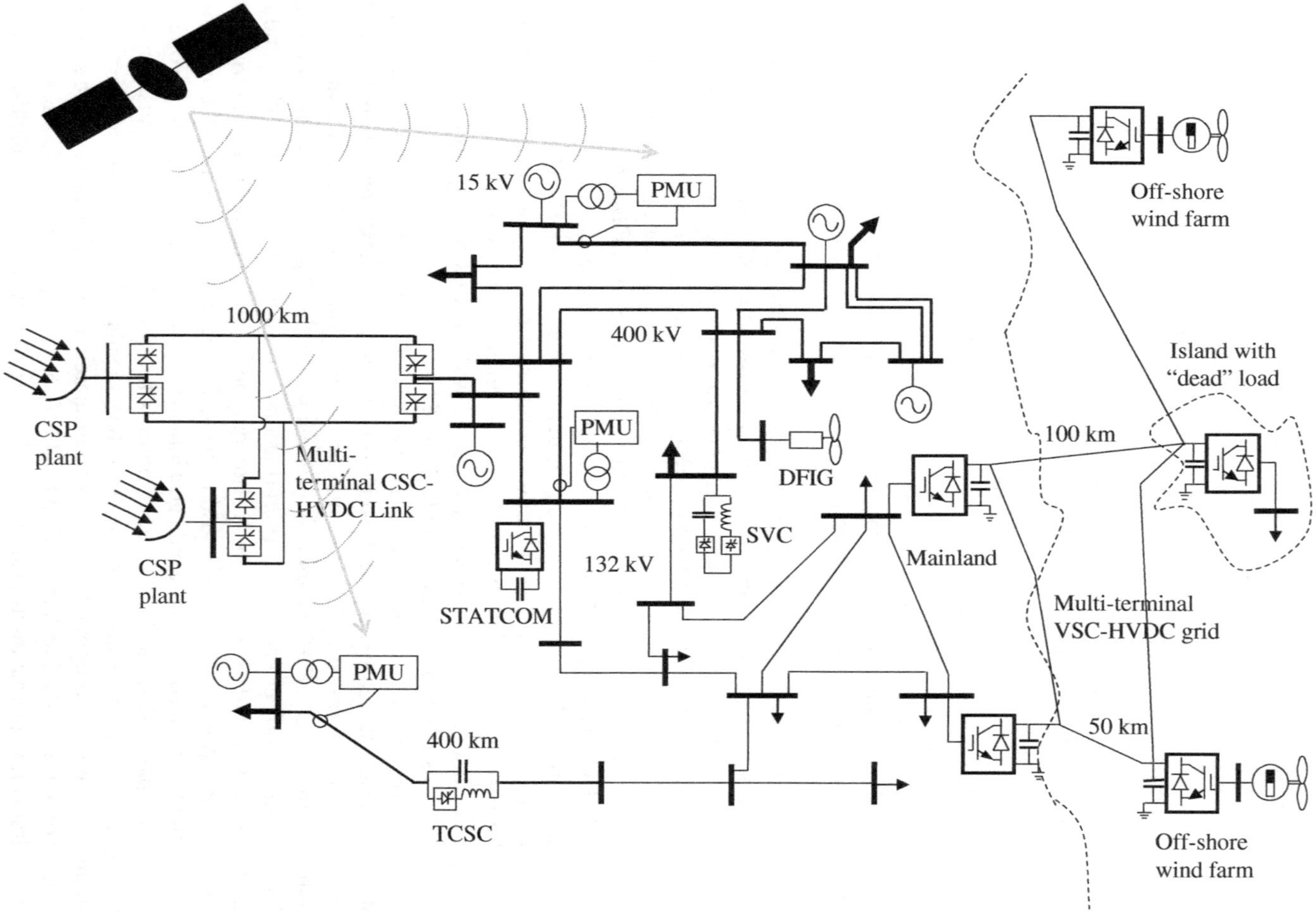

Figure 1.1 Flexible transmission system with renewable energy sources.

1.2 Classification of Flexible Transmission System Equipment

Many of the ideas upon which the foundation of FACTS rests evolved over a period of many decades. Chiefly among them is:

- experience gained with HVDC transmission technology [4]
- experience gained with SVC technology [5]
- experience gained with electric drives for motor control [23].

In a nutshell, FACTS and HVDC equipment uses power semiconductor devices and advanced power electronics control techniques and methods to fulfil its task in a matter of a few milliseconds [24].

The wide range of modern equipment available is bundled together in this book under the umbrella title of FTS equipment. The range of functions that this technology can fulfil is very wide but it is equipment-dependent. Table 1.1 lists the equipment comprising the FTS technology and the respective areas of power systems application.

Various classifications of the FTS equipment are possible. It can be classified in terms of the power systems application, as outlined in Table 1.1: (i) voltage control, (ii) reactive power control, (iii) active power control, (iv) frequency control, and (v) AC systems interconnection. This list is not exhaustive by any means and other applications exist, such as loop flow control.

Alternative classifications may be drawn according to the number of power converters used or the way in which the equipment connects to the AC power grid or grids,

Table 1.1 FTS equipment and the respective areas of power systems applications.

Equipment	Voltage control	Reactive power control	Active power control	Frequency control	AC systems interconnection
SVC	✓	✓			
STATCOM	✓	✓			
BESS	✓	✓		✓	
SSTC	✓				
TSSC			✓		
TCSC			✓		
SSPS			✓		
SSSC		✓	✓		
UPFC	✓	✓	✓		
IPFC		✓	✓	✓	
VFT			✓	✓	✓
CSC-HVDC			✓	✓	✓
VSC-HVDC	✓	✓	✓	✓	✓

SVC: static var compensator; **STATCOM**: static compensator; **BESS**: battery energy storage system; **SSTC**: solid-state tap changer; **TSSC**: thyristor switched series capacitor; **TCSC**: thyristor-controlled series compensator; **SSPS**: solid-state phase shifter; **SSSC**: solid-state series compensator; **UPFC**: unified power flow controller; **IPFC**: interphase power flow controller; **VFT**: variable frequency transformer; **CSC-HVDC**: current source converter high-voltage direct current; **VSC-HVDC**: voltage source converter high-voltage direct current.

Table 1.2 Classification of FTS equipment according to converter type.

Converter type	Equipment
Thyristor and phase control	SVC, SSTC, TSSC, TCSC, SSPS, CSC-HVDC
IGBT or GTO and PWM control	STATCOM, BESS, SSSC, UPFC, IPFC, VSC-HVDC, VFT

namely, shunt, series, cascade and multi-terminal. The equipment can also be classified according to the type of semiconductor valves and switching control that the converters use: (i) thyristor valves and phase control, and (ii) insulated gate bipolar transistor (IGBT) or gate turn-off (GTO) valves and pulse width modulation (PWM) control [2]. This classification is shown in Table 1.2.

For any practical purpose, an IGBT valve outperforms a GTO valve in terms of its speed of response, better power loss performance and improved reliability; they have become the standard forced commutated valves used in converters aimed at power systems applications. They are driven by PWM control of various kinds [24].

In this book, the application of IGBT-based converters driven by sinusoidal-PWM control is given priority. In particular, the ubiquitous STATCOM and the VSC-HVDC. Other equipment which uses the STATCOM as the basic building block also receives attention in this chapter, such as the BESS, the solid-state series compensator (SSSC), the unified power flow controller (UPFC) and the interphase power flow controller (IPFC). As a means of emphasizing the much-increased functionality that the STATCOM has brought into the arena of electrical power systems, in general, the SVC and the current source converter high-voltage direct current (CSC-HVDC) are also covered in sufficient detail.

1.2.1 SVC

The SVC appeared on the power systems scene at least two decades before the FACTS initiative was put forward [5]. In some respects, the SVC may be considered the forebear of some of the equipment developed under the auspices of the FACTS initiative [1]. It comprises a bank of thyristor-controlled reactors (TCRs) in parallel with a bank of thyristor switched capacitors (TSCs). The one-line schematic representation of the SVC is shown in Figure 1.2.

The SVC is connected in shunt with the AC system through a step-up transformer. Its main function is to supply/absorb reactive power to support a specified voltage magnitude at the high-voltage side of its connecting transformer.

At the construction level, a criticism levelled at the SVC is its rather large footprint [8]. The inductor, in Figure 1.2a, is a bulky air-core reactor, to prevent saturation; the capacitor banks are also quite sizable and so are the tuned filters. As suggested in Figure 1.2b, the SVC performs like a variable susceptance.

The TCR consumes variable reactive power up to its design limit, governed by firing angle control, α, in the range: $\pi/2 \leq \alpha \leq \pi$. It achieves its fundamental frequency operating point at the expense of generating harmonic currents, which is an undesirable side effect. Hence, the three-phase TCR is connected in delta to prevent the triple harmonics from reaching the power system. In addition, passive filtering is required

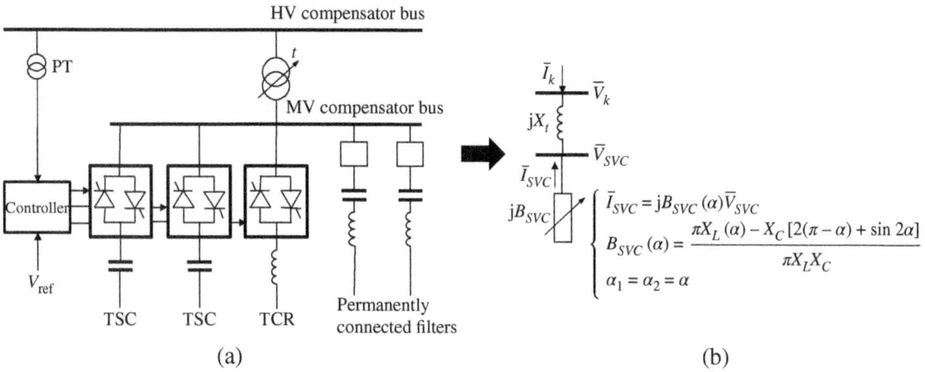

Figure 1.2 (a) An SVC (©MPCE, 2017) and (b) its equivalent circuit representation.

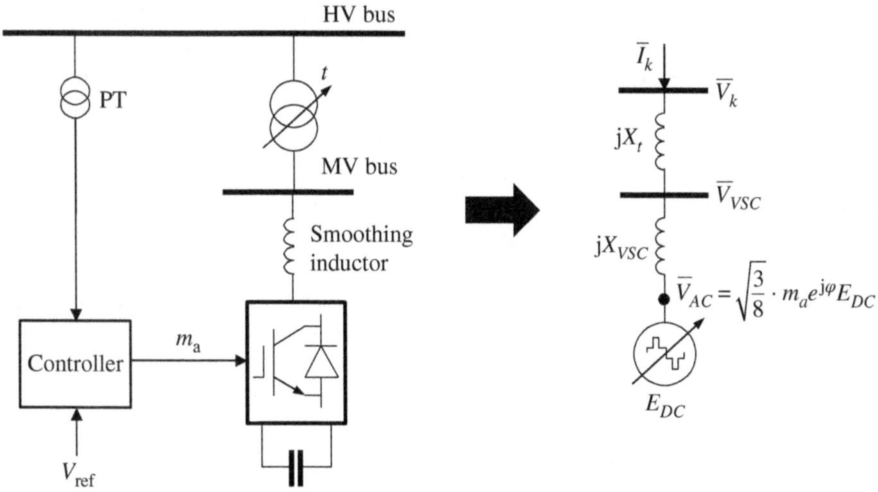

Figure 1.3 (a) A STATCOM (©MPCE, 2017) and (b) its equivalent circuit representation.

to mitigate the $6k \pm 1$ and $12k \pm 1$ harmonics generated by the 6-pulse and 12-pulse converter topologies, respectively, with $k = 1,2,3\ldots$ being the harmonic order. The TSC generates reactive power in a variable, discrete manner, with the thyristor pairs operating as switches (on/off); hence, during steady-state operation, no harmonic distortion is produced by the TSC [5].

1.2.2 STATCOM

The STATCOM is the modern counterpart of the SVC [8]. Coincidentally, its main operational task is to inject regulated volt-ampere reactives (VAR) to provide voltage support at the high-voltage side of its connecting transformer, but much more effectively. As illustrated in Figure 1.3, it connects in shunt with the AC power system. Its main elements are the VSC, the smoothing inductor, the interfacing transformer and the PWM control system.

Table 1.3 STATCOM operating modes.

Voltages	Operating mode	Functionality
$V_k > V_{AC}$	Consuming vars	Standard function
$V_k < V_{AC}$	Injecting vars	Standard function
$-\delta$	Consuming watts	Normal operation
$+\delta$	Injecting watts	Possible only with DC storage

Contrary to the SVC, the STATCOM does not use bulky inductors and banks of capacitors to absorb and to generate reactive power, respectively. This can be appreciated from Figure 1.3a. The reactive power production process is carried out entirely by the electronic processing of the voltage and current waveforms within the valves, to enable either leading or lagging VAR production to satisfy operational requirements [2]. The smoothing inductor laying between the VSC and the step-up transformer is used to eliminate the high-order harmonics produced by the action of the PWM control. The DC capacitor is an extruded capacitor of a small rating, employed only to support and stabilize the DC voltage to enable the converter operation – it does not play a significant part in the VAR generation process. The DC current in the capacitor in Figure 1.3a is taken to be zero during steady-state operation, as an indication that the capacitor is fully charged. During dynamic and transient operation, I_{DC} will differ from zero.

The STATCOM's operational behaviour is superior to that of the SVC because its operation mimics a variable voltage source as opposed to a variable susceptance [8]. This is illustrated in Figure 1.3b. In fact, the STATCOM's operational performance is closer to that of a rotating synchronous condenser but with a faster speed of response because it has no moving parts.

The basic operating principles of the STATCOM may be explained with reference to the complex voltages $\overline{V}_{AC} = V_{AC}\angle\delta$ and $\overline{V}_k = V_k\angle 0$ in Figure 1.3b, where the amplitude and the phase angle of the voltage drop across the reactances X_{SVC} and X_t can be controlled, to define the amount and direction of both active and reactive power flows.

For leading and lagging VARs, the STATCOM's active and reactive power flows are defined by the following fundamental expressions:

$$\left.\begin{aligned} P &= \frac{V_k V_{AC}}{X_{VSC} + X_t} \cdot \sin\delta \\ Q &= \frac{V_k^2}{X_{VSC} + X_t} - \frac{V_k V_{AC}}{X_{VSC} + X_t} \cdot \cos\delta \end{aligned}\right\} \tag{1.1}$$

To a large extent, the equation set (1.1) defines the STATCOM operating modes, which are summarized in Table 1.3.

The static voltage–current characteristics shown in Figure 1.4 correspond to the SVC and the STATCOM. The thick line is for the SVC (capacitor and thyristor-controlled inductor) at rated values; only one value of capacitor has been considered and the broken lines on the inductor side would be for various firing angle values of the TCR. As seen

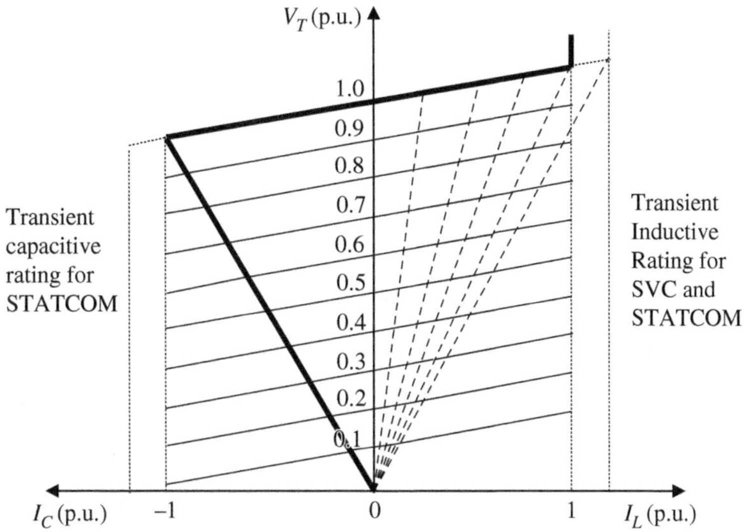

Figure 1.4 Static V-I characteristics of the VSC and the STATCOM.

from the figure, the characteristics of a STATCOM and a SVC of comparable ratings will coincide only at their rated values.

It can be seen from this characteristic that the SVC yields little capacitive current at low voltages. In contrast, the STATCOM is able to produce its full range of current, both inductive and capacitive, even when the voltage has dropped to about 10% of its nominal value. From the operational vantage, one of the main criticisms levelled at the SVC is that its ability to contribute reactive power becomes severely impaired in the presence of low system voltages in its vicinity, for instance in cases of voltage collapse. Conversely, the STATCOM's reactive power provision is system voltage-independent.

In spite of its superior technical performance, the STATCOM technology still carries a higher price tag than the SVC technology of comparable rating. Equipment manufacturer Alstom, Finland, has recently patented a new piece of VAR compensation equipment selecting the best attributes of the SVC and the STATCOM [25].

1.2.3 SSSC

This equipment may be seen as a series-connected STATCOM and serves the purpose of injecting a variable, controllable voltage to an incoming voltage to achieve a range of purposes, one of which is active power flow control through the power line and the injection of controlled reactive power at one of the two nodes [2].

An SSSC is normally a piece of equipment of small rating, with full control of the magnitude, phase and polarity of the injected series voltage. One may think of a small STATCOM which is connected to the secondary winding of an interfacing transformer whose primary winding is connected in series with the AC power grid, as illustrated schematically in Figure 1.5.

Note that the SSSC should be designed to carry the full line current, with the rated voltage being only a fraction of the rated line voltage.

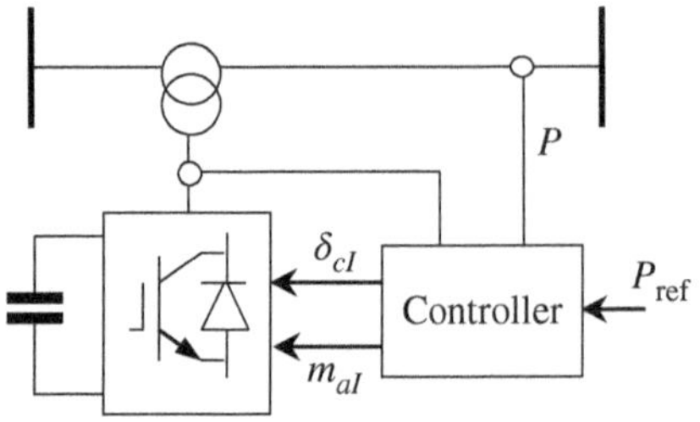

Figure 1.5 SSSC schematic representation (©MPCE, 2017).

1.2.4 Compound VSC Equipment for AC Applications

The VSC, which forms the kernel of the STATCOM and the SSSC, is a rather flexible device; not only does it connect easily in shunt or in series with the AC power system, it also combines rather well in any number and in any combination to suit a specific power systems application. This is illustrated in Figure 1.6.

The schematic diagrams of the four most popular compound FACTS devices – the UPFC, the IPFC, the back-to-back VSC-HVDC and the BESS – are shown in Figure 1.6. Their main salient characteristics are outlined below, according to the time at which they seem to have been conceptualized or emerged on the applications scene.

- The UPFC employs two VSCs, one connected in shunt and the other in series with the AC system. The two VSCs are connected back-to-back on their DC sides and have a unified control structure, giving rise to the UPFC, which has great operational functionality. Its schematic diagram is shown in Figure 1.6c. In this application, the shunt-connected VSC is rated at full capacity whereas the series-connected VSC is rated to carry only a fraction of the line current and have a fraction of the line voltage. Note that this contrasts with the SSSC, which is rated to carry the full line current [26].
- The UPFC combines the operational control capabilities of the STATCOM and the SSSC. It is capable of regulating, simultaneously, voltage magnitude at the high-voltage node of the shunt-connected VSC, active power flow arriving at the receiving node of the series-connected transformer – the opposite node to that where the shunt converter is connected – and the injection of reactive power at that node. Regulation of these parameters is limited by the ratings of the shunt and series converters. The UPFC plays the role, in any combination, of a STATCOM, a TCSC (thyristor-controlled series compensator), an SSSC, a phase-shifting transformer and a tap-changing transformer, simultaneously and in any combination. For all its operational flexibility and great expectations when it was conceptualized, the UPFC has not been a commercial success. Only two prototypes are known to exist in the world, in the USA [2] and Korea [27].
- If two or more transmission lines connecting to the same AC bus are fitted each with a SSSC then these can be made to share the same DC capacitor and to have a coordinated control system, giving rise to the IPFC [28]. Its schematic diagram is shown in Figure 1.6d. It has been designed to regulate active power flows between the various transmission lines by exchanging power through the common DC bus. No IPFC installation is known to exist at present but this is likely to change as more commercially-driven energy transactions take place between neighbouring transmission companies.

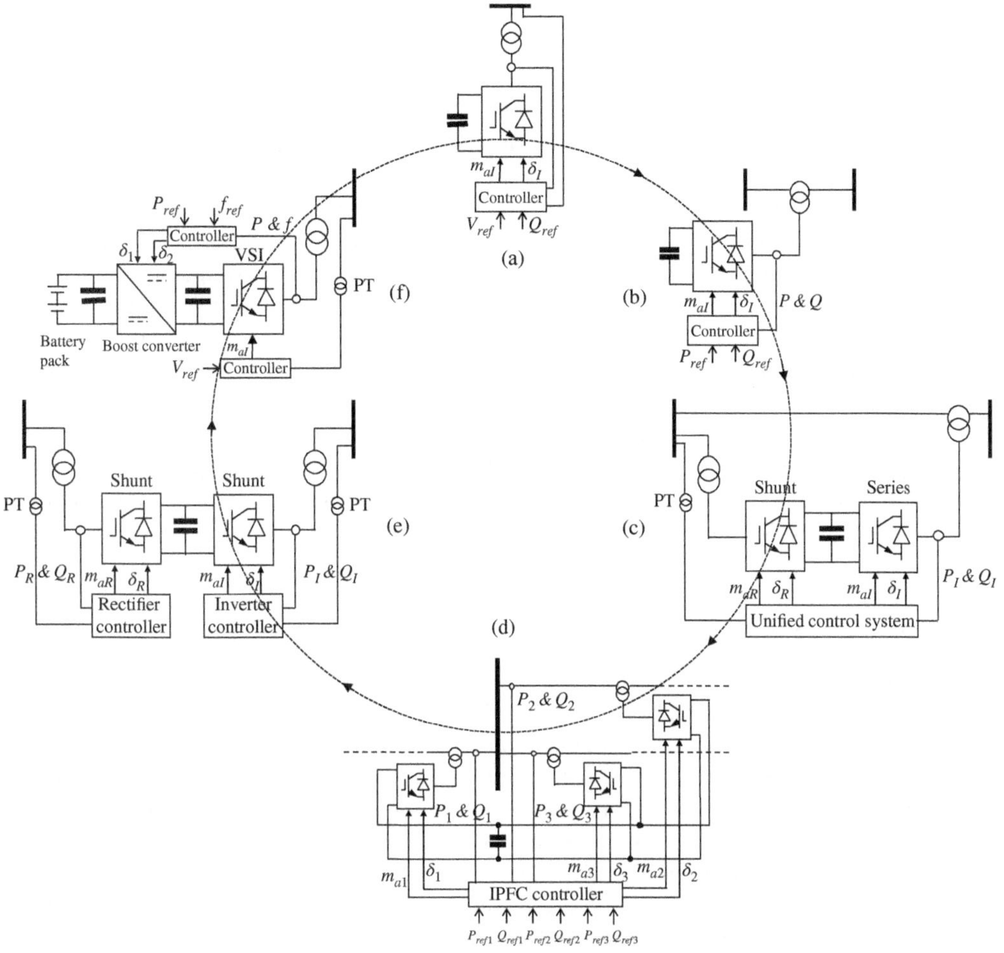

Figure 1.6 The most popular VSC-based equipment, placed, clockwise, in the order in which they appeared in the open literature: (a) STATCOM; (b) SSSC; (c) UPFC; (d) IPFC; (e) back-to-back VSC-HVDC: (f) BESS.

- Two VSCs connected in tandem form a back-to-back VSC-HVDC link, as shown in Figure 1.6e. The VSCs are connected in shunt at their respective AC systems and have a common DC bus [9]. There is some resemblance between the structures of the UPFC and the back-to-back VSC-HVDC. It may be argued that the operational functionality of the two controllers is comparable but the back-to-back VSC-HVDC achieves this at the expense of using two fully rated converters. Nevertheless, a key attribute of the VSC-HVDC link that the UPFC lacks is the ability to connect, in an asynchronous manner, two otherwise independent AC systems, which may have the same or different operating frequencies. Furthermore, the two VSCs of the HVDC link do not need to be connected back-to-back but instead may be linked by a cable and used to transport electrical power in DC form with less power loss than an AC transmission line of comparable rating and distance. Among all the VSC-based equipment, the VSC-HVDC link is the technology that has experienced the highest rate of growth since its introduction in 2000. Examples of VSC-HVDC installations in the world are given in Section 2.4 of Chapter 2.

- The VSC combines quite naturally with modern battery packs, such as lithium-ion batteries. The combined system is termed BESS, having the structure shown in Figure 1.6f. It should be noted that the battery pack connects to the VSC through a DC-DC converter to enable the smooth operation of the battery. Provided there is sufficient energy stored in the battery pack, the BESS is capable of injecting active power into the AC system, in a matter of milliseconds, to provide inertial and primary frequency support in the presence of synchronous generators frequency oscillations. Furthermore, the associated VSC acts as a source of VARs to enable effective voltage regulation at the AC bus. Indeed, not only can a modern BESS emulate the operation of a synchronous generator, it can also provide additional flexibility, such as adaptive time-varying behaviour during faults or perturbations. It is very likely that once BESS prices decrease further, this equipment will become ubiquitous in the power grid since it has a potentially major role to play in electrical energy retailing. BESSs commissioned in Chile and California [29, 30] in recent years are two good examples of where MW-size BESS have been installed.

1.2.5 CSC-HVDC Links

HVDC transmission using the classical six-pulse Graetz bridge, which uses thyristor valves and phase control, has been available, on a commercial basis, for more than half a century. Its main application has been in the area of bulk power transmission over long distances. The earliest schemes were point-to-point of the monopolar kind with ground return. The back-to-back and the point-to-point bipolar configurations were developed soon afterwards [4]. Today's classical HVDC transmission uses 12-pulse converters as opposed to 6-pulse converters. This type of HVDC technology is termed CSC-HVDC to distinguish it from the new HVDC technology which uses forced commutated valves and PWM control, namely VSC-HVDC [24].

The rectification and inversion processes are imperfect and the characteristic harmonics are produced in both the DC circuit and the AC circuits of the HVDC links. Hence, passive filtering is required to mitigate the $6k \pm 1$ and $12k \pm 1$ AC harmonic currents generated by the 6-pulse and 12-pulse converter topologies, respectively, with $k = 1,2,3\ldots$ being the harmonic order. On the DC side, filters may be required for the $6k$ order. The rectification process is achieved with thyristor firing angles in the range $0 \leq \alpha < \pi/2$ and the inversion process in the range $\pi/2 < \alpha \leq \pi$, although some margin would need to be left to avoid commutation failures. Moreover, the rectification/inversion process requires the provision of reactive power, which needs to be supplied locally to enable suitable operation of the link, with at least a part of this requirement being met by the passive harmonic filters in the installation. Contrary to the VSC, the Graetz bridge does not have VAR production capability and is unable to regulate AC voltage. Hence, connection to both AC grids is carried out through tap-changing transformers to enable AC voltage control.

Cases of a monopolar, back-to-back and bipolar, point-to-point HVDC schemes are exemplified in Figures 1.7a,b, respectively. For a full set of CSC-HVDC topologies, refer to [4].

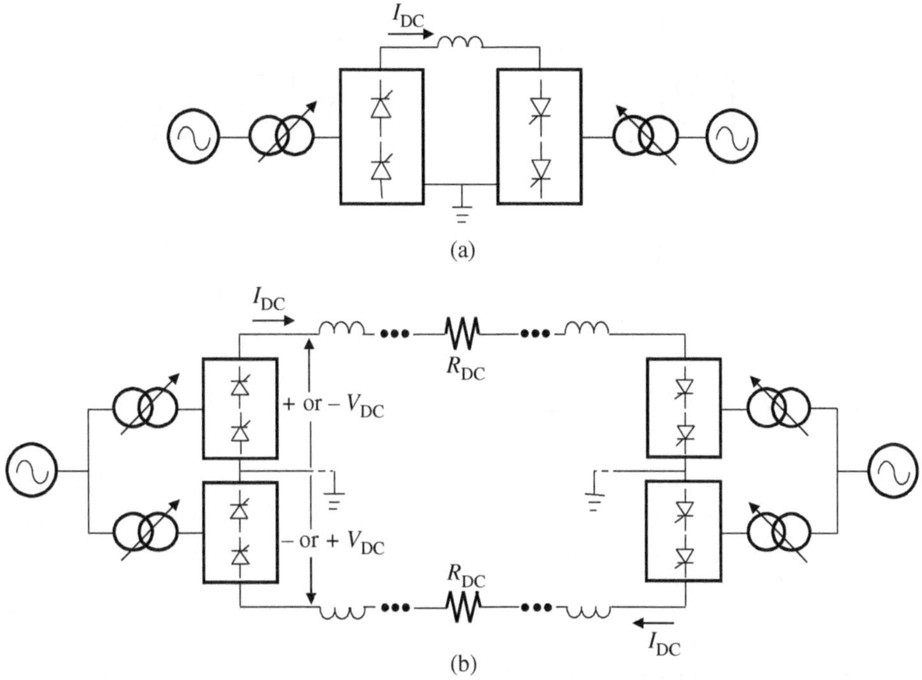

Figure 1.7 Two CSC-HVDC links: (a) back-to-back, monopolar HVDC; (b) point-to-point, bipolar HVDC.

Notice that the current flow is from the rectifier towards the inverter and so is the power flow when the voltage polarity is positive. Alternatively, the power follows an opposite direction to the current when the voltage polarity reverses, an operational characteristic achieved through firing angle control.

The two units of a back-to-back HVDC link have equal rating. They seem to be economical for voltage ratings as low as 50 kV. The bipolar link may be seen to comprise two monopolar links, one at positive and one at negative polarity with respect to ground. It is plausible to operate both monopolar links independently, each having its own ground return, but it is more effective to operate them together because their currents, being equal, cancel each other's ground return to zero. Indeed, the ground path is a valuable resource for cases when one pole is out of service due to a planned or an unplanned event [4].

This HVDC technology seems to have hit an intrinsic limitation when applied to multi-terminal HVDC systems because CSC-HVDC is based on current balances. Hence, only series, multi-terminal HVDC schemes seem to be realizable using this technology.

1.2.6 VSC-HVDC

In contrast to the UPFC, the two VSCs of an HVDC link are not constrained to be housed in the same substation [9]. They can be hundreds of kilometres apart and linked together by an overhead DC line or an underground or submarine DC cable, or a combination of these, to satisfy geographical, economical, technical and aesthetic requirements. Such

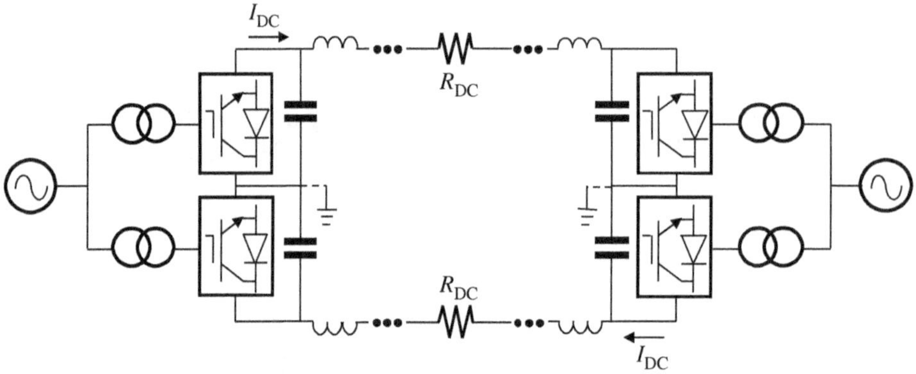

Figure 1.8 Point-to-point, bipolar VSC-HVDC system.

a VSC-HVDC link is termed point-to-point to differentiate it from the back-to-back VSC-HVDC link shown in Figure 1.6e. Since the point-to-point is aimed at bulk power transmission applications, it is normal to build it as a bipolar system as opposed to a monopolar one, as shown in Figure 1.8.

Note that the rectifier's AC system and the inverter's AC system do not necessarily need to have the same power frequency. The VSCs forming the bipole will be multi-level VSCs as opposed to the simpler two-level VSCs. Owing to the fast voltage-regulating capabilities afforded by the VSCs, it is likely that the connecting transformers will not need to have tap-changing facilities, saving on costs. Normally the rectifier is set to regulate power flow and the inverter is set to regulate DC voltage.

It is clear that the symmetrical bipole carries twice the rated power of a monopole HVDC link, with zero ground return. In the event of one of the poles being out of service because of maintenance or due to a contingency event, the link will remain in operation at 50% capacity. Tapping along the length of the DC line to pick up generation or to supply infeed points is carried out with ease. It requires only one extra converter station at each additional point. This is illustrated in the circuit diagram shown in Figure 1.9.

In some respects, the VSC-HVDC link in Figure 1.9 can be classified as a multi-terminal VSC-HVDC system [10], albeit of the radial type. As a matter of fact, the recent developments in VSC-HVDC technology are in the arena of multi-terminal VSC-HVDC systems; in particular, the kinds of multi-terminal systems which form meshed DC power grids. This is exemplified by the case of four VSCs interconnected in their DC sides through cables to make up a four-terminal VSC-HVDC system, as shown in Figure 1.10.

This is a generic concept that may be expanded to comprise n VSCs to link n AC systems of varying sizes, topologies and operational complexity. The DC cables may be overhead, underground or submarine, according to practical requirements.

The DC network may even comprise a single node – a common DC bus where the n VSCs would be sharing a DC capacitor, i.e. multi-terminal back-to-back configuration. Just as in meshed AC transmission systems there are transmission lines with non-regulated and regulated power flows, power flows sharing between neighbouring transmission lines, nodes with regulated and non-regulated voltages and so on, a meshed DC transmission system will have similar operational capabilities, as well as

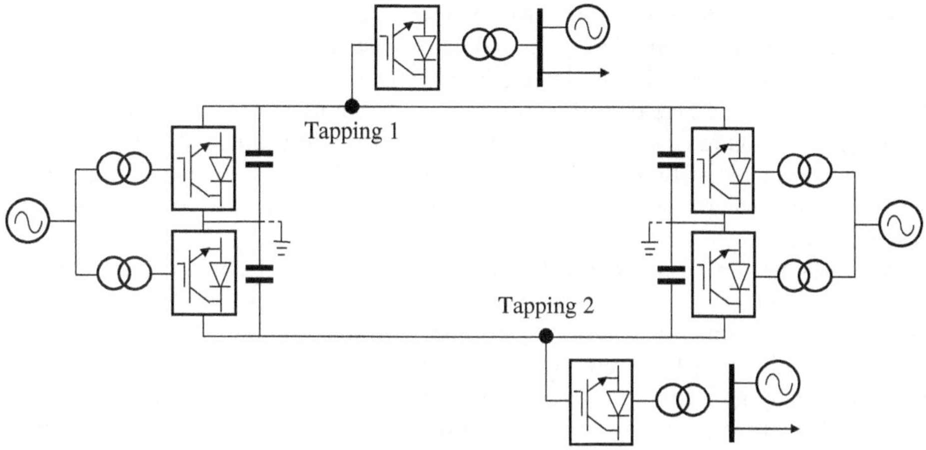

Figure 1.9 Radial, bipolar VSC-HVDC system with tappings.

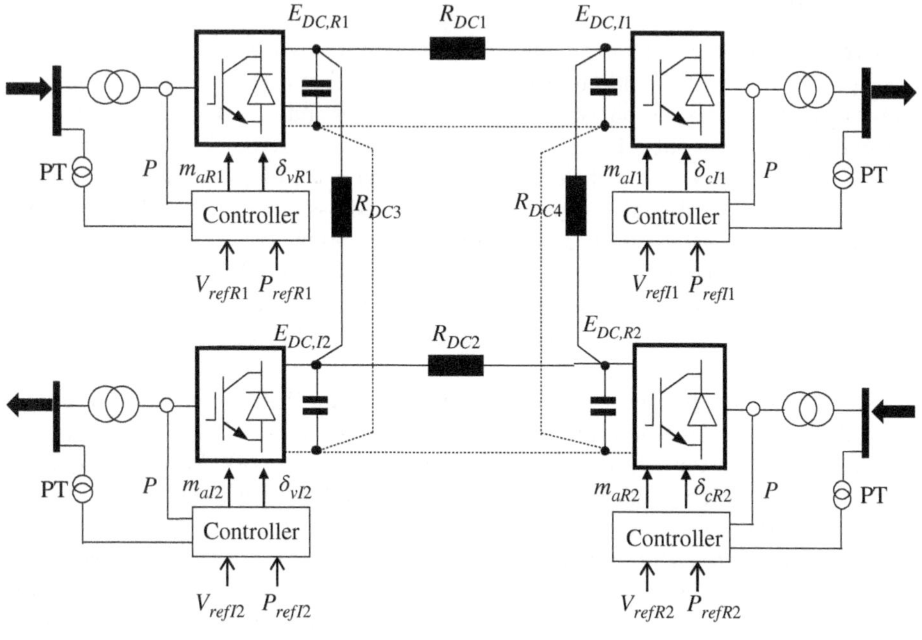

Figure 1.10 Four-terminal VSC-HVDC system.

suitable provisions for incorporating DC generation and DC loads directly. This will involve the use of DC-DC converters.

1.3 Flexible Systems Vs Conventional Systems

Conventional electrical energy systems have traditionally been divided into generation systems, transmission systems and distribution systems. The energy flows from the generation systems towards the distribution systems. It is assumed that generation exists neither in the transmission system nor in the distribution system. The three-phase voltage and currents waveforms are largely sinusoidal and symmetrical, with a current frequency of either 50 Hz or 60 Hz. In these AC systems, a transmission line is not normally loaded up to its thermal limit since angular stability limits take place at much lower values than the thermal limits. Furthermore, transmission lines longer than 300 km will normally require series and shunt VAR compensation to be able to operate in steady-state.

Power electronics valves, converter topologies switching techniques and control methods embedded in software have permeated all three sectors of the electricity supply industry, namely, generation, transmission and distribution. Power electronics has enabled the transformation of large, inflexible, inefficient, failure-prone and environmentally unfriendly power systems into flexible, efficient, reliable and environmentally benign power systems which may be large in size or a concatenation of microgrids.

1.3.1 Transmission

The large blocks of electrical energy which are moved from the large generating power plants to the cities and factories are transported at high voltages over long or extra-long distances. Three-phase AC transmission lines and HVDC transmission lines are employed to carry out this task. In particular, the latter option serves the purpose of transporting large amounts of electrical energy over extra-long distances, with lower power losses than a comparable high-voltage AC transmission line (HVAC) [4]. Further applications where HVDC outperforms the HVAC option, regardless of whether or not it uses FACTS upgrades, are submarine power transmission longer than 70 km and the interconnection of AC power grids exhibiting different operating frequencies or a substantial difference in network strength, i.e. connection of a weak network to a strong network [9].

It should be noted that the FACTS concept is based on the incorporation of power electronic devices and methods into the high-voltage side of the AC network to increase the control of power flows in the high-voltage side of the network during steady-state and transient conditions [31]. From the outset, the developers of the FACTS initiative emphasized that this was not intended to be a direct competitor to HVDC transmission but, rather, an initiative able to provide technical solutions to specific power transmission problems at a lower cost [1, 2], particularly when the AC transmission corridor already existed. In any case, the aim is to apply the FTS solution that carries the best technical performance and the best value for money for a specific power transmission problem [7]. By way of example, the following technical issues call for the application of FTS equipment and methods, either in AC form or in DC form:

- Higher power throughputs using the same right-of-way
- VAR compensation
- Frequency compensation/virtual inertia

1.3.1.1 HVAC Vs HVDC Power Transmission for Increased Power Throughputs

It is entirely feasible to have reinforced transmission systems using modern technology which perform their intended operating tasks in a rather smooth manner;

one which does not require more right-of-way than the original AC transmission system while increasing the power throughputs very substantially. Several options are available using FTS technology, one of which is to transport bulk power in DC form as opposed to AC form. The example presented in Figure 1.11 aims to illustrate this point.

Converting existing AC transmission systems to carry power in DC form, when there is a technical and economical justification for it, has been a longstanding aim [4, 32]. However, the technology associated with the DC transmission is still more expensive than the AC technology of comparable rating and the former, until quite recently, lacked the operational flexibility of the latter, although this paradigm is changing rapidly. The attractiveness of the conversion from AC to DC transmission is that at least a threefold increase in power-carrying can be achieved and with only a fraction of the investment that would otherwise be required. Certainly, this would be met with no need to widen the existing right-of-way, which nowadays is a major obstacle for building or expanding on existing transmission circuits.

As explained in [4], the insulation used in a standard AC transmission line to sustain the peak value of voltage to earth is entirely suitable to sustain a DC voltage to earth, particularly if atmospheric pollution is not a concern. Hence, the AC conductors become DC poles in the converted transmission system which may be operated up to their thermal rating. This contrasts with the AC conductors, where the transmission limiting factor is not the thermal rating of the cable but rather the stability limits associated with voltage phase angle. Accordingly, AC transmission limits have a great deal of slack capacity as far as their power-carrying potential is concerned.

In connection with the transmission systems shown in Figure 1.11, a conventional power transmission corridor is shown in Figure 1.11a, comprising three AC circuits running in parallel and delivering 800 MW of power, supplied mainly from hydro-power stations located 431 km away. There is intermediate and local generation and load points. An example of what was possible to achieve with the DC technology available in the early 1970s, the hybrid AC-DC transmission circuit, is shown in Figure 1.11b [32]. Two of the three-phase AC circuits are converted into two point-to-point HVDC links running in parallel. Each DC circuit is a bipole, one using two conductors per pole and the other using one conductor per pole, commensurate with the six conductors of the two three-phase circuits singled out for conversion. The power converters use thyristor valves and phase control since that was the standard DC technology available until the first decade of this millennium.

The CSC-HVDC technology does not allow for the tapping of intermediate load/generation points. Hence, the third AC circuit in Figure 1.1a is left untouched and used instead to supply the intermediate loads and to pick up the intermediate generation. The conversion of the two AC circuits into two bipolar DC circuits would increase almost fourfold the power-carrying capacity of the transmission scheme and with no compromises on operational flexibility. This is achieved at the expense of extra investment in power converters, additional transformers, compensation equipment, filters and switchgear, and protection equipment. Note that the rectification and inversion processes are imperfect and the characteristic harmonics are produced in both the DC circuit and the AC circuits of the HVDC links. Hence, AC filters are required for correct operation of the link and perhaps DC filters, if problems like interference in communications circuits are a cause for concern.

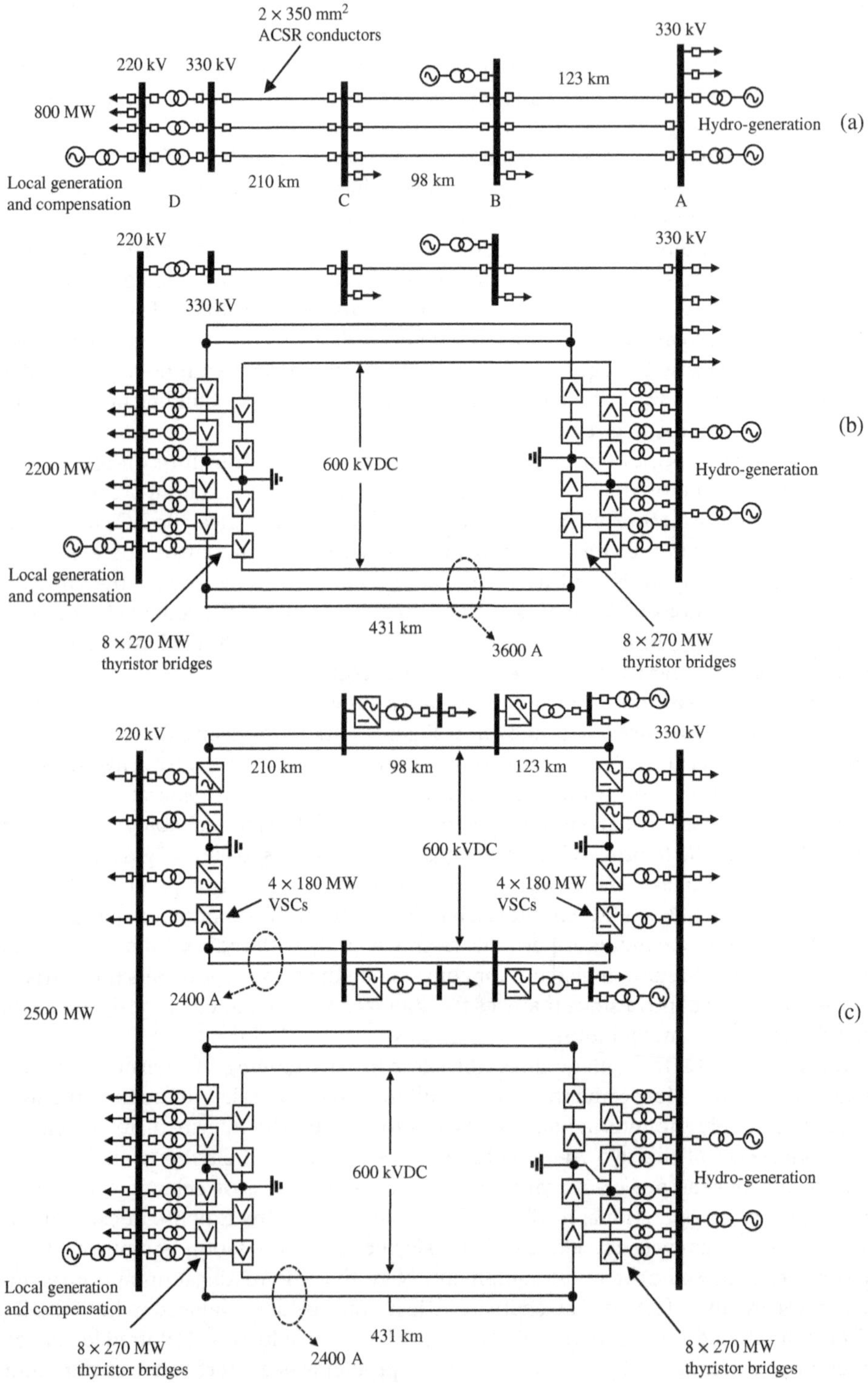

Figure 1.11 AC power transmission reinforcement/conversion using DC technology: (a) conventional AC transmission system [4]; (b) two AC circuits converted for DC transmission [4]; (c) three AC circuits converted for DC transmission.

A very important fact, from where we stand at this point in time, is that no extra land would be required for the upgraded right-of-way. Moreover, DC transmission systems do not contribute to fault level and this is normally considered to be an advantage, particularly in power grids where high levels of short-circuit ratio already have a problem. It should be noted that in bipolar HVDC systems, the poles may be operated independently when the need arises.

Reflecting on what is possible today, with the availability of a new generation of HVDC technology based on forced-commutated valves and modern switching valve techniques, it is possible to convert the AC transmission line in Figure 1.11b to transport electricity in only DC form. This requires the use of VSCs as opposed to CSCs. Contrary to the early HVDC technology, the new technology does allow for the tapping of intermediate load/generation points by installing as many inverter stations as required. In the all-HVDC transmission system shown in Figure 1.11c, the three-phase AC circuit has been converted into an HVDC bipole. Note that this is not a point-to-point scheme but rather a multi-terminal VSC-HVDC scheme, albeit of the radial form. Notice that one of the existing CSC-HVDC bipoles has been modified to include only three conductors as opposed to four. The fourth conductor is used to make up the new VSC-HVDC bipole.

The key point to bear in mind is that this upgrade fully meets the operational flexibility of the AC transmission system in Figure 1.11a while increasing more than fourfold the power-carrying capacity of the transmission corridor. Also, the use of VSCs increases the availability of dynamic reactive compensation at both the rectifier and the inverter sides of the hybrid HVDC transmission system, enabling a higher degree of system controllability. Equally important is the fact that the use of VSCs enables the supply of AC loads with no local synchronous generation, i.e. passive loads, such as the one connected to node C in Figure 1.11a.

1.3.1.2 VAR Compensation

Concerning shunt VAR compensation and the operational inflexibility associated with fixed capacitor banks and reactors, this issue has long been a concern for the power transmission industry and equipment manufacturers [5]. In the early days of AC power transmission, the requirements for dynamic reactive compensations were met with synchronous condensers, i.e. synchronous generators with no prime mover, which were a rather expensive and bulky option to provide dynamic VARs with. In the late 1970s, with the availability of affordable and reliable thyristor valve technology, TSC and TCR became commercially available to enable thyristor-fired VAR compensators, which had no moving parts, i.e. static, and exhibit dynamic reactive power capabilities, with an acceptable degree of operational performance. SVC designs improved within a decade and their prices dropped, displacing the synchronous condenser as the preferred option for providing dynamic reactive power support, except for very specialized applications. The use of shunt dynamic compensation became pervasive in high-voltage AC power transmission; it would be difficult to find a transmission system that does not have one or more of these compensators installed.

Towards the end of the millennium, a more powerful and versatile static VAR compensator emerged, which has been termed STATCOM. As discussed in Section 1.2, the STATCOM is essentially a VSC [8].

Steady-State Performance of Shunt VAR Compensation The steady-state power transfer concept, applied to the symmetrical transmission system shown in Figure 1.12, may be used to good effect to illustrate the effectiveness of adding a shunt compensator with dynamic capabilities, at the electrical midpoint [5].

In theory, the power transfer capability of a transmission line would double if an idealized shunt compensator were to be used at midpoint, as given by the following relationship:

$$P_{sh} = 2P_{\max} \cdot \sin\frac{\delta}{2} \tag{1.2}$$

where $P_{\max}$ is the ratio of the voltage magnitudes product at both ends of the transmission line, and its total inductive reactance. The angle δ is the phase angle difference of the two voltages at both ends of the line.

Having said that, some practical limitations should be borne in mind which decrease the effectiveness of the dynamic shunt controller. For instance, the compensator cannot respond instantaneously; the small positive slope in the characteristic of Figure 1.4, which accounts for the droop, and the inherent response delay in the controller combine to reduce the available area in the power-angle characteristic, as illustrated in Figure 1.13. Notice that the lost area is different for the SVC and the STATCOM, since the SVC will act as a fixed capacitor at low voltage.

Transient Performance of Shunt VAR Compensation By way of example, the simplified dynamic loops of the SVC and the STATCOM are shown in Figure 1.14 together with their simplified, schematic representation. The former takes the approach of modulating an equivalent susceptance in response to a voltage mismatch whereas the latter modulates directly the output AC voltage.

The SVC and the STATCOM will perform similarly when operating at around their design ratings. This statement can be extended to cover the dynamic range of operation. To show the effectiveness of the STATCOM to provide fault ride-through capability, its

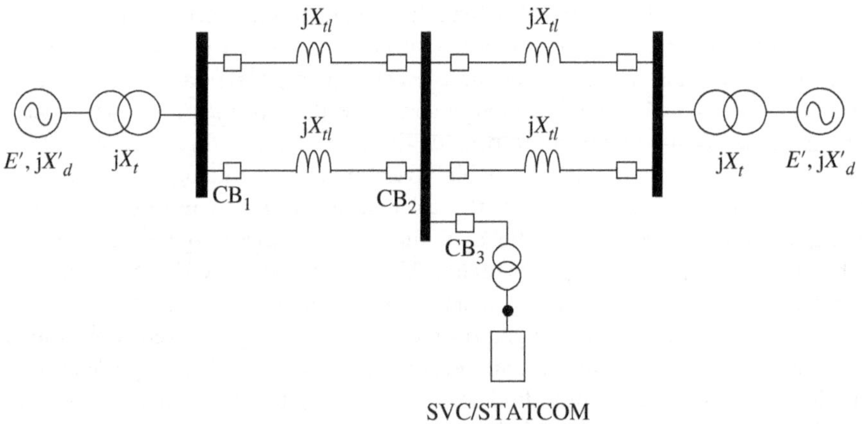

Figure 1.12 Symmetrical transmission system with midpoint dynamic shunt compensation.

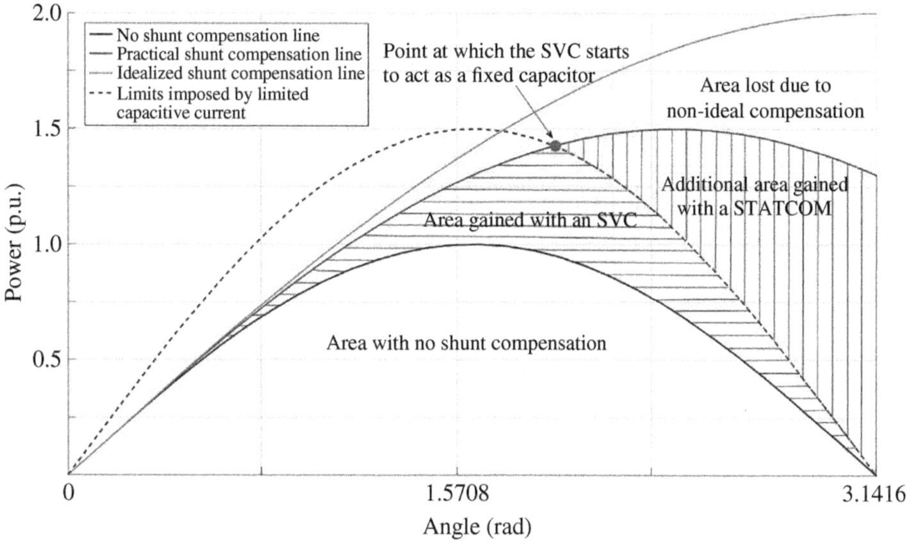

Figure 1.13 Power-angle characteristic for the transmission system shown in Figure 1.12.

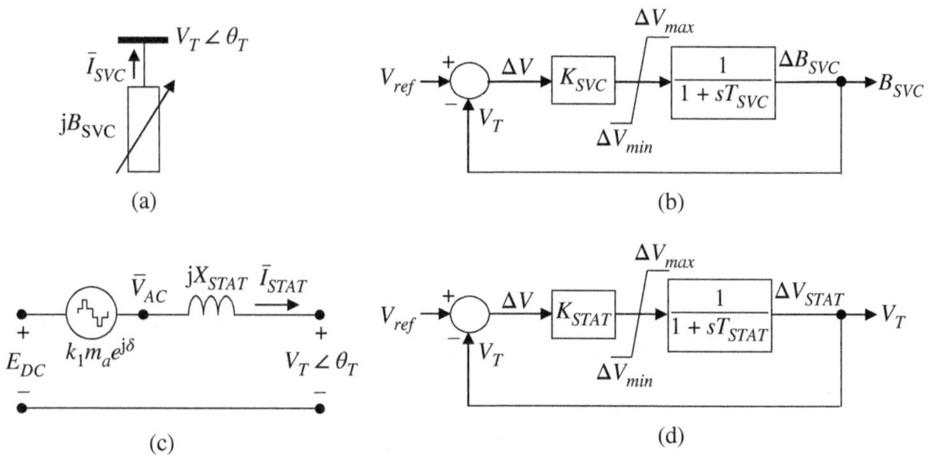

Figure 1.14 Dynamic SVC and STATCOM simplified representations: (a) schematic representation of the SVC; (b) dynamic control loop of the SVC; (c) schematic representation of the STATCOM; (d) dynamic control loop of the STATCOM.

voltage response to a three-phase fault applied to the electrical midpoint of the system shown in Figure 1.12 is given in Figure 1.15.

To visualize better the marked improvement that controllable-shunt compensation has on power systems dynamics, the classical equal area criterion for transient stability assessment is used below, as applied to the symmetrical transmission system with midpoint compensation, shown in Figure 1.12.

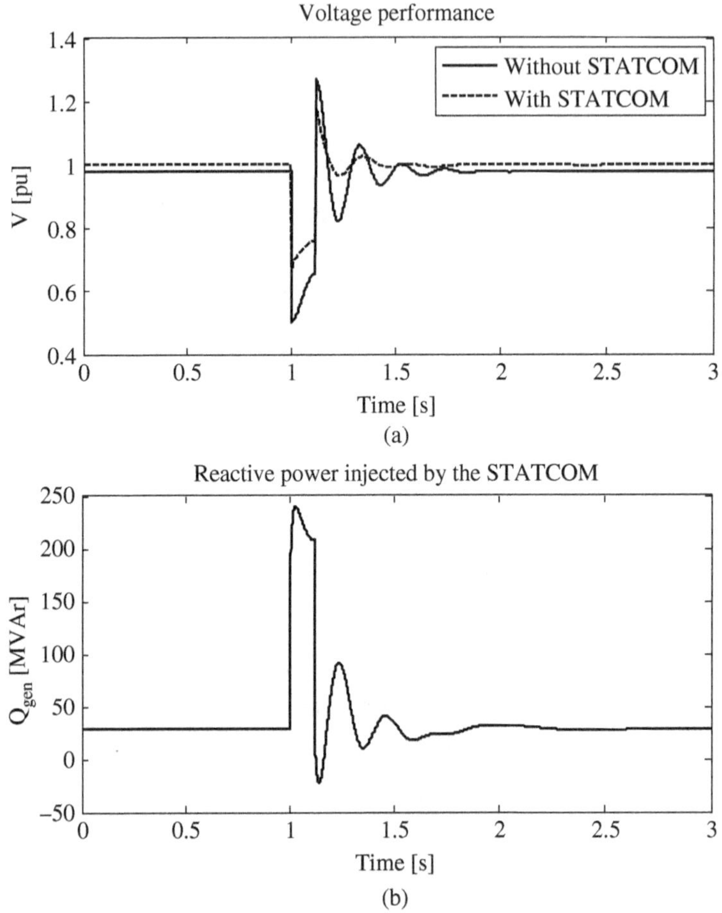

Figure 1.15 Fault ride-through capability of the STATCOM: (a) voltage performance at the electrical midpoint of the system shown in Figure 1.12; (b) reactive power generated by the STATCOM.

Let us consider first the case when CB_3 is in open position and a short-circuit fault takes place between CBs 1 and 2; the fault is cleared by the opening of CBs 1 and 2. The three power-angle curves associated with the three different states of the transmission system are shown in Figure 1.16a. The top curve corresponds to the healthy system – before the fault occurred. The bottom curve is for the period when the fault is in place, whereas the middle curve illustrates the situation when the faulted transmission line has been removed by the opening of CBs 1 and 2. To ensure that the overall system remains stable, these CBs should open at a time termed critical clearance time, t_c, which carries an associated critical clearance angle, d_c.

In connection with Figure 1.16a, due to the occurrence of the fault, the energy enclosed by area A_1 accelerate the rotors of the synchronous machines. Notice that before the occurrence of the fault, the system was operating in a stable equilibrium (P_1, d_1). Removal of the faulted transmission line results in rotor speed and angle

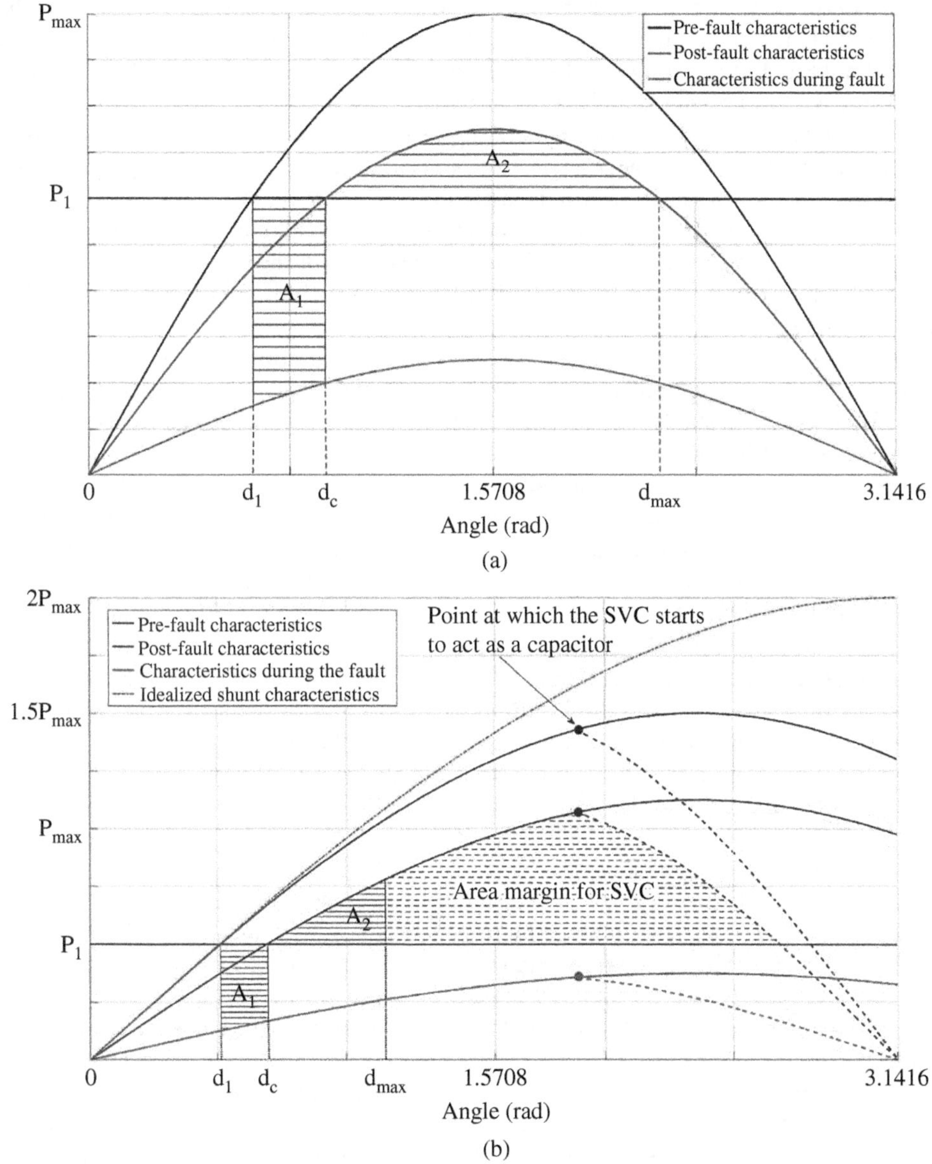

Figure 1.16 An application of the equal area criterion used to illustrate the stability margins gained by using midpoint shunt compensation in connection with the symmetrical transmission system of Figure 1.12: (a) power-angle characteristics with no compensator; (b) power-angle characteristics with dynamic shunt compensator (SVC/STATCOM) of large rating, suitable for power transmission applications.

oscillations. To ensure overall system stability, the area A_2 enclosed by the post-fault electrical characteristic and the unchanged, constant mechanical power between the angle limits d_c and d_{max} ought to be equal to area A_1. Otherwise, if area A_1 is greater than area A_2, the system would become unstable.

Let us consider now the previous operating scenario but with CB_3 in closed position. The corresponding three power-angle curves associated with the three different states of the transmission system but with the dynamic shunt compensator in place are shown in Figure 1.16b. The top curve corresponds to the pre-fault state. Notice that all three curves incorporate the limiting boundaries associated with the operation of a realistic dynamic shunt compensator, introduced in Figure 1.13. This applies to both the SVC and the STATCOM.

Let us examine now the case with dynamic shunt compensation, for the same pre-fault power as for the case with no compensation. Hence, area A_1 should be equal to area A_2 in order to ensure a stable system. It is observed from Figure 1.16b that this criterion is fulfilled rather handsomely; the available decelerating area enlarges as a result of the shunt compensator and is only partially used, increasing the transient stability margin (or limit) by a significant margin before the SVC starts behaving as a fixed capacitor. This means that the compensated transmission system could be made to work harder during steady-state operation than the system with no compensation, at least in principle, up to the point in which all the decelerating area is used up. Notice that the slack area is larger for the STATCOM than it is for the SVC.

Having said that, one has to bear in mind the rather simplifying nature of the equal area criterion for stability, such as the assumption of constant mechanical power and constant internal voltage of the synchronous generators. Detailed transient stability studies should be conducted using any of the industry-grade transient stability computer programs available.

1.3.1.3 Frequency Compensation

Large increases in wind and solar penetration in a power system present new challenges in its operation. This is intermittent generation, which may affect adversely the operation of the high-voltage transmission system in a number of ways. For one, the system-level power balance would become more difficult to attain, but also voltage control may become a significant problem under certain conditions [7]. Hence, the measures required to ameliorate the problems brought about by power fluctuations are worth investigating. For instance, the MW-level application of battery-BESSs is a case in point. It would be fair to say that interest in this application has been longstanding but the BESS technology had not matured sufficiently to warrant much interest from transmission system operators. However, this is changing rapidly with the advent of a new generation of BESS using Li-ion batteries and state-of-the-art power electronic converter topologies and control techniques [15]. Reports on operational experiences in two of the newest installations, using state-of-the-art technology, are rather encouraging [29, 30].

The structure of a modern BESS is shown in Figure 1.17, comprising several parts: (i) battery pack; (ii) DC-DC boost converter; (iii) VSC; (iv) step-up power transformer; (v) controls.

The DC-DC converter is responsible for active power control. It is designed to have bidirectional capabilities, with large enough capacitors to keep the ripple in its output

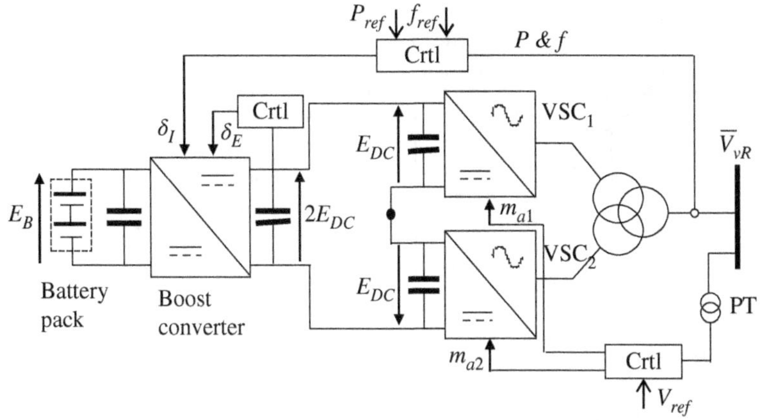

Figure 1.17 Structure of a BESS.

voltage as small as possible and to keep the voltage between the switches constant as it transfers power between the batteries and the VSC. The VSC injects/draws reactive power by controlling the output DC voltage of the inverter. Apart from enabling the connection of the battery, through the DC-DC converter, to the AC system, the main control objective of the VSC is to regulate the output voltage of the inverter and with it the reactive power flow.

In order to control the active power exchange between the storage system and the grid, the duty cycles of the DC-DC converter switches are adjusted. One switch controls the power drawn from the battery and another switch controls the power injected into the battery. Needless to say, the two switches cannot be closed at the same time. The switches also control the terminal voltage of the battery [15]. When the terminal voltage is higher or lower than the internal voltage of the battery, the converter injects or draws power from the battery, respectively.

It is well known that synchronous generators regulate frequency by adjusting the amount of power injected into the power grid. Following a system load change, all generators' rotors accelerate/decelerate to compensate the momentary power imbalance between the electrical and mechanical powers. This difference in power will be made up by the rotating mass (kinetic energy) of each generator. This is followed by the action of the speed-governing system, which acts to reduce the error between the nominal and the actual generator's speed by changing the input power to the turbine. This control system is termed load-frequency control [33].

It has been observed that a BESS possesses similar operating characteristics to those of a conventional power plant, even though it has no rotating parts. This makes the BESS much quicker to respond. The energy stored in the battery has the characteristics of inertia by acting as a stored kinetic energy. When the system load exceeds the power delivered by the synchronous generators, the BESS injects power into the grid, hence the battery's state of charge (SOC) decreases, and vice versa.

To emphasize further the operational resemblance between the BESS and a conventional power plant, an analogy between a BESS and a hydropower generator presented in [15] is reproduced in Figure 1.18.

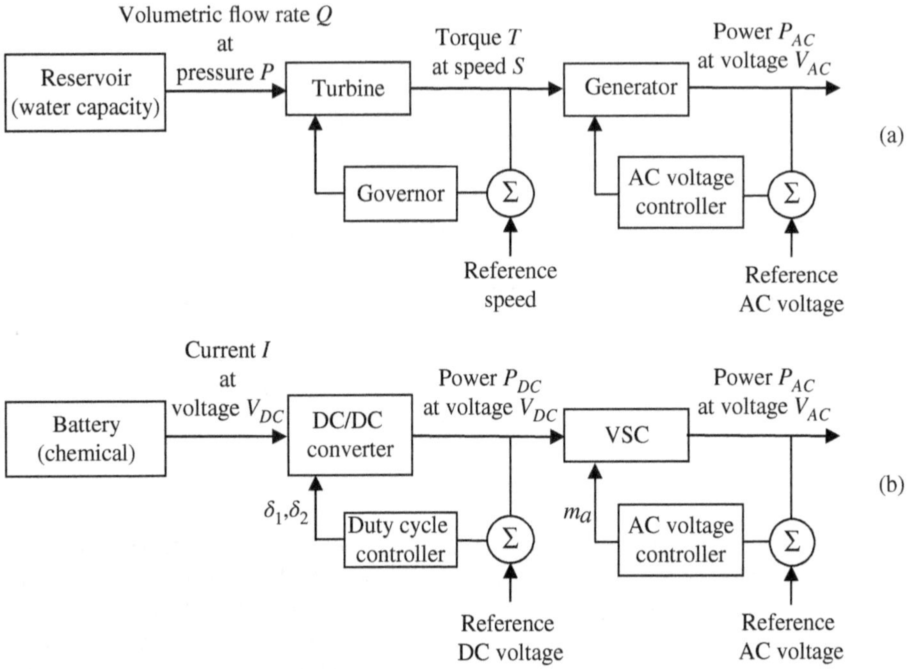

Figure 1.18 (a) A conventional turbine-governor control of a synchronous generator; (b) the BESS voltage and power controls (©IEEE, 2016).

The governor controls the inlet of water into the turbine (with the reservoir being discharged) and the duty cycle controller manages the rate of discharge of the battery. Note that the case of battery charging in this analogy would correspond to the case of hydro-pumped storage. The AC voltage controller in both cases is a regulated gain. At the outputs of the reservoir and battery blocks, the volumetric flow rate Q is akin to current I and pressure P is akin to voltage V_{DC}. Likewise, at the output of the turbine and DC-DC converter blocks, torque and speed are akin to power and AC voltage, respectively.

In order to illustrate the effectiveness of the BESS to impact on system frequency, we use the contrived test system in Figure 1.19. However, notice that this system is a simplified reflection of the practical BESS installation in Antofagasta in northern Chile [30]. The system responses to a load increase and a load decrease are presented with BESS and with no BESS in Figures 1.20 and 1.21, respectively.

The case of 100 MW load increase with BESS and with no BESS is discussed first. The positive contribution of the 22 MW BESS is self-evident in the system frequency plot in Figure 1.20, showing the result when the BESS is connected and when it is not. At a sudden increase in load, the frequency drop is smaller when the CB is in closed position.

The BESS not only improves the inertial response (i.e. first swing), it also reduces frequency oscillations and restores the system frequency to its 60 Hz nominal frequency (i.e. it provides secondary frequency support).

As expected, the case of 100 MW load decrease would also benefit from the use of the BESS, in very much the same manner as the case for the load increase, even though

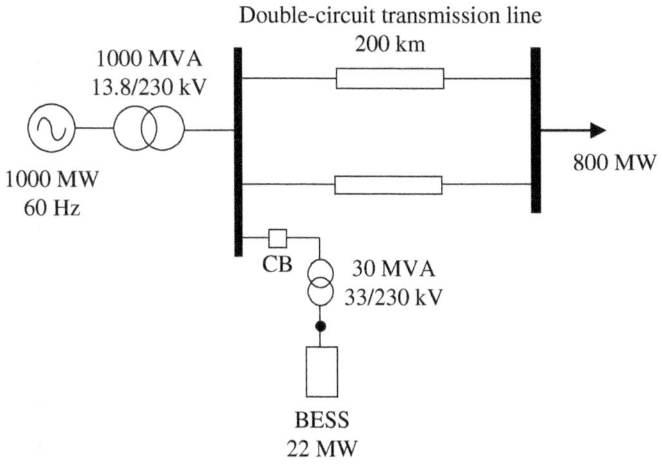

Figure 1.19 Test circuit to assess the frequency response of a BESS.

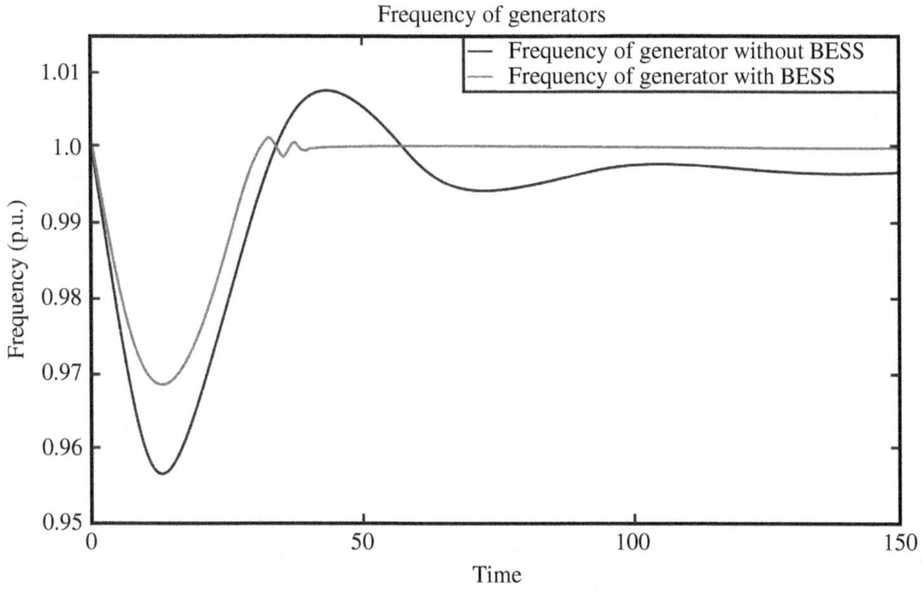

Figure 1.20 Frequency response of the BESS in the transmission circuit in Figure 1.19 – due to a load increase.

slightly more frequency oscillations are experienced in the BESS response, as seen from the frequency plots in Figure 1.21.

1.3.2 Generation

The vast majority of the electrical energy consumed today is produced by large, three-phase synchronous generators which use, as primary energy resources, the energy stored in huge water reservoirs, in fossil fuels and in nuclear materials [33].

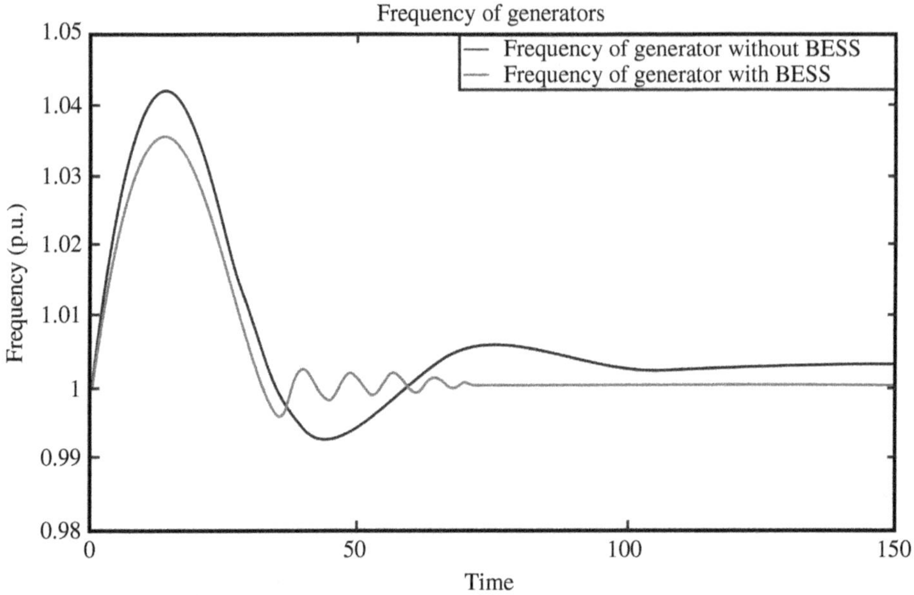

Figure 1.21 Frequency response of the BESS in the transmission circuit in Figure 1.19 – due to a load reduction.

However, there is widespread consensus that burning fossil fuels at the current rate is having adverse consequences for the global climate on a scale that poses a threat to many regions of the world.

A large reduction in greenhouse gas emissions seems not only desirable but essential in order to reign in global warming. For the last decade, there have been calls from many governments around the world to have in place a 60–80% cut in emissions before 2050. However, reducing global emissions is one of the greatest challenges facing mankind because environmental wellbeing and economic growth are heavily inter-linked.

Cost-effective, low-carbon electricity generation sources, together with effective demand-side and energy-saving measures, are pre-conditions for a low-carbon electric energy sector. Growth in wind and solar generation continues to outstrip all other forms of renewable generation, driven by concerns over climate change and energy diversity and by technological break-through in equipment and methods, and a better understanding of the wind and solar resources.

1.3.2.1 Wind Power Generation

The early commercial wind turbines were very basic in terms of their construction and their operation. They used a squirrel-cage induction generator (SCIG) driven above the synchronous speed by coupling the low-rotating shaft of the wind turbine and the high-speed shaft of the induction generator, using a gearbox [13]. This is illustrated in Figure 1.22a.

They run at almost constant speed (1–2% above synchronous speed) and that is the reason why they are known as fixed-speed induction generators (FSIGs). On the electrical side, the stator winding is connected to the secondary side of a step-up transformer, with the primary winding of the transformer connected to the AC power grid.

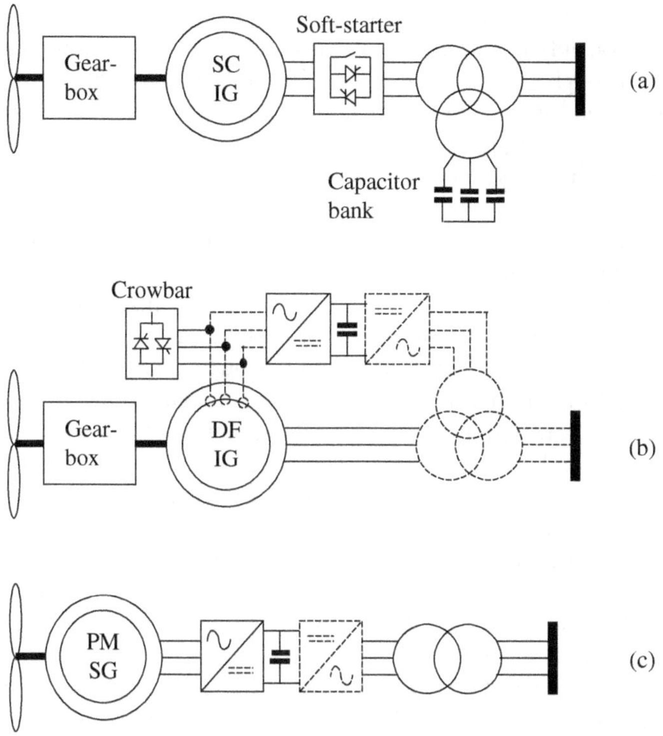

Figure 1.22 Schematic description of (a) FSIG, (b) DFIG and (c) PMSG with fully rated converters.

The reactive power required to flux the air-gap field of the induction machine was imported from the power grid or fixed bank of capacitors were installed at the wind farm for improved overall performance. Notwithstanding this, the following technical issues started to raise serious concerns among power system operators when the wind farms began to grow in capacity: (i) steady-state voltage control, (ii) reactive power support, (iii) transient/dynamic stability, (iv) fault ride-through capability and (v) power quality [34].

The reason for these technical issues was that the most likely location for an on-shore wind farm was in rural areas where the network was poorly developed (it had a low X/R ratio); hence, fluctuations in the wind speed induced power output changes which, in turn, led to fluctuations in the network voltage. Connection to a weak network also caused problems for the wind farm and network during transient conditions, which led to instability. These early wind generators were prone to instability following a network fault due to excessive reactive power absorption at high speeds. Nowadays, wind generators are required to have fault ride-through capability. From the power quality perspective, flickering was a concern.

Aiming to overcome these issues, power electronic-based compensators, such as the SVC and the STATCOM, were employed to provide dynamic reactive power to the wind farm. Particularly with the use of the STATCOM, those first-generation wind farms went a long way towards satisfying the more stringent low-voltage ride-through capability imposed by new grid codes [7].

Meanwhile, a new generation of wind turbines became available which were larger in size, able to capture the available wind resource in an optimal manner and with a much more enhanced ride-through capability, whose performance rivalled that of the conventional synchronous generators for any practical purpose relating to grid codes requirements. This was possible through the incorporation of power electronics technology. Suitable use of a back-to-back VSC-HVDC link enabled the electrical generators to have variable-speed capabilities.

Two quite distinct technologies emerged: the doubly-fed induction generator (DFIG) and the permanent magnet synchronous generator (PMSG). Both topologies are illustrated schematically in Figure 1.22b,c, respectively. In the former, a controlled flux is superimposed onto the rotor flux using a fractional VSC and slip rings. In the latter, a controlled voltage is injected directly into the stator winding through a fully rated VSC. In both cases, it has the effect of varying, in a finely regulated manner, the speed of rotation of the machines aiming at matching, in an optimal manner, the available wind resource. The DFIG employs an induction generator with a wound rotor whereas the PMSG employs multi-pole designs with permanent magnets to provide the required rotor excitation to enable the rotor to rotate at synchronous speed. However, the speed of rotation of the PMSG is very low; in fact, the speed of rotation is the same as that of the blades, hence no gear-box is required, as illustrated in Figure 1.22c. It follows that its current frequency is very low compared with the current frequency of the power grid and its connection to it is carried out through a back-to-back HVDC link, not restricted to a VSC-HVDC link. Sometimes, on economic grounds, a hybrid HVDC link has been suggested. It may be argued that a simple diode bridge rectifier could be used instead of the VSC connected to the stator winding of the PMSG in Figure 1.22c.

The induction generators of the kind used in wind power normally employ four poles and their speed of rotation must be quite high to produce a current frequency that matches that of the power grid; hence, the use of gear-boxes becomes mandatory. The FSIG, illustrated in Figure 1.15a, is driven above the synchronous speed to be able to perform as a generator. However, the DFIG, illustrated in Figure 1.15b, operates under a range of controllable speeds around the synchronous speed. In the presence of short-circuit faults in the vicinity of the installation, the crowbar quickly short-circuits the rotor windings to prevent large fault currents from entering into the rotor winding and causing terminal damage; the DFIG effectively becomes a FSIG for the duration of the short-circuit fault. This increases the ride-through fault capability of the DFIG technology, as shown in Figure 1.23. The DFIG is less expensive than the PMSG of comparable rating because its machine-side converter (rotor converter) is only a fraction of the machine-side converter (stator converter) of the PMSG. Moreover, its electrical generator uses standard rotating machinery designs compared with the PMSG designs employed in wind power applications.

1.3.2.2 Solar Power Generation

Solar power generation using the PV phenomenon is a growing industry around the world, particularly the case of grid-connected PV installations [14]. Since a PV generator produces voltage and current in DC form, connection to the AC power grid requires a power inverter. Early PV designs were able to track only the maximum amount of solar power available while maintaining a unit power factor at the connection point with the AC power grid. Modern PV designs incorporate a DC-DC boost converter,

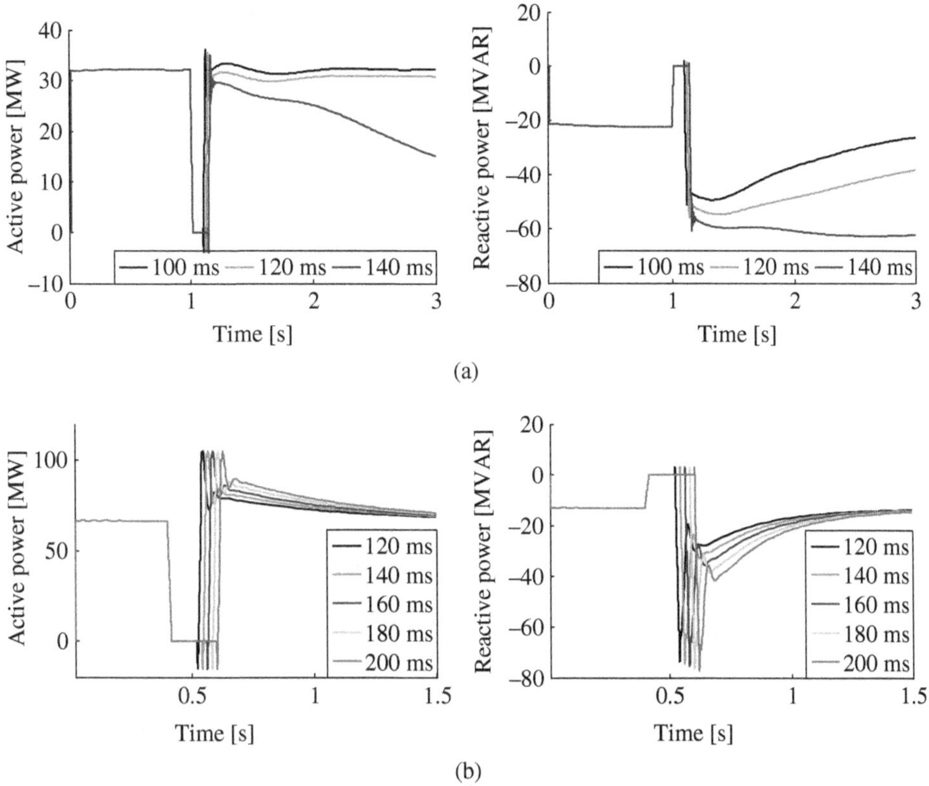

Figure 1.23 Fault ride-through capability tested for a (a) FSIG-based wind farm and (b) DFIG-based wind farm, for different time-fault durations.

which sits between the PV panels and the power inverter. This has enabled the new generation of PV installations to improve operational performance – they can perform voltage regulation at the point of common coupling with the AC power grid and provide frequency support, albeit this would be achieved at the expense of spilling out some of the solar resource available [35]. A schematic structure is shown in Figure 1.24.

To enable an even more powerful PV-based system, which would rival fully the operational flexibility of conventional, large synchronous generators, including the much after-sought characteristic of firm generation, the PV installation could be combined with local energy storage. In particular, BESS and superconductor magnetic energy storage (SMES) systems are most suitable because they have a relatively small footprint and use power electronics similar to those of the modern PV generator. The scheme, shown in Figure 1.25, would capture and store all the solar resource available, regardless of the grid-support function that it is required to perform by the grid operators [36].

Provided there is sufficient energy in the storage system, the associated controllers can quickly react by injecting active and reactive power into the AC system to provide frequency and voltage support in the event of power system disturbances.

One area of concern is that as renewable power continues to displace larger blocks of conventional generation, AC power grids are losing inertia and synchronizing power, with the ensuing detrimental effects on power system stability. To overcome

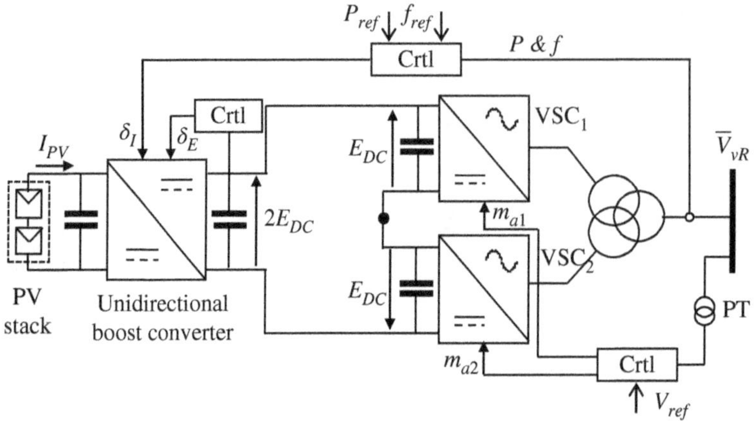

Figure 1.24 Structure of a PV generator.

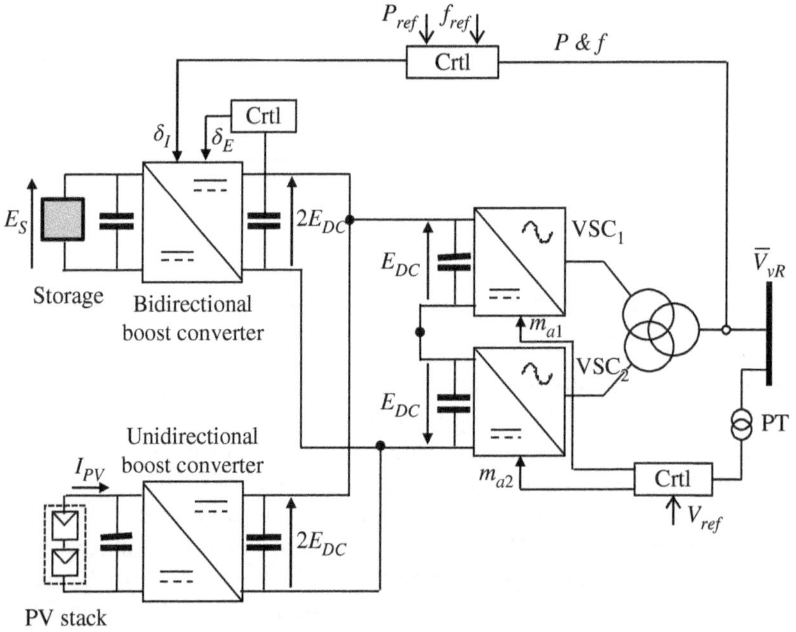

Figure 1.25 Structure of a PV generator with storage.

this undesirable side effect of increasing the generation share of the badly needed renewable generation, the power industry is engaged in the development of controllers for renewable generators and storage systems, to enable them to assist in primary frequency support by contributing synthetic inertia [7].

In order to illustrate the effectiveness of the BESS, to support the power output of a PV generator undergoing irradiance and temperature variations, the output powers shown in Figure 1.26 are self-evident. They show the case with no BESS and with BESS, respectively.

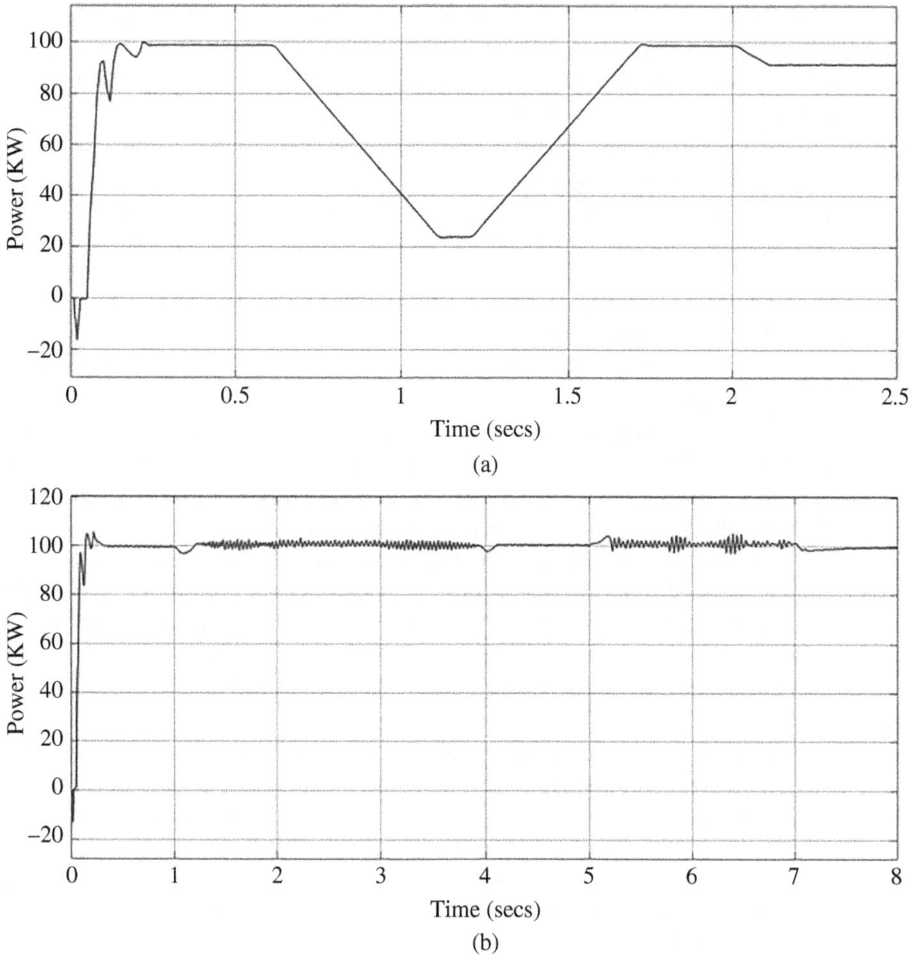

Figure 1.26 Power responses of (a) PV generator with varying solar irradiance conditions and (b) PV generator with BESS with varying solar irradiance conditions [36].

1.3.3 Distribution

It is widely acknowledged that the distribution system load is unbalanced and non-linear, in stark contrast to the bulk load points at the transmission level, which are taken to be balanced and linear. To limit the adverse effects that such loads may introduce into the distribution system, the industry has developed stringent grid codes aimed at maintaining high power-quality standards to increasingly demanding end users [7].

Moreover, the past two decades had seen a marked increase in the deployment of end-user equipment that was highly sensitive to poor power quality supply. In the early 1990s, several large industrial users were reported to have experienced significant financial losses as a result of even minor lapses in the quality of electricity supply [3]. A great many efforts were made to remedy the situation, where solutions based on the use of the latest power electronics technology figured prominently.

Indeed, custom power electronic technology, the low-voltage counterpart of the more widely known FACTS technology, aimed at high-voltage power transmission applications, emerged as a credible solution to solve many of the problems relating to continuity of supply at the end-user level. Both the original FACTS and custom power concepts were developed at EPRI [1, 3].

Several power electronics controllers have been developed as part of the custom power applications, such as the STATCOM for low-voltage applications, known as D-STATCOM, the dynamic voltage restorer (DVR) and the solid-state transfer switch (SSTS). These are system-level pieces of equipment which lies at the interface between the distribution system and industrial users, aiming at mitigating upstream disturbances [37], as indicated in Figure 1.27.

The D-STATCOM is aimed at load compensation whereas the DVR and SSTS provide dynamic voltage support.

To a greater or lesser extent, the short-circuit faults cause voltage sags in the neighbouring areas where they take place; in turn, voltage sags may impair industrial processes. Even though events like capacitor switching and voltage sags due to motor starting do play an important role in equipment tripping, by and large equipment tripping is due to short-circuit faults in the distribution system. A number of mitigation actions could be applied to reduce the number and severity of such unforeseen events. For instance: (i) to improve the power system design so that it reflects in less severe and frequent events; (ii) to increase the immunity of equipment; (iii) to use mitigation equipment at the interface between the utility company and users with sensitive equipment [37].

The Custom Power equipment, installed at the interface between a utility company and an industrial user with sensitive loads, is commercially available from several vendors. The following are some of the potential benefits of the Custom Power technology to end users with critical loads: practically no power interruptions, magnitude and duration of voltage dips/swells within specified, stringent limits, low harmonic distortion, low phase unbalance, acceptance in the neighbourhood of fluctuating, non-linear and low power factor loads [3].

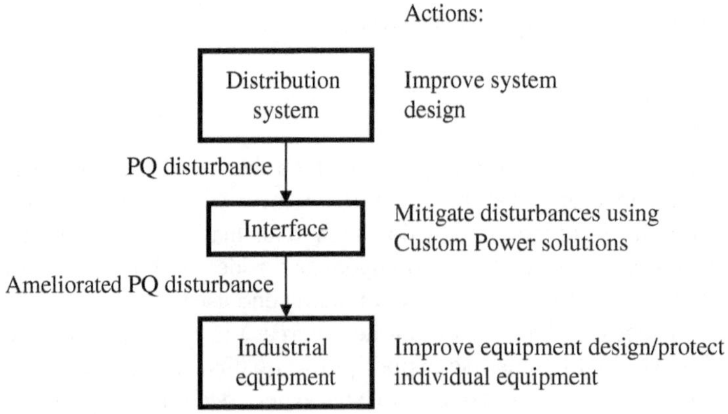

Figure 1.27 Upstream PQ disturbances and actions.

1.3.3.1 Load Compensation

The D-STATCOM is the conventional solution used to overcome power quality (PQ) phenomena such as voltage imbalances, poor power factor and harmonic distortion. If the D-STATCOM's control possesses harmonic cancellation capabilities, it is termed active power filter (APF). If the D-STATCOM is fitted with an energy source on its DC bus, such as a battery pack, it is termed BESS. Concerning the amelioration of the above PQ phenomena, these devices inject a set of suitable compensating currents, which restore the operating currents to their ideal, balanced and sinusoidal form.

1.3.3.2 Dynamic Voltage Support

Voltage sags/swells have the potential to impair industrial processes, with entire production lines shutting downs [3]. To a larger or greater extent, these power quality disturbances are the result of upstream short-circuit faults. Design improvements on the power grid and intelligent equipment maintenance programs are preconditions for reducing the number and severity of short-circuit faults; however, it is highly unlikely that these will be eradicated for good [37]. In view of this, a practical alternative for industrial users with sensitive equipment and processes, such as paper mills, is to shield their installations against the adverse voltage phenomena resulting from upstream short-circuit faults, namely voltage sags/swells. Possible options to protect their loads include using a DVR or an SSTS. The DVR injects a series voltage in phase with the incoming voltage to restore the voltage waveform to what it was prior to the fault. The SSTS ensures continuous high-quality power supply to sensitive loads by transferring, within milliseconds, the load from the faulted bus to an auxiliary healthy bus [3].

The test system shown in Figure 1.28 contains a sensitive load connected to Node 2. Nodes 1 and 2 are linked through the primary winding of a series-connected transformer. A DVR connects to the secondary winding of the transformer through a switch S_1. Note that this test system is an abstraction of Figure 8 in reference [38].

Two scenarios are considered below, one with no DVR and the other with DVR. The former scenario assumes that a three-phase-to-ground short-circuit through a fault resistance occurs at Node 1 in the time period 300–600 ms of the simulation. The RMS voltage profile in Figure 1.29a shows the expected behaviour of the voltage as measured

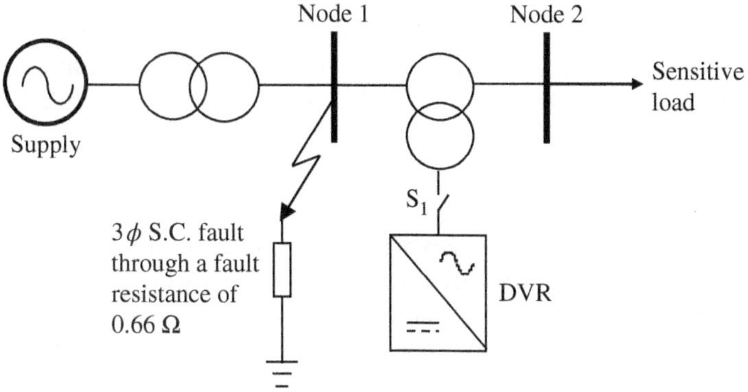

Figure 1.28 Equivalent circuit of a shielded sensitive load against upstream short-circuit faults.

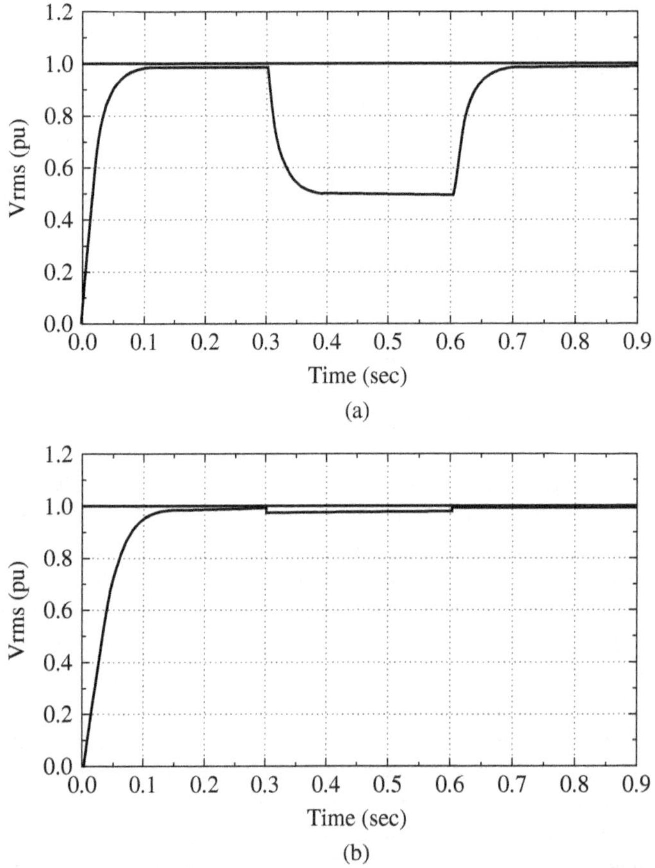

Figure 1.29 RMS voltage response at Node 2 of Figure 1.28 when a three-phase-to-ground short-circuit fault takes place at Node 1 through a fault resistance of a 0.66 Ω, (a) with no DVR and (b) with DVR (©IEEE, 2002).

at Node 2. The voltage sag at the load point is 50% with respect to the reference voltage. The simulation experiment is repeated but now with switch S_1 connected during the time of the fault. The new RMS voltage profile is shown in Figure 1.29b. The last simulation result shows the effectiveness of the DVR to provide dynamic voltage support.

1.3.3.3 Flexible Reconfigurations

Distribution systems may be classified as either radial or meshed networks, according to their structure. However, they are normally operated as if they were radial networks, owing to the simpler operation and cheaper protection equipment of radial networks. Heretofore, this practice has served well the aims and objectives of the electrical power utilities worldwide; however, the widespread installation of small-scale generators of various kinds at the customers' premises is challenging this long-standing practice. This array of new sources of electrical energy is termed distributed generation (DG) and is said to be turning the otherwise passive distribution systems into active distribution systems. One distinct characteristic of the passive distribution systems is that the power flows are unidirectional, i.e. from the utility company towards the customers, whereas in the active distribution systems the power flows are, at least in principle, bidirectional [7].

With the advent of reliable and affordable power converters for low-voltage, low-power applications, this technology is finding an increasing number of uses. For instance, the concept of switching centres has been put forward in [39]. This concept is illustrated with reference to Figure 1.30, where a four-terminal switching centre is installed at the far end of four three-phase, AC distribution feeders. It should be noticed that in all respects the switching centre is a back-to-back, multi-terminal VSC-HVDC system. It enables controlled power exchanges between the feeders through a common DC bus. The DC bus is also a useful resource for incorporating a wide range of storage and electrical energy sources, which produce power in DC form.

It is argued in [39] that key advantages of a flexible bridge, with respect to mechanical switches, are:

- The DC link is capable of carrying out scheduled exchanges of active power through the common DC bus, in a continuous fashion.
- Controlled injections of reactive power at the far end of the feeders are carried out by their respective VSCs. This is particularly useful when the switching centre is located at the far end of the feeder since the voltage drops will be more pronounced at such points in passive feeders. The opposite may occur in feeders containing distributed generation where the risk of over-voltages is high.
- The VSC technology is very effective in overcoming adverse PQ environments, i.e. poor voltage regulation and power factor, voltage imbalances, low-order harmonics.
- Anomalous phenomena such as short-circuit currents are not passed on between feeders owing to the almost instantaneous speed of response of the VSCs, which act to block the flow of anomalous currents.

One concern with the use of this technology that springs to mind is the high investment costs required, compared with those of an electromechanically based switching centre. However, it is conjectured that this technology may be cost effective when accompanied by the substantial integration of DERs [40].

1.3.3.4 AC-DC Distribution Systems

A conventional three-phase AC distribution feeder may be converted to carry DC current instead of AC current, with the use of suitable power electronic converters, as exemplified by the AC and DC distribution systems shown in Figure 1.31a,b, respectively. In the DC system, the conventional three-phase loads may be served in a more controllable manner while enabling the direct connection of DC distributed energy resources (DERs), such as battery packs, fuel cell stacks, PV stacks and EV recharging stations. Furthermore, if the amount of energy produced by the in-situ generation and storage is sufficient to meet the local demand plus the associated power losses, then the DC feeders may act as an independent microgrid, isolated from the HV utility supply, with the AC connection points taking the character of backup supply points.

It is stated that key concerns in today's distribution network development are cost effectiveness and system reliability [21]. In Finland, the distribution voltage levels were 20/0.4 kV until the benefits of an intermediate 1 kV voltage level was established, i.e.

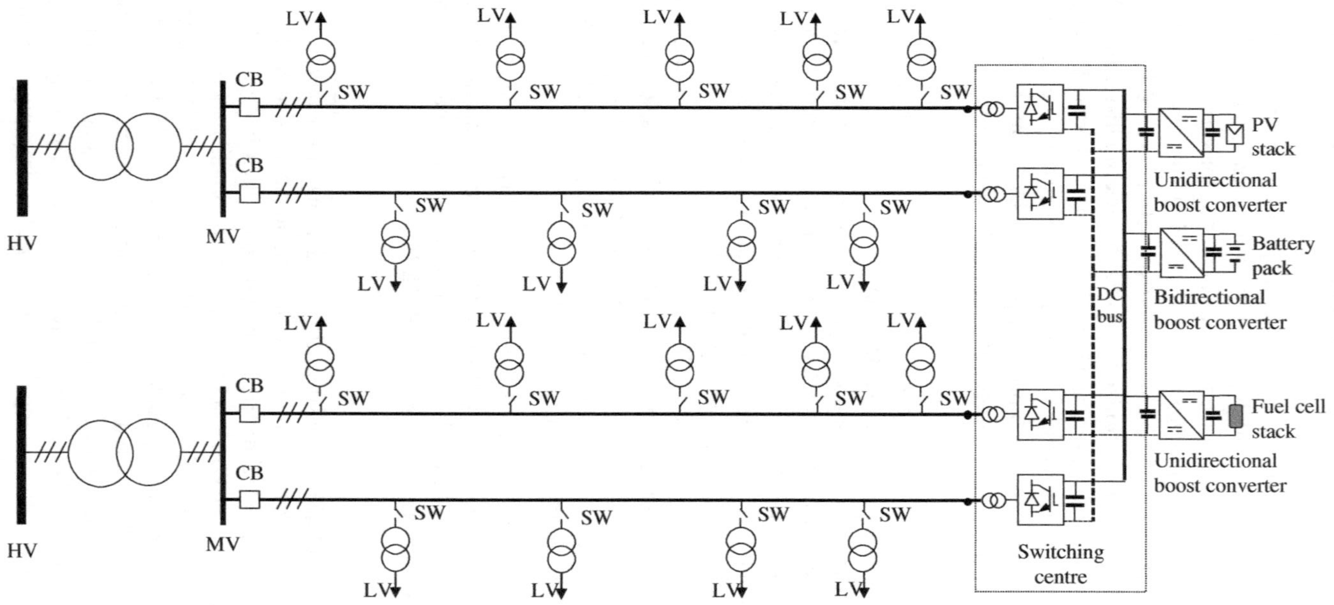

Figure 1.30 Distribution feeders with enhanced functionality with the use of power electronic converters and DC DERs.

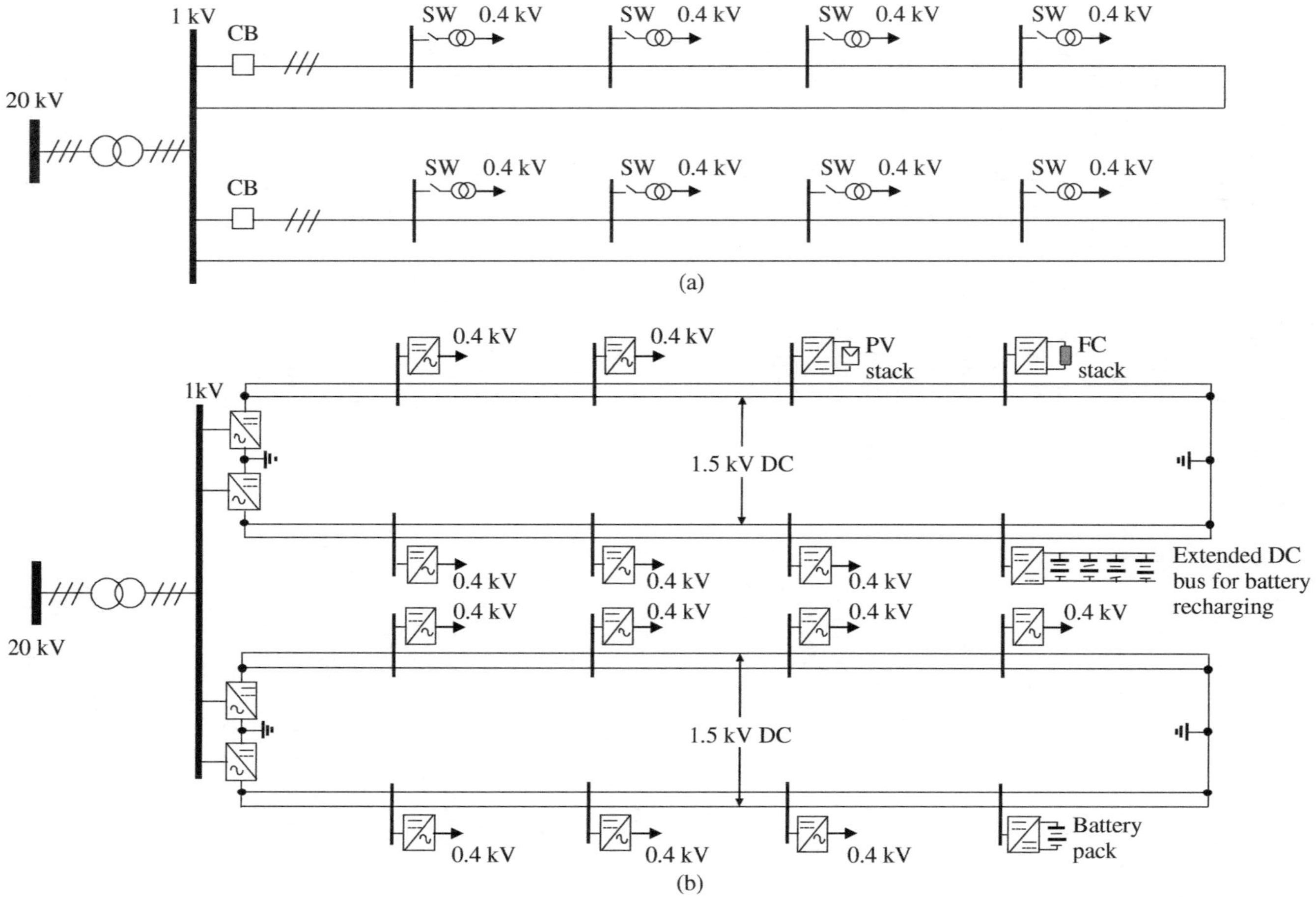

Figure 1.31 (a) Example of 1 kV AC distribution feeders and (b) LVDC bipoles incorporating DC DERs.

20/1/0.4 kV. This has proven to be a profitable solution but with application possibilities limited to small transmission powers and short transmission distances which, nonetheless, are suitable for the power distribution conditions prevailing in Finland. However, this is beginning to change as the challenges imposed by the introduction of DG are given due consideration [41]. A solution which has been put forward within Finnish academic circles is to transform the 20/1/0.4 kV AC distribution system into a distribution system operating at 1.5 kV DC.

The study carried out in [21] for transforming an AC distribution system into a DC distribution system explores a number of monopolar and bipolar DC topologies, aiming at reusing as many AC conductors as possible. A rather interesting result shows that for a distance of 500 m and a 6% voltage drop, the maximum transmitted power in a 400 V AC system is 20.5 kW. In contrast, for the same distance and voltage drop, the maximum transmitted power in a unipolar 1500 V DC system is 270.5 kW. The same comparison for the bipolar ±1500 V DC system gives 345 kW of transmitted power.

The authors conclude that for the Finnish distribution system, the low-voltage DC distribution system enables higher transmission powers and longer transmission distances compared with those of their traditional low-voltage system. Customers' voltage quality improves when voltage dips, and fluctuations and short time voltage drops can be eliminated using power electronic devices. They foresee more economical and reliable distribution networks than those we have today. However, for this to happen, power electronic converters prices should decrease even further and functionality should increase.

1.3.3.5 DC Power Grids with Multiple Voltage Levels

Power transformers have been used systematically in AC systems to suitably interconnect power circuits with different voltage levels. As a matter of fact, these devices were instrumental in the early developments of the AC power systems, vastly surpassing the applicability of the DC power systems at the time. However, over the past two decades, the DC technology has developed at an unprecedented pace and the AC power system has to integrate more and more DC subsystems and components, employing a variety of voltage levels, from LV to UHV. This has created a need for efficiently performing voltage level conversion at the points of interconnection using DC-DC converters. However, the kind of converters used in low power applications [42] may not be suitable for the handling of high power and high voltages.

A suitable solution involves the use of DC-AC converters employing two-level or multilevel VSCs, as illustrated in Figure 1.32 [7]. Note that the frequency on the AC transformer can be much higher than the power frequency, aiming at reducing the transformer size and, consequently, that of the DC-AC converters.

1.3.3.6 Smart Grids

Besides the incorporation of power electronic-based solutions to ameliorate PQ phenomena [3], to achieve higher power throughputs [6] and to enhance the operational flexibility of distribution power grids [39], the electricity industry is embarking on unprecedented changes to be able to cope with major challenges arising from an ageing infrastructure, market liberalization and the incorporation of renewable generation, affordable storage and smart metering [18]. It is said that tomorrow's power

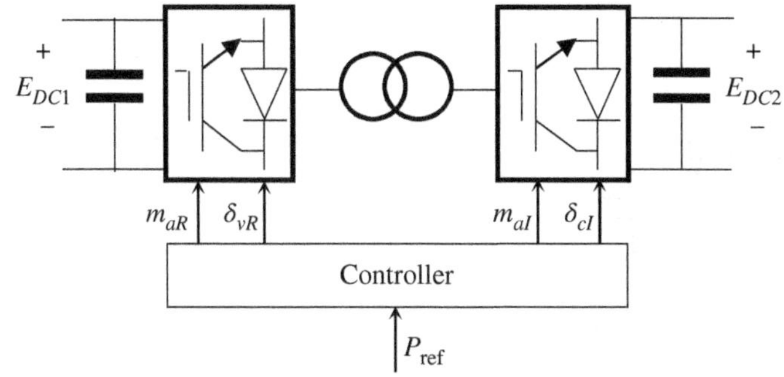

Figure 1.32 A DC-DC converter based on VSCs (©MPCE, 2017).

grids must ensure secure and sustainable electricity supplies with low energy losses and low CO_2 emissions.

In Europe, these power grids should comply with new policy imperatives and changing business frameworks, and should incorporate the state-of-the-art information technology, communications technology and the latest generation of electrical equipment. The confluence of these technologies and policies has given rise to the smart grid concept [19]. Paramount in this array of technologies is the ubiquitous power electronic converter [43]. In the near future, this would pave the way for the existence of AC-DC smart grids as opposed to today's AC smart grids [17].

AC-DC smart grids would be amenable to higher energy yields than AC smart grids, reducing very considerably carbon footprints and using fewer material resources. This would be rather significant in the face of ever escalating prices of metals such as copper and iron, which are a major cost factor in equipment manufacturing and in the electricity supply industry.

The design and operation of future AC-DC smart grids call for the development of models, methods and control techniques embedded in software. These developments in software would be invaluable in making informed recommendations concerning the transformation of traditionally driven, energy-intensive processes such as in the mining industry into a more energy-efficient, reliable operation with a much-reduced carbon footprint.

It is expected that the trend of CO_2 emissions reduction will accelerate when a full deployment of the smart grid technology takes place in earnest. It should be noted that the smart grid technology is likely to have a significant influence on the next top contributor of CO_2 emissions, namely the transport sector – which is reported to contribute to as much as 22% of total global emissions. Additional key drivers of the smart grid technology are its user-centric approach; electricity networks renewal due to ageing infrastructure; enhanced security of supply owing to the underground nature of its power cabling system and self-healing properties; the liberalized markets philosophy; the interoperability of European electricity networks due to standardization; the distributed generation and renewable energy sources (RES) approach; central generation environmental issues; the demand response and DSM policies; European regulatory and social aspects [19].

The multi-terminal VSC-HVDC systems (MT-VSC-HVDC) are very well placed to be the transmission structures that will be used in the next generation of smart grids, both

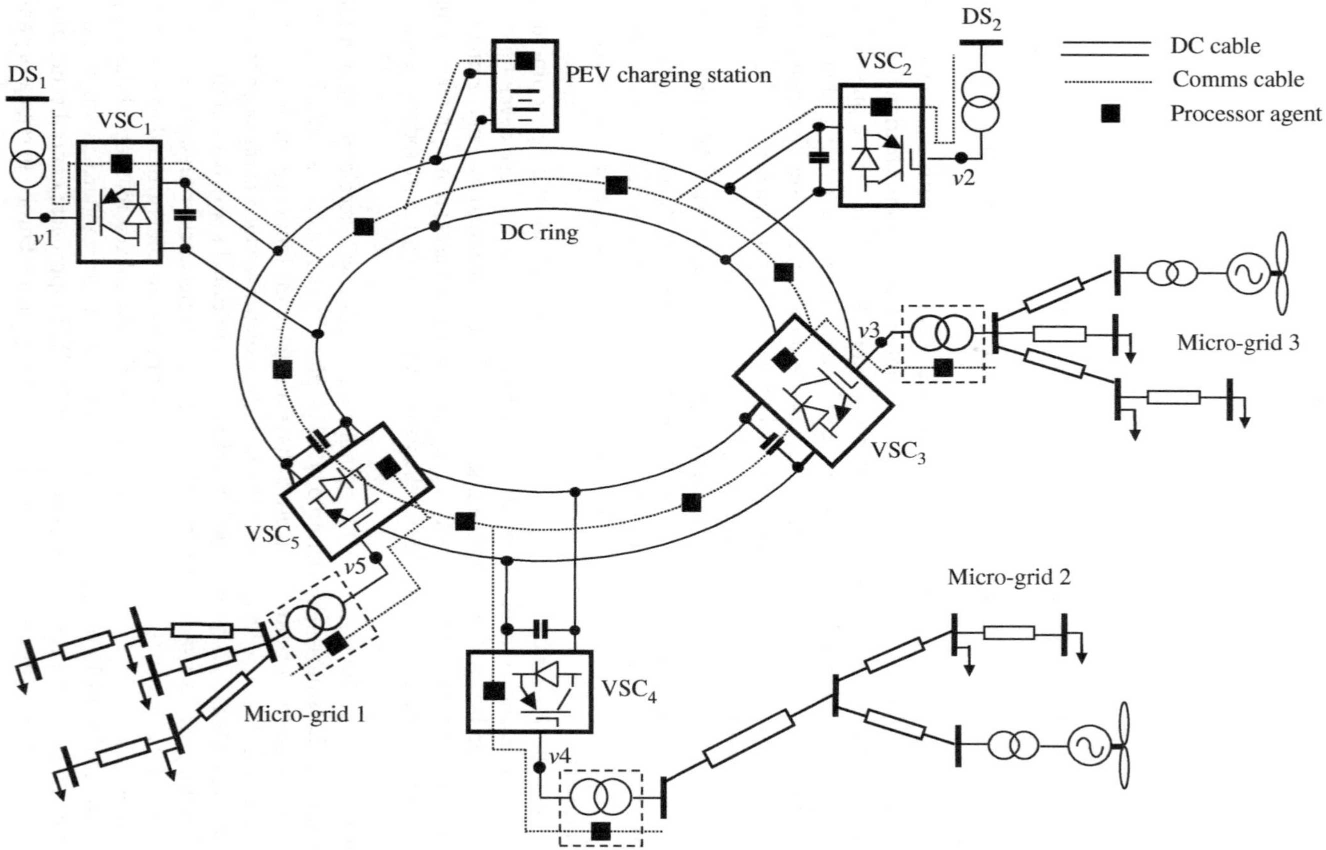

Figure 1.33 Structure of a DC microgrid (©PSCC, 2016).

at the transmission level and at the distribution level [44]. This may be a system within a larger AC system, as exemplified in Figure 1.1. Alternatively, the core system may be a MT-VSC-HVDC system interconnecting a number of AC systems, such as low-voltage, low-power microgrids, as illustrated by Figure 1.33.

It was suggested earlier in the chapter that we should call the equipment of FACTS and HVDC, collectively, FTS equipment. Heretofore, FTS equipment has been used to solve only one or more problems at the local level as opposed to the system level; FTS equipment has been endowed neither with intelligence nor with advanced ICT capabilities.

The current challenge is to make FTS equipment have a system reach within the power grid and to act either alone or in a coordinated fashion with other FTS equipment to ameliorate disturbances arising anywhere in the power grid, in an optimal, smart fashion. The challenge is in devising mechanisms that would provide FTS equipment with the required intelligence that would enable smart grids to be fault-tolerant [7] and, in the longer term, 'self-healing' [19]. To such an end, as illustrated in Figure 1.33, the individual equipment would be fitted with embedded processing agents [22], which would be linked by a dedicated communications network running in parallel with the AC-DC power grid.

1.4 Phasor Measurement Units

The developments of the PMUs are strongly linked to the development of distance relay using symmetrical components algorithms in the early 1970s [45]. By the early 1980s, GPS (Global Positioning System) satellites had been deployed in sufficient numbers and the great potential of digital relays to provide an instantaneous picture of the state of the power system had been realized. The basic idea was to use GPS time signals as inputs to the sampling clocks in the measurement system of the digital relays. More recently, the concept has evolved to encompass the calculation of phasors in real-time synchronized to an absolute time reference – the term 'synchrophasor' has been coined to describe such a concept.

Phasors have been used to describe AC power circuit voltages and currents since their introduction by Charles P. Steinmetz in 1893 using complex quantities [46]. In theory, phasors are applicable only to AC circuits operating in a 'true' steady-state condition, but in practice they have been used to good effect to describe the behaviour of AC power systems when their operations depart from the steady-state. For instance, when a power system undergoes electro-mechanical oscillations, the voltage and current waveforms will not observe constant amplitudes over the whole range of the oscillation period and the power system frequency itself will depart from its nominal operating value. Nevertheless, since the voltage and current variations are relatively slow, it makes sense to group a number of waveform cycles and to treat those as a steady-state condition corresponding to a given time interval; hence, the whole range of the oscillation period may be treated as a series of connected, steady-state conditions. Indeed, it may be argued that the limiting point of such an approach would be when the 'steady-state' is made up of one full cycle of the voltage and current waveforms. In practice, there are applications, such as relaying and power electronic control, where it is not uncommon to use phasors of voltage and current over a half-cycle observation window [47].

In a digital measuring system, samples of the waveform are collected for one full period of the fundamental frequency, starting at $t = 0$. If the interest is the fundamental frequency component in the Fourier series, then the following relation, derived from the Discrete Fourier Transform (DFT), may be used:

$$\overline{X} = \frac{\sqrt{2}}{N} \sum_{k=1}^{N} X_k \cdot e^{\frac{j2k\pi}{N}} \tag{1.3}$$

where N is the number of samples in one period, $\overline{X}$ is the phasor and X_k is the waveform samples.

The phasor expression (1.3) yields the correct value of the fundamental frequency component even in the presence of transient components. As expected, if the incoming signal frequency differs from the nominal frequency, the magnitude and phase angle of the phasor will incur some error. However, as it turns out, the phasor error may be used to determine the frequency of the incoming signal [48].

Other sources of error may not be as useful and ought to be minimized as much as practicable. The processing is based on the use of the DFT and frequencies that are not harmonics of the fundamental frequency will introduce error in the phasor calculation. Moreover, any frequency term above the Nyquist rate ($Nf_s/2$) will not be processed correctly by the DFT – this being the so-called aliasing phenomenon – which requires suitable data filtering to bring it down to acceptable levels.

It was recognized very early on in the area of synchrophasor applications that all the voltages and currents measured in a power system ought to be measured at exactly the same point in time to ensure the existence of a common reference for all the measured quantities. This represented a formidable challenge in network-wide applications, which was resolved satisfactorily only when fully fledged satellite systems were deployed and opened to business for civilian applications. More than one option is available but the preferred system is the Navstar GPS satellite transmission – a system originally designed for navigational purposes which is capable of providing a common-access timing pulse at any location on earth with an accuracy of up to 1 μs. The Navstar system is made up of a constellation of 24 satellites orbiting the earth at 10 000 miles. For accurate acquisition of the timing pulse, only one of the satellites needs to be visible to the antenna which would normally be mounted on the roof of a substation control house. Nevertheless, with the current array of 24 satellites, any point on earth is visible by more than one satellite all the time.

Satellites systems are not the only option open to phasors synchronization. In fact, most of the communication systems normally available in electrical utilities – namely, leased lines, microwave transmission, AM radio broadcasts and fibre optic links – have been used in the past with varying degrees of success. Apart from the dedicated fibre optic links, the other non-satellite communications means have proved too coarse to be of any practical use in this application. Even multiplexed fibre channels may not be suitable – they are reported to incur errors of up to 100 μs. Other satellite systems, such as the Geostationary Operational Environmental Satellite (GOES), have also been used for the purpose of phasors synchronization but their performance has not been sufficiently accurate [48].

A PMU may take many forms but it is widely agreed that its origins may be traced back to the field of computer relaying and, in particular, to the pioneering work of Phadke and Thorp at the Virginia Polytechnic Institute in the early 1970s. Indeed, evidence of the earliest PMU prototypes that used GPS transmission to synchronize their sampling clocks can be found in the Power Systems Research Laboratory at Virginia Tech. It is reported that in the early 1980s the GPS receivers were quite expensive because GPS satellites were few in number and crystal clocks were used as part of the receivers' circuitry to keep time accurately until the next satellite came into viewing. A full deployment of the satellite system and its opening to commercial use solved the logistical problems of having to allocate dedicated fibre optic links to distribute timing pulses of sufficient accuracy and to rely on expensive crystal clocks inside the GPS receivers to keep accurate timing – the chip set of today's GPS receivers is quite affordable.

A modern PMU uses one-pulse-per-second signals provided by the GPS receiver fed to a phase-locked oscillator (PLO) which is tasked to generate a sequence of high-speed timing pulses to sample waveforms from which synchrophasor information is extracted. A functional block diagram of the PMU and associated equipment is shown in Figure 1.34.

The GPS receiver's function is twofold: to provide the one-pulse-per-second signal to the PLO and to provide a time-tag to each extracted synchrophasor. The time-tag comprises the year, day, hour, minute and second, which may be provided with reference to the local time where the PMU is located or in Universal Time Coordinated (UTC) units. The PLO's input from the GPS receiver is split up into a predefined sequence of high-speed timing pulses for sampling of the analogue signals. These signals, which are derived from the secondary outputs of potential and current transformers, are conditioned with anti-aliasing and surge filtering to remove from the incoming waveforms any trace of frequency components above the Nyquist rate. The microprocessor executes all the required DFT phasor calculations, and the timing message from the GPS receiver together with the sample number at the start of a data window is assigned to the calculated phasor as its identifying tag – a process that gives rise to a synchrophasor. A string of positive-sequence synchrophasors is transmitted to one or more remote collection

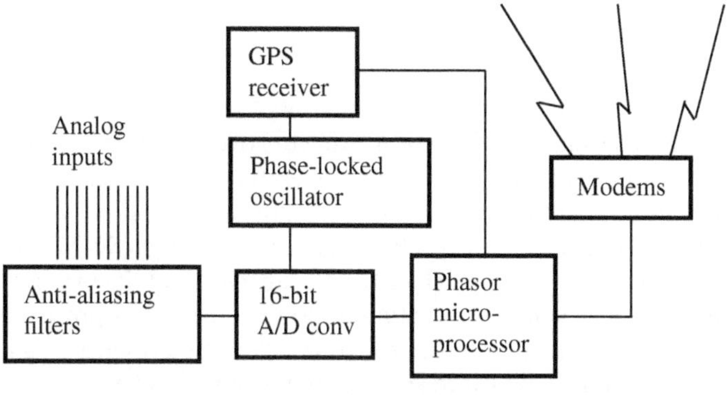

Figure 1.34 Functional block diagram of the elements in a phasor measurement unit. The general structure is similar to many power system relays and digital fault recorders (©IEEE, 2006).

devices known as phasor data concentrators (PDC) over a dedicated communication line using modems.

Note that for a purely sinusoidal waveform, the phasor expression (1.3) has the following representation:

$$\overline{X} = \frac{X_m}{\sqrt{2}} \cdot e^{j\varphi} \tag{1.4}$$

where $\overline{X}$ is now termed synchrophasor, X_m is the waveform amplitude and φ is its argument.

Alternatively, the polar expression (1.4) may be expressed in rectangular form:

$$\overline{X} = X \cdot \cos\varphi + jX \cdot \sin\varphi \tag{1.5}$$

where $X = X_m/\sqrt{2}$ is the rms value of the synchrophasor $\overline{X}$.

Owing to the growing use of synchrophasors in power systems state estimation, the most popular PMU measurements are synchrophasors of nodal voltages, branch currents and nodal current injections. This is elaborated on in Chapter 5.

1.5 Future Developments and Challenges

Current research efforts in the power electronics area concentrate on the development of modular, multilevel power electronic converters, SiC semiconductor valves, self-monitoring and fault-tolerant converters, converters with embedded processor agents and ICTs to add intelligence [7]. This would result in more efficient, scalable, reliable and inexpensive converters with longer lifetimes and improved performances and, in the fullness of time, would produce transformer-less, grid-connected power converters.

It is not difficult to agree that current and future developments in the power electronics industry will drive, to a large extent, the direction and structure of future electric energy grids. These will impact all sectors of the electrical energy industry, namely generation, transmission and distribution.

1.5.1 Generation

Sustained growth of wind and PV solar installations is expected over the next decades. This include both MW-level installations and micro-generation for domestic and commercial applications. Wind power plants are likely to be large offshore installations connected to mainland AC power grids in an asynchronous manner using VSC-HVDC links. The individual wind turbines may reach 10 MW capacity and incorporate superconductor materials. PV solar generators naturally produce electrical power in DC form and require a VSC for connection to the AC power grid. In any case, this is intermittent generation, much of it installed at the distribution system level, which may affect adversely the operation of the high-voltage transmission system in a number of ways: for one, the system level power balance would become more difficult to attain but also voltage control may become a significant problem. Both issues are raising concerns among power system operators, since power exchange between countries ought to be free of frequency deviations. Hence, the search for measures to ameliorate present and future problems caused by intermittent power generation points in the direction of MW-level BESSs. As with PV generators, the BESSs connect to the AC power grid through a VSC. It is also likely that the conventional synchronous generators, those driven by steam and waterfalls, will be replaced by permanent magnet synchronous generators and connected to the power grid through VSCs. This would open the possibility of operating the power grid at frequencies higher than the customary 50 Hz or 60 Hz power frequencies, leading to more compact power transmission grids.

A higher share of renewables in the generation mix calls for more advanced controllers than those we have at present, controllers able to mimic or even improve the operational behaviour of conventional power plants [49]. The very large penetration of distributed and renewable resources would require a mammoth effort to develop new standards, addressing communication and protection issues, as well as the grid codes, imposing stringent connection requirements [50].

All this is poised to increase the share of non-synchronous generation in AC power grids, bringing new challenges to power system operators and power systems application software developers. It is in this tenor that this book is aiming to contribute, presenting new models, methods and control techniques embedded in software, with which to plan and operate the future electrical power systems, which will integrate more and more DC subsystems and components, connected to a diversity of voltage levels, from LV to UHV.

1.5.2 Transmission

It is quite clear that the current major challenges in power transmission are in the area of DC transmission as opposed to AC transmission, such as increasing the capacity of VSC-HVDC to enable higher voltage and power levels. The VSC-HVDC option stands at 2000 MW and ± 380 kV, as exemplified by the INELFE and Ultranet VSC-HVDC transmission systems on the Spain–France border and in Germany, respectively – see Tables 2.3 and 2.4. As a yardstick, the classical HVDC option is currently rated at 7200 MW and ± 800 kV, as exemplified by the Jinping-Sunan HVDC link in East China, spanning a distance of 2100 km – see Table 2.5. However, current limitations in the VSC-HVDC technology seem to be in making further progress in power electronic valves and converters as opposed to any lack of fundamental knowledge of the VSC-HVDC technology, at least as far as the point-to-point VSC-HVDC links are concerned. Indeed, the VSC-HVDC technology has reached such a good level of development that converting some of the existing long-distance, AC transmission corridors into point-to-point VSC-HVDC links may be a profitable option, from the economic and technical vantages.

A different proposition is the multi-terminal HVDC technology, where the classical HVDC option does not present a challenge to the VSC-HVDC technology, owing to the intrinsic flexibility of the former. However, practical problems still need to be resolved before fully fledged multi-terminal VSC-HVDC technology becomes available [51].

For instance, a simple and cost-effective DC circuit breaker has yet to be developed [52, 53]. In point-to-point HVDC links, short-circuit faults are cleared by AC switches. However, AC switches cannot be applied in multi-terminal VSC-HVDC systems without compromising the integrity of the neighbouring AC systems. Also, suitable mechanisms of power flow control in meshed DC power grids still need to be realized. Having said that, it is likely that an option similar to the IPFC, designed for power flow control in AC power grids, may be devised for power flow control in DC power grids. Furthermore, as the DC power grid develops further, it is likely that different voltage levels will co-exist and suitable DC power transformers, capable of handling large power throughputs, would need to be developed.

The investigation of some of these outstanding issues in DC transmission may benefit from using, as a springboard, the theory presented in this book. The material in the various chapters is entirely relevant to gain insight into the current and future issues that are likely to dominate the scene of DC power transmission circuits for the foreseeable future.

One key technological development pertaining to power transmission is high temperature superconductivity (HTSC). It is thought that when this technology is fully developed, it will have a profound impact on both AC and DC transmission systems. In particular, DC circuits stand to benefit from HTSC not only because of the near-zero power loss that HTSC cables incur but also because these cables may have an embedded current-blocking characteristic, which is temperature-triggered, hence resolving the outstanding issue of lack of DC circuit breakers.

1.5.3 Distribution

It is thought that low-voltage distribution networks will continue to experience vast improvements in a number of areas as a result of the widespread use of power electronics controllers and storage. Power quality is an area which is clearly benefiting from the availability of advanced power electronic equipment and control methods [2, 3]. However, there is more to come: the massive penetration of new agents with power electronic components, such as electric vehicles, distributed generation and storage, will lead to a massive number of distributed energy resources which can be controlled in a coordinated fashion to optimize the operation of utilities and large users [16].

It is highly likely, once power electronic converter prices decrease further, that power distribution systems which currently are three-phase radial AC systems will migrate to become multi-terminal, bipolar, VSC-DC systems [21, 54]. This would be due to higher energy throughputs, lower energy losses and smaller footprints of the DC equipment and cables [55]. On the DC side of the system, the conventional three-phase loads may be served in a more controllable manner while enabling the direct connection of DC DERs, such as PV battery packs, fuel cell stacks, PV stacks and EV recharging stations. Furthermore, if the amount of energy produced by the in-situ generation and storage is sufficient to meet the local demand plus the associated power losses, then the DC feeders may act as an independent microgrid, isolated from the HV utility supply, with the AC connection points taking the character of backup supply points [39]. In the meantime, an issue worth investigating, particularly in urban areas, is the suitability of current AC infrastructure for hosting DC power grids in order to reduce operating costs and to combat the growing problem of high short-circuit levels in AC power grids.

References

1 Hingorani, N.G. (1988). High power electronics and flexible AC transmission systems. *IEEE Power Engineering Review* 8 (7): 3–4.
2 Hingorani, N.G. and Gyugyi, N. (1999). *Understanding FACTS: Concepts and Technology of Flexible AC Transmission Systems.* Wiley–IEEE Press.
3 Hingorani, N.G. (1995). Introducing custom power. *IEEE Spectrum* 32 (6): 41–48.
4 Arrillaga, J. (1998). *High Voltage Direct Current Transmission.* IET.
5 Miller, T.J.E. (1982). *Reactive Power Control in Electric Systems.* Wiley.
6 Hingorani, N.G. (1993). Flexible AC transmission. *IEEE Spectrum* 30 (4): 44–45.
7 Maza-Ortega, J.M., Acha, E., Garcia, S., and Gomez-Exposito, A. (2017). Overview of power electronics technology and applications in power generation, transmission and distribution. *Journal of Modern Power Systems and Clean Energy* 5, (4): 499–514.
8 Gyugyi, L. (1994). Dynamic compensation of AC transmission lines by solid-state synchronous voltage sources. *IEEE Transactions on Power Delivery* 9 (2): 904–911.
9 Asplund, G. (2000). Application of HVDC light to power systems enhancement. *IEEE PES Winter Meeting* 4: 2498–2503.
10 Acha, E. and Castro, L.M. (2016). A generalized frame of reference for the incorporation of multi-terminal VSC-HVDC systems in power flow solutions. *Electric Power Systems Research* 136: 415–424.
11 Ekanayake, J.B., Jenkins, N., Liyanage, K. et al. (2012). *Smart Grid: Technology and Applications.* Wiley.
12 Ribeiro, P., Johnson, B.K., Crow, M.L. et al. (2001). Energy storage systems for advanced power applications. *Proceedings of the IEEE* 89 (12): 1744–1756.
13 Anaya-Lara, O., Jenkins, N., Ekanayake, J. et al. (2009). *Wind Energy Generation: Modelling and Control.* Wiley.
14 Xiao, W. (2017). *Photovoltaic Power Systems: Modelling, Design and Control.* Wiley.
15 J. Servotte, *The Impact of a Distributed Battery Energy Storage System on Transmission and Distribution Power Grids*, MSc Thesis, Tampere University of Technology, 2013.
16 Acha, S. (2013). *Modelling Distributed Energy Resources in Energy Service Networks.* London: IET.
17 P. Järventausta, P. Verho, J. Partanen and D. Kronman, "Finnish Smart Grids – A Migration From Version One to the Next Generation ", 21st Int. Conf. & Exhib. on Electricity Distribution, CIRED, Frankfurt, Germany, 6–9 June 2011.
18 Momoh, J. (2012). *Smart Grid: Fundamentals of Design and Analysis.* Wiley–IEEE Press.
19 European Commission, "Smart Grids: Vision and Strategy for Europe's Electricity Networks of the Future", Directorate-General for Research, Sustainable Energy Systems, 2006.
20 Z.K. Rather, Z. Chen and P. Thϕrgersen, "Challenges of Danish Power System and Their Solutions", IEEE Int. Conf. on Power Systems Technology (POWERCON), 30 October–2 November 2012, Auckland, New Zealand.
21 T. Kaipia, P. Salonen, J. Lassila and J. Partanen, "Possibilities of the Low Voltage DC Distribution Systems", Nordic Distribution Automation Conference (NORDAC), Stockholm, August 2006.

22 S.M. Amin and B.F. Wollenberg, "Towards a Smart Grid", IEEE Power & Energy Magazine, pp. 34–41, September–October 2005.
23 Miller, T.J.E. (1989). *Brushless Permanent Magnet and Reluctance Motor Drives*, Monographs in Electrical Engineering. Oxford Press.
24 Acha, E., Agelidis, V.G., Anaya-Lara, O., and Miller, T.J.E. (2001). *Power Electronic Control in Electrical Systems*, Newnes Power Engineering Series. Elsevier.
25 O. Törhönen, Benefits of Main Reactor-Based SVC in Utility Applications, MSc Thesis, Tampere University of Technology, 2016.
26 Gyugyi, L., Rietman, T.R., Edris, A. et al. (1995). The unified power flow controller: a new approach to power transmission control. *IEEE Transactions on Power Delivery* 10 (2): 1085–1097.
27 Y.S. Han, I.Y. Suh, J.M. Kim, H.S. Lee, J.B. Choo and B.H. Chang, "Commissioning and Testing of the KangJin UPFC in Korea". Paper presented at the 2004 CIGRE Conference, Paris, session B4–211, August 2004.
28 Gyugyi, L., Kalyan, K., and Schauder, C.D. (1999). The interline power flow controller concept: a new approach to power flow management in transmission systems. *IEEE Transactions on Power Delivery* 14 (3): 1115–1123.
29 Miller, N.W., Zrebiec, R.S., Hunt, G., and Deimerico, R.W. (1996). Design and commissioning of a 5 MVA, 2.5 MW-h battery energy storage system. In: *Proc. of IEEE Transmission and Distribution Conference*, 339–345.
30 "AES Combines Advanced Battery-Based Energy Storage with a Traditional Power Plant", Available: https://www.businesswire.com/news/home/20120503005472/en/AES-Combines-Advanced-Battery-Based-Energy-Storage-Traditional 25.11.2018.
31 Acha, E., Fuerte-Esquivel, C.R., Ambriz-Perez, H., and Angeles-Camacho, C. (2004). *FACTS Modelling and Simulation in Power Networks*. Wiley.
32 Jones, K.M. and Kennedy, M.W. (1973). Existing AC transmission facilities converted for use with DC. In: *IEE Conf. on High-Voltage DC and/or AC Power Transmission*, 253–260. London.
33 Kundur, P. (1993). *Power System Stability and Control*. McGraw-Hill.
34 Jenkins, N., Allan, R., Crossley, P. et al. (2000). *Embedded Generation*. IET.
35 A. Pazynych, *A study of the harmonic content of distribution power grids with distributed PV systems*, MSc Thesis, Tampere University of Technology, 2014.
36 M.S. Rahman, *Modelling and Simulation of Combined PV-BESS systems*, MSc Thesis, Tampere University of Technology, 2017.
37 Bollen, M.H. (1999). *Understanding Power Quality Problems: Voltage Sags and Interruptions*. Wiley–IEEE Press.
38 Anaya-Lara, O. and Acha, E. (2002). Modeling and analysis of custom power systems by PSCAD/EMTDC. *IEEE Transactions on Power Delivery* 17 (1): 266–272.
39 Romero-Ramos, E., Gomez-Exposito, A., Marano-Marcolini, A. et al. (2011). Assessing the load ability of active distribution networks in the presence of DC controllable links. *IET Generation, Transmission & Distribution* 5 (11): 1105–1113.
40 Maza-Ortega, J.M., Gomez-Exposito, A., Barragan-Villarejo, M. et al. (2012). Voltage source converter-based topologies to further integrate renewable energy sources in distribution systems. *IET Renewable Power Generation* 6 (6): 435–445.
41 J. Lohjala, T. Kaipia, J. Lassila, J. Partanen, P. Järventausta and P. Verho, "Potentiality and Effects of the 1 kV Low Voltage Distribution System", IEEE Future Power Systems Conference, 18 November 2005, Amsterdam, Netherlands.

42 Mohan, N., Undeland, T.M., and Robbins, W.P. (2003). *Power Electronics, Converters, Applications, and Design*. Wiley.

43 Suntio, T., Messo, T., and Puukko, J. (2018). *Power Electronic Converters: Dynamics and Control in Conventional and Renewable Energy Applications*. Wiley–VCH.

44 E. Acha, T. Rubbrecht and L.M. Castro, "Power Flow Solutions of AC/DC Micro-grid Structures", 19th Power Systems Computation Conference (PSCC'16), Genoa, Italy, 19–24 June 2016.

45 A.G. Phadke and J.S. Thorp, "History and Applications of Phasor Measurements", Power Systems Conference and Exposition, 29 October–1 November 2006, Atlanta, GA.

46 Cayemitee, F.I. (2000). *Contribution of the Theory of Complex Numbers to the Field of Electrical Engineering Education and Practice*. NY, USA: Columbia University.

47 Phadke, A.G. and Thorp, J.S. (2017). *Synchronized Phasor Measurements and Their Applications*. Springer International Publishing AG.

48 Phadke, A.G. (1993). Synchronized phasor measurements in power systems. *IEEE Computer Applications in Power* 102 (2): 10–15.

49 M. Seyedi and M. Bollen, "The Utilization of Synthetic Inertia From Wind Farms and Its Impact on Existing Speed Governors and System Performance", Vindforsk Project Report V-369, Part 2. ELFORSK, Stockholm, Sweden, 2013.

50 Mohseni, M. and Islam, S.M. (2012). Review of international grid codes for wind power integration: diversity, technology and a case for global standard. *Renewable Sustainable Energy Review* 16 (6): 3876–3389.

51 P. Rodriguez and K. Rouzbehi, "Multi-terminal DC grids: challenges and prospects", Journal of Modern Power Systems and Clean Energy, Springer Link, Open Access, 8 July 2017.

52 Y.L. Li, X.J. Shi and F. Wang, "DC Fault Protection of Multi-terminal VSC-HVDC System With Hybrid DC Circuit Breaker", Proc. 2016 IEEE Energy Conversion Congress and Exposition (ECCE'16), Milwaukee, WI, USA, 18–22 September 2016.

53 Liu, X., Wang, P., and Loh, P.C. (2011). A hybrid AC/DC micro-grid and its coordination control. *IEEE Transactions on Smart Grid* 2 (2): 278–286.

54 Antoniou, D., Tzimas, A., and Rowland, S.M. (2015). Transition from alternating current to direct current low voltage distribution networks. *IET Generation, Transmission & Distribution* 9 (12): 1391–1401.

55 Justo, J.J., Mwasilu, F., Lee, J., and Jung, J.W. (2013). ACM-microgrids versus DC-microgrids with distributed energy resources: a review. *Renewable and Sustainable Energy Reviews* 24 (C): 387–405.

2

Power Electronics for VSC-Based Bridges

2.1 Introduction

Over the past decade there have been huge developments in power semiconductors. The main features of such developments are an increase in power capability, easier control, higher switching frequencies, reduced switching losses, improved packaging and cost reduction.

These technology improvements have resulted in power semiconductors that resemble, more and more, ideal switches. One consequence is that more simplified models and approaches are justified when studying power semiconductor circuits, particularly if they are aimed at power grid applications. New converter topologies and applications have emerged, such as voltage source converters (VSCs) of various kinds, which are already ubiquitous in electrical power networks. They have turned out to be key elements in the harnessing of offshore wind power, solar power and the enablement of energy storage with fast speed of response.

The first part of this chapter deals with the basic power semiconductors and their main features, while the fundamentals of VSCs and the most popular configurations are explained in the second part. The various VSC multilevel topologies and their advantages, compared to the traditional two-level VSCs, are presented here. The last part of the chapter is devoted to the application of VSC topologies to HVDC systems. Furthermore, the chapter contrasts the main advantages of the VSC-HVDC technology and the classical, CSC-HVDC technology, and gives a comprehensive list of the current VSC-HVDC installations around the world, their ratings, year of installation and type of converters.

2.2 Power Semiconductor Switches

The technology of semiconductor materials intended for use in high-current high-power applications continues to make steady progress in terms of developing a variety of switches that are able to withstand larger breakdown voltages, lower on-state losses, faster turn-on and turn-off capabilities and to handle larger amounts of power. However, these properties tend to be mutually exclusive as opposed to mutually inclusive, and the various kinds of switches are endowed with only a varying degree of such physical

VSC-FACTS-HVDC: Analysis, Modelling and Simulation in Power Grids, First Edition.
Enrique Acha, Pedro Roncero-Sánchez, Antonio de la Villa Jaén, Luis M. Castro and Behzad Kazemtabrizi.

Companion website: www.wiley.com/go/acha_vsc_facts

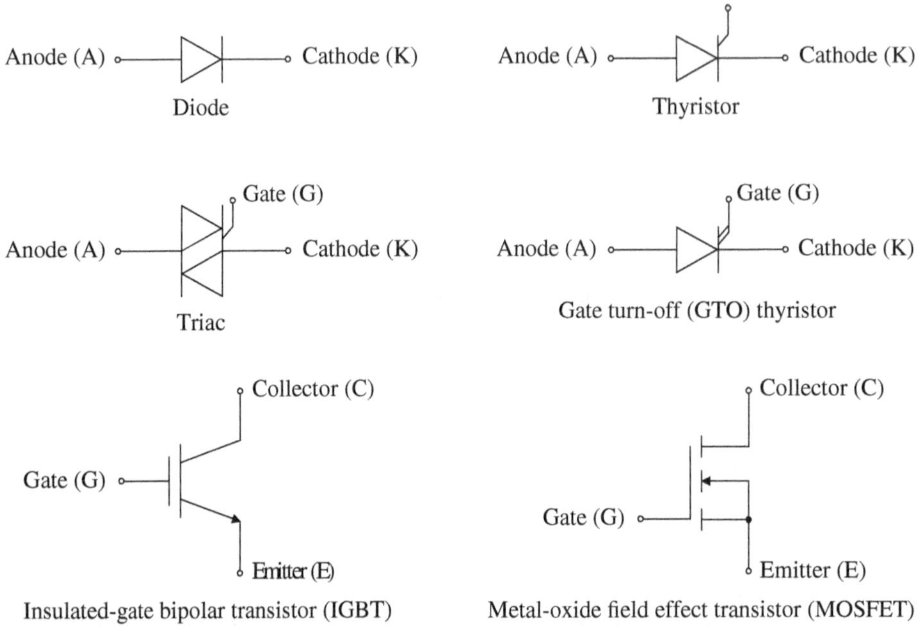

Figure 2.1 Electronic symbols of the most popular semiconductor switches.

attributes – for all the progress made, there is no single semiconductor device that dominates and is fit for all applications. We are thus in the realm of the elusive ideal semiconductor switch.

The main semiconductor switches used in power electronics applications and their symbols are shown in Figure 2.1.

In general, there is a trade-off between break-down voltages and on-state losses, and in the bipolar-type devices there is also a trade-off between on-state losses and switching speeds. The choice of a particular power semiconductor device thus tends to be application-dependent, and more often than not a number of individual semiconductor switches of the same kind are combined to make up series and parallel combinations in order to meet specified voltage and current levels for the bridge. The exception rather than the rule is the diode, which combines rather well with a number of the switching valves either to provide some form of protection or to aid the process of commutation. The combined symbols shown in Figure 2.2 are quite popular in the power electronics literature.

The semiconductor valves can be classified according to their switching characteristics:

1. *Uncontrolled.* The on and off states in these devices are governed by the external circuit operating conditions. It is possible that the diode is the only device in this category. The diode is a two-terminal device, as indicated by its electronic symbol in Figure 2.1, in which these terminals are termed as anode (A) and cathode (K), respectively.
2. *Semi-controlled.* If provisions are made for a third terminal in the device, which is termed as a gate (G), then it becomes possible to use this terminal to inject an

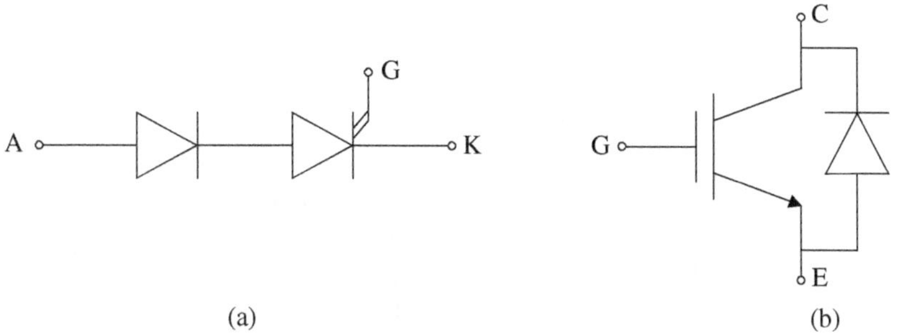

Figure 2.2 Compound symbols. (a) GTO with series diode. One limitation of the GTO thyristor is its reduced reverse break-down voltage in comparison with a conventional thyristor, a drawback that can be improved on by using a series diode. (b) IGBT with anti-parallel diode.

avalanche of electrons at predetermined time periods in order to accurately control the turning on of the device. The avalanche of electrons through the gate takes the form of electric pulses of current or light and in many ways may be thought of as a catalytic process. As with the diode, no such control exists for the turning off of the device, which is governed by the external circuit operating conditions. The thyristor and the triac belong to this category.

3. *Fully controlled.* An extended gate circuit may be used not only to inject positive pulses of current to control the turning on of the device but also to enable the injection of negative pulses of currents in order to turn the device off before it reaches its natural turning-off point. The gate turn-off (GTO) thyristor embodies such a description rather well, and at the macroscopic level it can be viewed as a thyristor with added functionality. This is an area in which power semiconductor technology has made great progress over the last quarter of a century, during which a great many devices have been improved as regards their design and operational capabilities, although new devices have also emerged. In addition to the current driven devices, such as the GTO, voltage driven devices are widely available in the form of power transistors, such as the bipolar junction transistor (BJT) and the metal-oxide-semiconductor field-effect transistor (MOSFET). In the quest for the ideal semiconductor device, attention has turned to hybrid devices, such as the insulated-gate bipolar transistor (IGBT) and the MOS-controlled thyristor (MCT).

2.2.1 The Diode

The diode is the most basic electronic switch whose on- and off-states are determined by the power circuit. The electrical symbol of the diode is shown in Figure 2.3a, while its ideal steady-state *i*-*v* characteristic is plotted in Figure 2.3b. The diode is forward-biased when the applied voltage v_{AK} is zero and the anode current is positive. When the voltage v_{AK} becomes negative, the diode does not conduct current through it and exhibits open circuit behaviour. Nevertheless, the actual *i*-*v* characteristics depicted in Figure 2.3c show that a positive voltage across the diode (the forward voltage drop) is necessary if the diode is to begin to conduct.

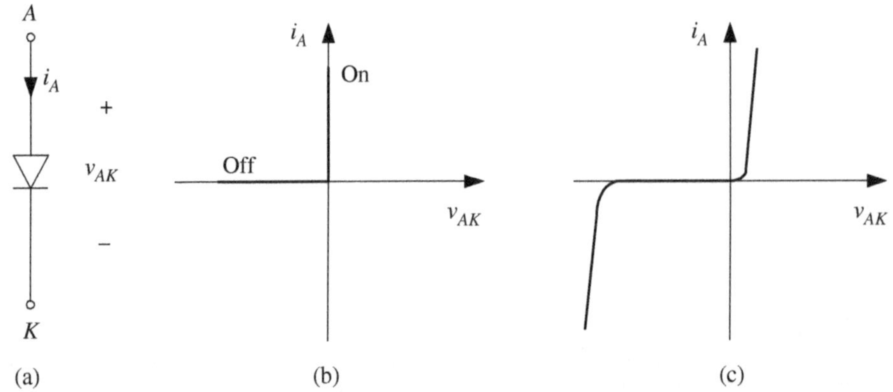

Figure 2.3 Diode (©Wiley, 2002), (a) electrical symbol; (b) ideal *i-v* characteristics; and (c) *i-v* characteristics.

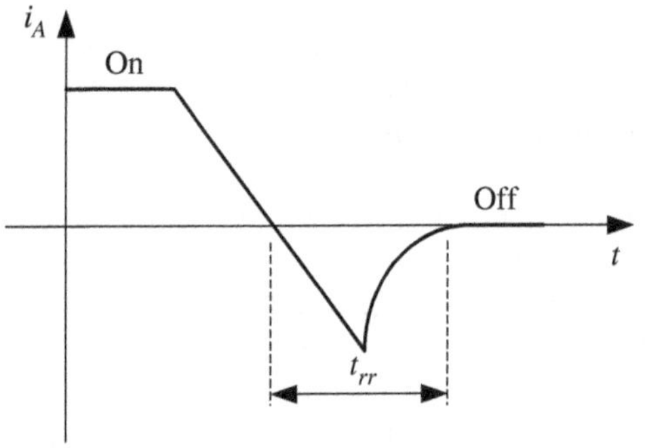

Figure 2.4 Diode reverse-recovery time (©Wiley, 2002).

The switch-on and switch-off processes can be considered as instantaneous if the ideal model is used. However, when the real model is considered, the diode has a dynamic response at switch-off: the current becomes negative for a time interval, known as reverse-recovery time t_{rr}, before becoming completely zero, as shown in Figure 2.4. During the reverse-recovery time, the current can produce overvoltages in inductive circuits [1].

The reverse-recovery time is a key factor in high-frequency applications where fast-recovery diodes, which guarantee a small t_{rr}, are normally used.

Schottky diodes are used in applications which require a low forward voltage drop: these diodes typically have a voltage drop of 0.3 V when forward-biased [2]. Line-frequency diodes have larger reverse-recovery times, signifying that they are not suitable for high-frequency applications and are normally used in grid applications, which require only a frequency of 50 Hz or 60 Hz.

2.2.2 The Thyristor

The thyristor is an electronic switch whose turn-on state is controlled, but the turn-off is determined by the power circuit to which the thyristor is connected. It is, therefore, a semi-controlled device which has three terminals: anode (A), cathode (K) and gate (G). Although the most well-known thyristor is the silicon-controlled rectifier (SCR), other

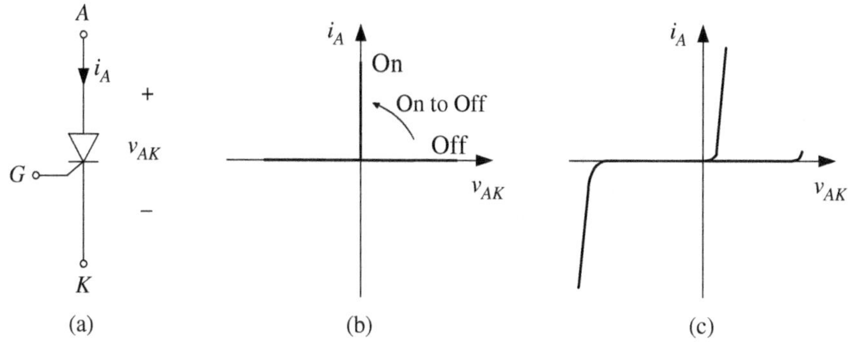

Figure 2.5 Thyristor (a) electrical symbol of the SCR; (b) ideal *i-v* characteristics; and (c) actual *i-v* characteristics.

electronic devices are included in the group of thyristors, such as the GTO thyristor or the MCT, among others.

Figure 2.5a shows the electrical symbol of an SCR, while Figures 2.5b,c show the ideal and real *i-v* characteristics, respectively. The SCR starts to conduct if the voltage v_{AK} is positive, and a positive pulse of current is simultaneously applied to the gate. Once the SCR has begun to conduct, the current pulse applied to the gate is no longer necessary. When the anode current attempts to become negative, the SCR turns off and the current is zero. In order to switch the SCR on again, it is necessary to repeat the process of applying a positive current pulse to the gate.

The SCR undergoes a forward voltage drop when it starts to conduct, as shown in Figure 2.5c. The value of this voltage typically varies between 1 V and 3 V [1].

Unlike the diode, the SCR is able to block forward and reverse voltages v_{AK}, and the rating values of these voltages are normally the same.

Several types of thyristors exist depending on the different applications. The two best known types are:

- Phase-control thyristors. These are normally used as rectifiers of the grid voltage and in the control of large DC motors. They are slow devices and do not therefore support large voltage- or current-time derivatives. They are manufactured to conduct an average current of up to 4 kA and to block voltages of up to 7 kV [1].
- Light-triggered thyristors. Rather than applying a current pulse to the gate in order to start to conduct, these thyristors are triggered by means of a light pulse that is transmitted by means of optic fibre. The main advantages of these thyristors are: (i) galvanic isolation between the control circuit and the power circuit to which the thyristor is connected; and (ii) control-circuit immunity to electromagnetic interferences (EMI) since light signals are not affected by them [3]. This galvanic isolation permits the series connection of the various thyristors required in high-voltage applications. As the potential of each thyristor increases with the numbers of devices, it is difficult to trigger these devices when using conventional thyristors.

2.2.3 The Bipolar Junction Transistor

The BJT is a current-controlled device which can operate in three different modes: cut-off mode, active mode and saturation mode. The active mode is used only if the BJT

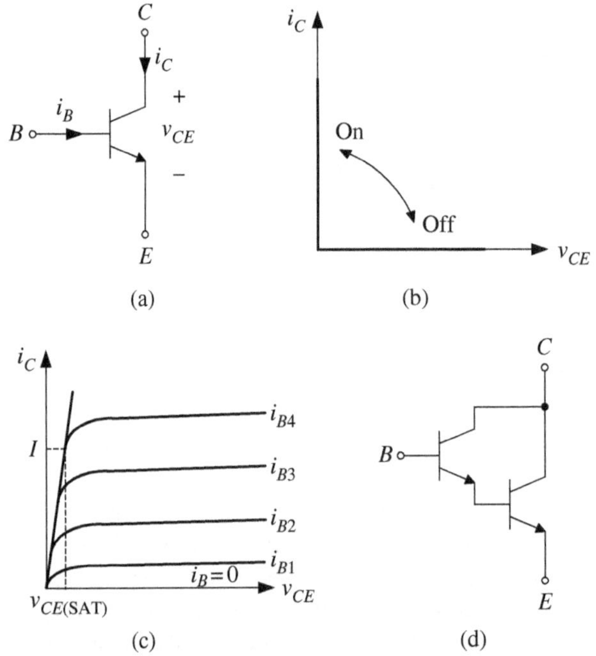

Figure 2.6 Bipolar transistor (NPN) (©Wiley, 2002) (a) electrical symbol; (b) ideal *i-v* characteristics; (c) actual *i-v* characteristics; and (d) Darlington topology.

is used as a linear device, i.e. amplifier applications, whereas the cut-off and saturation modes are used for switching applications [4].

Figure 2.6a shows the electrical symbol of an NPN BJT: the three terminals of the transistor are the emitter (*E*), the collector (*C*) and the base (*B*). The ideal and actual *i-v* characteristics are plotted in Figure 2.6b,c, respectively. The off-state is achieved when the BJT operates in the cut-off mode, which implies that the base current is zero. When the base current is sufficiently large, the transistor operates in the saturation mode, which determines the on-state.

The condition needed to saturate the transistor is:

$$I_B > \frac{I_C}{\beta} \tag{2.1}$$

where the parameter β is the DC current gain whose value is normally between 5 and 10 for high-power transistors, unlike a signal transistor whose DC current gain is usually 100.

The transistor undergoes a drop voltage $v_{CE(SAT)}$ when conducting, as shown in Figure 2.6c, which typically varies between 1 V and 2 V. The dissipated power during the on-state is therefore small.

In order to increase the DC current gain, Darlington configurations using two or more transistors can be employed, as shown in Figure 2.6d, and can be built either using a discrete transistor or as an integrated device. This avoids the need to use large values of the current base to saturate the transistor.

In the past, power transistors were frequently used in power applications. However, increased power capabilities and the ease of control of new devices signify that they have gradually been substituted by MOSFETs and IGBTs.

2.2.4 The Metal-Oxide-Semiconductor Field-Effect Transistor

The MOSFET is a voltage-controlled transistor device that is capable of operating at high switching frequencies, i.e. up to the range of megahertz. It cannot handle as much power as the BJT and is therefore used in low-power applications. Its voltage ratings are in the range of 1000–1500 V for small current values and current ratings of up to 600 A for small voltage values. Figure 2.7a shows the electrical symbol of an N-channel MOSFET. It contains three terminals: the gate (*G*), the drain (*D*) and the source (*S*). The ideal *i-v* characteristics of the MOSFET are shown in Figure 2.7b, whereas the actual *i-v* characteristics are depicted in Figure 2.7c. The off-state is determined by a gate-to-source voltage v_{GS} that is lower than a threshold value, and it is necessary to continuously apply a voltage v_{GS} that is higher than an appropriate value in order to achieve the on-state. When the MOSFET is in the on-state, it can be modelled as a small resistance between the drain and the source $r_{DS(\mathrm{ON})}$ of a few milliohms.

As the gate impedance is almost infinite, the power required to control the on- and off-states is very small.

2.2.5 The Insulated-Gate Bipolar Transistor

The IGBT combines some of the advantages of the MOSFET and the BJT. On the one hand, it is a voltage-controlled device and has as high a gate impedance as the MOSFET, signifying that it is easy to implement the control circuit of the switch and it requires little power. The IGBT has, on the other hand, similar on-state characteristics to those of the BJT, with a small forward drop voltage.

The electrical symbol of the IGBT is shown in Figure 2.8a: the three terminals are the gate (*G*), the collector (*C*) and the emitter (*E*). Figure 2.8b shows the ideal *i-v* characteristics: unlike the BJT or the MOSFET, the IGBT can be designed to block negative voltages. The typical *i-v* characteristics are plotted in Figure 2.8c.

BJTs have progressively been replaced by IGBTs in applications such as AC motor drives. They can block voltages of up to 3.3 kV and their current ratings are as large as 1200 A. [5]. Commercial IGBTs can operate at switching frequencies of up to 80 kHz [6].

2.2.6 The Gate Turn-Off Thyristor

Unlike the conventional SCR in which only the on-state can be controlled, the GTO thyristor is a fully controlled device. The symbol of the GTO is shown in

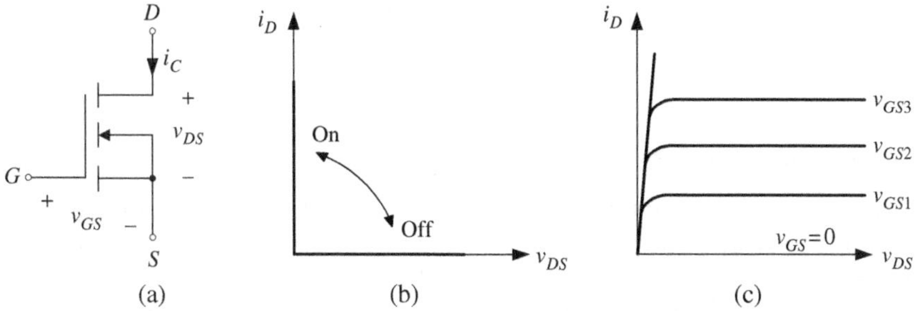

Figure 2.7 N-channel MOSFET (©Wiley, 2002) (a) electrical symbol; (b) ideal *i-v* characteristics; and (c) actual *i-v* characteristics.

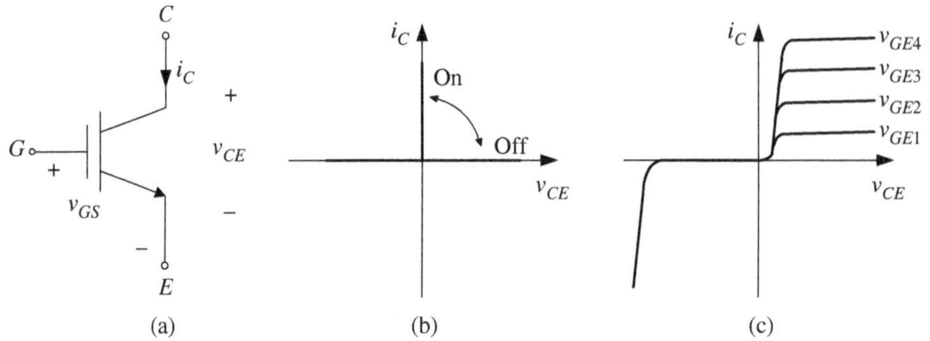

Figure 2.8 An IGBT (©Wiley, 2002) (a) electrical symbol; (b) ideal *i*-*v* characteristics; and (c) actual *i*-*v* characteristics.

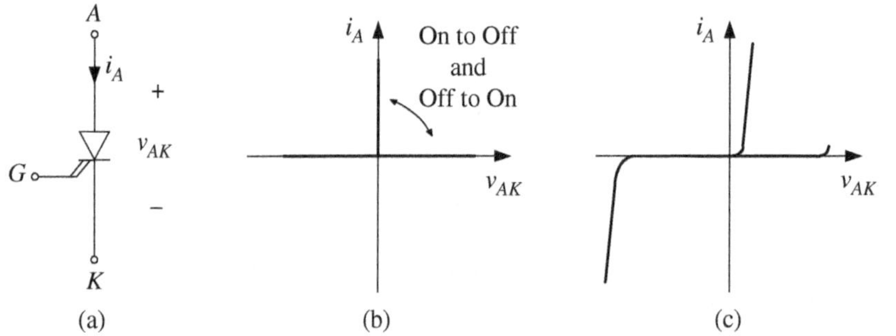

Figure 2.9 A GTO (©Wiley, 2002) (a) electrical symbol; (b) ideal *i*-*v* characteristics; and (c) actual *i*-*v* characteristics.

Figure 2.9a, whereas the ideal and actual *i*-*v* characteristics are plotted in Figure 2.9b,c, respectively.

As in the case of the SCR, the on-state of the GTO is obtained by applying a positive current pulse to the gate when the voltage v_{AK} is positive. Furthermore, the GTO can be switched off with a negative gate current for a short period of time (the turn-off time of the GTO, which is typically a few microseconds). However, the amplitude of the gate current needed to turn off the GTO can be up to one third of the anode current and complex gate control circuits are therefore required to process this current value.

The GTOs can block negative voltages and are slow devices in comparison with other switches such as IGBTs or MOSFETs, i.e. the switching frequency is lower than 1 kHz [3], while the maximum voltage and current ratings for commercial GTOs are 6.5 kV and 4.5 kA.

2.2.7 The MOS-Controlled Thyristor

The MCT is a device which combines features of MOS and thyristor technologies. It is a voltage-controlled switch whose electrical symbols are shown in Figure 2.10a,b for an N-channel MCT and a P-channel MCT, respectively.

The *i*-*v* characteristics are very similar to those of the GTO, as can be seen in Figure 2.10b,c, with a small voltage drop in the on-state. Nonetheless, the on- and

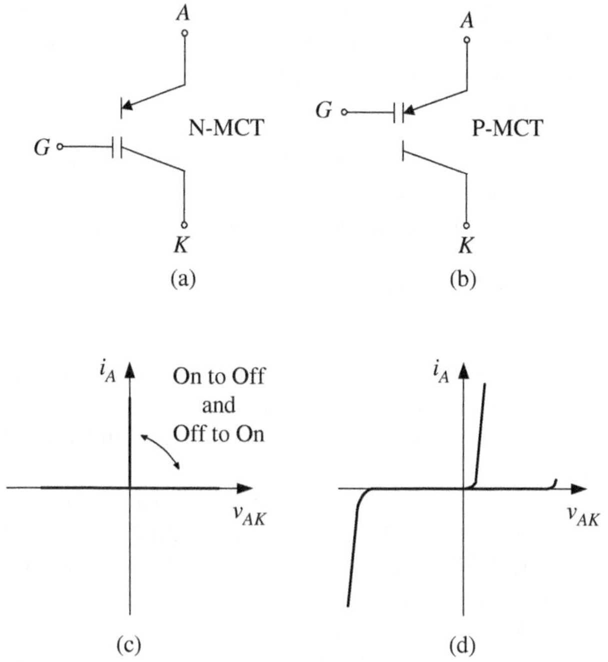

Figure 2.10 An MCT (©Wiley, 2002) (a) N-channel MCT electrical symbol; (b) P-channel MCT electrical symbol; (c) ideal *i-v* characteristics; and (d) actual *i-v* characteristics.

off-states are controlled by means of short voltage pulses that are applied to the gate, unlike the GTO which is controlled by the gate current. Moreover, the MCT is capable of operating at higher switching frequencies than the GTO. These are the main advantages of the MCT over the GTO. Commercial MCTs have voltage ratings of 1500 V at current ratings in the range of 50 A and a few hundred amperes [1].

2.2.8 Considerations for the Switch Selection Process

In the selection process of a switch, the device chosen depends on the type of application and the main factors to take into consideration are the voltages and currents required, necessary switching frequency, maximum acceptable value of power losses, device price and degree of controllability.

2.3 Voltage Source Converters

HVDC systems use power electronic converters to control the electrical energy flow between the DC side and the AC side in a bidirectional manner. Two technologies can be employed [7]:

1. Current source converters use a DC current source as their input source. The load current can be controlled and the load voltage depends on the nature of the load. Mercury-arc rectifiers and thyristors have been employed in this type of converter using line-commutated control schemes.

2. In voltage source converters, the input source is a voltage source, the voltage across the load can be controlled and the current through the load depends on its nature. The power switches that are most frequently used in VSCs are IGBTs and GTOs. These converters are self-commutated.

CSCs have been used in HVDC systems since the mid-1950s; they represent a mature technology [8]. With the recent advances in power semiconductors, VSCs are on the ascendency over the CSC technology. The following sections explain the working principles of the VSC.

According to [1], it is possible to classify VSCs as follows:

1. Pulse width-modulated VSCs. In these converters, the output voltage is controlled by using a PWM scheme in order to obtain an output voltage that is close to a sinusoidal waveform with a small distortion, keeping the voltage constant on the DC side. The switching frequency used in these converters is much higher than the fundamental frequency of the output voltage.
2. Square-wave VSCs. The VSC can be operated in square-wave mode. In this mode, the switching frequency is equal to the fundamental harmonic of the output voltage and the magnitude of this voltage is controlled by modifying the value of the DC-side voltage. In these converters, the output voltage is a square waveform with odd harmonic components and all the switches on the converter have a duty ratio of 50%.
3. Single-phase VSCs operated by phase-shift control. These can be operated in square-wave mode and the output voltage is controlled by overlapping the two leg voltages at a given angle. As in the previous case, since the converter works in square-wave mode, only odd harmonics are present in the output voltage and the switching frequency is equal to the fundamental frequency of the output voltage.

2.3.1 Basic Concepts of Pulse Width Modulated-Output Schemes and Half-Bridge VSC

PWM provides a method with which to control the amplitude of the output voltage in a VSC, thus maintaining a low distortion of that voltage. In order to fully understand the PWM process, some concepts will be defined first.

Any PWM scheme uses two waveforms to operate: a carrier signal, which is usually a triangular waveform in the case of the VSC control, whose frequency is the switching frequency of the converter, f_{sw}, and a modulating signal, whose frequency f_1 is the fundamental frequency of the VSC output voltage. In the case of VSCs, the modulating signal is a sinusoidal waveform.

Let $\hat{V}_{mod}$ and $\hat{V}_{tri}$ be the amplitudes of the modulating signal and the carrier, respectively. The amplitude modulation index is defined as:

$$m_a = \frac{\hat{V}_{mod}}{\hat{V}_{tri}} \tag{2.2}$$

while the frequency modulation index is:

$$m_f = \frac{f_{sw}}{f_1} \tag{2.3}$$

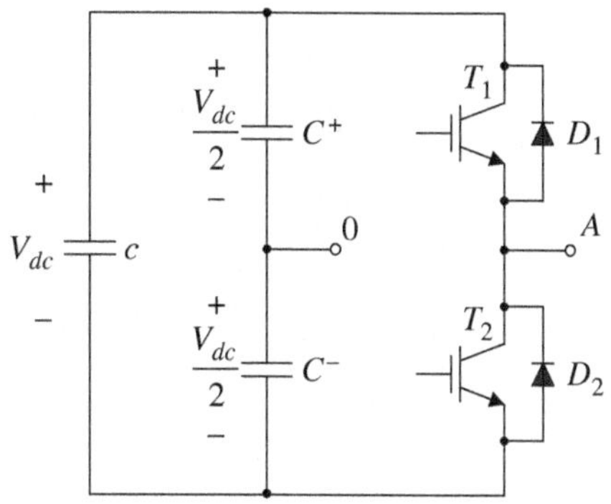

Figure 2.11 Single-phase half-bridge VSC.

Let us consider a basic topology of the VSC: the single-phase half-bridge VSC, which is plotted in Figure 2.11, in which two capacitors are connected in series and split the DC voltage of the input in half. The VSC also contains a leg with two switches, T_1 and T_2. The output voltage v_o is measured between the terminal A and the midpoint of the two series capacitors.

The on- and off-states of the two switches are determined by comparing a modulating signal v_{mod} (sinusoidal) and a carrier signal v_{tri} (triangular). Bearing in mind that the two switches of the leg cannot be simultaneously on since this would cause short-circuits, and they cannot be off since the output voltage must be defined with regard to the VSC circuit, the valve T_1 will be on when $v_{\text{mod}} > v_{tri}$, with $v_o = \frac{V_{dc}}{2}$, and the valve T_2 will be switched on when $v_{\text{mod}} < v_{tri}$, and therefore $v_o = -\frac{V_{dc}}{2}$.

When the PWM process is applied to control the output voltage of the VSC, the waveforms plotted in Figure 2.12 are obtained. Figure 2.12a shows the sinusoidal signal and the carrier waveform when $m_a = 0.8$ and $m_f = 15$: the frequency of the carrier signal is the switching frequency of the valves T_1 and T_2. It should be noted that if $m_a \leq 1$, then the amplitude of the fundamental harmonic of the output voltage is [1]:

$$\hat{V}_{o1} = m_a \frac{V_{dc}}{2} \tag{2.4}$$

This can be seen in Figure 2.12b. Furthermore, if the frequency modulation index is sufficiently high (typically $m_f \geq 9$), the frequencies f_h of the harmonic components of the output voltage are distributed around the switching frequency and its integer multiples, as shown in Figure 2.12c, as follows:

$$f_h = (m_f \pm k) f_1 \tag{2.5}$$

with

$$\begin{aligned} k &= 1, 3, 5, \ldots \text{ when } m_f = 0, 2, 4, \ldots \\ k &= 2, 4, 6, \ldots \text{ when } m_f = 1, 3, 5, \ldots \end{aligned} \tag{2.6}$$

What is more, the amplitudes of the different harmonic components are almost independent of m_f. If the frequency modulation index is chosen to be an odd integer number, the resulting output voltage will have odd symmetry and no even harmonic will be present.

When $m_a \leq 1$, the PWM process has a linear behaviour according to (2.4), as shown in Figure 2.12b. Nevertheless, the amplitude modulation index can be greater than 1.0, and

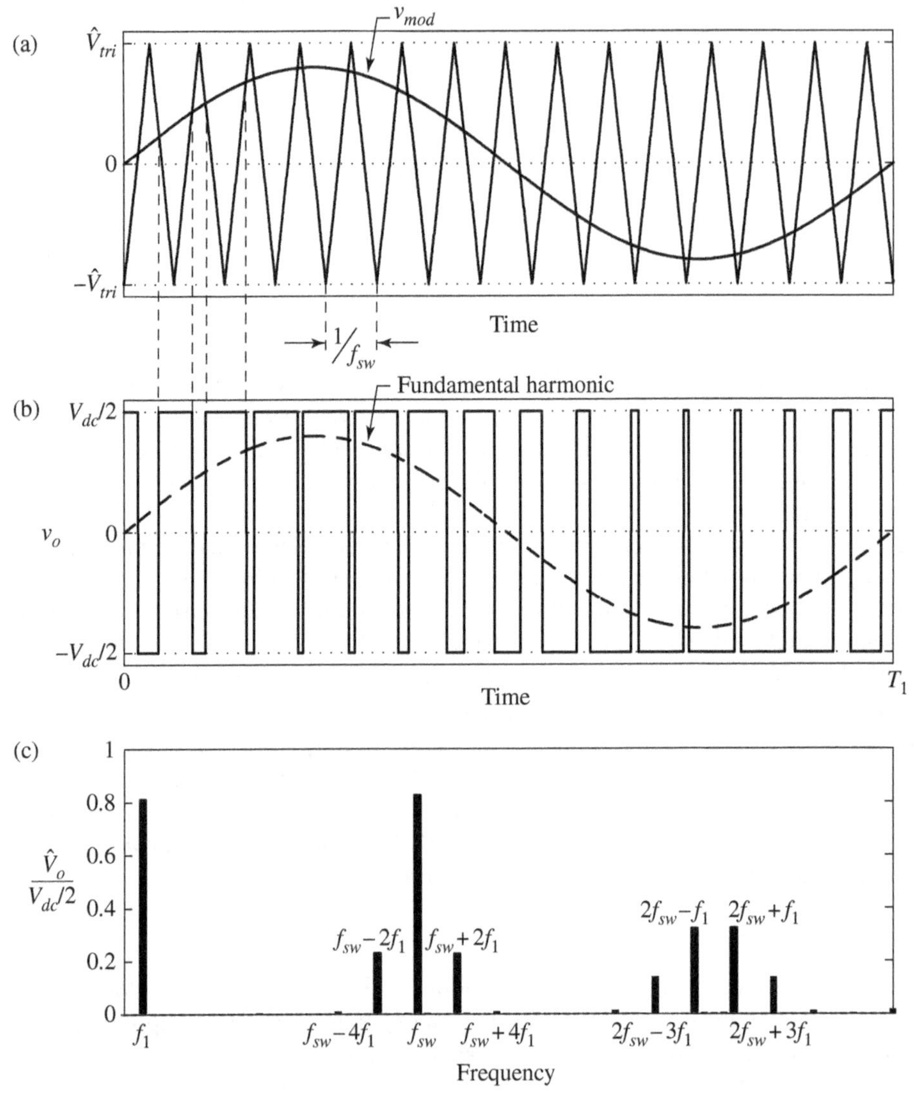

Figure 2.12 PWM method for a half-bridge VSC (©Wiley, 2002), (a) modulating and carrier waveforms for $m_a = 0.8$ and $m_f = 15$; (b) output voltage of the VSC v_o(–) and the fundamental harmonic (– –); and (c) normalized harmonic components of the output voltage.

the PWM operates in the so-called overmodulation region. In this case, the amplitude of the fundamental harmonic does not vary linearly with m_a and low-frequency components are present in the output voltage. Figure 2.13 shows the results obtained when using a PWM scheme with $m_a = 2$ and $m_f = 15$. As shown in Figure 2.13a, the amplitude of the modulating signal is twice the amplitude of the carrier waveform. Nevertheless, the amplitude of the fundamental harmonic does not vary linearly, as Figures 2.13b,c show. Furthermore, the harmonic spectrum does not satisfy the relationship (2.5) previously explained, and low-order odd harmonics are now present in the output voltage.

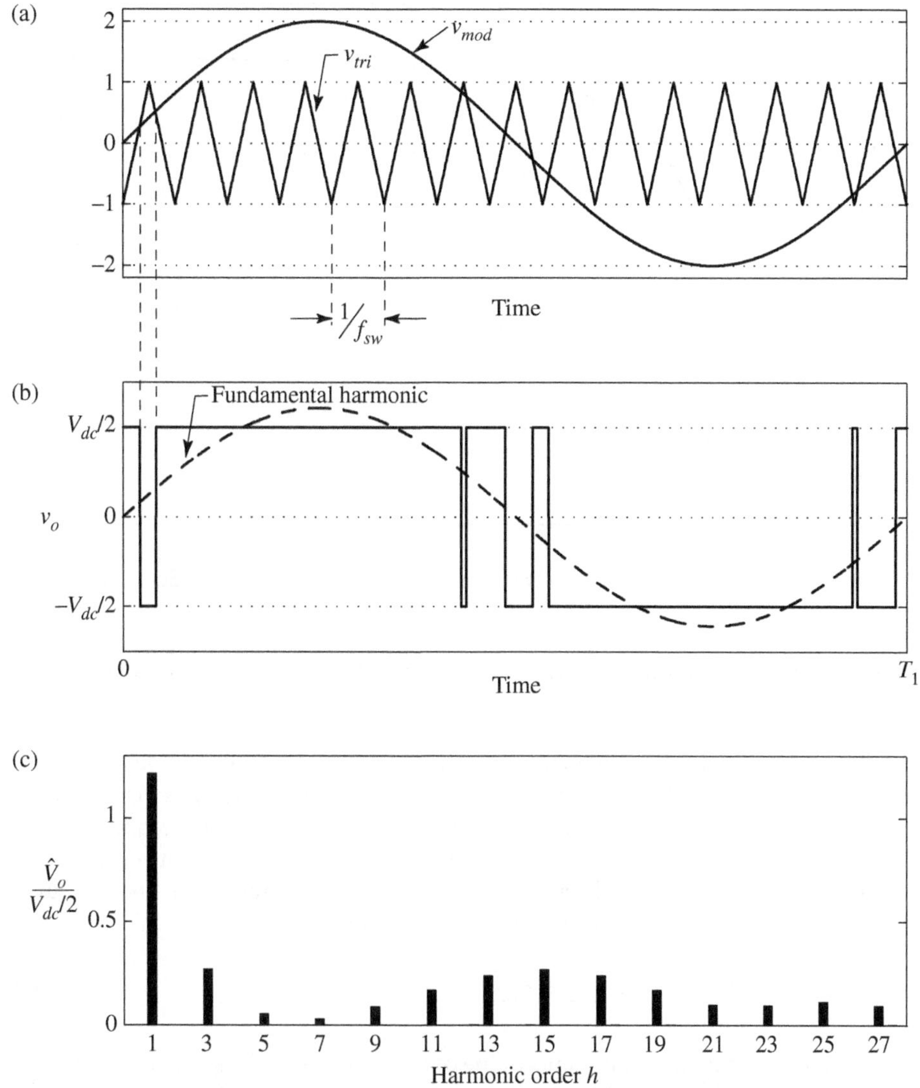

Figure 2.13 PWM scheme operating in the overmodulation region, (a) modulating and carrier waveforms for $m_a = 2$ and $m_f = 15$; (b) output voltage of the VSC v_o(−) and the fundamental harmonic (− −); and (c) normalized harmonic components of the output voltage (©Wiley, 2002)

If the value of the amplitude modulation index continues to increase, the number of intersections between the modulating signal and the carrier waveform is reduced. These intersections are eventually reduced to the zero crossing instants of the modulating signal, with a maximum normalized value of the amplitude of the fundamental component equal to $4/\pi$. In this situation, the VSC is working in the square-wave operation mode and each switch of the leg is turned on the 50% of the fundamental period of the output voltage. Figure 2.14 shows the evolution of the normalized amplitude of the fundamental

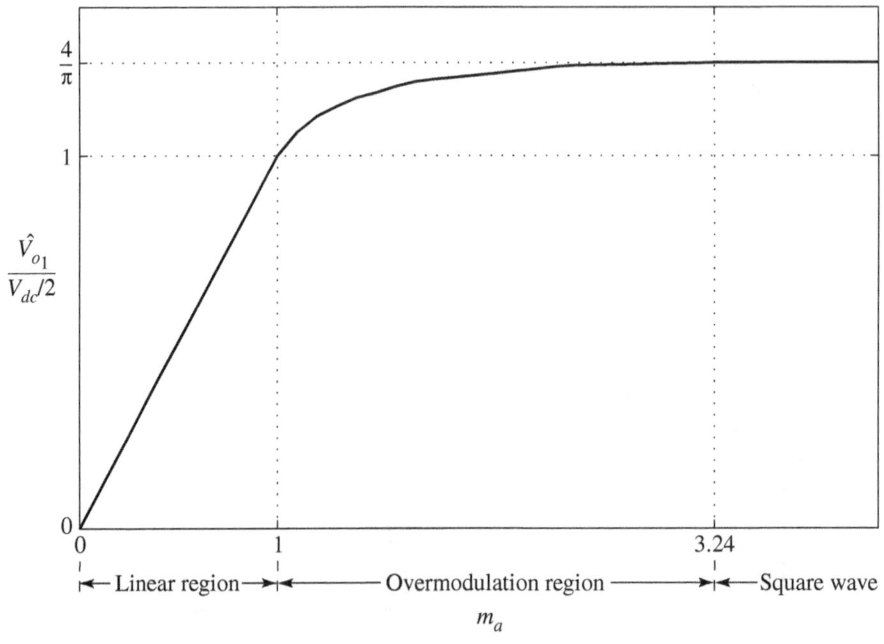

Figure 2.14 Normalized amplitude of fundamental component of the output voltage vs. m_a, for $m_f = 15$.

component of the output voltage with m_a. This evolution has been divided into three zones: the linear region, the overmodulation region and the square-wave operation.

When the VSC operates using the square-wave scheme, the output frequency is the fundamental frequency and the output voltage is a square wave with a duty cycle of 50%, as shown in Figure 2.15a. The harmonic spectrum can be obtaining by using Fourier analysis and the amplitudes of the different components are:

$$\frac{\hat{V}_{o1}}{V_{dc}/2} = \frac{4}{\pi}; \frac{\hat{V}_{oh}}{\hat{V}_{o1}} = \frac{1}{h}, h = 1, 3, 5, 7, \ldots \tag{2.7}$$

The different harmonic components of the output voltage can be seen in Figure 2.15b. When the VSC works using the square-wave scheme, the switching frequency is the fundamental frequency of the output voltage, which may be an advantage in high-power applications in which the power devices cannot commutate at high frequencies.

2.3.2 Single-Phase Full-Bridge VSC

The scheme of the single-phase full-bridge VSC is depicted in Figure 2.16. In this case, the VSC comprises two legs (*A* and *B*) and the output voltage is obtained as $v_o = v_{AN} - v_{BN}$. The full-bridge VSC allows a maximum output voltage equal to the voltage of the DC side, unlike the half-bridge VSC whose maximum output voltage is half of the DC voltage.

There are two PWM alternatives with which to control the full-bridge VSC: PWM with bipolar switching and PWM with unipolar switching.

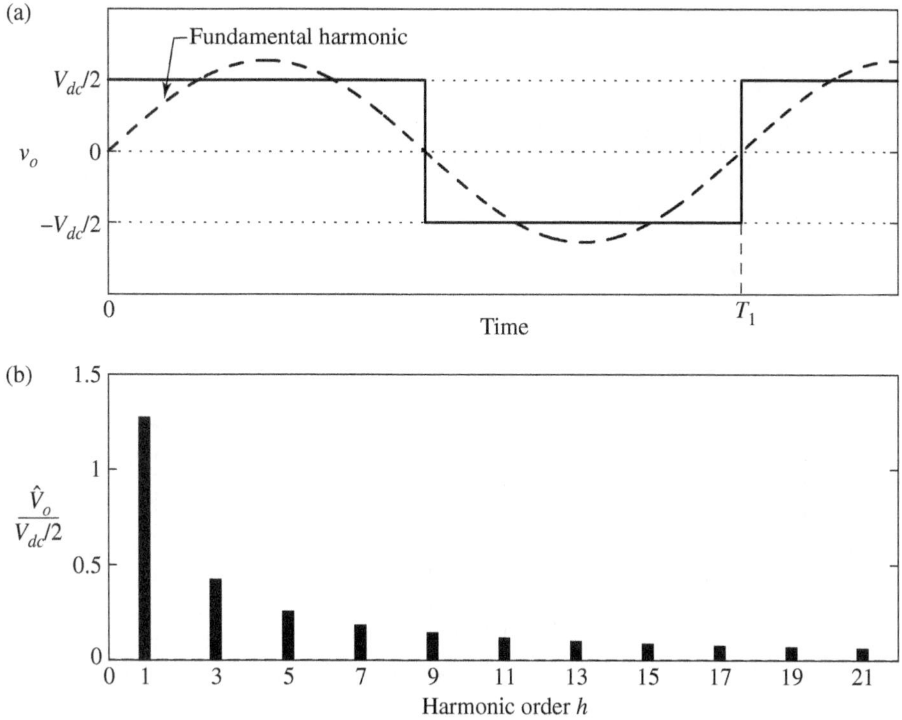

Figure 2.15 Square-wave operation of the VSC, (a) output voltage of the VSC v_o(–) and the fundamental harmonic (– –); and (b) normalized harmonic components of the output voltage.

Figure 2.16 Single-phase full-bridge VSC.

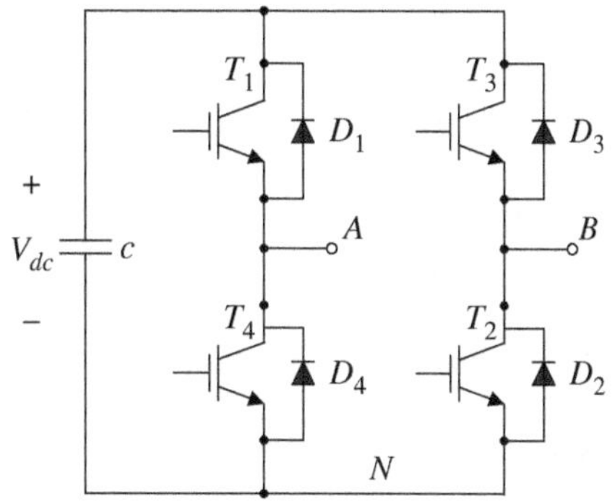

2.3.2.1 PWM with Bipolar Switching

In this scheme, the switches are gathered in pairs (T_1,T_2) and (T_3,T_4) and their on- and off-states are controlled by means of the intersection of one modulating signal with a carrier signal. When $v_{mod} > v_{tri}$, both members of the pair (T_1,T_2) are on while both members of the pair (T_3,T_4) are off, resulting in $v_{AN} = V_{dc}$, $v_{BN} = 0$ and $v_o = V_{dc}$. On the contrary, when $v_{\mathrm{mod}} < v_{tri}$, the switches (T_1,T_2) are both off and the devices (T_3,T_4) are on and, therefore, $v_{AN} = 0$, $v_{BN} = V_{dc}$ and $v_o = -V_{dc}$.

Figure 2.17a shows the sinusoidal signal and the carrier waveform when $m_a = 0.8$ and $m_f = 15$. Since $m_a \le 1$, the VSC operates in the linear region and the amplitude of the

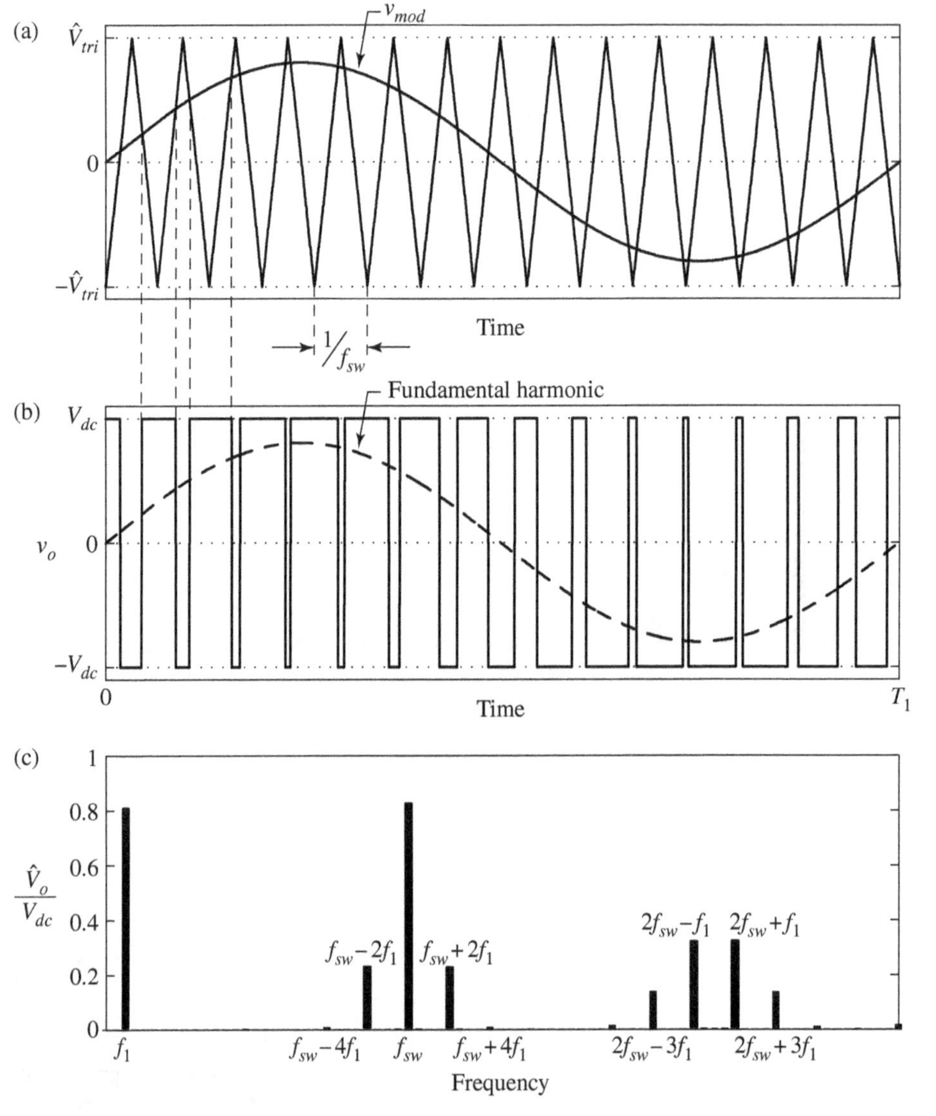

Figure 2.17 PWM with bipolar switching for a full-bridge VSC (©Wiley, 2002), (a) modulating and carrier waveforms for $m_a = 0.8$ and $m_f = 15$; (b) output voltage of the VSC v_o(–) and the fundamental harmonic (– –); and (c) normalized harmonic components of the output voltage.

fundamental harmonic of the output voltage is:

$$\hat{V}_{o1} = m_a V_{dc} \tag{2.8}$$

Figure 2.17b shows the output voltage and the fundamental component of this voltage and it will be noted that this component is proportional to m_a. In the bipolar switching scheme, the output is switched from V_{dc} to $-V_{dc}$, as shown in Figure 2.17b. The harmonic components are plotted in Figure 2.17c: the distribution of these components also fulfils (2.5) and (2.6), as in the case of the half-bridge VSC operated with PWM in the linear region.

2.3.2.2 PWM with Unipolar Switching

If the unipolar PWM method is used, the voltage output is switched from either V_{dc} to zero or $-V_{dc}$ to zero. This scheme is therefore called PWM with unipolar switching. In this scheme, there are two modulating signals,v_{mod} and $-v_{\text{mod}}$, with which to individually control the switches of the legs A and B, signifying that the modulating signals are shifted 180° with regard to each other. Recall that the switches of the pairs (T_1,T_4) and (T_3,T_2) cannot be simultaneously on, and that when $v_{mod} > v_{tri}$ the switch T_1 is on and $v_{AN} = V_{dc}$ and when $v_{mod} < v_{tri}$ the switch T_4 is on, and therefore $v_{AN} = 0$. Similarly, when $-v_{mod} > v_{tri}$ the switch T_3 is on with $v_{BN} = V_{dc}$, and when $-v_{\text{mod}} < v_{tri}$ the switch T_2 is on, and therefore $v_{BN} = 0$. The comparison process between the modulating signals and the carrier signal is shown in Figure 2.18a for $m_a = 0.8$ and $m_f = 15$. The output voltage is shown in Figure 2.18b: since $m_a < 1$, the amplitude of the fundamental component of the output voltage again fulfils (2.8).

The main advantage of this PWM scheme is that the effective switching frequency is twice the actual switching frequency. This is owing to the phase-shift of 180° of the modulating signals [3]. The harmonic spectrum of the output voltage shown in Figure 2.18c exhibits the fundamental component, and the switching components around twice the switching frequency and its even multiples.

2.3.2.3 Square-Wave Mode

As in the case of the half-bridge VSC, the full-bridge VSC can be operated in order to obtain a square waveform with odd harmonic components in which all the switches on the VSC have a duty ratio of 50%. The magnitude of the first harmonic of the output voltage is:

$$\hat{V}_{o1} = \frac{4}{\pi} V_{dc} \tag{2.9}$$

whereas the magnitudes of the remaining harmonic components can be obtained as:

$$\frac{\hat{V}_{oh}}{\hat{V}_{o1}} = \frac{1}{h}, h = 1, 3, 5, 7, \ldots \tag{2.10}$$

where h is the harmonic order.

2.3.2.4 Phase-Shift Control Operation

In this mode, the converter is operated in square-wave mode and the output voltage is controlled by overlapping the two leg voltages v_{AN} and v_{BN} at an angle of α, as shown in Figure 2.19. When the two leg voltages are overlapped, the output voltage is zero since either switches T_1 and T_3 are on or switches T_2 and T_4 are on. Since the converter operates in square-wave mode, as Figures 2.19a,b show, only odd harmonics are present in the output voltage. This can also be observed in Figure 2.19c in which the output-voltage waveform is an odd function and hence contains only odd components.

Let β be defined as:

$$\beta = \frac{\pi - \alpha}{2} \tag{2.11}$$

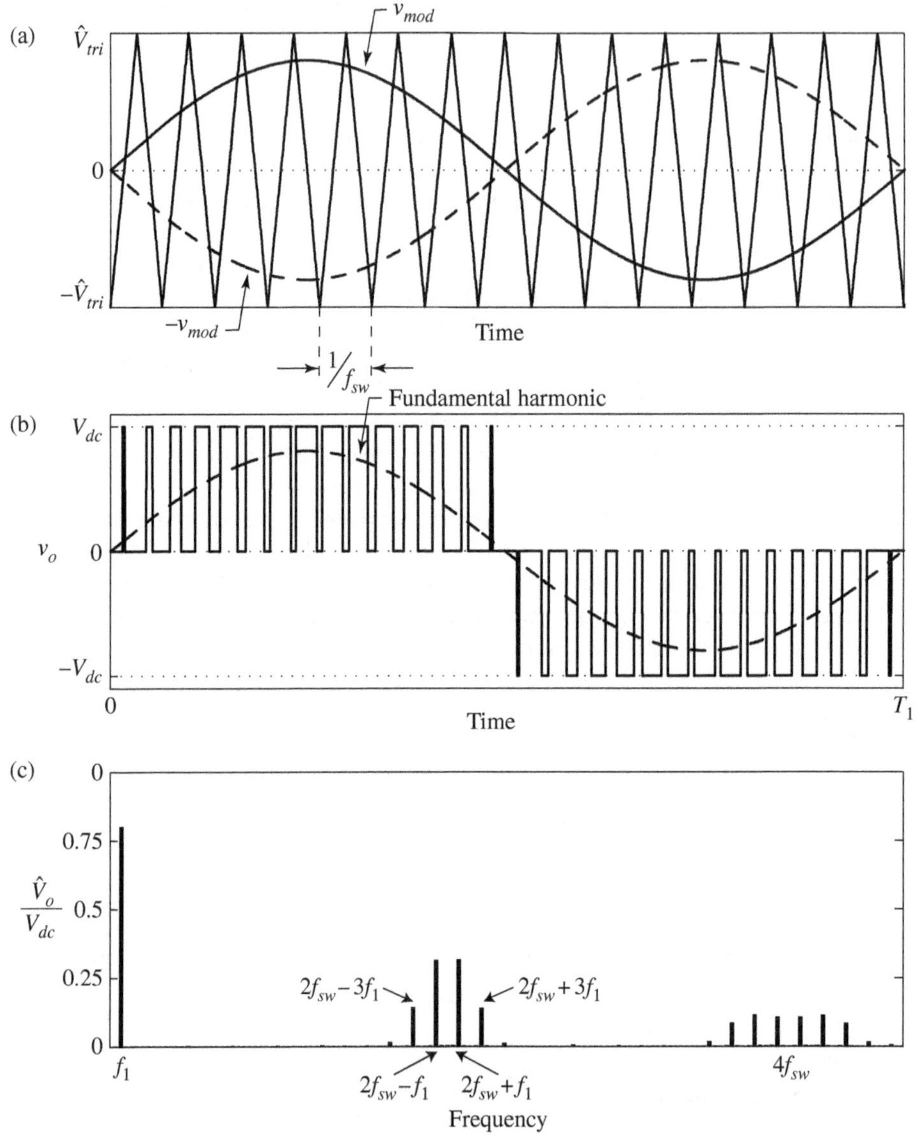

Figure 2.18 PWM with unipolar switching for a full-bridge VSC (©Wiley, 2002), (a) modulating and carrier waveforms for $m_a = 0.8$ and $m_f = 15$; (b) output voltage of the VSC v_o(–) and the fundamental harmonic (– –); and (c) normalized harmonic components of the output voltage.

The magnitude of the harmonic of order h of the output voltage can be easily obtained by using Fourier analysis as in [1]:

$$\hat{V}_{oh} = \frac{2}{\pi}\left|\int_{-\pi/2}^{\pi/2} v_o \cos(h\tau)d\tau\right| = \frac{2}{\pi}\left|\int_{-\beta}^{\beta} v_o \cos(h\tau)d\tau\right| = \frac{4}{\pi h}V_{dc}\sin(h\beta) \tag{2.12}$$

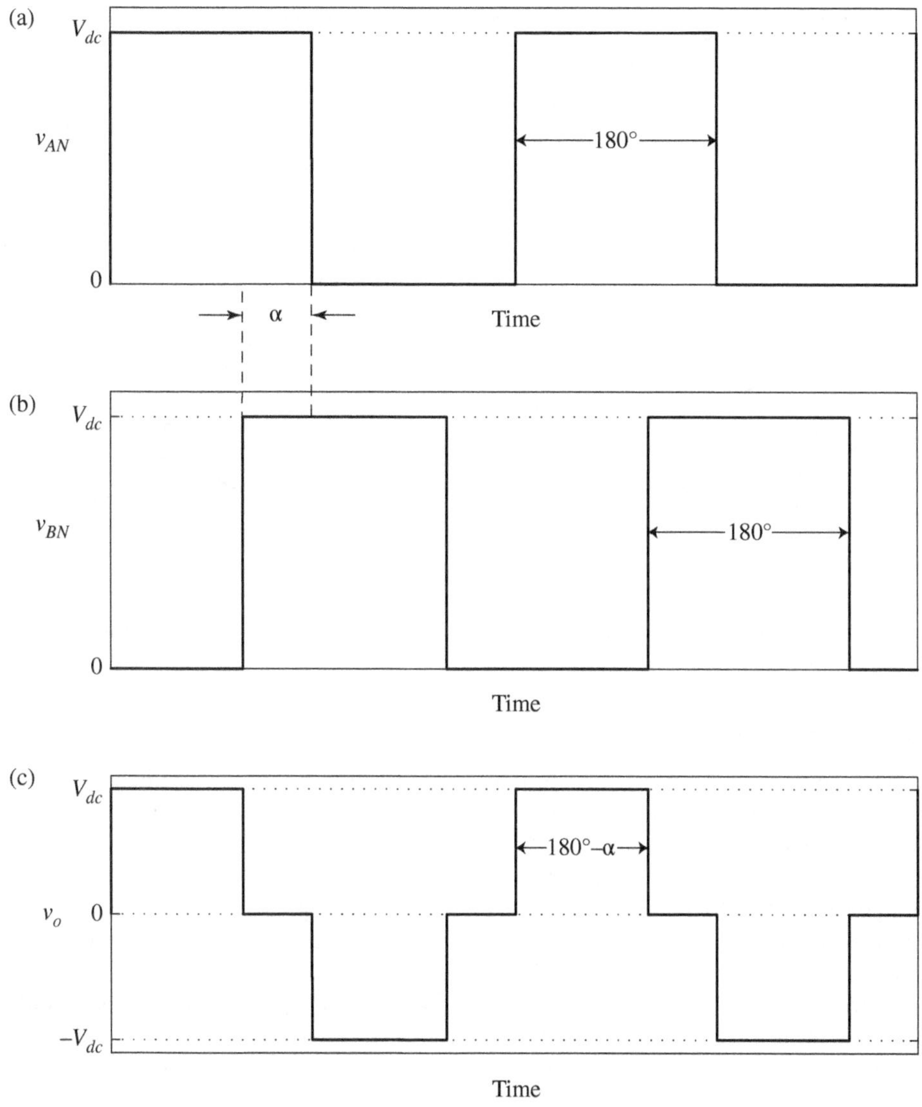

Figure 2.19 Waveforms obtained in the VSC using the phase-shift control (©Wiley, 2002), (a) v_{AN}; (b) v_{BN}; and (c) output voltage v_o.

The value required for the first harmonic component can therefore be obtained by simply varying the overlap angle α.

Figure 2.20 shows the amplitudes of the different harmonic components as a function of the angle α. The results show that the angle α can be chosen either to control the amplitude of the fundamental frequency or to cancel a particular harmonic component (e.g. if $\alpha = 60°$, the third harmonic is not present in the output voltage).

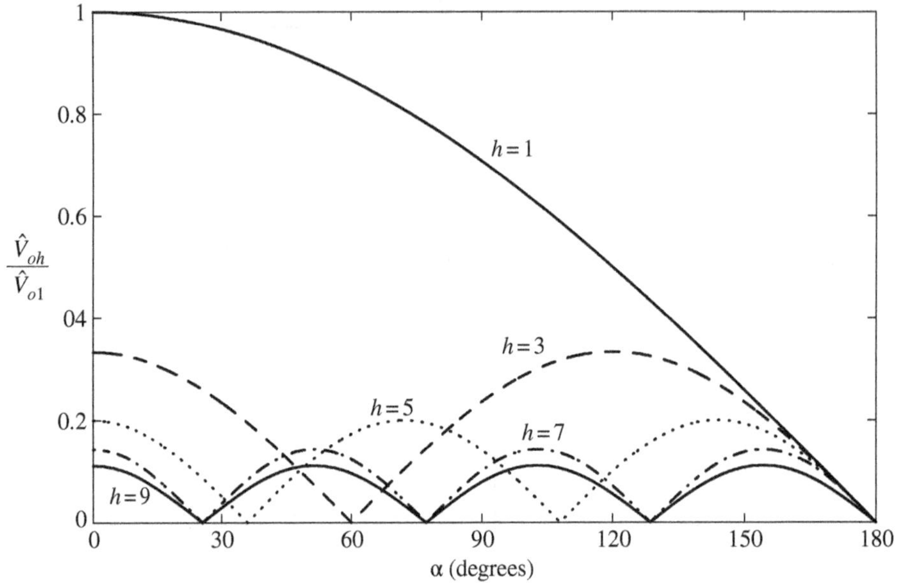

Figure 2.20 Normalized harmonic components of the output voltage (from $h=1$ to $h=9$) as a function of the overlap angle α (©Wiley, 2002).

2.3.3 Three-Phase VSC

In high-power applications, three-phase VSCs are normally used rather than the single-phase configuration. The best-known topology of the three-phase VSC is depicted in Figure 2.21, in which the VSC consists of three independent legs (A, B and C), with two switches per leg. As in the case of the full-bridge VSC, the three-phase configuration allows a maximum output voltage per leg equal to the DC voltage to be obtained, and both switches on the same legs cannot be either on or off simultaneously.

The three-phase VSC can be controlled using a PWM scheme. In this case, three sinusoidal waveforms $v_{mod\,A}$, $v_{mod\,B}$ and $v_{mod\,C}$, i.e. the modulating signals with a phase-shift of 120° with regard to each other, are compared with the triangular signal v_{tri} in order to produce the three independent voltages v_{AN}, v_{BN} and v_{CN}, as shown in Figure 2.22a

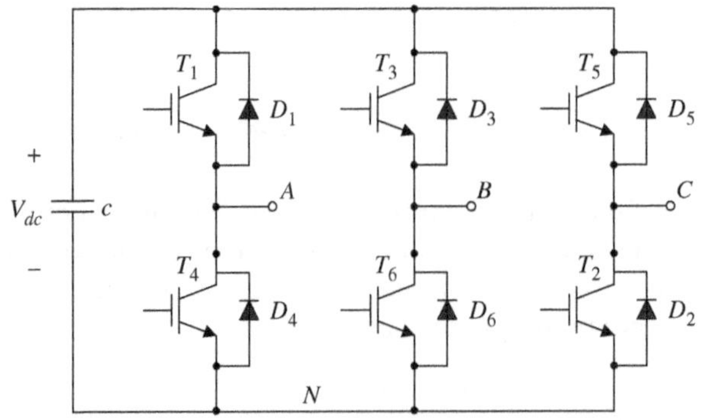

Figure 2.21 Three-phase two-level VSC.

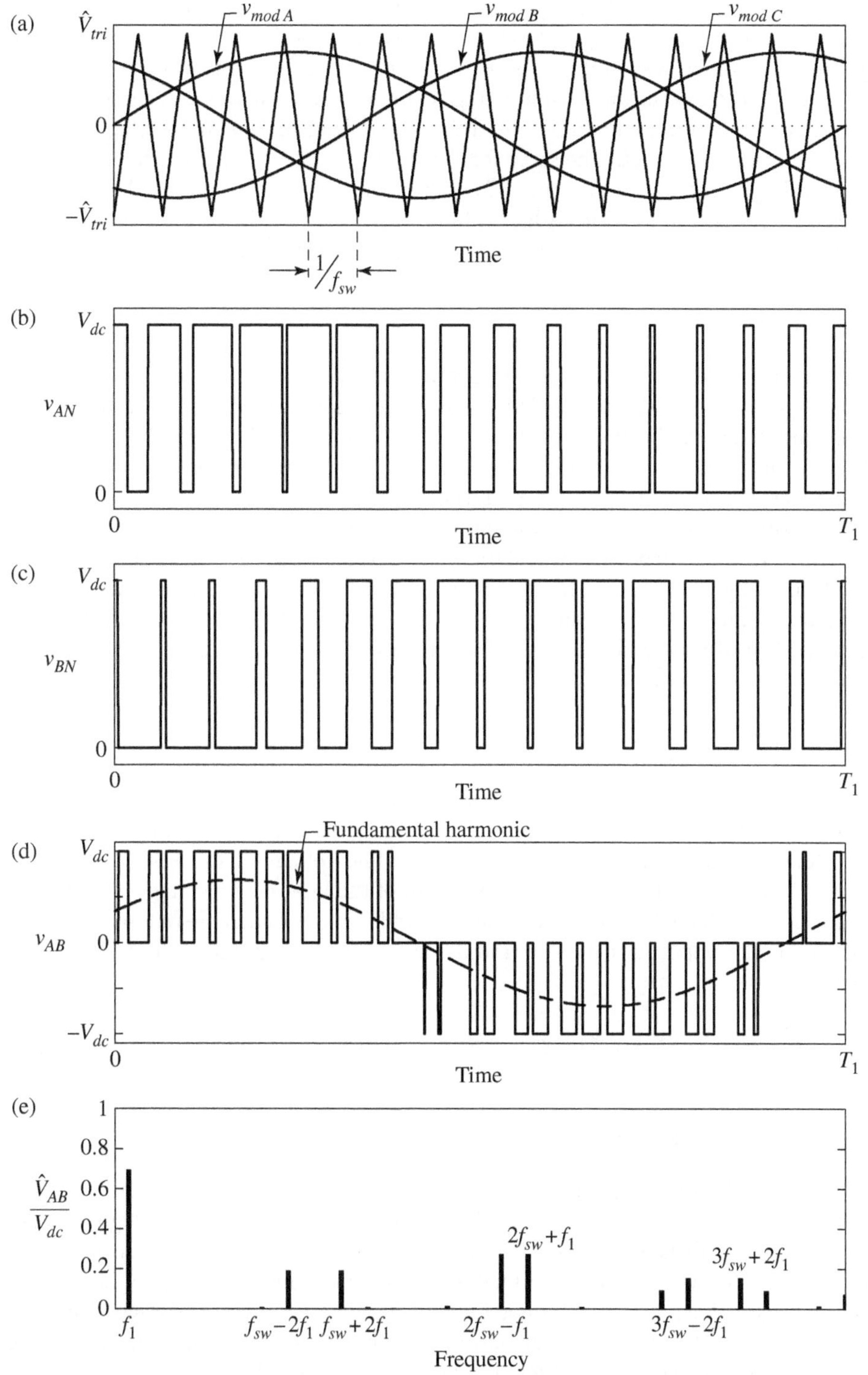

Figure 2.22 PWM process for a three-phase VSC (©Wiley, 2002), (a) modulating and carrier waveforms for $m_a = 0.8$ and $m_f = 15$; (b) output voltage v_{AN}; (c) output voltage v_{BN}; (d) line-to-line output voltage v_{AB} (–) and the fundamental harmonic (– –); and (e) normalized harmonic components of the line-to-line output voltage.

for $m_a = 0.8$ and $m_f = 15$. When $v_{mod\,A} > v_{tri}$, the switch T_1 is on and therefore $v_{AN} = V_{dc}$. On the contrary, if $v_{mod\,A} < v_{tri}$, the valve T_4 is on and therefore $v_{AN} = 0$. The process is repeated in a similar way in the other two legs involving the modulating signals $v_{mod\,B}$ and $v_{mod\,C}$, and the resulting output voltages are plotted in Figures 2.22b,c for the v_{AN} and v_{BN}, respectively. It should be noted that the output voltage of each leg has a DC component which is exactly the same for the three legs, signifying that the line-to-line output voltage does not contain any DC component, as shown in Figure 2.22d. The harmonic components of the line-to-line voltage are plotted in Figure 2.22e: in the linear modulation region, the fundamental component amplitude $\hat{V}_{AB_1}$ is [1]:

$$\hat{V}_{AB_1} = \frac{\sqrt{3}}{2} m_a V_{dc} \tag{2.13}$$

With regard to the remaining harmonic components, they are odd components if m_f is chosen to be uneven and they are placed at the sidebands of the switching frequency f_{sw} and its multiples. Furthermore, owing to the 120° phase shift between the modulating signals, if m_f is chosen as an odd integer multiple of 3, the harmonic components at frequencies nf_{sw}, $n = 1, 3, 5\ldots$ are not present [1].

The maximum value of the fundamental-harmonic amplitude is obtained when $m_a = 1$ in (2.13). This value can be increased if the VSC operates in the overmodulation region at the expense of generating low-frequency harmonics and obtaining a nonlinear control of the amplitude of the fundamental harmonic. This nonlinear behaviour is shown in Figure 2.23, in which it will be observed that the maximum amplitude is obtained at the square-wave region and is:

$$\hat{V}_{AB_1} = \frac{2\sqrt{3}}{\pi} V_{dc} \tag{2.14}$$

When the VSC operates in the square-wave mode, all switches have a 50% duty cycle and the period of the output voltage is the fundamental period. Figures 2.24a–c show the voltages of the three legs, while the line-to-line voltage of the VSC is shown in Figure 2.24d. The resulting harmonic spectrum of the line-to-line voltage is plotted in Figure 2.24e, in which it will be noted that only odd harmonics that are not multiples of 3 are present. The amplitudes of the harmonics are inversely proportional at the harmonic order:

$$\frac{\hat{V}_{AB_h}}{\hat{V}_{AB_1}} = \frac{1}{h}, h = 1, 5, 7, 11, 13, \ldots \tag{2.15}$$

2.3.4 Three-Phase Multilevel VSC

Power electronic converters are very often connected to a medium-voltage grid, and the use of a conventional two-level VSC is not appropriate owing to the high voltages that the devices valves must block. Multilevel VSC topologies have been developed as an alternative to two-level VSCs, aimed at high-power applications. They have a low harmonic distortion and are well suited for applications such as reactive power compensation and HVDC transmission. Furthermore, as the various levels allow the switching frequency to be reduced, the switching losses are also reduced.

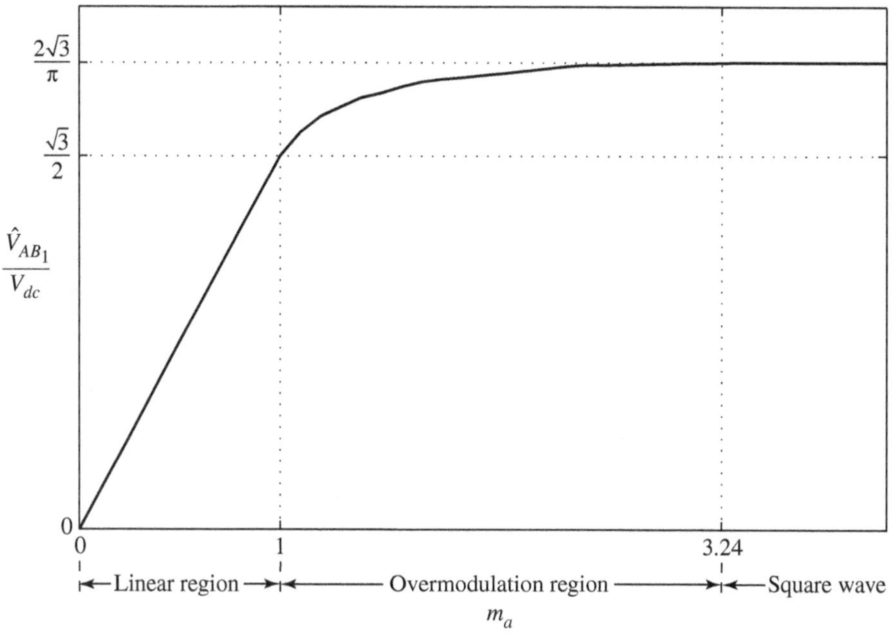

Figure 2.23 Normalized fundamental-component amplitude of the line-to-line output voltage vs. m_a, for $m_f = 15$.

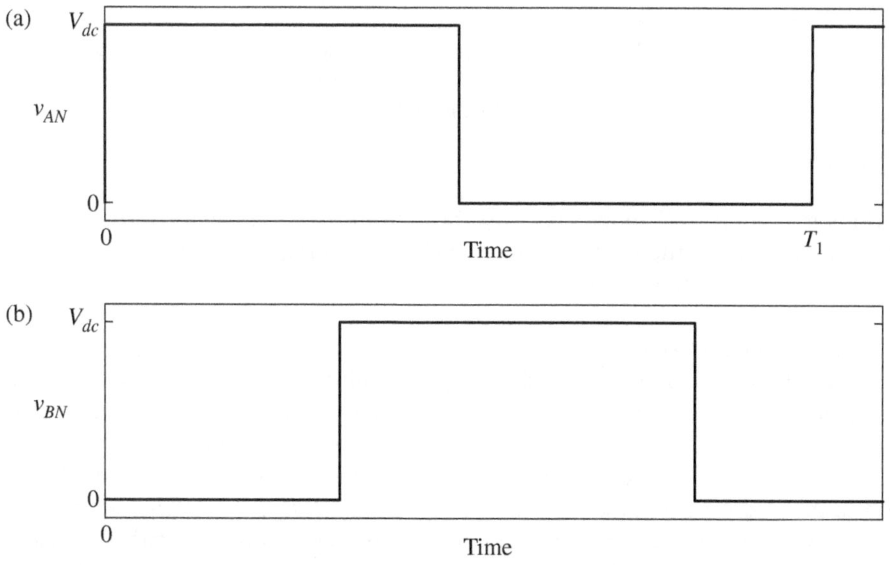

Figure 2.24 Square-wave operation of a three-phase VSC, (a) output voltage v_{AN}; (b) output voltage v_{BN}; (c) output voltage v_{CN}; (d) line-to-line output voltage v_{AB} (–) and the fundamental harmonic (– –); and (e) normalized harmonic components of the line-to-line output voltage.

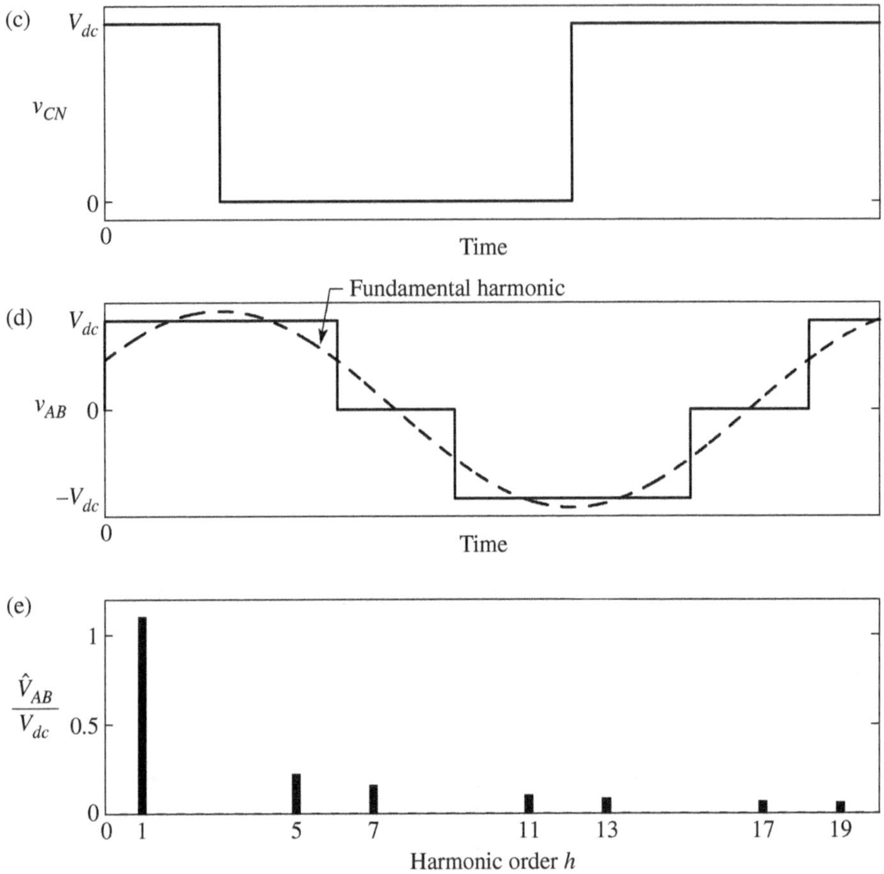

Figure 2.24 *(Continued)*

Various multilevel VSC topologies exist, but the most well-known are the neutral-point-clamped (NPC), the flying capacitor (FC), the H-bridge converter, the modular multilevel converter, and the hybrid multilevel converter.

2.3.4.1 The Multilevel NPC VSC

Figure 2.25 shows the simplest multilevel NPC topology for a three-phase VSC: the three-level NPC configuration in which each leg comprises four switches and four diodes. In order to explain the operation of the converter, the DC voltage has been split in half by means of two capacitors and the output voltage of each leg is measured with regard to the midpoint of these two capacitors, e.g. v_{A0}. A positive voltage is obtained when the two upper switches are on, i.e. $v_{A0} = V_{dc}/2$ for leg A, while a negative voltage is produced if the two lower switches are on, i.e. $v_{A0} = -V_{dc}/2$ for leg A. Besides these two voltage levels, this topology can also generate zero voltage (the third voltage level) when the switches T_{x2} and T_{x3} are on simultaneously, where the subscript x stands for the A, B and C legs and the devices T_{x1} and T_{x4} are off; the diodes D_{cx1} and D_{cx2} allow the current to flow in both directions when the converter works in this zero-voltage region: for positive current value, the devices T_{x2} and D_{cx1} conduct, whereas if the current is

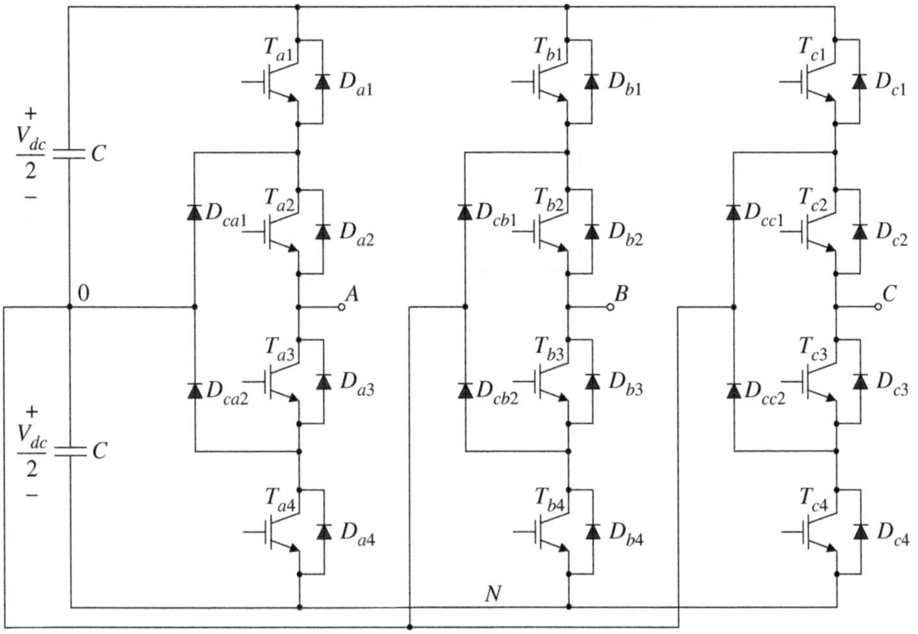

Figure 2.25 Three-level three-phase NPC VSC.

negative, the semiconductors T_{x3} and D_{cx2} are on. In order to avoid short-circuits, it is obvious that when the switches T_{x1} and T_{x2} are on, the devices T_{x3} and T_{x4} must be off, and vice versa. These restrictions imply that the firing signals of both switches T_{x1} and T_{x3} and switches T_{x2} and T_{x4} must be complementary. The resulting output voltage for the on- and off-states of the different devices is summarized in Table 2.1.

The operation of the converter can be carried out by using either a square-wave scheme or a PWM process. The PWM operation will be explained later, as it can be applied in a similar way to other multilevel topologies. The square-wave scheme is identical to the phase-shift control method applied to the full-bridge single-phase VSC: the fundamental harmonic and the harmonic content of the output voltage can be controlled by varying the angle α. Figure 2.26 shows the resulting output voltage for a specified angle α: the output voltage of each leg exhibits the three levels explained above, as shown in Figures 2.26a,b. Nonetheless, the line-to-line output voltage v_{AB} has five levels (two additional levels), which implies a lower harmonic content.

Table 2.1 Three-level NPC VSC: output voltage as a function of the switch states.

Switch state				
T_{x1}	T_{x2}	T_{x3}	T_{x4}	**Output voltage**
On	On	Off	Off	$V_{dc}/2$
Off	Off	On	On	$-V_{dc}/2$
Off	On	On	Off	0

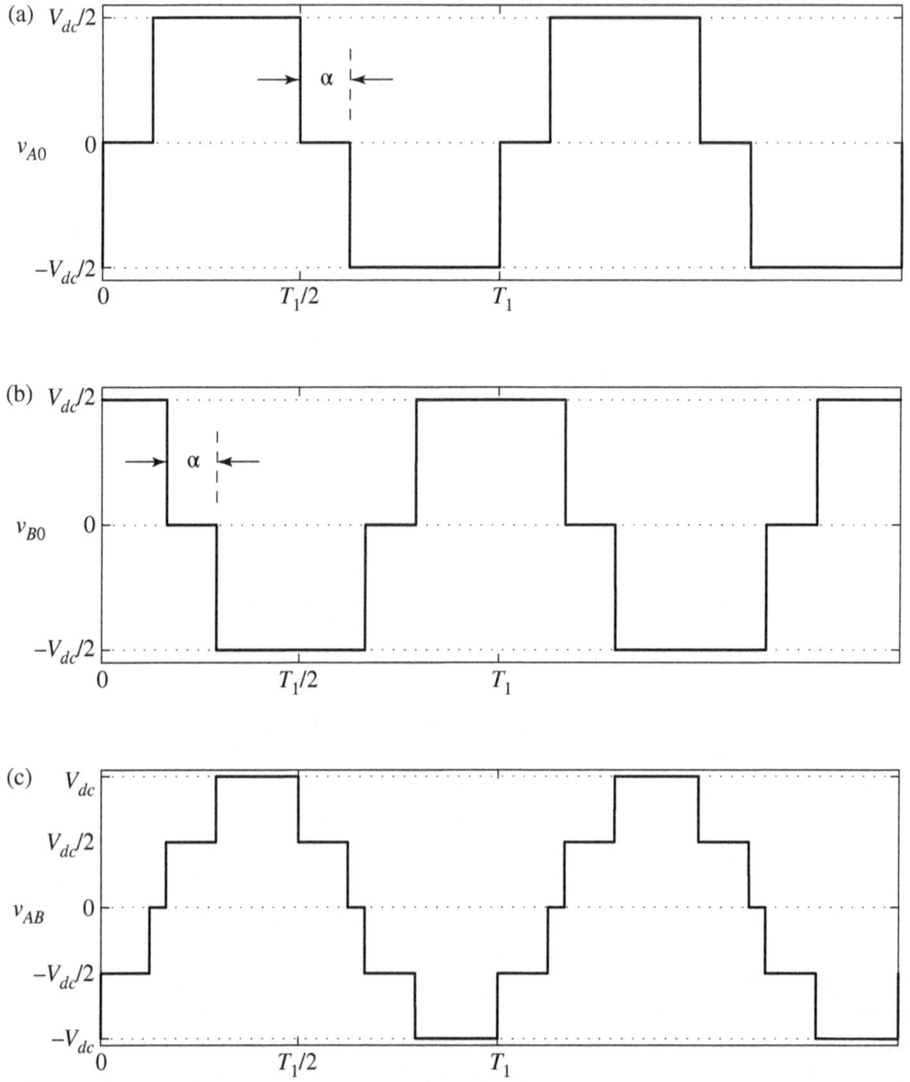

Figure 2.26 Output voltage of a three-level NPC VSC using square-wave operation, (a) output voltage v_{A0}; (b) output voltage v_{B0}; and (c) line-to-line output voltage v_{AB}.

The number of levels can be increased in order to reduce not only the harmonic content but also the time derivative of the voltage. If the number of levels of a VSC is m, then the converter will require $m-1$ capacitors on the DC side and $2 \times (m-1)$ controllable switches per leg.

Figure 2.27 shows the DC side and one leg of a five-level NPC VSC, where it can be seen that there are four capacitors on the DC side and eight switches per leg. Six

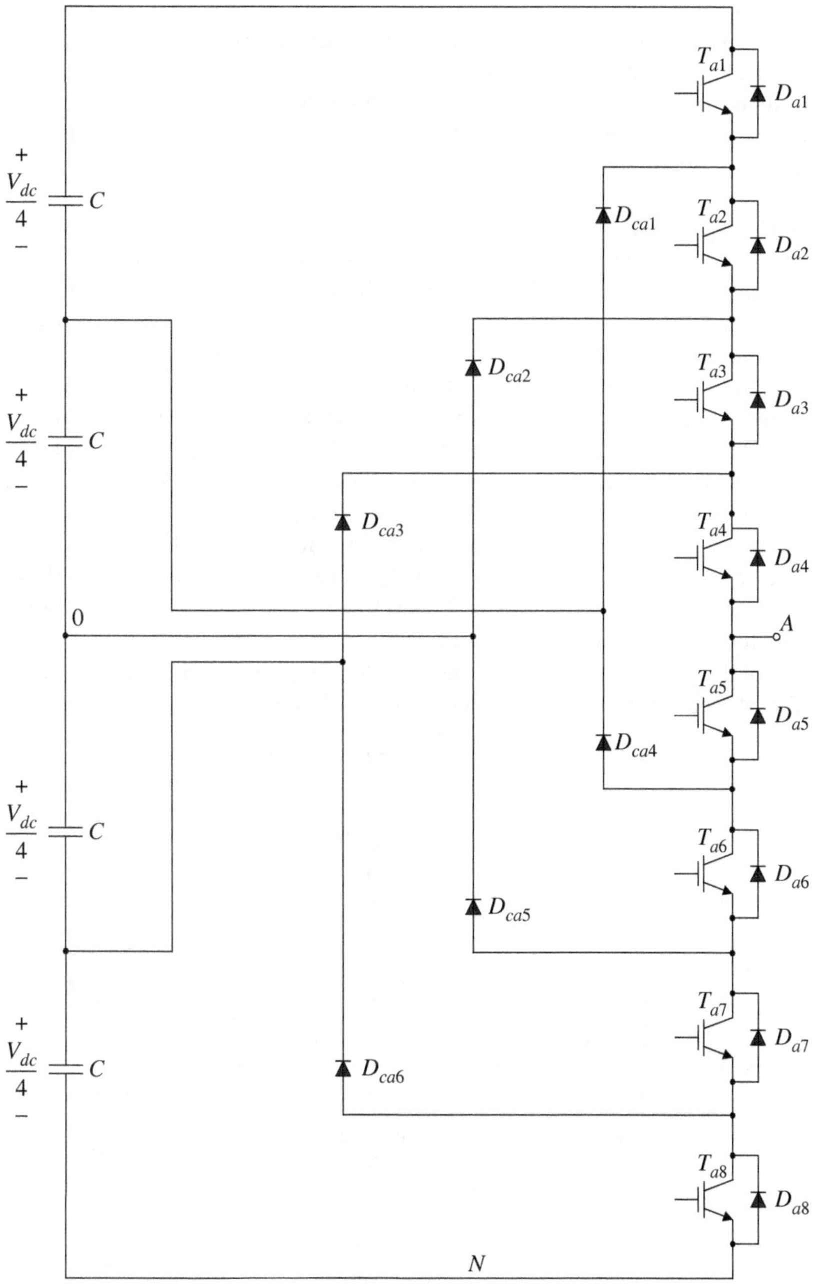

Figure 2.27 Five-level NPC configuration.

clamping diodes are also required per leg: for given m levels. Although the controllable switches block a voltage equal to $V_{dc}/(m-1)$, the clamping diodes must block a different voltage [9] and the number of necessary clamping diodes is almost proportional to the square of the number of levels [10]. The main drawbacks of the NPC configuration are the great difficulties present in keeping the voltages of the DC capacitors balanced and the unequal power rating of the different switches.

2.3.4.2 The Multilevel FC VSC

Rather than using diodes, the multilevel FC VSC uses capacitors to clamp the switch voltage to the capacitor voltage level [11]. The most basic structure of an FC converter is the three-level FC VSC, shown in Figure 2.28: the voltage of the capacitor C_1 must be $v_{C1} = V_{dc}/2$ and the converter provides three output-voltage levels: $V_{dc}/2$, 0 and $-V_{dc}/2$. As in the case of the three-level NPC VSC, if the switches T_{x1} and T_{x2} are on, the output voltage of the leg x is $v_{x0} = V_{dc}/2$, whereas when the devices T_{x3} and T_{x4} are on, the output voltage is $v_{x0} = -V_{dc}/2$. In order to obtain zero output voltage, either switches T_{x1} and T_{x3} or switches T_{x2} and T_{x4} must be on. The flying capacitor C_1 is charged when the pair T_{x1}- T_{x3} is on, while it is discharged when the pair T_{x2}-T_{x4} is on. Table 2.2 shows the different states of each switch and the corresponding output voltage.

When T_{x1} and T_{x2} are on, the devices T_{x3} and T_{x4} must be off, and vice versa. These restrictions imply that the firing signals of both switches T_{x1} and T_{x3} are complementary as well as the firing signals of switches T_{x2} and T_{x4}.

As in the case of the multilevel NPC VSC, the multilevel FC converter can be operated using either PWM techniques or a phase-shift control scheme.

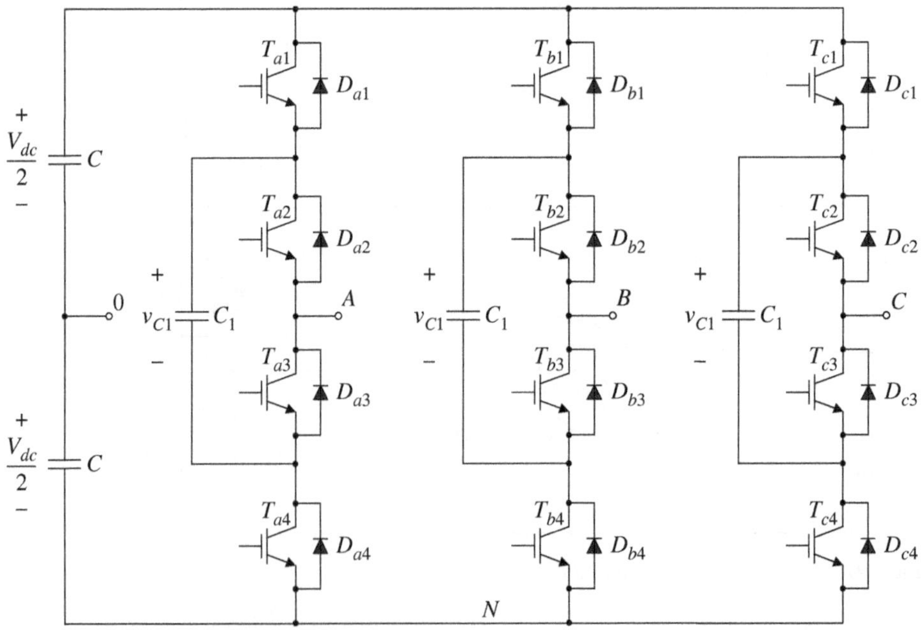

Figure 2.28 Three-level three-phase FC VSC.

Table 2.2 Three-level FC VSC: output voltage as a function of the switch states.

Switch state				
T_{x1}	T_{x2}	T_{x3}	T_{x4}	Output voltage
On	On	Off	Off	$V_{dc}/2$
Off	Off	On	On	$-V_{dc}/2$
On	Off	On	Off	0
Off	On	Off	On	0

One of the main drawbacks of this topology is the large number of flying capacitors when the number of voltage levels increases. For example, Figure 2.29 shows a five-level FC topology for the leg A, where the voltages of the different capacitors must be $v_{C1} = 3V_{dc}/4$, $v_{C2} = V_{dc}/2$ and $v_{C3} = V_{dc}/4$. As in the NPC case, the final number of flying capacitors is almost proportional to the square of the number of voltage levels [10]. Furthermore, the ratings of the flying capacitors must be high since the full load current flows through them, and although the flying capacitor voltages are, in theory, balanced, in practice there are many factors that lead to asymmetrical operation, involving voltage imbalances in the flying capacitors. A method with which to balance the voltages of the flying capacitors is therefore required [12].

The main advantage of this topology is that it provides more flexibility in the choice of switching combinations, allowing accurate control of the flying capacitor voltages. Although it requires a large number of capacitors, the implementation of a three-level FC converter is less onerous than the NPC topology [13]. In this multilevel topology, all the devices have the same switching frequency.

2.3.4.3 The Cascaded H-Bridge VSC

This topology uses several single-phase full-bridge converters with independent DC voltage sources to build a multilevel VSC. Each leg of the multilevel VSC is obtained by connecting the different single-phase converters in series, as shown in Figure 2.30. The resulting output voltage is the sum of the individual output voltage of each single-phase converter: $v_{AN} = v_{VSC1} + v_{VSC2} + v_{VSC3}$. In general, for an n given number of full-bridge converters, the resulting number of levels in the output voltage is $m = 2n + 1$. Each individual converter can be controlled by using the phase-shift control technique, and the harmonic content of the output voltage v_{AN} can be controlled by adjusting the different angle α of each single-phase converter.

Figure 2.31 shows the voltage waveforms obtained by applying the phase-shift control technique to the VSC in Figure 2.30: the output voltages of each single-phase converter are plotted in Figures 2.31a–c, respectively; all these output voltages have the same three levels (V_{dc}, 0 and $-V_{dc}$). The resulting output voltage v_{AN} is a staircase waveform with a low distortion, which is highly sinusoidal and has seven different voltage levels: $3V_{dc}$, $2V_{dc}$, V_{dc}, 0, $-V_{dc}$, $-2V_{dc}$ and $-3V_{dc}$, as shown in Figure 2.31d.

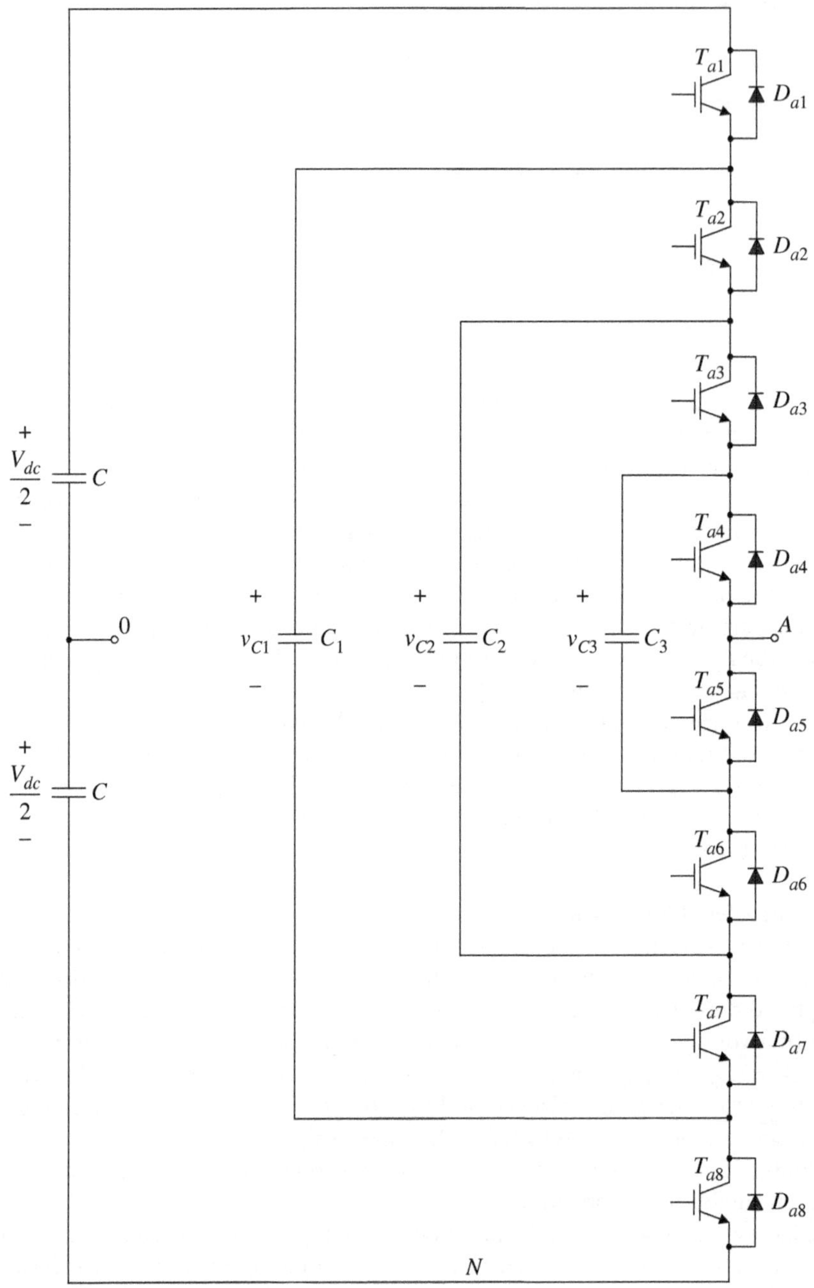

Figure 2.29 Five-level FC configuration.

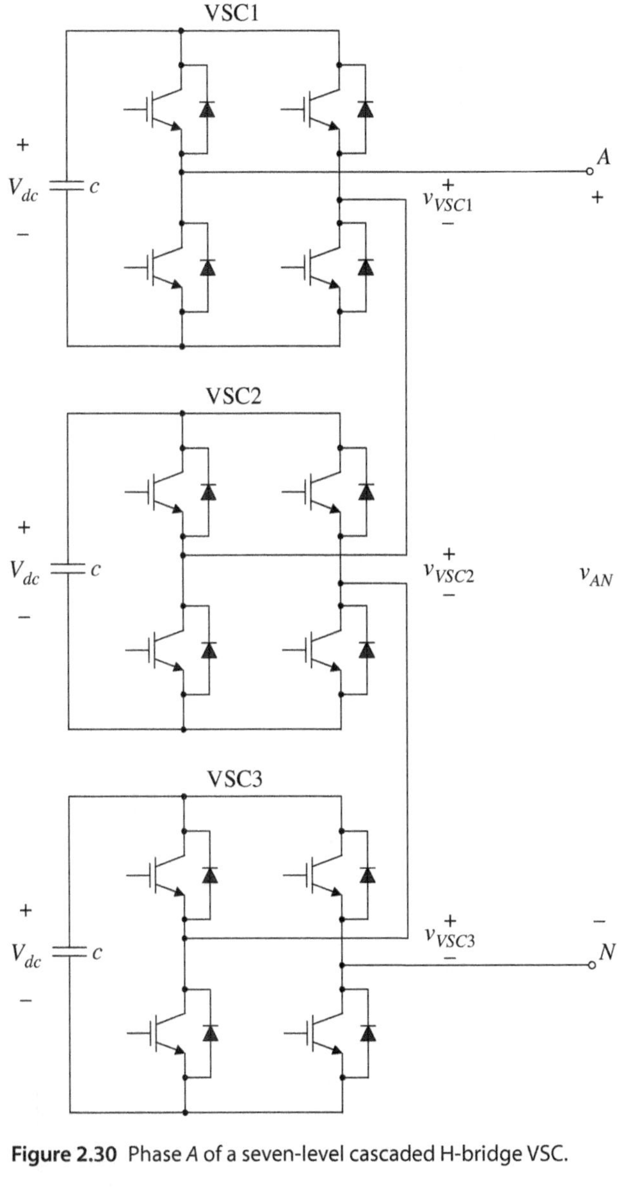

Figure 2.30 Phase *A* of a seven-level cascaded H-bridge VSC.

The use of PWM techniques is very suitable for the case of the H-bridge topology as it allows more harmonics to be cancelled than other multilevel topologies [3].

Until not so long ago, it was considered that the drawbacks of this topology were the large number of individual VSCs and the number of isolated DC voltage sources required as the number of levels increases [14]. However, in today's context, this seems to be a very interesting technology, which may be considered the forebear of modern multilevel converters presented in Section 2.3.4.5.

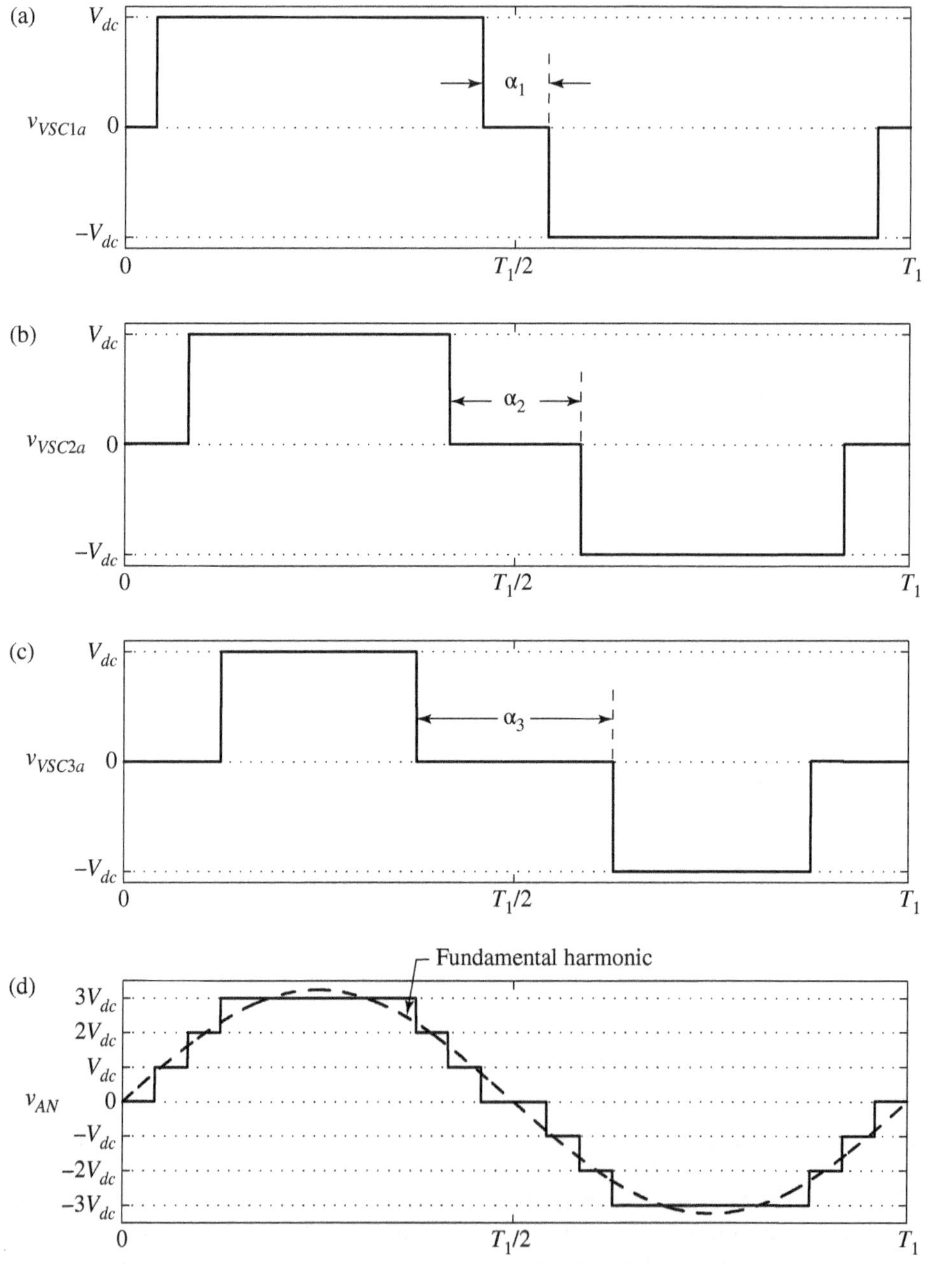

Figure 2.31 Output voltage of a seven-level cascaded H-bridge VSC using the phase-shift control technique, (a) output voltage of VSC1; (b) output voltage of VSC2; (c) output voltage of VSC3; and (d) resulting output voltage v_{AN} for the phase A (–) and its fundamental harmonic (– –).

2.3.4.4 PWM Techniques for Multilevel VSCs

The previous sections have presented multilevel topologies that use fundamental switching techniques, such as phase-shift control, in order to control the multilevel VSC.

If a high switching frequency is required, one of the most frequently used techniques is that of the phase-shifted SPWM (sinusoidal pulse width modulation) switching method: for a given number of levels m and the number of individual VSC cell voltages n necessary to generate the m levels, this scheme employs a sinusoidal modulating signal which is compared with a number of triangular carrier signals equal to n [9]. These triangular signals are shifted to an angle $\theta = 2\pi/n$ and the results of these comparisons are used to turn the different switches on and off. One of the advantages of this technique is that the output voltage has an equivalent switching frequency of n times the frequency of each carrier signal f_{sw}, i.e. the switching frequency of each individual cell. This allows the frequency of the carrier signals to be decreased, thus reducing the switching losses [15].

The three-level, three-phase FC VSC of Figure 2.28 has been operated by using the phase-shifted SPWM technique. In this case, only two carrier signals are necessary and the resultant angle θ is π radians. The two carrier signals and the three modulating waveforms are plotted in Figure 2.32a. The output voltage v_{A0} has three levels, as shown in Figure 2.32b, and according to Figure 2.32c, the main harmonic components are placed around $2f_{sw}$. The line-to-line voltage v_{AB} contains five levels: V_{dc}, $V_{dc}/2$, 0, $-V_{dc}/2$ and $-V_{dc}$, as can be observed in Figure 2.32d. Finally, the harmonic content of the line-to-line voltage shows that the effective switching frequency is twice the frequency of the carrier signals.

The phase-shifted SPWM switching technique is applied not only to multilevel VSCs but also to multiconverter applications [16] and DC-DC interleaved converters [17].

2.3.4.5 An Alternative Multilevel Converter Topology

The need to improve on the voltage waveform and reduce switching losses has led to the development of multilevel converters. The classical multilevel VSC configurations – the NPC, the flying capacitor and the H-bridge – have been described in the sections above. Each topology presents its advantages and disadvantages with respect to the others, but the current consensus is that in spite their many advantages over the two-level VSCs for low- and medium-power applications, they are difficult to escalate for more than three or five voltage levels and hence are rendered unsuitable for high-voltage and high-power applications [18]. Some of these drawbacks have been stated at the end of Sections 2.3.4.1, 2.3.4.2 and 2.3.4.3.

Aiming at reducing design circuit complexity, increasing modularity, lowering switching losses and achieving a more sinusoidal waveform of the output voltage, the modular multilevel converter (MMC) was put forward in [19]. Depending on the number of modules (levels−1) used, the MMC may synthesize a near-sinusoidal voltage waveform. This is achieved by the incremental switching of each module in turn to shape up as much as possible a sinusoidal waveform, with frequencies in the range of 100–150 Hz. This contrasts with the switching frequencies employed by two-level and three-level VSCs, which are in the range of 1–2 kHz. One outcome is that MMCs incur power losses of around 1% as opposed to 2% incurred by the two-level and three-level VSCs. This is a rather significant achievement, but it should be noted that the losses incurred by MMCs are still higher than the losses incurred by CSCs, which stand at around 0.7%.

There are currently three vendors of the MMC VSC technology: ABB, Siemens and Alstom. Each has its own proprietary designs and trademark names: HVDC *Light*, HVDC *Plus* and *MaxSine* HVDC, respectively. The HVDC *Light* and the HVDC *Plus* are very similar and based on the design put forward in [19]. Interestingly, the HVDC *Light* has also been referred to as a cascaded two-level converter [20]. Let us say that these two proprietary designs are pure MMC technologies. Meanwhile, *MaxSine* HVDC uses a hybrid VSC-HVDC converter concept, involving a two-level converter as the main switching component with a low switching frequency and a multilevel converter of small rating to cancel out, in an active fashion, the harmonics produced by the main, two-level converter.

The MMC is built by cascading a large number of half-bridge submodules (SMs), as illustrated in Figure 2.33, where two such SMs are connected in series.

Appropriate switching of S_1 and S_2 in each SM will enable to add up the two capacitances, or to bypass both of them, or to use either of them, to achieve the target output voltage level. In each SM, when switch S_1 is on, switch S_2 is off, and the other way around. When both switches S_1 are on, the output voltage is $V_{out} = V_{sm1} + V_{sm2}$. When both switches S_2 are on the output, voltage is nil, i.e. $V_{out} = 0$. These switching operations already represent two levels in the converter operation and a third level arises when S_1 in either SM is on, e.g. $V_{out} = V_{sm1}$.

The schematic representation of a three-phase MMC is shown in Figure 2.34 where $2n$ SMs are employed to make up a phase leg, as illustrated in the figure.

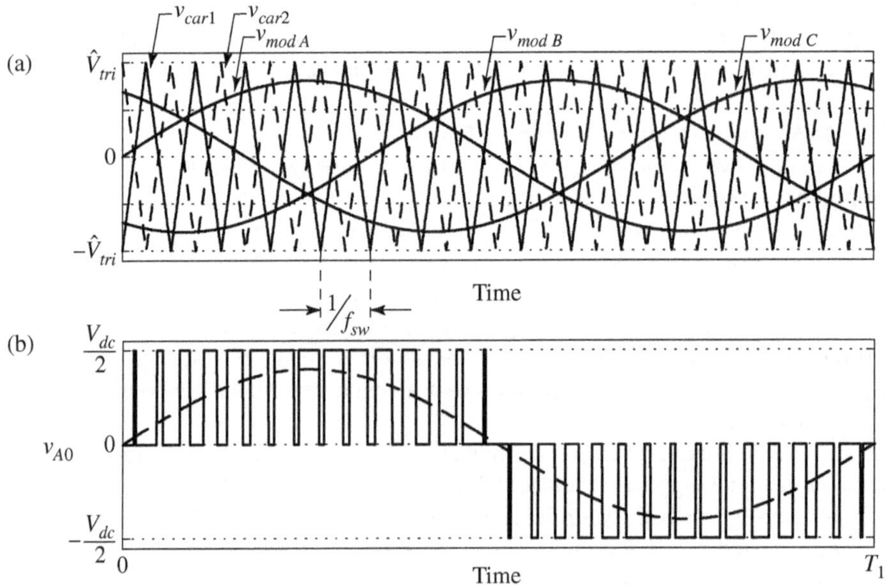

Figure 2.32 SPWM for the three-level three-phase FC VSC for $m_a = 0.8$ and $m_f = 15$, (a) modulating signals and carrier waveforms shifted π radians; (b) output voltage v_{A0} (–) and its fundamental harmonic (– –); (c) normalized harmonic components of v_{A0}; (d) line-to-line output voltage v_{AB} (–) and its fundamental harmonic (– –); and (e) normalized harmonic components of v_{AB} for the phase A (–) and its fundamental harmonic (– –).

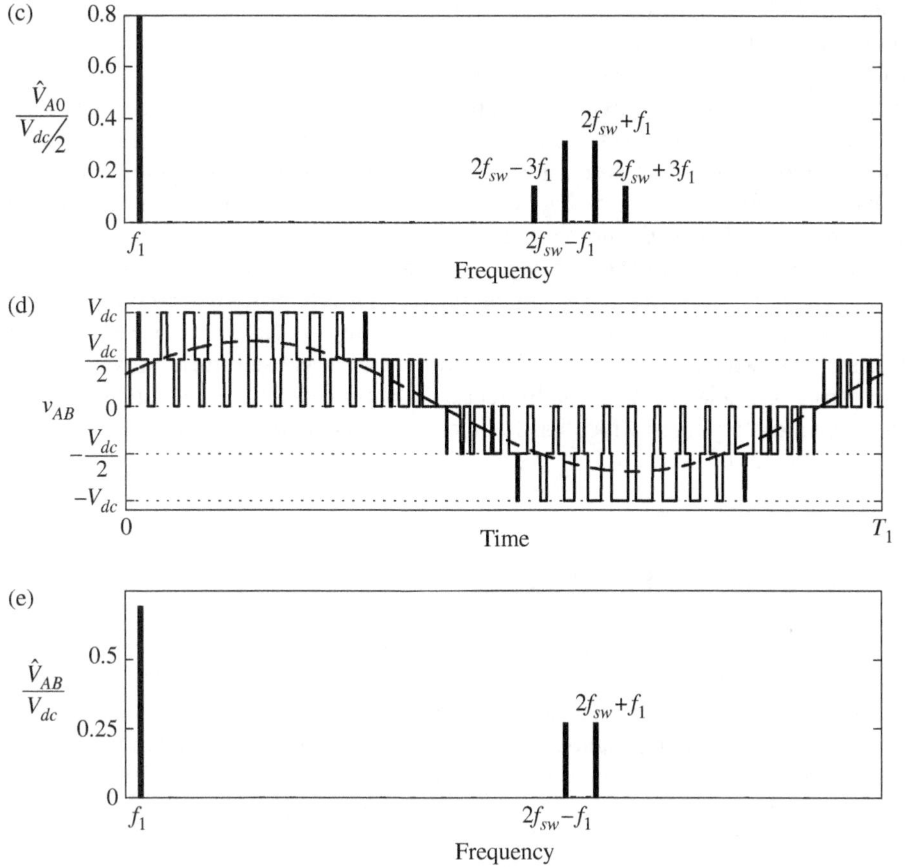

Figure 2.32 (*Continued*)

Notice that the phase legs are connected in parallel in the DC side and ideally, each phase leg contributes one third to the DC current. Nonetheless, balancing or circulating current will flow between the phase legs since a small natural unbalance exists between the voltages of the three phase legs. The function of the phase reactors is to limit such circulating currents, but they also limit the current rise in cases of DC faults, to enable the IGBTs to be turned off relatively safely. More specifically, the circulating currents are suppressed using a parallel resonant filter tuned at the second harmonic and in the case of the ABB's proprietary design, placed at the centre of the phase reactors. The filter also serves to eliminate the third harmonic [18]. The two companies seem to take very different approaches when it comes to the number of SMs employed in their MMCs. For instance, in the Trans Bay Cable installation, Siemens used 200 SMs in each phase arm of the converter. This VSC-HVDC link is rated at ±200 kV [21]. In contrast, ABB employs only 38 SMs per phase arm for a DC link of ±320 kV [20]. Owing to the rather large number of SMs, the former design requires no filter whereas the latter design employs a small filter.

The hybrid VSC is shown schematically in Figure 2.35. It comprises an H-bridge connected in parallel with a wave-shaping circuit in each phase. In turn, a wave-shaping circuit comprises half-bridge SMs – see Figure 2.33.

In this configuration the three phases are connected in series at their DC sides; so, there are no circulating currents and no need for the use of inductors as is the case in the MMC topology. We shall use the acronym HMC for this converter topology. Another key advantage of this hybrid topology is that the wave-shaping circuit lies outside the main current path; hence, the number of SMs is very much reduced.

The working principle of the H-bridge has already been explained in Section 2.3.2. The function of the wave-shaping circuit on the DC side is to produce a rectified voltage sinewave, which is transferred onto the AC side, resulting in a sinusoidal waveform with little distortion, which is a function of the number of SMs used in the wave-shaping circuit. An undesirable byproduct of this rectification process is the appearance of a large sixth harmonic component which has to be removed using filters or an active filtering technique [18], [22].

2.4 HVDC Systems Based on VSC

In Section 1.2 it was illustrated that two power converters connected in tandem may be used to interconnect two AC systems to exchange electrical power in DC form. This involves one converter acting as a rectifier and the other as an inverter, according to a pre-specified command. In principle, any pair of the VSC types presented in Section 2.3 may be used to such an aim, giving rise to single-phase and three-phase HVDC links. It should be noted that diode and thyristor-based bridges may also be used to form HVDC links or may even be combined with VSCs to form hybrid HVDC links. Moreover, we may have two-level and multilevel HVDC links of various kinds. In bulk power transmission applications, only three-phase HVDC links are employed owing to reasons of energy efficiency and costs.

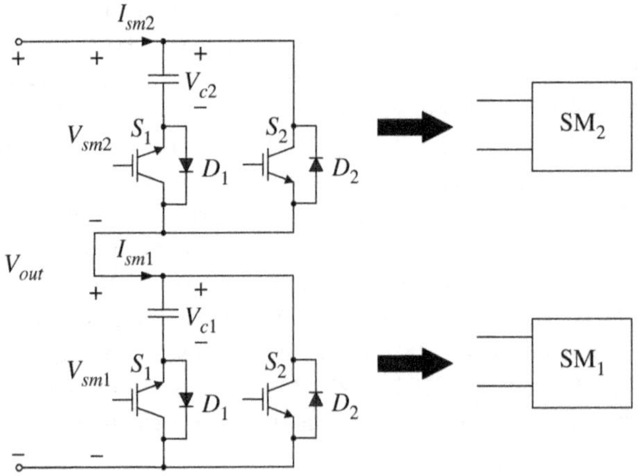

Figure 2.33 Two SMs connected in series.

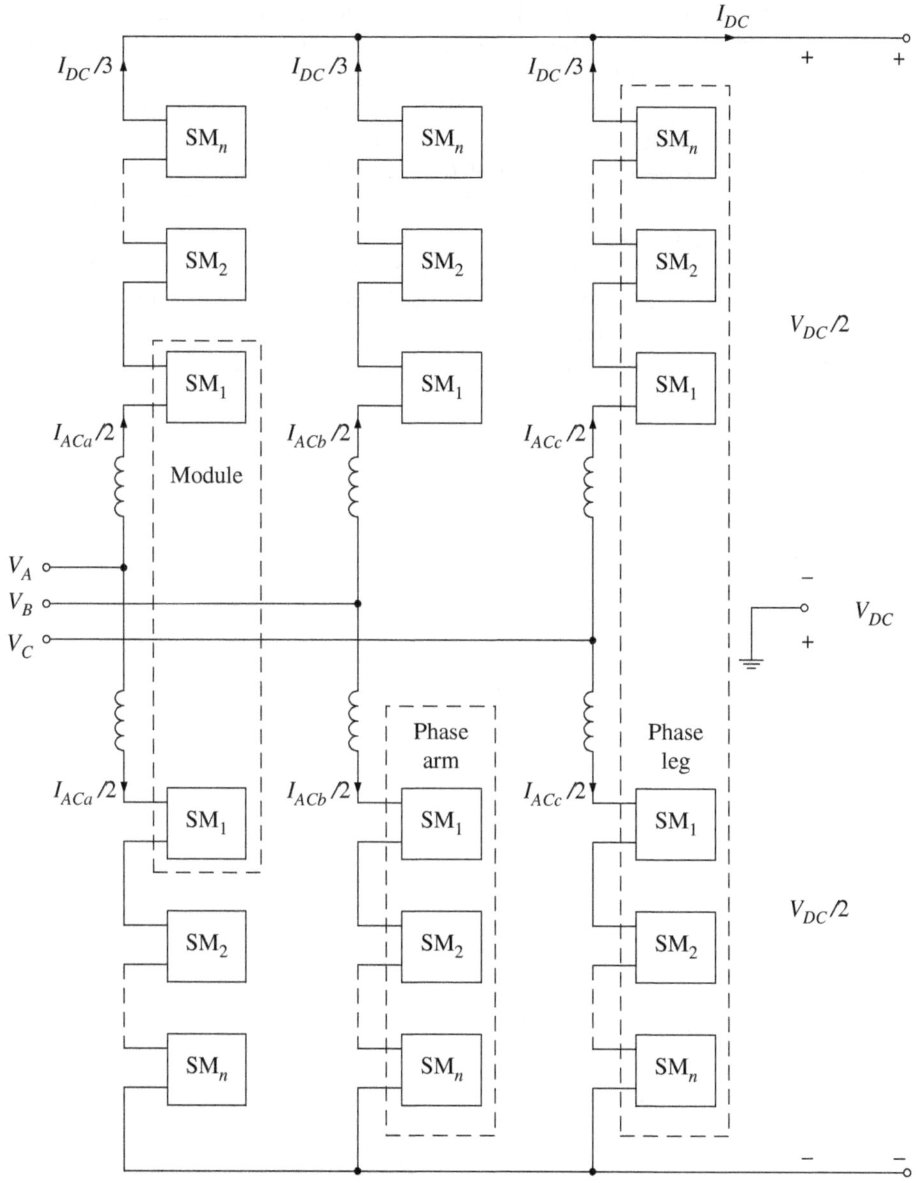

Figure 2.34 Schematic representation of a three-phase MMC.

Reputedly, the first commercial HVDC facility was installed to connect the island of Götland and mainland Sweden in 1954 [23]. Technology has evolved since then, and modern power semiconductors such as IGBTs have substituted the old mercury arc valves and the classical thyristors in that installation.

Classical HVDC systems employ line-commutated CSC thyristor bridges. This mature technology enables the transmission of power in the region of 8 GW and perhaps more if the requirement were to arise [24] – note that in 2012, the Jinping-Sunan link in

East China came into operation, rated at 7200 MW and ±800 kV, spanning a distance of 2100 km. These HVDC links may be classified, depending on the function and location of the CSCs, in back-to-back HVDC systems, monopolar HVDC systems, bipolar HVDC systems and series, multi-terminal HVDC systems. A comprehensive description of these configurations can be found in [25]. In contrast, current development of power semiconductors permits the use of IGBT-based VSCs for power levels up to 1000 MW per circuit, as in the case of the 2000 MW INELFE DC link between Baixas, France and Santa Llogaia, Spain, commissioned in 2015.

Figure 1.8 shows an HVDC system based on the VSC technology. There are two VSC stations, which can be of the following kinds: the basic two-level, the multilevel or the more advanced modular multilevel. In any case, the active power, in DC form, can flow through the DC link in either direction, according to command, whereas the respective

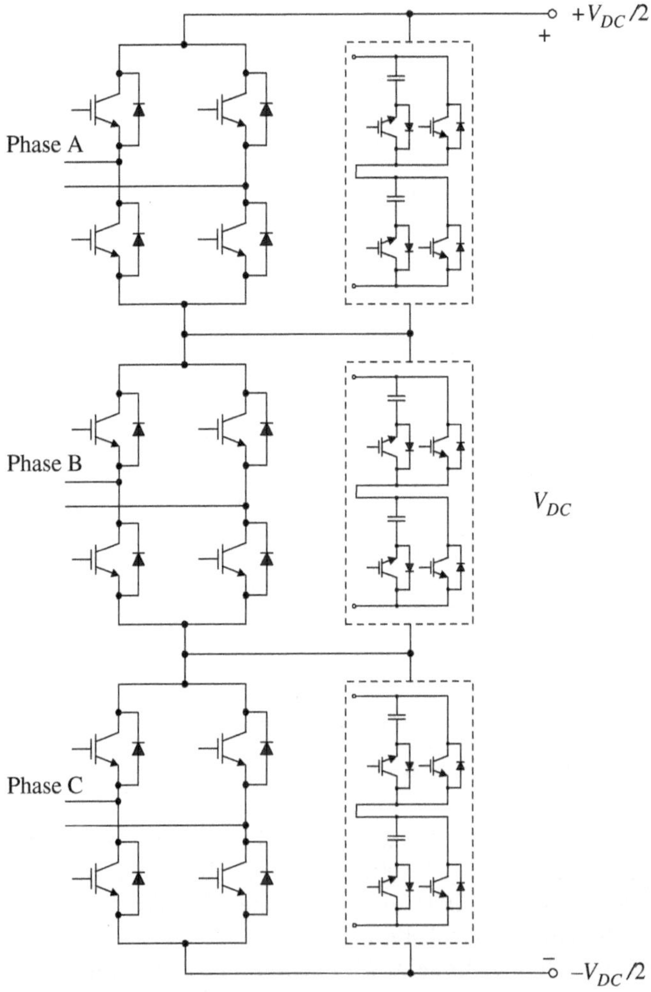

Figure 2.35 Schematic representation of a three-phase hybrid VSC.

reactive powers can be controlled to flow in and out of both VSCs, or even to operate at unity power factors. Note that no reactive power exists in a DC circuit.

The use of an HVDC system based on the VSC technology compared with the CSC technology exhibits several advantages, some of which are listed below:

- VSCs are four-quadrant operation converters in terms of their active and reactive powers, as opposed to two-quadrant operation, which is the case with CSC converters [7, 26].
- Unlike CSCs, there are no commutation failures in VSCs, which incidentally cause voltage sags and waveform distortion in the AC power grid.
- VSC-HVDC links may be used to interconnect strong and weak grids (those exhibiting low short-circuit ratios) [26]. This option is very much curtailed with CSC-HVDC links.
- VSC-HVDC links are ideally suited to connect wind and solar power parks to the main grid – in particular, offshore wind parks – as well as to supply oil and gas rigs from the main power grid [27].
- VSC-HVDC links, like CSC-HVDC systems, are economically feasible solutions if the transmission distance exceeds a threshold value, termed the 'break even distance' [26], [18]. However, there are many instances where the distance does not even come into consideration, such as the back-to-back (or zero-distance) schemes and submarine applications, the latter being an application where the VSC-HVDC outperforms the CSC-HVDC.
- Unlike CSCs, VSCs exhibit very fast control of the active and reactive powers owing to the high PWM switching frequency.
- Since the power semiconductors used in VSCs are fully controlled, there is no need for bulky transformers to assist in the commutation process as is the case in CSCs [7].
- CSC-HVDCs produce characteristic harmonics, which are low-order harmonics. In contrast, two-level VSC-HVDCs may produce no low-order harmonics but at the expense of substantial converter power losses (switching losses) [26]. However, the new breed of MMC VSCs have much more reduced power losses than the two-level converters, which are in the order of those incurred by the CSC-HVDCs, and a much-improved harmonic profile. In fact, the MMC produces only the second and third harmonics, which may necessitate filtering out and the use of a transformer with its secondary winding connected in delta. However, if the number of SMs per phase arm is large enough (e.g. 200), such as in the case of the Siemens design, then the second harmonic is low in value and does not require filtering. The hybrid multilevel design produces a large sixth harmonic which needs to be filtered out. The high-order harmonics are negligibly low in the three MMC designs presented in Section 2.3.4.5.
- Each terminal of a VSC-HVDC system can operate as the STATCOM when the active power in the DC link is zero. Such a feature is not an attribute of the CSC-HVDC technology.
- VSC-HVDCs, but not CSC-HVDCs, have the capability to start up in the case of a dead network [28].
- Unlike CSC-HVDC, VSC-HVDC enables meshed DC power grids. However, no multi-terminal VSC-HVDC systems has yet been built, although it is thought that one would enable the electrification of the North Sea, bringing many technical and economic advantages to the whole of Northern Europe [29].

Table 2.3 VSC-HVDC installations in operation.

Installation name	Installed	Rated power (MW)	Rated voltage (kV)	Manufacturer	Converter topology
Hälsjön	1997	3	±10	ABB	2-level
Gotland	1999	50	±80	ABB	2-level
Tjaereborg	2000	7	±9	ABB	2-level
Terranora (Directlink)	2000	180	±80	ABB	2-level
Eagle Pass	2000	36	±15.9	ABB	2-level
Murraylink	2002	220	±150	ABB	3-level
Cross Sound Cable	2002	330	±150	ABB	3-level
Troll A 1&2	2005	88	±60	ABB	2-level
Estlink1	2006	350	±150	ABB	2-level
BorWin1	2009	400	±150	ABB	2-level
Trans Bay	2010	400	±200	Siemens	MMC
Caprivi Link	2010	300	+350	ABB	2-level
Valhall	2011	78	+150	ABB	2-level
East West Interconnector	2013	500	±200	ABB	2-level
INELFE	2013	2000	±320	Siemens	MMC
Mackinac	2014	200	±71	ABB	2-level
Skagerrak4	2014	700	+500	ABB	MMC
BorWin2	2015	800	±300	Siemens	MMC
HelWin1	2015	576	±250	Siemens	MMC
HelWin2	2015	690	±320	Siemens	MMC
Troll A 3&4	2015	100	±60	ABB	2-level
Åland	2015	100	±80	ABB	2-level
Nordbalt	2015	700	±300	ABB	MMC
DolWin1	2015	800	±320	ABB	MMC
DolWin2	2017	900	±320	ABB	MMC
EWIC	2017	500	±200	ABB	2-level
Caithness-Moray	2018	1200	±320	ABB	MMC
Maritime Link	2018	500	±200	ABB	MMC

Given that the VSC-HVDC technology is barely 20 years old, having started in 1997 with the Hälsjön prototype in Sweden, a healthy number of installations are already in operation around the world, as shown in Table 2.3. The names, ratings, year of commissioning, converter manufacturer and converter topology are presented for the most relevant VSC-HVDC installations. Table 2.4 shows similar information but for VSC-HVDC installations that are in the construction stage and due for commissioning in the period 2019–2020. The tables show that at the time of writing there are three major vendors of the VSC-HVDC technology: ABB, Siemens and Alstom.

Table 2.4 VSC-HVDC installation due for commissioning in the period 2019–2020.

Installation name	Rated power (MW)	Rated voltage (kV)	Manufacturer	Converter topology
South-West Link	1440	±300	Alstom	HMC
Italy–France link	1200	±320	Alstom	HMC
DolWin3	900	±320	Alstom	HMC
NEMO	1000	±400	Siemens	MMC
SylWin	864	±320	Siemens	MMC
Cobra Cable	700	±320	Siemens	MMC
BorWin3	900	±320	Siemens	MMC
Ultranet	2000	±380	Siemens	MMC

As illustrated in both tables and except for the specific low-power applications aimed at the power supply of oil and gas rigs, the upward trend is the use of VSC-HVDC systems of higher ratings, a task for which the multilevel topologies are very well suited, as described previously. The reason is that multilevel topologies extend the advantages of medium-power VSCs to high-power applications, which necessarily calls for higher voltages. Furthermore, the use of multilevel converters reduces not only the total harmonic voltage distortion but also the switching losses, hence reducing filtering to a bare minimum – normally, there is no need to filter out any of the characteristic harmonics, 5th, 7th, 11th, 13th and above. Currently, the total power losses of two-level VSCs and MMC are in the region of 2% and 1%, respectively. Note that they are still higher than those of thyristor-based converters, which are in the region of 0.7% [18].

The following are some of the main applications of the VSC-HVDC technology:

- Grid connection of offshore wind farms, with offshore wind power projects dominating the scene in Europe. There are sufficient technical and economic justifications for offshore wind parks to be connected to the main power grid using a VSC-HVDC link as opposed to an AC cable when they lie more than 70 km away from the shore [18]. The former option helps to improve the grid transient stability since the VSCs are natural sources of fast reactive power control [26, 30]. The layout of the submarine cable is also simpler with the former option since there is no need for shunt reactors to be connected at regular intervals along the length of the cable, as would be the case with the latter option. Also, power losses will be less using a VSC-HVDC submarine link compared with a submarine AC cable of comparable length.
- Power distribution in densely populated areas. It is highly likely that multi-terminal VSC-HVDC systems will be used in future distribution applications in large cities [8]. A generic four-terminal VSC-HVDC system is shown in Figure 1.10. One of the key advantages of multi-terminal topologies is the ability to connect a number of power grids in an asynchronous manner; they can therefore contribute towards delivering the electricity produced by different kinds of distributed power sources to customers, with no increase in the troublesome short-circuit ratio. Moreover, they would help to improve on the power quality [31].
- VSC-HVDC systems for bulk power transmission. Unidirectional bulk power transmission has been one of the main applications of HVDC systems [26]. If a CSC

Table 2.5 The 10 largest CSC-HVDC installations in the world.

Installation name	Rated power (MW)	Rated voltage (kV)	Installed	Distance (km)	Country
Jinping-Sunan	7200	±800	2012	2090	China
Xiangjiaba-Shanghai	6400	±800	2010	1980	China
Xiluodo-Guangdong	6400	±500	2014	1223	China
Yunnan-Guangdong	5000	±800	2010	1418	China
Ningxia-Shandong	4000	±660	2010	1348	China
Itaipu 2	3150	±600	1984	805	Brazil
Itaipu 1	3150	±600	1984	785	Brazil
Pacific DC Intertie	3100	±500	1970	1362	USA
Three Gorges-Changzhou	3000	±500	2003	860	China
Guizhou-Guangdong	3000	±500	2004	980	China

solution is used, power ratings above 6.4 MW can be achieved with no difficulty, as illustrated in Table 2.5, containing the names, ratings, distances and year of commissioning of the 10 largest CSC-HVDC links in the world. At present, as Tables 2.3 and 2.4 show, the VSC technology reaches values of 1000 MW (2000 MW using two circuits), such as the case of the INELFE and Ultranet links. Some existing HVAC systems are under consideration to be replaced or reconverted to HVDC systems using the VSC option. Furthermore, hybrid solutions combining CSC-HVDC and VSC-HVDC already exist in practice, as in the case of the Skagerrak 4. In fact, the Skagerrak HVDC transmission system comprises four HVDC links, with a total combined capacity of 1700 MW. It goes from Kristiansand in southern Norway to Tjele on Denmark's Jutland peninsula. Skagerrak 1 and 2 rated 500 MW are CSC-HVDC links commissioned in 1976–1977 and so is Skagerrak 3, commissioned in 1993. Skagerrak 4, commissioned in 2014, is a VSC-HVDC link and operates in a bipole mode with Skagerrak 3, which is a CSC-HVDC link. The latter required a control system upgrade to enable this mode of operation.

2.5 Conclusions

The evolution that has taken place in the power semiconductor field over the past two decades has enabled new developments in power-electronics applications. Although the power semiconductor with the highest power capability is the thyristor, the emergence of the GTO first and the IGBT later on, as fully controlled semiconductors, with reasonable high switching frequencies and medium voltage and medium power capabilities, has made possible their increasing use in power electronics applications. This is particularly the case with the IGBT, which outperforms the GTO in terms of power losses and exhibits a much higher switching frequency range.

A clear example of its growing popularity in power grid applications is the ubiquitousness of VSCs that employ IGBTs. They are used in the generation, transmission, distribution and industrial sectors. Modern control switching techniques allow multi-level VSCs to be operated with a very low output voltage distortion and power losses of around 1%. The majority of the early commercial solutions were based on the two-level configuration. However, there is now a variety of well-developed multilevel topologies such as the NPC and the FC converters, which are key elements in high-voltage applications.

The number and rating of installations using VSC systems which employ the IGBT technology have risen since 2013, with the majority using multilevel converter topologies. The current facilities are in the range of several hundred megawatts [32]. VAR compensation systems based on VSC technology such as the STATCOM have several advantages over the classical static VAR compensators using thyristor-switched capacitors and inductor-controlled reactors. Among them is the faster speed of response, smaller footprint and the ability to contribute reactive power when the power system is undergoing a voltage collapse. A further, more general feature of the VSC is its four-quadrant operation, which enables the connection of a DC energy source or an energy storage system onto its DC bus. This feature is extended even further when two or more VSCs are employed to make up an HVDC system, representing a newer form of DC power transmission that exhibits numerous advantages over the conventional CSC-HVDC systems. For instance, VSC-HVDC enables the connection of strong and weak AC power grids; each VSC converter is able to regulate voltage magnitude at the respective AC nodes and exerts fast control of active and reactive powers, according to a pre-specified control operating philosophy. An unrivalled feature of the VSC-HVDC technology over the CSC-HVDC technology is its ability to form multi-terminal HVDC systems of a generic nature.

All these attributes of the VSC-HVDC technology have encouraged their use in an ever-growing number of applications. Further to the well-known long-distance power transmission application, they look unassailable in the grid connection of offshore wind generators, in energy supply from the mainland to offshore oil and gas rigs, and in underground power distribution in the city centres of large conurbations.

References

1 Mohan, N., Undeland, T.M., and Robbins, W.P. (2002). *Power Electronics: Converters, Applications, and Design*, 3e. Wiley.
2 Hart, D.W. (2011). *Power Electronics*. McGraw-Hill.
3 Acha, E., Agelidis, V.G., Anaya-Lara, O., and Miller, T.J.E. (2002). *Power Electronic Control in Electrical Systems*. Newness.
4 Sedra, A.S. and Smith, K.C. (1998). *Microelectronic Circuits*. Oxford University Press.
5 Mohan, N. (2012). *Power Electronics. A First Course*. Wiley.
6 Hu, A.P. (2009). *Wireless/Contactless Power Supply – Inductively Coupled Resonant Converter Solutions*. VDM Verlag.
7 Flourentzou, N., Agelidis, V.G., and Demetriades, G.D. (2009). VSC-based HVDC power transmission systems: an overview. *IEEE Transactions on Power Electronics* 592–602.
8 Sood, V.K. (2004). *HVDC and FACTS Controllers. Applications of Static Converters in Power Systems*. Boston, MA: Kluwer Academic Publishers.

9 Rodríguez, J., Lai, J.-S., and Peng, F.Z. (2002). Multilevel inverters: a survey of topologies, controls, and applications. *IEEE Transactions on Industrial Electronics* 49 (4): 724–738.
10 Arrillaga, J., Liu, Y.H., and Watson, N.R. (2007). *Flexible Power Transmission. The HVDC Options*. Wiley.
11 Lai, J.-S. and Peng, F.Z. (1996). Multilevel converters – a new breed of power. *IEEE Transactions on Industry Applications* 32 (3): 509–517.
12 Xu, L. and Agelidis, V.G. (May 2004). Active capacitor voltage control of flying capacitor multilevel converters. *IEEE Proceedings on Electric Power Applications* 151 (3): 313–320.
13 Feng, C., Liang, J., and Agelidis, V.G. (2007). Modified phase-shifted pwm control for flying capacitor multilevel converters. *IEEE Transactions on Power Electronics* 22 (1): 178–185.
14 Roncero-Sánchez, P. and Acha, E. (2009). Dynamic voltage restorer based on flying capacitor multilevel converters operated by repetitive control. *IEEE Transactions on Power Delivery* 24 (2): 951–960.
15 Liang, Y. and Nwankpa, C.O. (2000). A power-line conditioner based on flying-capacitor multilevel voltage-source converter with phase-shift SPWM. *IEEE Transactions on Industry Applications* 36 (4): 965–971.
16 Mwinyiwiwa, B., Ooi, B.-T., and Wolanski, Z. (1998). UPFC using multiconverter operated by phase-shifted triangle carrier SPWM strategy. *IEEE Transactions on Industry Applications* 34 (3): 495–500.
17 Choe, G.-Y., Kim, J.-S., Kang, H.-S., and Lee, B.-K. (2010). An optimal design methodology of an interleaved boost converter for fuel cell applications. *Journal of Electrical Engineering & Technology* 5 (2): 319–328.
18 J. Glasdam, J. Hjerrild, L.H. Kocewiak and C.L. Bak, "Review on multi-level voltage source converter based HVDC technologies for grid connection of large offshore wind farms," in Proceedings of the IEEE International Conference on Power System Technology (POWERCON), 2012, Auckland, 2012.
19 A. Lesnicar and R. Marquardt, "An innovative modular multilevel converter topology suitable for a wide power range" in 2003 IEEE Bologna Power Tech Conference Proceedings, Bologna, 2003.
20 B. Jacobson, P. Karlsson, G. Asplund and L. Harnefors, "VSC-HVDC transmission with cascaded two-level converters", in Proceedings CIGRE Conference, Paris, 2010.
21 H.J. Knaak, "Modular multilevel converters and HVDC/FACTS: A success story", in Proceedings of the 201 I-14th European Conference on Power Electronics and Applications (EPE 2011), Birmingham, 2006.
22 C.C. Davidson and D.R. Trainer, "Innovative concepts for hybrid multi-level converters for HVDC power transmission", in 9th IET International Conference on AC and DC Power Transmission (ACDC 2010), London, 2010.
23 Jacobson, B., Jiang-Härner, Y., Rey, P. et al. (2006). HVDC with voltage source converters and extruded cables for up to +/−300 kV and 1000 MW. In: *Cigre Session*. Canada: Montreal,.
24 Hingorani, N.G. (1996). High-voltage DC transmission: a power electronics workhorse. *IEEE Spectrum* 33 (4): 63–72.
25 Arrillaga, J. (1998). *High Voltage Direct Current Transmission*, 2e. London: The Institution of Electrical Engineers.

26 A.M. Abbas and P.W. Lehn, "PWM based VSC-HVDC systems — A review", in Proceedings of the IEEE Power & Energy Society General Meeting, 2009. PES '09, Calgary, 2009.

27 Z.R. Zhang, Z.D. Yin and F.X. Hu, "Research of Multi-Farms Transmission of Distributed Generation Based on HVDC Light", in Proceedings of the International Conference on Power System Technology, 2006. PowerCon 2006, Chongqing, 2006.

28 W. Pan, Y. Chang and H. Chen, "Hybrid multi-terminal HVDC system for large scale wind power", in Proceedings of the IEEE PES Power Systems Conference and Exposition, 2006, Atlanta, 2006.

29 Van Hertem, D. and Ghandhari, M. (2010). Multi-terminal VSC HVDC for the European supergrid: obstacles. *Renewable and Sustainable Energy Reviews* 14 (9): 3156–3163.

30 Haileselassie, T.M. (2008). *Control of Multi-Terminal VSC-HVDC, Master of Science in Energy and Environment*. Norwegian University of Science and Technology.

31 C. Zhao and Y. Sun, "Study on control strategies to improve the stability of multi-infeed HVDC systems applying VSC-HVDC", in Canadian Conference on Electrical and Computer Engineering. CCECE '06, Ottawa, 2006.

32 T.K. Vrana, "Review of HVDC component ratings: XLPE cables and VSC converters", in Proceedings of the 2016 IEEE International Energy Conference (ENERGYCON), Leuven, Belgium, 2016.

3

Power Flows

3.1 Introduction

The main objective of a power flow study is to assess the steady-state operating condition of an electrical power network operating under a fixed set of generation and load pattern. Such assessment may conveniently be carried out by determining the voltage magnitudes and phase angles in all buses of the network; the flow of active and reactive powers in all branches of the network; the active and reactive powers contributed by each generator; and the total network active and reactive power losses [1].

The planning and daily operation of modern power systems call for numerous power flow studies. If the study indicates that there are voltage magnitudes outside bounds at certain points of the network, then appropriate control actions become necessary in order to regulate voltage magnitude. Similarly, if the study predicts that the power flow in a given transmission line or transformer is beyond the power-carrying capability of the line or transformer, then a suitable control action is required. Voltage magnitude regulation is achieved by controlling the amount of reactive power generated/absorbed at key points of the network, as well as by controlling the flow of reactive power throughout the network. In this application it is common practice to represent the power network using only positive-sequence parameters, based on the assumption that the power network is perfectly balanced in terms of both network impedance and generator voltage excitation [2].

From the mathematical modelling point of view, the power flow exercise consists in solving a set of nonlinear, algebraic equations that describe the power network under steady-state operating conditions. Over the years, many approaches have been used to solve the nodal power flow equations; the early methods employed the Gauss-Seidel technique and nodal impedance matrix concepts. These were followed by applications of the Newton-Raphson method, with the nodal complex voltages expressed in either polar coordinates or rectangular coordinates. Simplified formulations of the Newton-Raphson method that exhibit a constant Jacobian matrix, such as the fast decoupled and second-order Newton-Raphson methods, were put forward and used with varying degrees of success. At the present time, with the benefit of modern computers with practically unlimited storage capacity, the power flow Newton-Raphson method may be considered the de facto standard in conventional power flow computations [3].

VSC-FACTS-HVDC: Analysis, Modelling and Simulation in Power Grids, First Edition.
Enrique Acha, Pedro Roncero-Sánchez, Antonio de la Villa Jaén, Luis M. Castro and Behzad Kazemtabrizi.

Companion website: www.wiley.com/go/acha_vsc_facts

3.2 Power Network Modelling

Conventional power flow formulations take the approach of balancing to zero the complex powers leaving and entering a given node of the electrical power network. To be commensurate with this principle, which stems from the fundamental law of energy conservation, all the network elements are taken to be connected to one or more nodes, depending on their nature. Power generators are represented as power injections into the node to which they are connected and load points are represented as power drains at the node to which they are connected. The latter may be made to reflect a degree of voltage dependency to widen the representation of electrical power-consuming equipment, such as electric motors of various kinds, electric lighting, heating and power electronics-based appliances.

For the purpose of electrical power transmission, the main pieces of equipment are overhead transmission lines and underground cables, power transformers, and series and shunt volt ampere reactive (VAR) compensating equipment of various kinds. Early VAR compensation equipment used iron-core inductors and switchable banks of capacitors with mechanical switches. Modern compensation equipment uses power electronic switches and its functionality has been extended to exert active power control as opposed to only reactive power control. This comes under the realm of the power technology area known as flexible alternating current transmission systems (FACTSs) [4].

3.2.1 Transmission Lines Modelling

In general, the magnetic and electric fields associated with an AC current-carrying overhead transmission lines and underground cables are suitably represented by series inductances and shunt capacitances which, in turn, may be represented by series impedances and shunt admittances at a given operating frequency. For the purpose of fundamental-frequency, power system application studies, the relationships between the nodal currents and nodal voltages at the sending end and the receiving end of these elements are well accommodated in a nodal transfer admittance matrix, as shown in (3.1):

$$\begin{bmatrix} \overline{I}_k \\ \overline{I}_m \end{bmatrix} = \begin{bmatrix} \overline{Z}_{series}^{-1} + \overline{Y}_{shunt}/2 & -\overline{Z}_{series}^{-1} \\ -\overline{Z}_{series}^{-1} & \overline{Z}_{series}^{-1} + \overline{Y}_{shunt}/2 \end{bmatrix} \begin{bmatrix} \overline{V}_k \\ \overline{V}_m \end{bmatrix} \tag{3.1}$$

where $\overline{Z}_{series} = R + j\omega L$ is the series combination of the cable's resistance and the inductive reactance at the fundamental frequency: $\omega = 2\pi f$ and, say, $f = 50$ Hz. The shunt admittance accounts for the cable's dielectric losses and capacitive effects: $\overline{Y}_{shunt} = G + j\omega C$. Notice that when the cable is an overhead conductor with no external insulation, the dielectric losses are practically zero, hence $\overline{Y}_{shunt}$ contains only the imaginary part.

3.2.2 Conventional Transformers Modelling

The transformer's operation at power frequencies is well represented by a series impedance linking the primary and secondary windings' nodes: k and m. This involves a per-unit representation of the primary and secondary windings' impedances with, say, the secondary winding impedance referred to as the transformer's primary side. In nodal admittance form, this representation takes the following form:

$$\begin{bmatrix} \overline{I}_k \\ \overline{I}_m \end{bmatrix} = \begin{bmatrix} \overline{Y}_{SC} & -\overline{Y}_{SC} \\ -\overline{Y}_{SC} & \overline{Y}_{SC} \end{bmatrix} \begin{bmatrix} \overline{V}_k \\ \overline{V}_m \end{bmatrix} \tag{3.2}$$

where $\overline{Y}_{SC} = \overline{Z}_{SC}^{-1} = 1/(R_t + jX_{SC})$ is the inverse of the series combination of the transformer's resistance and the short-circuit reactance at the fundamental frequency, derived from standard short-circuit tests.

3.2.3 LTC Transformers Modelling

Power transformers may be made to have a degree of voltage-regulating capabilities by fitting one of the main windings with an auxiliary winding having variable turns ratio capability; the auxiliary winding is placed in series with the main winding. It is normal to place the auxiliary winding in series with the primary winding since the current is lower in this winding.

The tap-changing transformer, with the tap T placed on the primary winding, is illustrated in Figure 3.1.

The following basic relationships may be established with reference to the circuit of Figure 3.1:

$$\overline{V}_p = T\overline{V}_m$$
$$T\overline{I}_k = -\overline{I}_m$$
$$\overline{V}_p = \overline{V}_k - \overline{I}_k\overline{Z}_{SC}$$

and suitable combination of these equations yields the following nodal transfer admittance matrix:

$$\begin{bmatrix} \overline{I}_k \\ \overline{I}_m \end{bmatrix} = \begin{bmatrix} \overline{Y}_{SC} & -T\overline{Y}_{SC} \\ -T\overline{Y}_{SC} & T^2\overline{Y}_{SC} \end{bmatrix} \begin{bmatrix} \overline{V}_k \\ \overline{V}_m \end{bmatrix} \tag{3.3}$$

3.2.4 Phase-Shifting Transformers Modelling

If the tap T is complex, with magnitude and phase angle of 1 and φ, respectively, the nodal admittance matrix would correspond to that of a phase-shifting transformer:

$$\begin{bmatrix} \overline{I}_k \\ \overline{I}_m \end{bmatrix} = \begin{bmatrix} \overline{Y}_{SC} & -\overline{Y}_{SC} \cdot (\cos\varphi + j\sin\varphi) \\ -\overline{Y}_{SC} \cdot (\cos\varphi - j\sin\varphi) & \overline{Y}_{SC} \end{bmatrix} \begin{bmatrix} \overline{V}_k \\ \overline{V}_m \end{bmatrix} \tag{3.4}$$

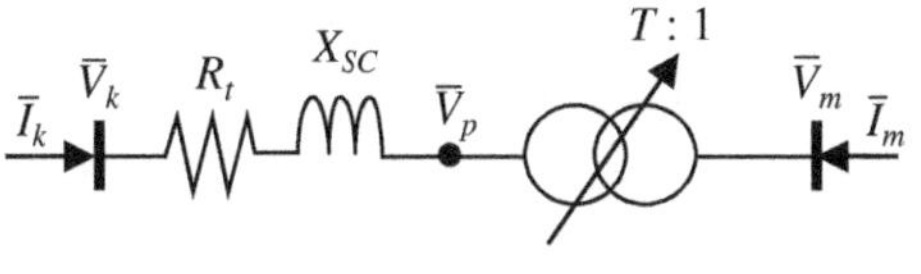

Figure 3.1 LTC representation, with the per-unit impedance in the primary side of the transformer.

3.2.5 Compound Transformers Modelling

It should be noted that the system of equations is asymmetrical and that the ensuing equivalent circuit is a non-reciprocal one, hence a passive circuit representation is not possible.

In a circuit representation of a transformer where the tap is complex and the magnitude is not 1 but T, the nodal admittance matrix would be as follows:

$$\begin{bmatrix} \overline{I}_k \\ \overline{I}_m \end{bmatrix} = \begin{bmatrix} \overline{Y}_{SC} & -\overline{Y}_{SC} \cdot T(\cos\varphi + \mathrm{j}\sin\varphi) \\ -\overline{Y}_{SC} \cdot T(\cos\varphi - \mathrm{j}\sin\varphi) & T^2\overline{Y}_{SC} \end{bmatrix} \begin{bmatrix} \overline{V}_k \\ \overline{V}_m \end{bmatrix} \tag{3.5}$$

3.2.6 Series and Shunt Compensation Modelling

For completeness, a series component such as an inductor or a capacitor has the following nodal transfer admittance representation:

$$\begin{bmatrix} \overline{I}_k \\ \overline{I}_m \end{bmatrix} = \begin{bmatrix} \overline{Y}_{\text{series}} & -\overline{Y}_{\text{series}} \\ -\overline{Y}_{\text{series}} & \overline{Y}_{\text{series}} \end{bmatrix} \begin{bmatrix} \overline{V}_k \\ \overline{V}_m \end{bmatrix} \tag{3.6}$$

where $\overline{Y}_{\text{series}} = 1/\mathrm{j}\omega L$ or $\overline{Y}_{\text{series}} = \mathrm{j}\omega C$ for the series reactor and series capacitor, respectively.

Likewise, a shunt inductor or capacitor has the following nodal admittance representation:

$$[\overline{I}_k] = [\overline{Y}_{\text{shunt}}][\overline{V}_k] \tag{3.7}$$

where $\overline{Y}_{\text{shunt}} = 1/\mathrm{j}\omega L$ or $\overline{Y}_{\text{shunt}} = \mathrm{j}\omega C$ for the shunt reactor and shunt capacitor, respectively.

3.2.7 Load Modelling

A composite power load point in the network, say load point k, may comprise a combination of three kinds of power loads, in various amounts. They are suitably represented by the power expressions given in (3.8) and (3.9):

$$P_{Lk} = P_{0k} \cdot \left[a_p \cdot \left(\frac{V_k}{V_{0k}} \right)^2 + b_p \cdot \left(\frac{V_k}{V_{0k}} \right) + c_p \right] \tag{3.8}$$

$$Q_{Lk} = Q_{0k} \cdot \left[a_q \cdot \left(\frac{V_k}{V_{0k}} \right)^2 + b_q \cdot \left(\frac{V_k}{V_{0k}} \right) + c_q \right] \tag{3.9}$$

The first, second and third terms represent those portions of the load that exhibit constant impedance, constant current and constant power behaviours, respectively. Notice that different coefficients are used for active power and reactive power.

3.2.8 Network Nodal Admittance

The various nodal transfer admittance matrices representing the individual components – Eqs. (3.1)–(3.9) – are easily assembled to represent an electrical power network of practically any size. The result would be the nodal admittance matrix of the entire electrical power system having the following structure for a network of n buses:

$$\begin{bmatrix} \overline{I}_1 \\ \overline{I}_2 \\ \vdots \\ \overline{I}_k \\ \vdots \\ \overline{I}_n \end{bmatrix} = \begin{bmatrix} \overline{Y}_{11} & \overline{Y}_{12} & \cdots & \overline{Y}_{1k} & \cdots & \overline{Y}_{1n} \\ \overline{Y}_{21} & \overline{Y}_{22} & \cdots & \overline{Y}_{2k} & \cdots & \overline{Y}_{2n} \\ \vdots & \vdots & \ddots & \vdots & \ddots & \vdots \\ \overline{Y}_{k1} & \overline{Y}_{k2} & \cdots & \overline{Y}_{kk} & \cdots & \overline{Y}_{kn} \\ \vdots & \vdots & \ddots & \vdots & \ddots & \vdots \\ \overline{Y}_{n1} & \overline{Y}_{n2} & \cdots & \overline{Y}_{nk} & \cdots & \overline{Y}_{nn} \end{bmatrix} \begin{bmatrix} \overline{V}_1 \\ \overline{V}_2 \\ \vdots \\ \overline{V}_k \\ \vdots \\ \overline{V}_n \end{bmatrix} \tag{3.10}$$

It should be noted that inclusion of the powers P_k and Q_k in expressions (3.8)–(3.9) requires conversion into admittances. This is done by dividing these powers by the square of the voltage magnitude at node k. Moreover, the power expression (3.9) is negated.

The entries along the main diagonal, $(\overline{Y}_{11}, \overline{Y}_{22}, \overline{Y}_{kk}, \overline{Y}_{nn})$, are termed self-elements of the nodal admittance matrix and the remaining entries are termed mutual terms. Most entries in the matrix equation (3.10) are complex and will comprise a real part and an imaginary part, say, $\overline{Y}_{kk} = G_{kk} + \mathrm{j}B_{kk}$ and $\overline{Y}_{kn} = G_{kn} + \mathrm{j}B_{kn}$. In general, any self-element of the nodal admittance matrix in (3.10) corresponds to the sum of all the branch admittances that connect to that node. Moreover, a mutual term of the nodal admittance matrix corresponds to the negative value of the branch admittance connecting two nodes, say, branch-linking nodes k and n will have the negative of its admittance value placed in locations kn and nk of the matrix. If no direct connection exists between two nodes, its corresponding entry in the matrix will be nil.

Sometimes it is convenient to represent the nodal admittance matrix equation (3.10) using real quantities as opposed to complex quantities, hence:

$$\begin{bmatrix} I_{\mathrm{Re},1} \\ I_{\mathrm{Im},1} \\ I_{\mathrm{Re},2} \\ I_{\mathrm{Im},2} \\ \vdots \\ I_{\mathrm{Re},k} \\ I_{\mathrm{Im},k} \\ \vdots \\ I_{\mathrm{Re},n} \\ I_{\mathrm{Im},n} \end{bmatrix} = \begin{bmatrix} G_{11} & -B_{11} & G_{12} & -B_{12} & \cdots & G_{1k} & -B_{1k} & \cdots & G_{1n} & -B_{1n} \\ B_{11} & G_{11} & B_{12} & G_{12} & \cdots & B_{1k} & G_{1k} & \cdots & B_{1n} & G_{1n} \\ G_{21} & -B_{21} & G_{22} & -B_{22} & \cdots & G_{2k} & -B_{2k} & \cdots & G_{2n} & -B_{2n} \\ B_{21} & G_{21} & B_{22} & G_{22} & \cdots & B_{2k} & G_{2k} & \cdots & B_{2n} & G_{2n} \\ \vdots & \vdots & \vdots & \vdots & \ddots & \vdots & \vdots & \ddots & \vdots & \vdots \\ G_{k1} & -B_{k1} & G_{k2} & -B_{k2} & \cdots & G_{kk} & -B_{kk} & \cdots & G_{kn} & -B_{kn} \\ B_{k1} & G_{k1} & B_{k2} & G_{k2} & \cdots & B_{kk} & G_{kk} & \cdots & B_{kn} & G_{kn} \\ \vdots & \vdots & \vdots & \vdots & \ddots & \vdots & \vdots & \ddots & \vdots & \vdots \\ G_{n1} & -B_{n1} & G_{n2} & -B_{n2} & \cdots & G_{nk} & -B_{nk} & \cdots & G_{nn} & -B_{nn} \\ B_{n1} & G_{n1} & B_{n2} & G_{n2} & \cdots & B_{nk} & G_{nk} & \cdots & B_{nn} & G_{nn} \end{bmatrix} \begin{bmatrix} V_{\mathrm{Re},1} \\ V_{\mathrm{Im},1} \\ V_{\mathrm{Re},2} \\ V_{\mathrm{Im},2} \\ \vdots \\ V_{\mathrm{Re},k} \\ V_{\mathrm{Im},k} \\ \vdots \\ V_{\mathrm{Re},n} \\ V_{\mathrm{Im},n} \end{bmatrix} \tag{3.11}$$

In the above, note the following complex numbers equivalence:

$$\begin{aligned} \overline{I} = \overline{Y}\,\overline{V} \;&\Leftrightarrow\; I_{\mathrm{Re}} + \mathrm{j}I_{\mathrm{Im}} = (G + \mathrm{j}B)(V_{\mathrm{Re}} + \mathrm{j}V_{\mathrm{Im}}) \\ &\Leftrightarrow\; \begin{bmatrix} I_{\mathrm{Re}} \\ I_{\mathrm{Im}} \end{bmatrix} = \begin{bmatrix} G & -B \\ B & G \end{bmatrix} \begin{bmatrix} V_{\mathrm{Re}} \\ V_{\mathrm{Im}} \end{bmatrix} \end{aligned} \tag{3.12}$$

3.3 Peculiarities of the Power Flow Formulation

In conventional AC power flow studies, buses can be of three different types: voltage-controlled, load and reference. If synchronous generation is available at the bus with sufficient reactive power provisions, the nodal voltage magnitude may be regulated at a specified value and the bus will be of the voltage-controlled type; if no synchronous generation exists at the bus, voltage magnitude cannot be regulated and the node will be of the load type. On the other hand, one synchronous generator bus in the network must be selected to act as reference for all the other buses in the network. In the parlance of power systems engineers the reference bus is termed slack bus or swing bus and its main function is to provide for transmission power losses and any shortfall in generation that the other generators in the network are unable to meet. In the power flow application, the power injected at a bus is well handled by using the concept of net power, which is the difference between the scheduled generation (positive) and the total load (negative) at the bus. If the bus contains neither generation nor load, its net nodal power carries a zero value.

Each bus in the network has associated two power equations and four nodal quantities: voltage magnitude, voltage phase angle, net active power and net reactive power. Two of these quantities are specified a priori and the other two are calculated using the two available power equations – one is for active power and the other is for reactive power. In a voltage-controlled node, the two specified quantities are the voltage magnitude and the net active power injected into the bus; in a load node, the two specified quantities are the net active power injected into the bus and the reactive power injected into the bus; in the reference node, the two specified quantities are the voltage magnitude and the voltage phase angle. As a complement of the specified nodal variables, the calculated variables are the net reactive power injected into the bus and the voltage phase angle in a voltage-controlled node; the voltage magnitude at the bus and the voltage phase angle in a load node; the net active power injected into the bus and the net reactive power injected into the bus in the slack node. Because of the nature of the specified variables in the voltage-controlled bus and the load bus, they are also known as PV-type bus and PQ-type bus, respectively. Notice that the nodal power injections may be either positive or negative.

Three main control systems directly affect the turbine-generator set: the boiler's firing control, the governor control and the excitation system control. The excitation system consists of the exciter and the automatic voltage regulator (AVR). The latter regulates the generator terminal voltage by controlling the amount of current supplied to the field winding by the exciter. For the purpose of steady-state analysis, it is assumed that the three control systems act in an idealized manner, enabling the synchronous generator to produce constant power output, to run at synchronous speed and to regulate voltage magnitude at the generator's terminal with no delay and up to its reactive power design limits. These are the reasons why in a voltage-controlled bus the nodal voltage magnitude and the active power are selected to be known quantities (and specified a priori) in the power flow solution. On the other hand, in a load-type bus, the active power and the reactive power are taken to be available quantities derived from measurements, say supervisory control and data acquisition (SCADA) measurements. In network expansion studies, the active and reactive powers in load type buses may correspond to forecasted values determined from load growth assessments. In the slack bus, the voltage magnitude is specified because in essence this is a generator-type bus or the strongest point in the network (with the largest reserve of reactive power) and its phase voltage angle provides a reference for the phase voltage angles of all the other buses in the network – it is quite normal to select a value of zero for the phase voltage angle of the slack bus but practically any other value may be selected instead.

It should be noted that within the context of power flows in DC power circuits, no complex variables exist, i.e. voltages and currents are real quantities and reactive power does not exist. Apart from this simplification in variables, no other difference seems to exist between the power flow solutions of an AC power circuit and a DC power circuit. As a matter of fact, the power flow solution of DC circuits may be carried out with no difficulty using a conventional power flow computer program which is normally intended for solving AC power circuits. The results furnished by an AC power flow algorithm when solving a DC power circuit show nodal voltages with zero phase angle voltages at all nodes and zero reactive power flows throughout the DC power circuit. It is quite clear that for this to happen, the slack's phase angle voltage ought to be initialized at zero value. Moreover, it is common sense to interpret such complex voltages (provided by the AC power flow algorithm) with phase angle voltages of zero to be real quantities, as one would expect to see in DC power circuits.

Combined AC and DC power circuits employ frequency converter equipment of various kinds at the interface between AC and DC parts of the power circuit. These power converters require suitable modelling representation within the combined AC-DC power flow solution. Strictly speaking, the frequency does not enter as state variable in conventional power flow solutions, but it is clear that a HVDC link decouples, frequency-wise, the AC power circuits connected to the rectifier and the AC power circuit connected to the inverter. As a consequence of this frequency decoupling, each AC circuit has its own slack node and no relationship exists between the nodal phase angle voltages of the two AC power circuits [5]. The above points apply also to multi-terminal HVDC systems where each AC power circuit connected to a terminal of the HVDC link will require its own slack node. In even more specialized power flow studies, such as the voltage source converter VSC-HVDC-infeed of micro-grids where conventional synchronous generation is not readily available, the inverter's AC internal angle plays the role of slack node. This concept also applies to the case of offshore wind farms where the energy is transferred to the mainland AC system using VSC-HVDC transmission.

3.4 The Nodal Power Flow Equations

The equation describing the complex power injection at bus l is the starting point for deriving nodal active and reactive power flow equations suitable for the Newton-Raphson power flow algorithm. This may be established by appealing to Kirchhoff's Current Law (KCL), where the current balance at bus l is exemplified in Figure 3.2a and its extension to complex powers balance is exemplified in Figure 3.2b.

The current balance in Figure 3.2a may be expressed as follows:

$$\overline{I}_l = \sum_{m=1}^{n} \overline{I}_{lm} = \sum_{m=1}^{n} \overline{Y}_{lm}\overline{V}_m \tag{3.13}$$

Extending this to include all n nodes of the network yields the nodal admittance matrix expression (3.10), which in compact form may be written as follows:

$$\overline{\mathbf{I}} = \overline{\mathbf{Y}}\overline{\mathbf{V}} \tag{3.14}$$

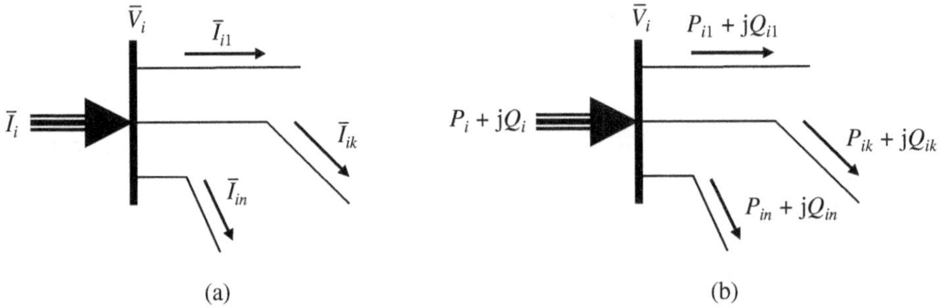

Figure 3.2 Current and power balances in bus *I*.

The nodal complex voltages $\overline{\mathbf{V}}$ may be expressed in either rectangular coordinates, i.e. $\overline{\mathbf{V}} = \mathbf{e} + \mathbf{jf}$, or polar coordinates, i.e. $\overline{\mathbf{V}} = \mathbf{V}\mathbf{e}^{\mathbf{j}\boldsymbol{\theta}}$. Similarly, the admittances may be expressed in either rectangular coordinates, i.e. $\overline{\mathbf{Y}} = \mathbf{G} + \mathbf{jB}$, or polar coordinates, i.e. $\overline{\mathbf{Y}} = \mathbf{Y}\mathbf{e}^{\mathbf{j}\boldsymbol{\gamma}}$. Using rectangular notation as suggested in (3.12), Eq. (3.14) transforms into the following one:

$$\begin{bmatrix} \mathbf{I_{re}} \\ \mathbf{I_{im}} \end{bmatrix} = \begin{bmatrix} \mathbf{G} & -\mathbf{B} \\ \mathbf{B} & \mathbf{G} \end{bmatrix} \begin{bmatrix} \mathbf{e} \\ \mathbf{f} \end{bmatrix} \tag{3.15}$$

Moreover, the nodal complex power equations in rectangular form may be obtained by the matrix product of the nodal complex voltages and the complex conjugate of the nodal currents:

$$\begin{bmatrix} \mathbf{P} \\ \mathbf{Q} \end{bmatrix} = \begin{bmatrix} \mathbf{e} & -\mathbf{f} \\ \mathbf{f} & \mathbf{e} \end{bmatrix} \begin{bmatrix} \mathbf{I_{re}} \\ \mathbf{I_{im}} \end{bmatrix}^* = \begin{bmatrix} \mathbf{eI_{re}} + \mathbf{fI_{im}} \\ \mathbf{fI_{re}} - \mathbf{eI_{im}} \end{bmatrix} \tag{3.16}$$

In a more detailed form:

$$\begin{bmatrix} \mathbf{P} \\ \mathbf{Q} \end{bmatrix} = \begin{bmatrix} \mathbf{e} & -\mathbf{f} \\ \mathbf{f} & \mathbf{e} \end{bmatrix} \cdot \left(\begin{bmatrix} \mathbf{G} & -\mathbf{B} \\ \mathbf{B} & \mathbf{G} \end{bmatrix} \begin{bmatrix} \mathbf{e} \\ \mathbf{f} \end{bmatrix} \right)^* = \begin{bmatrix} \mathbf{e(Ge - Bf) + f(Be + Gf)} \\ \mathbf{f(Ge - Bf) - e(Be + Gf)} \end{bmatrix} \tag{3.17}$$

3.5 The Newton-Raphson Method in Rectangular Coordinates

The nonlinear set of algebraic equations in (3.17) may be used to calculate the nodal voltages that exist at all nodes of the power network, corresponding to a given set of generation and load patterns. These calculated powers will match the specified net powers up to a specified tolerance. In turns, the net active power and net reactive power at a given node correspond to the difference between the generated power and the consumed power at that node. This gives rise to the nodal power mismatches which are solved by iteration using any suitable numerical method, such as the Newton-Raphson:

$$\mathbf{\Delta P} = (\mathbf{P^{gen}} - \mathbf{P^{load}}) - \mathbf{P^{cal}} = \mathbf{P^{net}} - \mathbf{P^{cal}} \tag{3.18}$$

$$\mathbf{\Delta Q} = (\mathbf{Q^{gen}} - \mathbf{Q^{load}}) - \mathbf{Q^{cal}} = \mathbf{Q^{net}} - \mathbf{Q^{cal}} \tag{3.19}$$

where $\mathbf{P^{gen}}$, $\mathbf{Q^{gen}}$, $\mathbf{P^{load}}$, $\mathbf{Q^{load}}$ are vectors of specified active and reactive powers at all nodes of the network; $\mathbf{P^{net}}$, $\mathbf{Q^{net}}$ are the corresponding difference vectors between nodal generation and demand; and $\mathbf{P}^{cal}$, $\mathbf{Q}^{cal}$ are vectors of calculated powers as determined by (3.17). It is very unlikely that the nodal voltages used in (3.17) in the first instance to calculate $\mathbf{P}^{cal}$, $\mathbf{Q}^{cal}$ will yield sufficient accuracy to fulfil the following condition: $\Delta\mathbf{P} \leq 10^{-12}$ and $\Delta\mathbf{Q} \leq 10^{-12}$. Hence, successive approximations (iterations) are normally required.

3.5.1 The Linearized Equations

The expansion of (3.18)–(3.19) in a Taylor series form and retaining only the first-order derivative terms, expressed in compact matrix notation, yields:

$$\begin{bmatrix} \Delta\mathbf{P} \\ \Delta\mathbf{Q} \end{bmatrix} = -\begin{bmatrix} \partial\Delta\mathbf{P}/\partial\mathbf{e} & \partial\Delta\mathbf{P}/\partial\mathbf{f} \\ \partial\Delta\mathbf{Q}/\partial\mathbf{e} & \partial\Delta\mathbf{Q}/\partial\mathbf{f} \end{bmatrix} \begin{bmatrix} \Delta\mathbf{e} \\ \Delta\mathbf{f} \end{bmatrix} \tag{3.20}$$

The linearized system of Eq. (3.20) is solved for the vector of nodal voltage increments $[\Delta\mathbf{e}\ \ \Delta\mathbf{f}]^t$ and with it the nodal voltages from the previous iteration, say $(r-1)$, are updated:

$$\mathbf{e}^{(r)} = \mathbf{e}^{(r-1)} + \Delta\mathbf{e}^{(r)} \tag{3.21}$$

$$\mathbf{f}^{(r)} = \mathbf{f}^{(r-1)} + \Delta\mathbf{f}^{(r)} \tag{3.22}$$

where r is the iteration counter.

The partial derivatives may be obtained in a number of ways. For instance, the form developed below makes for rather clean implementations and straightforward solutions. Starting from the voltage derivatives of the product voltage times the conjugate of the current:

$$\frac{1}{\partial\mathbf{e},\partial\mathbf{f}} \cdot \begin{bmatrix} \Delta\mathbf{P} \\ \Delta\mathbf{Q} \end{bmatrix} = \frac{1}{\partial\mathbf{e},\partial\mathbf{f}} \cdot \begin{bmatrix} \mathbf{P}^{net} - \mathbf{P}^{cal} \\ \mathbf{Q}^{net} - \mathbf{Q}^{cal} \end{bmatrix} = -\frac{1}{\partial\mathbf{e},\partial\mathbf{f}} \cdot \left(D_V \begin{Bmatrix} \mathbf{e} & -\mathbf{f} \\ \mathbf{f} & \mathbf{e} \end{Bmatrix} \times \begin{bmatrix} \mathbf{I}_{\mathbf{Re}} \\ \mathbf{I}_{\mathbf{Im}} \end{bmatrix}^* \right) \tag{3.23}$$

where $\mathbf{P}^{net}$ and $\mathbf{Q}^{net}$ are constant entries; $D_V\{\cdot\}$ is a block-diagonal matrix of nodal voltages and the current vector accommodates contiguous values of the real and imaginary parts of the current of a given node.

Elaborating further the expression (3.23):

$$\begin{aligned}
\frac{1}{\partial\mathbf{e},\partial\mathbf{f}} \cdot \begin{bmatrix} \Delta\mathbf{P} \\ \Delta\mathbf{Q} \end{bmatrix} &= -\frac{1}{\partial\mathbf{e},\partial\mathbf{f}} \cdot D_V \begin{Bmatrix} \mathbf{e} & -\mathbf{f} \\ \mathbf{f} & \mathbf{e} \end{Bmatrix} \times \begin{bmatrix} \mathbf{I}_{\mathbf{Re}} \\ \mathbf{I}_{\mathbf{Im}} \end{bmatrix}^* - D_V \begin{Bmatrix} \mathbf{e} & -\mathbf{f} \\ \mathbf{f} & \mathbf{e} \end{Bmatrix} \\
&\quad \times \frac{1}{\partial\mathbf{e},\partial\mathbf{f}} \cdot \begin{bmatrix} \mathbf{I}_{\mathbf{Re}} \\ \mathbf{I}_{\mathbf{Im}} \end{bmatrix}^* \\
&= -D_U \begin{Bmatrix} \mathbf{1} & \mathbf{0} \\ \mathbf{0} & \mathbf{1} \end{Bmatrix} \begin{bmatrix} \partial\mathbf{e}/\partial\mathbf{e} \\ \partial\mathbf{f}/\partial\mathbf{f} \end{bmatrix} \times D_{I^*} \begin{Bmatrix} \mathbf{I}_{\mathbf{Re}} & \mathbf{I}_{\mathrm{Im}} \\ -\mathbf{I}_{\mathrm{Im}} & \mathbf{I}_{\mathbf{Re}} \end{Bmatrix} - D_V \begin{Bmatrix} \mathbf{e} & -\mathbf{f} \\ \mathbf{f} & \mathbf{e} \end{Bmatrix} \\
&\quad \times \left(\begin{bmatrix} \mathbf{G} & -\mathbf{B} \\ \mathbf{B} & \mathbf{G} \end{bmatrix}^* \times \frac{1}{\partial\mathbf{e},\partial\mathbf{f}} \cdot D_{V^*} \begin{Bmatrix} \mathbf{e} & -\mathbf{f} \\ \mathbf{f} & \mathbf{e} \end{Bmatrix}^* \right) \\
&= -D_{I^*} \begin{Bmatrix} \mathbf{I}_{\mathbf{Re}} & \mathbf{I}_{\mathrm{Im}} \\ -\mathbf{I}_{\mathrm{Im}} & \mathbf{I}_{\mathbf{Re}} \end{Bmatrix} - D_V \begin{Bmatrix} \mathbf{e} & -\mathbf{f} \\ \mathbf{f} & \mathbf{e} \end{Bmatrix} \\
&\quad \times \begin{bmatrix} \mathbf{G} & \mathbf{B} \\ -\mathbf{B} & \mathbf{G} \end{bmatrix} \times D_{U^*} \begin{Bmatrix} \mathbf{1} & \mathbf{0} \\ \mathbf{0} & -\mathbf{1} \end{Bmatrix} \begin{bmatrix} \partial\mathbf{e}/\partial\mathbf{e} \\ \partial\mathbf{f}/\partial\mathbf{f} \end{bmatrix} \\
&= -D_{I^*} \begin{Bmatrix} \mathbf{I}_{\mathbf{Re}} & \mathbf{I}_{\mathrm{Im}} \\ -\mathbf{I}_{\mathrm{Im}} & \mathbf{I}_{\mathbf{Re}} \end{Bmatrix} - D_V \begin{Bmatrix} \mathbf{e} & -\mathbf{f} \\ \mathbf{f} & \mathbf{e} \end{Bmatrix} \times \begin{bmatrix} \mathbf{G} & -\mathbf{B} \\ -\mathbf{B} & -\mathbf{G} \end{bmatrix}
\end{aligned} \tag{3.24}$$

The following slightly modified result is preferred for implementation because it makes use of the already available nodal admittance matrix in (3.11):

$$\begin{bmatrix} \partial \mathbf{\Delta P}/\partial \mathbf{e} & \partial \mathbf{\Delta P}/\partial \mathbf{f} \\ \partial \mathbf{\Delta}/\partial \mathbf{\Delta e} & \partial \mathbf{\Delta Q}/\partial \mathbf{f} \end{bmatrix} = -D_{I^*}\left\{ \begin{matrix} \mathbf{I}_{Re} & \mathbf{I}_{\mathrm{Im}} \\ -\mathbf{I}_{\mathrm{Im}} & \mathbf{I}_{Re} \end{matrix} \right\} - D\left\{ \begin{matrix} \mathbf{e} & \mathbf{f} \\ \mathbf{f} & -\mathbf{e} \end{matrix} \right\} \begin{bmatrix} \mathbf{G} & -\mathbf{B} \\ \mathbf{B} & \mathbf{G} \end{bmatrix} \tag{3.25}$$

Further to the block-diagonal matrix of nodal voltages, $D_V\{\cdot\}$ in (3.24), the conjugated nodal currents and voltages are also arranged in the same format, $D_{I^*}\{\cdot\}$ and $D_{V^*}\{\cdot\}$, respectively. Notice also that the differentiation process in (3.24) yields a block-diagonal unit matrix and a block-diagonal matrix resulting from the derivative of the conjugated voltage, $D_U\{\cdot\}$ and $D_{U^*}\{\cdot\}$, respectively.

The Jacobian matrix of the Newton-Raphson power flow method in (3.25) involves the matrix product of a band-diagonal matrix of voltages and the nodal admittance matrix added to a band-diagonal matrix of nodal currents. Notice that the Jacobian matrix yields a negative term which with the negative term of linearized expression (3.20) becomes an overall positive Jacobian term. Notice also that the arrangement of the band-diagonal matrix of nodal voltages, $D\{\cdot\}$, in (3.25) differs from both $D_V\{\cdot\}$ and $D_{V^*}\{\cdot\}$.

Eqs. (3.17)–(3.22) and (3.25) are valid for cases when all $n-1$ buses in the network are load type – it should be remembered that one bus is selected to be the slack bus. Voltage-controlled buses, where the generator has not reached its reactive power ceiling, require different equations from those of load-type buses. For a given voltage-controlled node, say node l, the mismatch voltage ΔV_l in (3.26) replaces the entry corresponding to reactive power mismatch ΔQ_l in (3.20):

$$\Delta U_l = U_l^{spec} - \sqrt{e_l^2 + f_l^2} \tag{3.26}$$

where U_l^{spec} is the specified voltage to be attained, and e_l and f_l are the real and imaginary components of the calculated voltage at a given iteration (r).

The corresponding Jacobian entries for the voltage-controlled buses are:

$$\frac{\partial U_l}{\partial e_l} = -\frac{e_l}{\sqrt{e_l^2 + f_l^2}} \tag{3.27}$$

$$\frac{\partial U_l}{\partial f_l} = -\frac{f_l}{\sqrt{e_l^2 + f_l^2}} \tag{3.28}$$

The various equations used to program the Newton-Raphson power flow algorithm in rectangular coordinates may be summed up by the flow diagram presented in Figure 3.3.

Owing to the nature of these voltage equations, all locations in the Jacobian matrix corresponding to the row where ΔU_l has replaced ΔQ_l will yield zero entries except the two self-locations corresponding to e_l and f_l.

3.5.2 Convergence Characteristics of the Newton-Raphson Method

The convergence characteristics of the Newton-Raphson power flow method in either rectangular or polar coordinates are known to be very strong, particularly if the problem at hand is suitably initialized – it is said to yield quadratic convergence. The reason for such a flawless performance may be better appreciated in connection with the

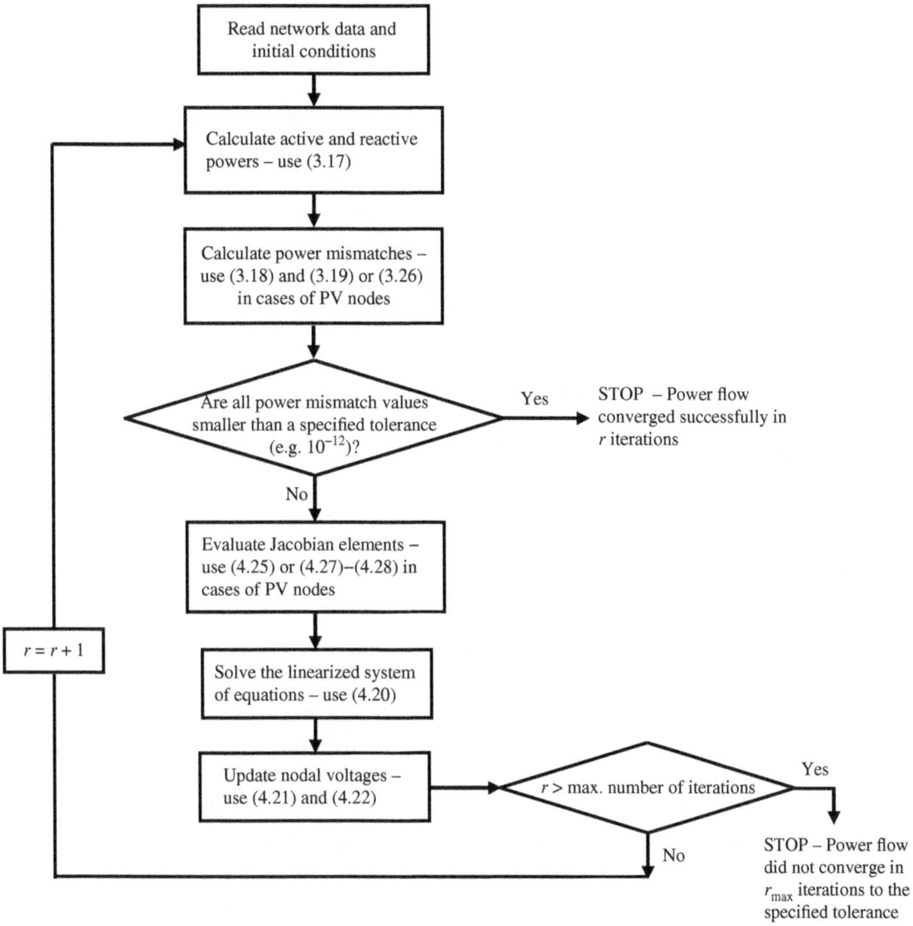

Figure 3.3 Flow diagram for the Newton-Raphson power flow in rectangular coordinates.

performance towards the convergence of a single variable function, say $f(x)$, as shown in Figure 3.4.

However, if the iterative solution is poorly initialized, convergence may be impaired and the method may even diverge.

3.5.3 Initialization of Newton-Raphson Power Flow Solutions

Owing to practical reasons, the electrical power network is designed to have very different rated voltages at the generation, transmission, distribution and utilization systems – the various voltage levels in the overall network are separated by step-up and step-down power transformers, according to requirement. Power engineers have wisely used the per-unit system of units. In essence, the per-unit system is a normalized system of units where the rated values of the interface transformers may be taken to be the base voltage values against which all voltage values in their respective zone of influence are normalized – so, in a power network with no load attached, neglecting charging currents

Quadratic convergence, e.g.:

$\Delta f^{(0)} = 1$

$\Delta f^{(1)} = 0.1$

$\Delta f^{(2)} = 10^{-2}$

$\Delta f^{(3)} = 10^{-4}$

$\Delta f^{(4)} = 10^{-8}$

$\Delta f^{(5)} = 10^{-16}$

$\Delta x^{(1)} = -\Delta f^{(0)}/f'^{(0)}$

$\Delta x^{(2)} = -\Delta f^{(1)}/f'^{(1)}$

$\Delta x^{(3)} = -\Delta f^{(2)}/f'^{(2)}$

$\Delta x^{(4)} = -\Delta f^{(3)}/f'^{(3)}$

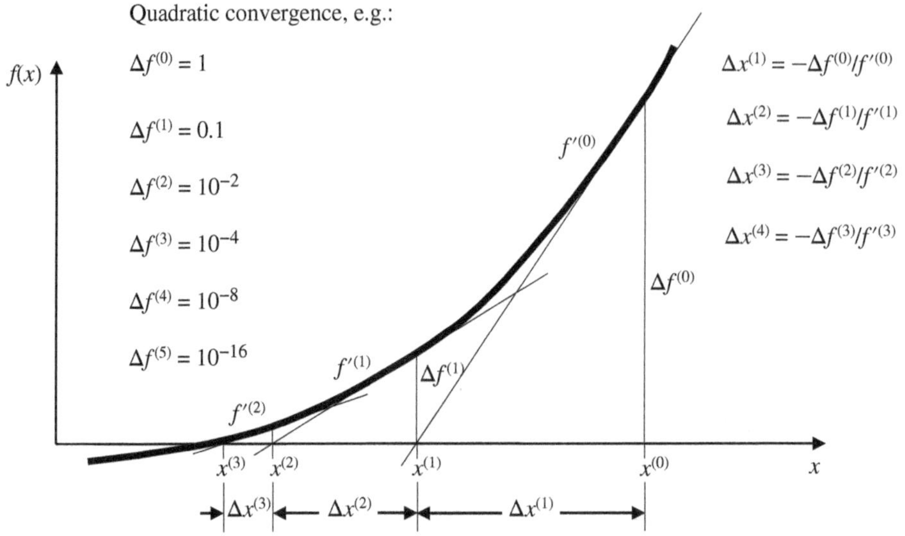

Figure 3.4 Convergence characteristic of the Newton-Raphson method.

in equipment and with all synchronous generators operating at their rated voltages, the voltage profile throughout the network will be one per-unit. When the power network is operating under normal loading conditions, the nodal voltages will depart from their rated 1 p.u. values to some degree but healthy operation of the power systems will call for maximum deviations of 0.06 around the rated 1 p.u. value. Concerning power, which is the other standard base value to be specified, a judicious single value of power is selected for the whole power system under analysis – in academic circles, a base power of 100 MVA is normally selected for high-voltage power networks whereas a base value of 100 kVA looks more amenable for low-voltage distribution systems.

The working practice of using per-unit values in a number of power system applications is rather fortuitous for cases when the Newton-Raphson method is used to solve the associated nonlinear system of equations – which is carried out by iteration. It has been observed that the Newton-Raphson is at its best when the state variables are within the same order of magnitude and when they are initialized to be within close range of the final solution. Thanks to the per-unit practice, all nodal voltage magnitudes that enter the Newton-Raphson formulation as variables may be initialized at 1 p.u. Notice that there will be some nodal voltage magnitudes that will be entered as control variables and that they may not necessarily be specified to keep a 1 p.u. value – generator buses may be cases in point.

Phase angle voltages are handled in radians and it is customary to set them to zero at the start of the iterative solution. In power systems applications where a reference bus is mandatory – for instance, the slack bus in a power flow study – its phase angle is normally taken to be zero, a value that does not change throughout the iterative solution because the phase angle of the slack bus is only a reference against which all other phase angle voltages in the network are measured. It should be remarked that the zero phase angle at the slack is only a convenience practice and that any other value might work just as well – upon convergence to the same tolerance, the only discernible difference between two solutions that use different phase angle voltages at the slack bus would be the phase angle voltages throughout the network. However, their relative values, with respect to the slack bus, would still coincide.

The discussion above impinges directly onto state variable initialization of the Newton-Raphson method in rectangular coordinates – where the real and the imaginary parts of most nodal voltages are set to 1 and 0 p.u. at the start. Such initialization, known as flat voltage profile, may cause the Newton-Raphson method in rectangular coordinates to crash during the first iteration if all the branch elements connecting to a bus of the network contain no resistance – the reason is that one or more entries in the diagonal of the Jacobian will become zero. However, if this is a problem, the Jacobian and vector of incremental state variable values in the right-hand-side of Eq. (3.20) may be arranged differently to prevent this possibility.

The voltage magnitude and phase angle of the series voltage sources, such as the one found in the unified power flow controller (UPFC), requires special care at the initialization stage – its magnitude and phase angle values are likely to be around 0.1 p.u. and $\pi/2$ rad, respectively. Initializing at 1 p.u. and 0 rad would be seen by the Newton-Raphson solutions as being too far away from the final solution; at best, impairing its quadratic convergence and at worst, causing the algorithm to diverge.

3.5.4 Incorporation of PMU Information in Newton-Raphson Power Flow Solutions

The widespread adoption of the phasor measurement unit (PMU) technology may change the practice of initializing nodal phase angle voltages at zero values. The reason is that a PMU is designed to measure the voltage magnitude and the phase angle as well as the complex current at a point in the network – this is actual information, as it exists at the point in time at which the measurement was carried out [6]. It has been observed that the measured phase angles are normally very far away from zero – perhaps in the region of $-\pi/2$ rad. Values measured by PMU would remain constant throughout the iterative power flow solution, with the remaining phase angle voltages being pulled in their direction. Hence, customary initialization of the non-PMU phase angle voltages may be seen as being too far away from the final solution values, resulting in an impaired convergence. However, it should be borne in mind that the phase angle difference between adjacent nodes remains relatively small, just as in the conventional power flow solution. Thus, it makes common sense to initialize the phase angle voltages of nodes with no PMW measurements at values close to those neighbouring nodes with available PMU measurements.

Overall, the new information calls for a slight reformulation of the linearized system of Eq. (3.20), which is explicitly illustrated with specific reference to Eqs. (3.29) and (3.30). The first equation shows the case when an n-node system comprises the slack node (taken to be node 1), PQ-type nodes and PV-type nodes. The slack node is essentially a PV-type bus but one where the voltage phase angle is known a priori. Hence the voltage magnitude and phase angle in this node remain constant from iteration to iteration, i.e. the real and imaginary part of the nodal voltage. This node would normally contain the largest reserve of reactive power in the system and in conventional power flow solutions is singled out to cater for the network power losses. We shall refer to this slack node as a 'physical' slack node.

If PMU information becomes available, say at node k, then V_k and θ_k are known a priori and e_k and f_k remain constant throughout the iterative solution. Hence, node k changes its status from being a PQ-type node to being a 'mathematical' slack node where the voltage magnitude and phase angle remain fixed as given by the PMU measurement; the corresponding implementation is shown in Eq. (3.30).

$$\begin{bmatrix} \Delta P_1 \\ \Delta Q_1 \\ \vdots \\ \Delta P_k \\ \Delta Q_k \\ \vdots \\ \Delta P_n \\ \Delta V_n \end{bmatrix} = - \begin{bmatrix} 1\ 0 & \cdots & 0\ 0 & & \cdots & 0\ 0 & \\ 0\ 1 & \cdots & 0\ 0 & & \cdots & 0\ 0 & \\ \vdots\ \vdots & \ddots & \vdots\ \vdots & & \ddots & \vdots\ \vdots & \\ 0\ 0 & \cdots & \partial \Delta P_k/\partial e_k & \partial \Delta P_k/\partial f_k & \cdots & \partial \Delta P_k/\partial e_n & \partial \Delta P_k/\partial f_n \\ 0\ 0 & \cdots & \partial \Delta Q_k/\partial e_k & \partial \Delta Q_k/\partial f_k & \cdots & \partial \Delta Q_k/\partial e_n & \partial \Delta Q_k/\partial f_n \\ \vdots\ \vdots & \ddots & \vdots & \vdots & \ddots & \vdots & \vdots \\ 0\ 0 & \cdots & \partial \Delta P_n/\partial e_k & \partial \Delta P_n/\partial f_k & \cdots & \partial \Delta P_n/\partial e_n & \partial \Delta P_n/\partial f_n \\ 0\ 0 & \cdots & 0 & 0 & \cdots & \partial \Delta V_n/\partial e_n & \partial \Delta V_n/\partial f_n \end{bmatrix} \times \begin{bmatrix} \Delta e_1 = 0 \\ \Delta f_1 = 0 \\ \vdots \\ \Delta e_k \\ \Delta f_k \\ \vdots \\ \Delta e_n \\ \Delta f_n \end{bmatrix} \tag{3.29}$$

$$\begin{bmatrix} \Delta P \\ \Delta Q_1 \\ \vdots \\ \Delta P_k \\ \Delta Q_k \\ \vdots \\ \Delta P_n \\ \Delta V_n \end{bmatrix} = - \begin{bmatrix} 1\ 0 & \cdots & 0\ 0 & \cdots & 0\ 0 & \\ 0\ 1 & \cdots & 0\ 0 & \cdots & 0\ 0 & \\ \vdots\ \vdots & \ddots & \vdots\ \vdots & \ddots & \vdots\ \vdots & \\ 0\ 0 & \cdots & 1\ 0 & \cdots & 0\ 0 & \\ 0\ 0 & \cdots & 0\ 1 & \cdots & 0\ 0 & \\ \vdots\ \vdots & \ddots & \vdots\ \vdots & \ddots & \vdots\ \vdots & \\ 0\ 0 & \cdots & 0\ 0 & \cdots & \partial \Delta P_n/\partial e_n & \partial \Delta P_n/\partial f_n \\ 0\ 0 & \cdots & 0\ 0 & \cdots & \partial \Delta V_n/\partial e_n & \partial \Delta V_n/\partial f_n \end{bmatrix} \begin{bmatrix} \Delta e_1 = 0 \\ \Delta f_1 = 0 \\ \vdots \\ \Delta e_k = 0 \\ \Delta f_k = 0 \\ \vdots \\ \Delta e_n \\ \Delta f_n \end{bmatrix} \tag{3.30}$$

If PMU information becomes available in more than one bus, we repeat the procedure carried out in Eq. (3.30), with the ensuing system of linearized equations containing the actual or physical slack node and as many mathematical slack nodes as there are PMU voltage measurements in the system. It should be noted that some of the newly created slack nodes may not even contain synchronous generation, such as the PQ-type buses. In these buses, steps must be taken to suitably constrain the net reactive power. Meanwhile, most PV-type buses will contain generation and the phasor voltages furnished by the PMU measurement at the corresponding bus will correctly account for the net power injected at that bus.

3.6 The Voltage Source Converter Model

The STATCOM model comprises the series connection of a VSC and a transformer whose primary winding is shunt-connected with the AC power network [7]. Physically, the VSC is built as a two-level or a multilevel inverter that uses a converter bridge made up of self-commutating switches driven by is pulse width modulation (PWM) control

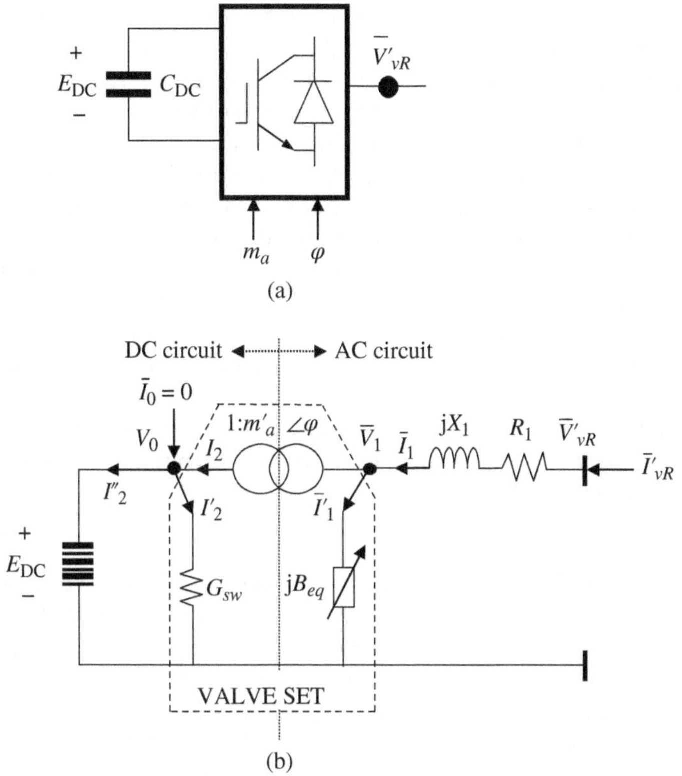

Figure 3.5 (a) VSC schematic representation, (b) VSC equivalent circuit. Source: ©IEEE, 2013.

[8]. It uses a small capacitor bank on its DC side to support and stabilize the DC voltage to enable converter operation. The converter keeps the capacitor charged to the required voltage level by making its output voltage lag the AC system voltage by a small phase angle. The DC capacitor bank of value C_{DC} is shown schematically in Figure 3.5a. It should be stated that C_{DC} is not used per se in the VAR generation/absorption process. Instead, this process is carried out by action of the PWM control, which shifts the voltage and current waveforms within the VSC to yield either leading or lagging VAR operation to satisfy operational requirements.

It is noted that the VSC has practically no inertia, it does not significantly alter the existing system impedance and it can internally generate reactive (both capacitive and inductive) power. For the purpose of fundamental frequency analysis, the VSC's electronic processing of the voltage and current waveforms is well synthesized by the notional variable susceptance, B_{eq}, which connects to the ac bus of the ideal complex tap-changing transformer – see Figure 3.5b. Note that the notional B_{eq} is responsible for the whole of the reactive power production in the valve set of the VSC.

3.6.1 VSC Nodal Admittance Matrix Representation

The fundamental frequency operation of the VSC shown schematically in Figure 3.5a may be modelled by means of electric circuit components, as shown in Figure 3.5b.

From the conceptual point of view, the central component of this VSC model is the ideal tap-changing transformer with complex tap, which, in the absence of switching losses, may be seen to act as a *nullator* that constrains the source current to zero, with the source being the capacitor C_{DC} [9]. In this context, the associated *norator* is the variable susceptance B_{eq}. Indeed, in steady-state operation the DC capacitor may be thought of as a battery that yields voltage E_{DC} and draws no current. The dynamic behaviour of the STATCOM's DC capacitor is covered in Chapter 6. Notice that the winding connected to node 1 is an AC node internal to the VSC and that the winding connected to node 0 is a notional DC node. Two elements connect to the VSC's DC bus: the source, E_{DC}, and the current-dependent resistor, G_{sw}. Hence, the ideal tap-changing transformer is the element that provides the interface for the VSC's AC and DC circuits, as illustrated in Figure 3.4b. It should be emphasized that no reactive power flows through it, only *real* power which is akin to DC power.

The following expression, which is the root mean square (RMS) counterpart of Eq. (2.13) in Chapter 2, is the cornerstone of the fundamental-frequency, positive-sequence STATCOM model, developed in this chapter:

$$\overline{V}_1 = m'_a \, e^{j\varphi} \, E_{\mathrm{DC}} \tag{3.31}$$

where $\overline{V}_1$ corresponds to the fundamental-frequency component of the VSC's output line-to-line voltage in RMS form. In this expression, the angle φ is the phase angle of the complex voltage $\overline{V}_1$ relative to the system phase reference; E_{DC} is the DC bus voltage and $m'_a = k_1 m_a$ is the inverter's amplitude modulation coefficient, where in the linear range of modulation, the amplitude modulation index m_a takes values within bounds: $0 < m_a < 1$. For the case of a two-level, three-phase VSC, the constant k_1 takes the form $k_1 = \sqrt{3}/2\sqrt{2} \cdot (E_{B,DC}/V_{B,AC}^{LL})$, where $V_{B,AC}^{LL}$ and $E_{B,DC}$ are the base voltages in the AC and DC sides, respectively. In the remainder of this chapter and the book, the base voltages of the AC side and the DC side will be considered to have the same values. Hence, in this chapter the constant k_1 takes the value $\sqrt{3/8}$. This value is explained in terms of $1/\sqrt{2}$ being the relationship between an instantaneous quantity and the RMS value of a sinusoidal quantity; $\sqrt{3}$ being the relationship between a phase-like quantity and a line-to-line quantity and 1/2 representing the fact that only half of the peak-to-peak AC voltage induced by a DC voltage is useful for the purpose of extracting the RMS value. Note that this differs from the case in [5, 7] where the following base voltages are used: $V_{B,AC}^{LL}$ and $\sqrt{2}E_{B,DC}$.

Another element of the electric circuit shown in Figure 3.45b is the series impedance, which is connected to the ideal transformer's AC side. The series reactance X_1 represents the VSC's interface magnetics. The series resistor R_1 accounts for the ohmic losses which are proportional to the AC terminal current squared. Note that the secondary winding current $I_2 = I_{DC}$, which is always a real quantity, splits into I'_2 and I''_2. The latter current is always zero during steady-state operation.

As one would expect, the complex power conservation property of the ideal transformer in Figure 3.5b stands, but note that there is no reactive power flowing through it, since all the reactive power requirements of the VSC model (generation/absorption) are met by the shunt branch B_{eq} connected at node 1. The power relationships between nodes 1 and 0, which account for the full VSC model, are:

$$E_{DC}I_{DC} = \overline{V}_1(\overline{I}_1 - \overline{I}'_1)^* = \overline{V}_1\overline{I}_1^* + jB_{eq}V_1^2 \quad \Rightarrow \quad E_{DC}I_{DC} = \mathrm{Re}\{\overline{V}_1\overline{I}_1^*\} \tag{3.32}$$

The switching loss model corresponds to a constant resistance (conductance) G_0, which under the presence of constant DC voltage and constant load current would yield constant power loss for a given switching frequency of the PWM converter. Admittedly, the constant resistance characteristic may be inaccurate because although the DC voltage is kept largely constant, the load current will vary according to the prevailing operating condition. Hence, it is proposed that the resistance characteristic derived at rated voltage and current be corrected by the quadratic ratio of the actual current to the nominal current,

$$G_0 \cdot \left(\frac{I_1^{\mathrm{act}}}{I_1^{\mathrm{nom}}}\right)^2 \Rightarrow G_{sw} \tag{3.33}$$

where G_{sw} would be a resistive term exhibiting a degree of power behaviour.

Expressing the fundamental equation (3.31) as a voltage ratio and using this information into (3.32) yields the following two relationships:

$$\frac{\overline{V}_1}{E_{DC}} = \frac{k_1 m_a \angle\varphi}{1} \text{ and } \frac{I_{DC}}{(\overline{I}_1 - \overline{I}'_1)} = \frac{k_1 m_a \angle -\varphi}{1} \tag{3.34}$$

which corresponds well with the voltage and current ratios which are the hallmark of an ideal, complex tap-changing transformer.

The current through the admittance connected between nodes vR and 1 is:

$$\overline{I}_1 = \overline{Y}_1(\overline{V}'_{vR} - \overline{V}_1) = \overline{Y}_1\overline{V}'_{vR} - k_1 m_a \angle\varphi\ \overline{Y}_1 E_{DC} = \overline{I}'_{vR} \tag{3.35}$$

where $\overline{Y}_1 = 1/(R_1 + jX_1)$.

At node 0, the following relationship holds:

$$\begin{aligned} \overline{I}_0 &= -I_2 + I'_2 \\ &= -k_1 m_a \angle -\varphi\ (\overline{I}_1 - \overline{I}'_1) + G_{sw}E_{DC} \\ &= -k_1 m_a \angle -\varphi\, \overline{Y}_1\, \overline{V}'_{vR} + G_{sw}E_{DC} + k_1^2 m_a^2(\overline{Y}_1 + jB_{eq})E_{DC} \end{aligned} \tag{3.36}$$

Combining (3.35) and (3.36) and incorporating constraints from the electric circuit in Figure 3.5b:

$$\begin{pmatrix} \overline{I}'_{vR} \\ \overline{I}_0 = 0 \end{pmatrix} = \begin{pmatrix} \overline{Y}_1 & -k_1 m_a \angle\varphi\, \overline{Y}_1 \\ -k_1 m_a \angle -\varphi\, \overline{Y}_1 & G_{sw} + k_1^2 m_a^2(\overline{Y}_1 + jB_{eq}) \end{pmatrix} \begin{pmatrix} \overline{V}'_{vR} \\ E_{\mathrm{DC}} \end{pmatrix} \tag{3.37}$$

Notice that this expression represents the VSC equivalent circuit in Figure 3.3b in steady-state, with the capacitor effect represented by the DC voltage E_{DC}.

3.6.2 Full VSC Station Model

As suggested in [10] and shown in Figure 3.6 for a generic VSC station, i connected to node k using a load tap changer (LTC) transformer. A complete VSC station comprises

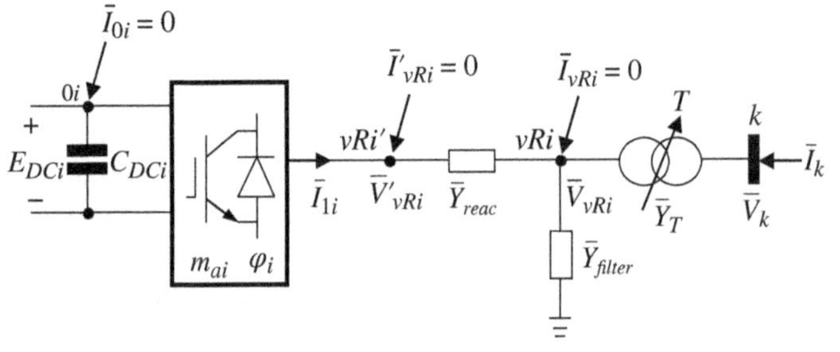

Figure 3.6 VSC station i with ancillary elements connected to node k through a LTC transformer. Source: ©ELSEVIER, 2016.

the AC-DC converter with its AC and DC buses explicitly represented, phase reactors and AC filters. The connecting LTC transformer is added for the sake of generality, but its role will be discussed in Section 3.7.

The smoothing line reactor and the shunt filter are essential elements of the VSC station in order to enable harmonic-free currents and voltages at the secondary winding of the transformer. It should be noted that the resistive components of these elements are rather small, but they are represented here as admittances $\overline{Y}_{reac}$ and $\overline{Y}_{filter}$ for the sake of generality.

The smoothing line reactor and the shunt filter are added below to the basic model of the VSC given by Eq. (3.37). Owing to the modular philosophy of the modelling approach, this can be done in two different ways: (i) by simply extending the nodal admittance matrix to include one more node in the model, namely vR; (ii) by developing a more compact model where node vR' is eliminated mathematically. Developing these two points and dropping the subscript i in Figure 3.6:

$$\begin{pmatrix} \overline{I}_{vR} \\ \overline{I}'_{vR} \\ \overline{I}_0 \end{pmatrix} = \begin{pmatrix} \overline{Y}_{reac} + \overline{Y}_{filter} & -\overline{Y}_{reac} & 0 \\ -\overline{Y}_{reac} & \overline{Y}_{reac} + \overline{Y}_1 & -k_1 m_a \angle\varphi\, \overline{Y}_1 \\ 0 & -k_1 m_a \angle -\varphi\, \overline{Y}_1 & k_1^2 m_a^2 (\overline{Y}_1 + \mathrm{j}B_{eq}) + G_{sw} \end{pmatrix} \begin{pmatrix} \overline{V}_{vR} \\ \overline{V}'_{vR} \\ E_{DC} \end{pmatrix} \tag{3.38}$$

Moreover, since the external injected current at node vR' is nil, the more compact representation of Eq. (3.39) is arrived at:

$$\begin{pmatrix} \overline{I}_{vR} \\ \overline{I}_0 \end{pmatrix} = \begin{pmatrix} \overline{Y}_{vRvR} & -\overline{Y}_{vR0}\angle\varphi \\ -\overline{Y}_{vR0}\angle -\varphi & (G_{sw} + \mathrm{j}k_1^2 m_a^2 B_{eq}) + \overline{Y}_{00} \end{pmatrix} \begin{pmatrix} \overline{V}_{vR} \\ E_{DC} \end{pmatrix} \tag{3.39}$$

where

$$\overline{Y}_{vRvR} = G_{vRvR} + \mathrm{j}B_{vRvR} = \overline{Y}_{filter} + \overline{Y}_1 \overline{Y}_{reac} / (\overline{Y}_1 + \overline{Y}_{reac})$$
$$\overline{Y}_{vR0} = G_{vR0} + \mathrm{j}B_{vR0} = k_1 m_a\, \overline{Y}_1\, \overline{Y}_{reac} / (\overline{Y}_1 + \overline{Y}_{reac})$$
$$\overline{Y}_{00} = G_{00} + \mathrm{j}B_{00} = k_1^2 m_a^2 \cdot \overline{Y}_1 \overline{Y}_{reac} / (\overline{Y}_1 + \overline{Y}_{reac})$$

Both representations, Eq. (3.38) and Eq. (3.39), correctly account for all the elements existing in the electric circuit of Figure 3.5. However, it should be noted that the latter option is preferred with the use of the Newton-Raphson algorithm because voltage regulation using m_a will be applied at a point of the VSC where the voltage signal has already been cleaned by action of the smoothing reactor and filter, namely node vR.

3.6.3 VSC Nodal Power Equations

The complex power model is derived from the nodal admittance matrix:

$$\begin{pmatrix} \overline{S}_{vR} \\ \overline{S}_0 \end{pmatrix} = \begin{pmatrix} \overline{V}_{vR} & 0 \\ 0 & E_{DC} \end{pmatrix} \begin{pmatrix} \overline{I}_{vR}^* \\ \overline{I}_0^* \end{pmatrix}$$
$$= \begin{pmatrix} (e_{vR} + \mathrm{j}f_{vR}) & 0 \\ 0 & E_{DC} \end{pmatrix}$$
$$\times \left\{ \begin{pmatrix} (G_{vRvR} - \mathrm{j}B_{vRvR}) & -(\cos\varphi - \mathrm{j}\sin\varphi)(G_{vR0} - \mathrm{j}B_{vR0}) \\ -(\cos\varphi + \mathrm{j}\sin\varphi)(G_{vR0} - \mathrm{j}B_{vR0}) & (G_{sw} - \mathrm{j}k_1^2 m_a^2 B_{eq}) + (G_{00} - \mathrm{j}B_{00}) \end{pmatrix} \begin{pmatrix} (e_{vR} - \mathrm{j}f_{vR}) \\ E_{DC} \end{pmatrix} \right\} \tag{3.40}$$

Following some arduous algebra, the following nodal active and reactive power expressions are arrived at:

$$\begin{aligned} P_{vR} &= G_{vRvR}(e_{vR}^2 + f_{vR}^2) - E_{DC}(e_{vR}[G_{vR0}\cos\varphi - B_{vR0}\sin\varphi] + \\ &\quad f_{vR}[G_{vR0}\sin\varphi + B_{vR0}\cos\varphi]) \\ Q_{vR} &= -B_{vRvR}(e_{vR}^2 + f_{vR}^2) - E_{DC}(-e_{vR}[G_{vR0}\sin\varphi + B_{vR0}\cos\varphi] + \\ &\quad f_{vR}[G_{vR0}\cos\varphi - B_{vR0}\sin\varphi]) \end{aligned} \tag{3.41}$$

and

$$\begin{aligned} P_0 &= (G_{sw} + G_{00})E_{DC}^2 - E_{DC}(e_{vR}[G_{vR0}\cos\varphi + B_{vR0}\sin\varphi] \\ &\quad + f_{vR}[G_{vR0}\sin\varphi - B_{vR0}\cos\varphi]) \\ Q_0 &= -(k_1^2 m_a^2 B_{eq} + B_{00})E_{DC}^2 - E_{DC}(e_{vR}[G_{vR0}\sin\varphi - B_{vR0}\cos\varphi] \\ &\quad - f_{vR}[G_{vR0}\cos\varphi + B_{vR0}\sin\varphi] \end{aligned} \tag{3.42}$$

where

$$\begin{aligned} G_{vRvR} &= G_{filter} + \Delta G \text{ and } B_{vRvR} = B_{filter} + \Delta B \\ G_{vR0} &= k_1 m_a \Delta G \text{ and } B_{vR0} = k_1 m_a \Delta B \\ G_{00} &= k_1^2 m_a^2 \Delta G \text{ and } B_{00} = k_1^2 m_a^2 \Delta B \\ \Delta G &= \frac{(G_1 G_{reac} - B_1 B_{reac})(G_1 + G_{reac}) + (G_1 B_{reac} + B_1 G_{reac})(B_1 + B_{reac})}{(G_1 + G_{reac})^2 + (B_1 + B_{reac})^2} \\ \Delta B &= \frac{(G_1 B_{reac} + B_1 G_{reac})(G_1 + G_{reac}) - (G_1 G_{reac} - B_1 B_{reac})(B_1 + B_{reac})}{(G_1 + G_{reac})^2 + (B_1 + B_{reac})^2} \end{aligned}$$

Note that G_{vR0}, B_{vR0}, G_{00} and B_{00} are functions of m_a.

3.6.4 VSC Linearized System of Equations

These equations are nonlinear and their solution, for a predefined set of generation and load pattern ($P_{vR,\,\mathrm{net}}$, $Q_{vR,\,\mathrm{net}}$, $P_{0,\,\mathrm{net}}$, $Q_{0,\,\mathrm{net}}$), may be carried out using the Newton-Raphson method. This involves repeated linearization of the nodal power equations. Their initial evaluation requires an informed guess of the state variable values ($e_{vR}^{(0)}, f_{vR}^{(0)}, B_{eq}^{(0)}, m_a^{(0)}, \varphi^{(0)}$), when the aim is to regulate voltage magnitude at bus vR using the VSC's amplitude modulation ratio m_a and to keep E_{DC} at a constant value. In practice, the latter is possible due to the DC capacitor's action. The linearized system of equations is:

$$\begin{bmatrix} \Delta P_{vR} \\ \Delta Q_{vR} \\ \Delta U_{vR} \\ \Delta Q_{0-vR} \\ \Delta P_{0-vR} \end{bmatrix} = -\begin{bmatrix} \partial\Delta P_{vR}/\partial e_{vR} & \partial\Delta P_{vR}/\partial m_a & \partial\Delta P_{vR}/\partial f_{vR} & 0 & \partial\Delta P_{vR}/\partial\varphi \\ \partial\Delta Q_{vR}/\partial e_{vR} & \partial\Delta Q_{vR}/\partial m_a & \partial\Delta Q_{vR}/\partial f_{vR} & 0 & \partial\Delta Q_{vR}/\partial\varphi \\ \partial\Delta U_{vR}/\partial e_{vR} & 0 & \partial\Delta U_{vR}/\partial f_{vR} & 0 & 0 \\ \partial\Delta Q_{0-vR}/\partial e_{vR} & \partial\Delta Q_{0-vR}/\partial m_a & \partial\Delta Q_{0-vR}/\partial f_{vR} & \partial\Delta Q_{0-vR}/\partial B_{eq} & \partial\Delta Q_{0-vR}/\partial\varphi \\ \partial\Delta P_{0-vR}/\partial e_{vR} & \partial\Delta P_{0-vR}/\partial m_a & \partial\Delta P_{0-vR}/\partial f_{vR} & 0 & \partial\Delta P_{0-vR}/\partial\varphi \end{bmatrix} \times \begin{bmatrix} \Delta e_{vR} \\ \Delta m_a \\ \Delta f_{vR} \\ \Delta B_{eq} \\ \Delta\varphi \end{bmatrix} \quad (3.43)$$

Subsequent evaluations of the nodal power equations are carried out using the improved set of values being furnished by the iterative process: ($e_{vR}^{(1,2\ldots)}, f_{vR}^{(1,2\ldots)}, B_{eq}^{(1,2\ldots)}, m_a^{(1,2\ldots)}, \varphi^{(1,2\ldots)}$). The entries making up the Jacobian matrix in (3.43) are the following:

$$\frac{\partial P_{vR}}{\partial e_{vR}} = CR_{vR} + e_{vR}\frac{\partial CR_{vR}}{\partial e_{vR}} + f_{vR}\frac{\partial CI_{vR}}{\partial e_{vR}} \qquad \frac{\partial Q_{vR}}{\partial e_{vR}} = -CI_{vR} + f_{vR}\frac{\partial CR_{vR}}{\partial e_{vR}} - e_{vR}\frac{\partial CI_{vR}}{\partial e_{vR}}$$

$$\frac{\partial P_{vR}}{\partial f_{vR}} = CI_{vR} + e_{vR}\frac{\partial CR_{vR}}{\partial f_{vR}} + f_{vR}\frac{\partial CI_{vR}}{\partial f_{vR}} \qquad \frac{\partial Q_{vR}}{\partial f_{vR}} = CR_{vR} + f_{vR}\frac{\partial CR_{vR}}{\partial f_{vR}} - e_{vR}\frac{\partial CI_{vR}}{\partial f_{vR}}$$

$$\frac{\partial P_{vR}}{\partial \varphi} = e_{vR}\frac{\partial CR_{vR}}{\partial \varphi} + f_{vR}\frac{\partial CI_{vR}}{\partial \varphi} \qquad \frac{\partial Q_{vR}}{\partial \varphi} = f_{vR}\frac{\partial CR_{vR}}{\partial \varphi} - e_{vR}\frac{\partial CI_{vR}}{\partial \varphi}$$

$$\frac{\partial P_{vR}}{\partial m_a} = e_{vR}\frac{\partial CR_{vR}}{\partial m_a} + f_{vR}\frac{\partial CI_{vR}}{\partial m_a} \qquad \frac{\partial Q_{vR}}{\partial m_a} = f_{vR}\frac{\partial CR_{vR}}{\partial m_a} - e_{vR}\frac{\partial CI_{vR}}{\partial m_a}$$

$$\frac{\partial P_{vR}}{\partial E_{DC}} = e_{vR}\frac{\partial CR_{vR}}{\partial E_{DC}} + f_{vR}\frac{\partial CI_{vR}}{\partial E_{DC}} \qquad \frac{\partial Q_{vR}}{\partial E_{DC}} = f_{vR}\frac{\partial CR_{vR}}{\partial E_{DC}} - e_{vR}\frac{\partial CI_{vR}}{\partial E_{DC}}$$

$$\frac{\partial P_{0-vR}}{\partial e_{vR}} = E_{DC}\frac{\partial CR_0}{\partial e_{vR}} \qquad \frac{\partial Q_{0-vR}}{\partial e_{vR}} = -E_{DC}\frac{\partial CI_0}{\partial e_{vR}}$$

$$\frac{\partial P_{0-vR}}{\partial f_{vR}} = E_{DC}\frac{\partial CR_0}{\partial f_{vR}} \qquad \frac{\partial Q_{0-vR}}{\partial f_{vR}} = -E_{DC}\frac{\partial CI_0}{\partial f_{vR}}$$

$$\frac{\partial P_{0-vR}}{\partial \varphi} = E_{DC}\frac{\partial CR_0}{\partial \varphi} \qquad \frac{\partial Q_{0-vR}}{\partial \varphi} = -E_{DC}\frac{\partial CI_0}{\partial \varphi}$$

$$\frac{\partial P_{0-vR}}{\partial B_{eq}} = E_{DC}\frac{\partial CR_0}{\partial B_{eq}} \qquad \frac{\partial Q_{0-vR}}{\partial B_{eq}} = -E_{DC}\frac{\partial CI_0}{\partial B_{eq}}$$

$$\frac{\partial P_{0-vR}}{\partial m_a} = E_{DC}\frac{\partial}{\partial m_a}CR_0 \qquad \frac{\partial Q_{0-vR}}{\partial m_a} = -E_{DC}\frac{\partial CI_0}{\partial m_a}$$

$$\frac{\partial P_{0-vR}}{\partial E_{DC}} = CR_0 + E_{DC}\frac{\partial CR_0}{\partial E_{DC}} \qquad \frac{\partial Q_{0-vR}}{\partial E_{DC}} = -CI_0 - E_{DC}\frac{\partial CI_0}{\partial E_{DC}}$$

The various current derivatives are given explicitly in Appendix 3.A at the end of this chapter.

Similar to the power mismatch terms (3.18), (3.19) and (3.26), relevant mismatch terms for the VSC model, which has been assumed to be connected between buses vR and 0, are given below:

$$\begin{aligned}
\Delta P_{vR} &= P_{vR,\text{net}} - P_{vR,\text{cal}} = (P_{vR,\text{gen}} - P_{vR,\text{load}}) - P_{vR,\text{cal}} \\
\Delta Q_{vR} &= Q_{vR,\text{net}} - Q_{vR,\text{cal}} = (Q_{vR,\text{gen}} - Q_{vR,\text{load}}) - Q_{vR,\text{cal}} \\
\Delta P_{0} &= P_{0,\text{net}} - P_{0,\text{cal}} = (P_{0,\text{gen}} - P_{0,\text{load}}) - P_{0,\text{cal}} \\
\Delta Q_{0} &= Q_{0,\text{net}} - Q_{0,\text{cal}} = (Q_{0,\text{gen}} - Q_{0,\text{load}}) - Q_{0,\text{cal}} \\
\Delta P_{0-vR} &= -P^{\text{sch}}_{0-vR} + \Delta P_0 \\
\Delta Q_{0-vR} &= 0 + \Delta Q_0
\end{aligned} \tag{3.44}$$

where P^{sch}_{0-vR} is the active power which is scheduled to arrive at node 0 and the negative sign is to signify that it flows from node vR towards node 0. This value is normally set to zero but it could be the value of any suitable DC load. On the other hand, the scheduled value for the reactive power flow arriving at node 0 is always set to zero, which correctly implies that no reactive power exists at the DC bus.

Similar to Eqs. (3.21) and (3.22), the state variable increments calculated at iteration (r) are the difference between the state variable value at that iteration and their values at the previous iteration:

$$\begin{aligned}
\Delta e^{(r)}_{vR} &= e^{(r)}_{vR} - e^{(r-1)}_{vR} \\
\Delta f^{(r)}_{vR} &= f^{(r)}_{vR} - f^{(r-1)}_{vR} \\
\Delta B^{(r)}_{eq} &= B^{(r)}_{eq} - B^{(r-1)}_{eq} \\
\Delta m^{(r)}_{a} &= m^{(r)}_{a} - m^{(r-1)}_{a} \\
\Delta \varphi^{(r)} &= \varphi^{(r)} - \varphi^{(r-1)}
\end{aligned} \tag{3.45}$$

3.6.5 Non-Regulated Power Flow Solutions

If no voltage regulation at node vR applies because either the amplitude modulation index limit has been reached or one explicitly wishes to do so, m_a is removed from the list of state variables and the mismatch voltage ΔV_{vR} becomes deactivated in the linearized expression (3.43). Notice that no new Jacobian terms are required:

$$\begin{bmatrix} \Delta P_{vR} \\ \Delta Q_{vR} \\ \Delta Q_{0-vR} \\ \Delta P_{0-vR} \end{bmatrix} = - \begin{bmatrix} \partial \Delta P_{vR}/\partial e_{vR} & \partial \Delta P_{vR}/\partial f_{vR} & 0 & \partial \Delta P_{vR}/\partial \varphi \\ \partial \Delta Q_{vR}/\partial e_{vR} & \partial \Delta Q_{vR}/\partial f_{vR} & 0 & \partial \Delta Q_{vR}/\partial \varphi \\ \partial \Delta Q_{0-vR}/\partial e_{vR} & \partial \Delta Q_{0-vR}/\partial f_{vR} & \partial \Delta Q_{0-vR}/\partial B_{eq} & \partial \Delta Q_{0-vR}/\partial \varphi \\ \partial \Delta P_{0-vR}/\partial e_{vR} & \partial \Delta P_{0-vR}/\partial f_{vR} & 0 & \partial \Delta P_{0-vR}/\partial \varphi \end{bmatrix} \begin{bmatrix} \Delta e_{vR} \\ \Delta f_{vR} \\ \Delta B_{eq} \\ \Delta \varphi \end{bmatrix} \tag{3.46}$$

There are instances outside the scope of the STATCOM application in which it may be desirable to relax the VSC's DC voltage, such as in some HVDC applications. In such cases, a rearrangement of state variables in the linearized expression (3.43) also takes place:

$$\begin{bmatrix} \Delta P_{vR} \\ \Delta Q_{vR} \\ \Delta U_{vR} \\ \Delta Q_{0-vR} \\ \Delta P_{0-vR} \\ \Delta P_{0} \end{bmatrix} = -\begin{bmatrix} \partial \Delta P_{vR}/\partial e_{vR} & \partial \Delta P_{vR}/\partial m_a & \partial \Delta P_{vR}/\partial f_{vR} & 0 & \partial \Delta P_{vR}/\partial \varphi & \partial \Delta P_{vR}/\partial E_{DC} \\ \partial \Delta Q_{vR}/\partial e_{vR} & \partial \Delta Q_{vR}/\partial m_a & \partial \Delta Q_{vR}/\partial f_{vR} & 0 & \partial \Delta Q_{vR}/\partial \varphi & \partial \Delta Q_{vR}/\partial E_{DC} \\ \partial \Delta U_{vR}/\partial e_{vR} & 0 & \partial \Delta U_{vR}/\partial f_{vR} & 0 & 0 & 0 \\ \partial \Delta Q_{0-vR}/\partial e_{vR} & \partial \Delta Q_{0-vR}/\partial m_a & \partial \Delta Q_{0-vR}/\partial f_{vR} & \partial \Delta Q_{0-vR}/\partial B_{eq} & \partial Q_{\Delta 0-vR}/\partial \varphi & \partial \Delta Q_{0-vR}/\partial E_{DC} \\ \partial \Delta P_{0-vR}/\partial e_{vR} & \partial \Delta P_{0-vR}/\partial m_a & \partial \Delta P_{0-vR}/\partial f_{vR} & 0 & \partial \Delta P_{0-vR}/\partial \varphi & \partial \Delta P_{0-vR}/\partial E_{DC} \\ \partial \Delta P_{0}/\partial e_{vR} & \partial \Delta P_{0}/\partial m_a & \partial \Delta P_{0}/\partial f_{vR} & 0 & \partial \Delta P_{0}/\partial \varphi & \partial \Delta P_{0}/\partial E_{DC} \end{bmatrix} \begin{bmatrix} \Delta e_{vR} \\ \Delta m_a \\ \Delta f_{vR} \\ \Delta B_{eq} \\ \Delta \varphi \\ \Delta E_{DC} \end{bmatrix} \tag{3.47}$$

3.6.6 Practical Implementations

3.6.6.1 Control Strategy

As illustrated in Figure 3.5b, the VSC is assumed to be connected between a sending bus vR and a receiving bus 0, with the former being the VSC's AC bus and the latter being the VSC's DC bus. The voltage E_{DC} is kept constant by the action of a small DC capacitor bank with rated capacitance C_{DC}, which in steady-state draws no current. In the Newton-Raphson power flow solution, the voltage magnitude V_{vR} is regulated within system-dependent maximum and minimum values, afforded by the following basic relationship:

$$V_{vR} = k_1 m_a E_{DC} - \left(\sqrt{R_1^2 + X_1^2} + \left|\overline{Z}_{reac}\right| \right) \left|\overline{I}_1\right| \tag{3.48}$$

Note that in the VSC's linear range of modulation, the index m_a takes values within the bounds:

$$0 < m_a < 1 \tag{3.49}$$

However, in power systems reactive power control applications, it is unlikely that values of m_a lower than 0.5 will be used. The reason is that the voltage magnitude at the VSC's AC bus must be kept within practical limits because too high a voltage may induce insulation coordination failure at the point of connection with the power grid and too low a voltage may induce a condition of voltage collapse. To illustrate this point and using realistic values of $R_1 = 0.0002$ p.u., $X_1 = 0.002$ p.u., $R_{\mathrm{reac}} = 0$ p.u., $X_{\mathrm{reac}} = 0.02$ p.u. and $E_{DC} = 2$ p.u., and considering low-current operation, say 0.1 p.u., V_{vR} will take a value of 0.590 36 p.u. with $m_a = 0.5$.

3.6.6.2 Initial Parameters and Limits

Three VSC parameters require initialization. The amplitude modulation ratio m_a and its phase angle φ are normally set at 1 and 0, respectively. The VSC is assumed to operate within the linear region, whereas the phase angle φ is assumed to have no limits. The third parameter is the equivalent shunt susceptance B_{eq}, which is given an initial value that lies within the range B_{eq+} and B_{eq-}.

3.6.7 VSC Numerical Examples

A power flow computer program is an essential tool to carry out the steady-state solution of even simple electrical power networks. The computer program `VSC_Power_Flows`, written in **protected** Matlab code, is suitable to carry out the power flow solution of small and medium-size power systems; it resides in the following repository:

www.wiley.com/go/acha_vsc_facts

The program is general as far as the topology of the network is concerned; it caters for any number of *PV* and *PQ* buses. Any bus in the network can be designated to be a slack bus – contrary to convention, there can be more than one of these buses.

Models of the following conventional network elements are included: overhead transmission lines, underground cables, LTC transformers, constant-power loads and fixed shunt and series compensation. The model of the VSC is also included in this code. However, no checking of reactive power limits violations is carried out for generators or for VSCs.

For illustration purposes, the code of the main program, `VSC_Power_Flows`, is given below:

Program 3.1 A program written in Matlab to calculate positive sequence power flows using the Newton-Raphson method.

```
%***- - - Main Program
%Read system data
DataPowerFlowsR_1VSC;
%% Set up the nodal admittance matrix
[Ybus] = YBus(nbb,ntl,tlsend,tlrec,tlresis,tlreac,...
tlcond,tlsuscep,nsh,shbus,shresis,shreac);
%% Determine Vreal and Vimag by iteration, together with the VSC
%% and LTC parameters
[Vreal,Vimag,mVSC,phiVSC,BVSC,tftap]= VSC_NewtonRaphsonR↵
(tol,itmax,ngn,nld,nbb,bustype,genbus,loadbus,Pgen,Qgen,Pload,...
Qload,Ybus,Vreal,Vimag,nvsc,VSCsend,VSCrec,RVSC1,XVSC1,GVSC0,...
RVSCR,XVSCR,RVSCF,XVSCF,BVSC0,mVSC,phiVSC,VSCVCtrl,VSCVM,...
VSCPL,CVSC,ntf,tfsend,tfrec,tfresis,tfreac,tftap,TFVMT,TFVCtrl);
%% Calculate the power flows and power losses in transmission
%% lines
[PQtlsend,PQtlrec,PQtlloss,PQbus,PQgen,PQshunt,PQtotalloss]=...
PQflowsR(nbb,ngn,ntl,ntf,nsh,nld,genbus,tlsend,tlrec,...
tlresis,tlreac,tlcond,tlsuscep,tfsend,tfrec,Tfresis,Tfreac,...
tftap,shbus,shresis,shreac,loadbus,Pload,Qload,Vreal,Vimag);
%% Calculate the power flows and power losses in LTC
%% transformers
if ntf > 0 ; [PQbus,PQTFsend,PQTFrec,PQTFLoss]... = TF_PQflowsR...
(Vreal,Vimag,PQbus,ntf,tfsend,tfrec,tfresis,tfreac,tftap); end
%% Calculate the power flows in SVC
if nvsc > 0 ; [PQbus,PQVSCsend,PQVSCrec,PQVSC,PQVSCLoss]=...
VSC_PQflowsR(Vreal,Vimag,PQbus,nvsc,VSCsend,VSCrec,RVSC1,XVSC1,...
GVSC0,BVSC,RVSCR,XVSCR,RVSCF,XVSCF,mVSC,phiVSC,CVSC); end
%% End of Main Program
```

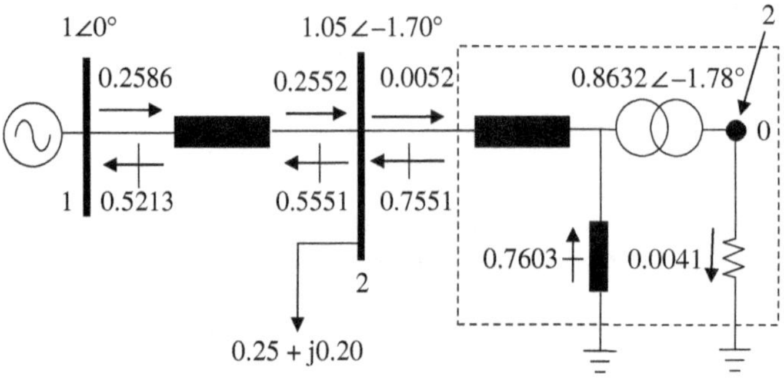

Figure 3.7 VSC providing voltage support at Node 2.

To illustrate the application of the theory so far covered and the usefulness of the computer program provided, the VSC model is applied in a rather contrived test case where the voltage source converter (VSC) is connected at the receiving end of a loaded transmission line. In this particular example, and in order to simplify matters, a connecting transformer is neglected and so are the impedances of the smoothing reactor and the shunt filter. Three cases are considered: (i) the VSC is used to provide reactive power; (ii) the VSC is used to draw reactive power; and (iii) the VSC is used to supply a DC power load.

Test Case 1. The three-node system shown in Figure 3.7 comprises one generator, one transmission line and one AC-DC converter (VSC), which is represented by the elements shown within the broken-line rectangle. The generator node is taken to be the slack bus where the voltage magnitude is kept at 1 p.u. and its phase angle provides a reference for all other phase angles in the network, excepting bus 0 since this is a DC bus and the voltage is always a real quantity.

The VSC consumes 0.0052 p.u. of active power from the system to account for its internal losses while supplying 0.7551 p.u. of reactive power to the system. The equivalent susceptance (in capacitive mode) produces 0.7603 p.u. of reactive power and its capacitive susceptance stands at $B_{eq} = 0.6803$ p.u. The VSC switching losses are 0.41%, corresponding to an actual conductance $G_0 \approx 0.1\%$.

The DC bus voltage is controlled at 2 p.u. and the voltage magnitude at bus 2 is kept at 1.05 with $m_a = 0.8632$. The phase shifter angle takes a value of $-1.78°$. The line current drawn by the VSC is $0.7192\angle{+87.91°}$.

The following parameters are used by the computer program in the function `DataPowerFlowsR_1VSC` to solve this system:

```
%% Bus data - notice that power generation and load are treated
%% as attributes of the node
% bustype = type of bus
% VM = nodal voltage magnitude in per-unit
% VA = nodal voltage phase angle in degrees
% Pgen and Qgen = nodal active and reactive power generation in
% per-unit
% Pload and Qload = nodal active and reactive power load in
% per-unit
bustype(1)=1; VM(1)=1; VA(1)=0; Pgen(1)=0; Qgen(1)=0;
Pload(1)=0; Qload(1)=0;
bustype(2)=3; VM(2)=1.05; VA(2)=0; Pgen(2)=0; Qgen(2)=0;
```

```
Pload(2)=0.25; Qload(2)=0.20;
bustype(3)=2; VM(3)=2; VA(3)=0; Pgen(3)=0; Qgen(3)=0;
Pload(3)=0; Qload(3)=0;
%% Transmission line data
% tlsend = sending end of transmission line
% tlrec = receiving end of transmission line
% tlresis = series resistance of transmission line in per-unit
% tlreac = series reactance of transmission line in per-unit
% tlcond = shunt conductance of transmission line in per-unit
% tlsuscep = shunt susceptance of transmission line in per-unit
tlsend(1)=1; tlrec(1)=2; tlresis(1)=0.01; tlreac(1)=0.10;
tlcond(1)=0; tlsuscep(1)=0;
%LTC transformer data
% tfsend = sending end of transformer - if tfsend(1)=0 then the
% number of transformers in the network are zero
% tfrec = receiving end of transformer
% tfresis = series resistance of transformer in per-unit
% tfreac = series reactance of transformer in per-unit
% tftap = transformer's tap
% TFVM = target voltage magnitude in per-unit
% TFVCtrl = voltage control with the transformer's tap. 1 and 2
% is for voltage control at the sending and receiving nodes,
% respectively, 0 is for no control.
tfsend(1)=0; tfrec(1)=0; Tfresis(1)=0; Tfreac(1)=0;
tftap(1)=0; TFVMT(1)=1.05; TFVCtrl(1)=1;
%% Shunt data
% shbus = shunt element bus number - if shbus(1)=0 then the
% number of shunt elements in the network are zero
% shresis = resistance of shunt element in per-unit
% shreac = reactance of shunt element in per-unit
% +ve for inductive reactance and -ve for capacitive reactance
shbus(1)=0; shresis(1)=0; shreac(1)=0;
%% VSC data
% Topology connection to the VSC's AC and DC buses- if
% VSCsend(1)=0 then the number of VSC in the network are zero
VSCsend(1)=2; VSCrec(1)=3;
% RVSC1 = series resistance in-per unit
% XVSC1 = series reactance in per-unit
% GVSC0 = shunt conductance in per-unit corresponding to
% switching losses at nominal current
% BVSC0 = shunt susceptance in per-unit corresponding to the
% VSC's nominal MVA
% RVSCR = resistance of the smoothing reactor in-per unit
% XVSCR = reactance of the smoothing reactor in-per unit
% RVSCF = resistance of the shunt filter in-per unit
% XVSCF = reactance of the shunt filter in-per unit
% VSCMVA = nominal power in per-unit
RVSC1(1)=0.002; XVSC1(1)=0.010; GVSC0(1)=0.002; BVSC0(1)=0.50;
RVSCR(1)=0; XVSCR(1)=0; RVSCF(1)=0; XVSCF(1)=0;
VSCMVA(1)=1;
% VSC Initial values
% mVSC = initial value of amplitude modulation ratio (0<m<1)
% phiVSC = initial value of phase shift (in degrees)
mVSC(1)=1; phiVSC(1)=0;
% VSC Control parameters
% VSCVCtrl = voltage control with the amplitude modulation
```

```
% index ma, with a theoretical range between 0 and 1. VSCVCtrl=1
% is for AC bus voltage control and 0 is for no control. No
% control of the DC bus is allowed.
% VSC_VM = target voltage in per-unit
% VSC_PL = a DC power load in per-unit
VSCV_Ctrl(1)=1; VSC_VM(1)=1.05; VSC_PL(1)=0;
% General parameters
itmax=10; %maximum number of iterations permitted
tol=1e-12; %tolerance criterion
%End of data
```

Test Case 2. The operating conditions of the power circuit in Test Case 1 are modified to force the VSC to draw reactive power from the slack generator connected at bus 1. As shown in Figure 3.8, the nodal voltage magnitude at node 2 (0.95 p.u.) is kept at a lower voltage magnitude than the voltage at node 1, which remains at 1 p.u. The relevant voltage changes are carried out in the node data section of the function `DataPowerFlowsR_1VSC`, previous to its computer execution. In explicit form:

```
bustype(2)=3; VM(2)=0.95; VA(2)=0; Pgen(2)=0; Qgen(2)=0;
Pload(2)=0; Qload(2)=0;
```

The VSC draws 0.0007 p.u. of active power and 0.2477 p.u. of reactive power. The equivalent susceptance absorbs 0.2470 p.u. of reactive power and its inductive susceptance stands at $B_{eq} = -0.2752$ p.u. The VSC switching losses are low, 0.05%, and the current drawn by the VSC is quite low, i.e. $0.2607\angle{-91.09}°$. The dc bus voltage is controlled at 2 p.u. and the voltage magnitude at bus 2 is kept at 0.95 with $m_a = 0.7735$. The phase shifter angle takes a value of −1.21°.

Test Case 3. Test Case 1 is expanded to incorporate a load in the DC side of the VSC in the form of a simple battery-type load, as shown in Figure 3.9.

This test network uses the same circuit parameters as in Test Case 1, but a second load is added in the form of a DC load which is being supplied through the VSC at 0.5 p.u. of power. The VSC is used to keep the voltage magnitude at 1.05 p.u. at bus 2. The required changes to carry out this VSC operation, in the function `DataPowerFlowsR_1VSC`, take place in the VSC data section. More explicitly:

```
VSC_PL(1) = 0.5 ;
```

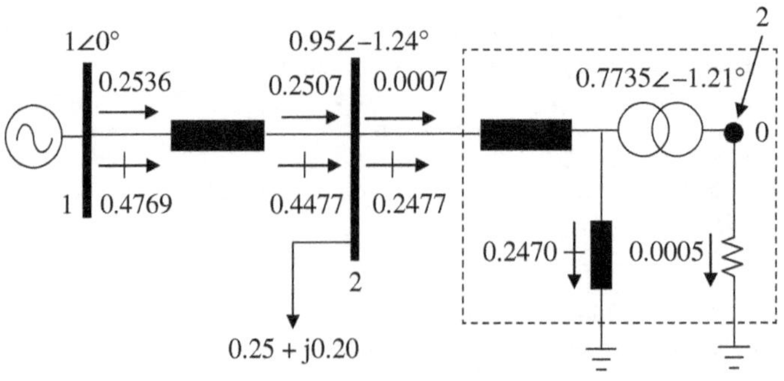

Figure 3.8 Test network uses the same circuit parameters as in Test Case 1 but the voltage magnitude at bus 2 is kept at 0.95 p.u. using m_a to force the reactive power flow into the VSC.

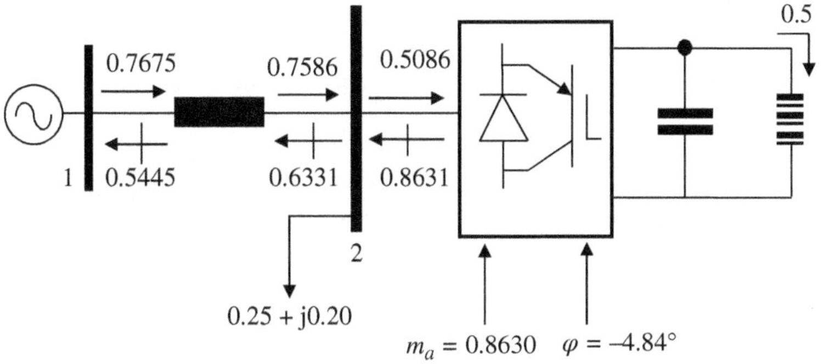

Figure 3.9 Test network with a battery load on its DC bus.

The total VSC active power loss stands at 0.86% p.u. where 0.69% corresponds to switching loss and 0.17% corresponds to ohmic loss. The VSC contributes 0.8631 p.u. to supply the reactive power load of 0.20 p.u. and the rest is exported to the slack generator. The VSC equivalent susceptance with a capacitive value of 0.7534 p.u. produces 0.8717 p.u. of reactive power. The SVC is set to regulate voltage magnitude at its AC bus at 1.05 p.u. and its actual complex modulation ratio is $0.8630\angle{-4.84^\circ}$. The current drawn by the VSC is $0.9296\angle{+54.10^\circ}$. The solution converges in five iterations to a tolerance of 10^{-12}.

3.7 The STATCOM Model

For studies at the fundamental frequency, the STATCOM may be seen to comprise a VSC and an interfacing transformer, which for the sake of modelling and programming flexibility may be taken to be an LTC, as shown in Figure 3.6. In practice, however, in STATCOM installations the interfacing transformers will not be of the LTC type but rather conventional transformers, owing to the slow action of their electromechanical tap compared with the very fast-acting nature of the VSC's amplitude modulation index.

The nodal admittance matrix of the LTC transformer is:

$$\begin{pmatrix} \overline{I}_k \\ \overline{I}_{vR} \end{pmatrix} = \begin{pmatrix} \overline{Y}_l & -T\overline{Y}_l \\ -T\overline{Y}_l & T^2\overline{Y}_l \end{pmatrix} \begin{pmatrix} \overline{V}_k \\ \overline{V}_{vR} \end{pmatrix} \tag{3.50}$$

where $\overline{Y}_l = \overline{Y}_{SC}$ and $vR = m$, with reference to the notation used in Eq. (3.3).

Inclusion of the STATCOM model in a power flow solution is straightforward. It requires only explicit representation of the nodal power flow equations of the VSC connected between, say, nodes 0 and vR, and the nodal power equations of the LTC transformer connected between, say, nodes k and vR. The nodal power equations and Jacobian entries for the LTC model in rectangular coordinates are given in Appendix 3.B.

Combining the individual nodal admittances of the LTC and VSC models yields the compound model representing the STATCOM model,

$$\begin{bmatrix} \overline{I}_k \\ \overline{I}_{vR} \\ \overline{I}_0 \end{bmatrix} = \begin{bmatrix} \overline{Y}_l & -T\overline{Y}_l & 0 \\ -T\overline{Y}_l & T^2\overline{Y}_l + \overline{Y}_{vRvR} & -(\cos\varphi + \mathrm{j}\sin\varphi)\overline{Y}_{vR0} \\ 0 & -(\cos\varphi - \mathrm{j}\sin\varphi)\overline{Y}_{vR0} & (G_{sw} + \mathrm{j}k_1^2 m_a^2 B_{eq}) + \overline{Y}_{00} \end{bmatrix} \begin{bmatrix} \overline{V}_k \\ \overline{V}_{vR} \\ E_{DC} \end{bmatrix} \tag{3.51}$$

Alternatively, a more compact set of power flow equations may be achieved by realizing that the interface point between the VSC and LTC circuits, namely node vR, receives a zero external (nodal) current injection. Then a mathematical elimination of node vR becomes an option.

$$\begin{bmatrix} \overline{I}_k \\ \overline{I}_0 = 0 \end{bmatrix} = \frac{1}{\Delta}\begin{bmatrix} \overline{Y}_l\overline{Y}_{vRvR} & -T(\cos\varphi + \mathrm{j}\sin\varphi)\overline{Y}_l\overline{Y}_{vR0} \\ -T(\cos\varphi - \mathrm{j}\sin\varphi)\overline{Y}_l\overline{Y}_{vR0} & \begin{matrix}[(G_{sw} + \mathrm{j}k_1^2 m_a^2 B_{eq}) + \overline{Y}_{00}] \\ \times(T^2 Y_l + Y_{vRvR}) - Y_{vR0}^2\end{matrix} \end{bmatrix} \begin{bmatrix} \overline{V}_k \\ E_{DC} \end{bmatrix} \tag{3.52}$$

where $\Delta = T^2\overline{Y}_l + \overline{Y}_1$.

Emulating the algebraic developments in Section 3.6.2, expressions for nodal active and reactive powers at the STATCOM terminals may be derived quite straightforwardly taking either Eq. (3.51) or (3.52) as the basis. However, it should be said that the reduced model would be attractive only if we were prepared to lose a degree of modelling flexibility, since this bus would not be explicitly available for the regulating action of either T or m_a. Instead, the combined regulating action would be taking place in the high-voltage side of the LTC transformer. Hence, this option is not pursued further in this chapter owing to the decrease in modelling flexibility.

Building on the result of Eq. (3.51), we have:

$$\begin{bmatrix} \overline{S}_k \\ \overline{S}_{vR} \\ \overline{S}_0 \end{bmatrix} = \begin{bmatrix} \overline{V}_k & 0 & 0 \\ 0 & \overline{V}_{vR} & 0 \\ 0 & 0 & E_{DC} \end{bmatrix} \begin{bmatrix} \overline{I}_k^* \\ \overline{I}_{vR}^* \\ \overline{I}_0^* \end{bmatrix}$$

$$= \begin{bmatrix} \overline{Y}_l^* V_k^2 - T\overline{Y}_l^*\overline{V}_k\overline{V}_{vR}^* \\ -T\overline{Y}_l^*\overline{V}_{vR}\overline{V}_k^* + (T^2\overline{Y}_l^* + \overline{Y}_{vRvR}^*)V_{vR}^2 - (\cos\varphi - \mathrm{j}\sin\varphi)\overline{Y}_{vR0}^*\overline{V}_{vR}E_{DC} \\ -(\cos\varphi + \mathrm{j}\sin\varphi)\overline{Y}_{vR0}^*E_{DC}\overline{V}_{vR}^* + [(G_{sw} + \mathrm{j}k_1^2 m_a^2 B_{eq}) + \overline{Y}_{00}]^* E_{DC}^2 \end{bmatrix} \tag{3.53}$$

Following some arduous algebra, the nodal active and reactive power expressions are arrived at:

$$\begin{aligned} P_k &= G_l(e_k^2 + f_k^2) - T(G_l e_{vR} - B_l f_{vR})e_k - T(G_l f_{vR} + B_l e_{vR})f_k \\ Q_k &= -B_l(e_k^2 + f_k^2) - T(G_l e_{vR} - B_l f_{vR})f_k - T(G_l f_{vR} + B_l e_{vR})e_k \end{aligned} \tag{3.54}$$

$$\begin{aligned} P_{vR} &= (T^2 G_l + G_{vRvR})(e_{vR}^2 + f_{vR}^2) - T(G_l e_k - B_l f_k)e_{vR} - T(G_l f_k + B_l e_k)f_{vR} \\ &\quad - (G_{vR0}\cos\varphi - B_{vR0}\sin\varphi)e_{vR}E_{DC} - (B_{vR0}\cos\varphi + G_{vR0}\sin\varphi)f_{vR}E_{DC} \\ Q_{vR} &= -(T^2 B_l + B_{vRvR})(e_{vR}^2 + f_{vR}^2) - T(G_l e_k - B_l f_k)f_{vR} + T(G_l f_k + B_l e_k)e_{vR} \\ &\quad - (G_{vR0}\cos\varphi - B_{vR0}\sin\varphi)f_{vR}E_{DC} + (B_{vR0}\cos\varphi + G_{vR0}\sin\varphi)e_{vR}E_{DC} \end{aligned} \tag{3.55}$$

$$\begin{aligned} P_0 &= (G_{sw} + G_{00})E_{DC}^2 - (G_{vR0}\cos\varphi + B_{vR0}\sin\varphi)e_{vR}E_{DC} \\ &\quad - (B_{vR0}\cos\varphi + G_{vR0}\sin\varphi)f_{vR}E_{DC} \\ Q_0 &= -(k_2^2 m_a^2 B_{eq} + B_{00})E_{DC}^2 - (G_{vR0}\cos\varphi - B_{vR0}\sin\varphi)f_{vR}E_{DC} \\ &\quad + (B_{vR0}\cos\varphi + G_{vR0}\sin\varphi)e_{vR}E_{DC} \end{aligned} \tag{3.56}$$

These equations are linearized around a base operating point, in a similar manner to the linearization process carried out in Section 3.6.4.

3.7.1 STATCOM Numerical Examples

Two test cases are presented to exemplify the performance of the STATCOM model. The first case relates to a contrived system which is, essentially, the same system as that used in Test Case 1, except that the STATCOM model replaces the VSC model. The second test case is a modified version of the IEEE 30-node system where two STATCOMs regulate voltage magnitude at two different points in the network.

Test Case 4. The power circuit in Test Case 1 is modified with the LTC's STATCOM now incorporated in Figure 3.10.

The test network uses the same circuit parameters as in Test Case 1 but with the parameters of the LTC transformer added to the circuit parameters: $R_T = 0.01$ p.u. and $X_T = 0.10$ p.u. The generator keeps the voltage magnitude at the slack node at 1 p.u. The STATCOM consumes 0.0081 p.u. of active power from the system to account for its internal losses while supplying 0.7555 p.u. of reactive power to the system. The VSC switching losses stand at 0.44% and the remaining 0.11% corresponds to ohmic losses in the LTC transformer and the VSC. The DC bus voltage is kept at 2 p.u. by action of the DC capacitor and this bus is treated in the power flow solution as a *PV* bus. The voltage magnitude at bus 2 is kept at 1.05 p.u. by action of the transformer tap, which stands at $T = 1.0342$. Likewise, the voltage magnitude at bus 3 is kept at 1.05 p.u. by action of the VSC's m_a, which stands at 0.8634. The current drawn by the STATCOM is $0.7442\angle{+87.67°}$.

Test Case 5. Two STATCOMs are assumed to be connected at two different locations of a larger network. The IEEE 30-node system is selected for this test case, with the relevant portions of the modified system shown in Figure 3.11.

The fixed banks of capacitors at nodes 10 and 24 in the original network are replaced with STATCOMs which are set to regulate voltage magnitudes at their points of connection with the power grid. Notice that in each case the STATCOMs comprise the VSC and an LTC interfacing transformer. Their respective DC voltages are kept at 2 p.u.

The voltage magnitudes at the compensated buses, namely 10 and 24, are compared in Table 3.1 to the case when conventional capacitor banks are connected to these nodes, and when no compensation is used.

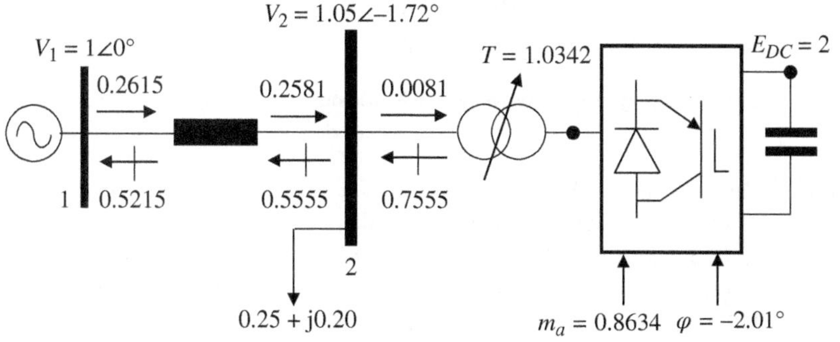

Figure 3.10 Upgraded network used in Test Case 1 to include the LTC transformer.

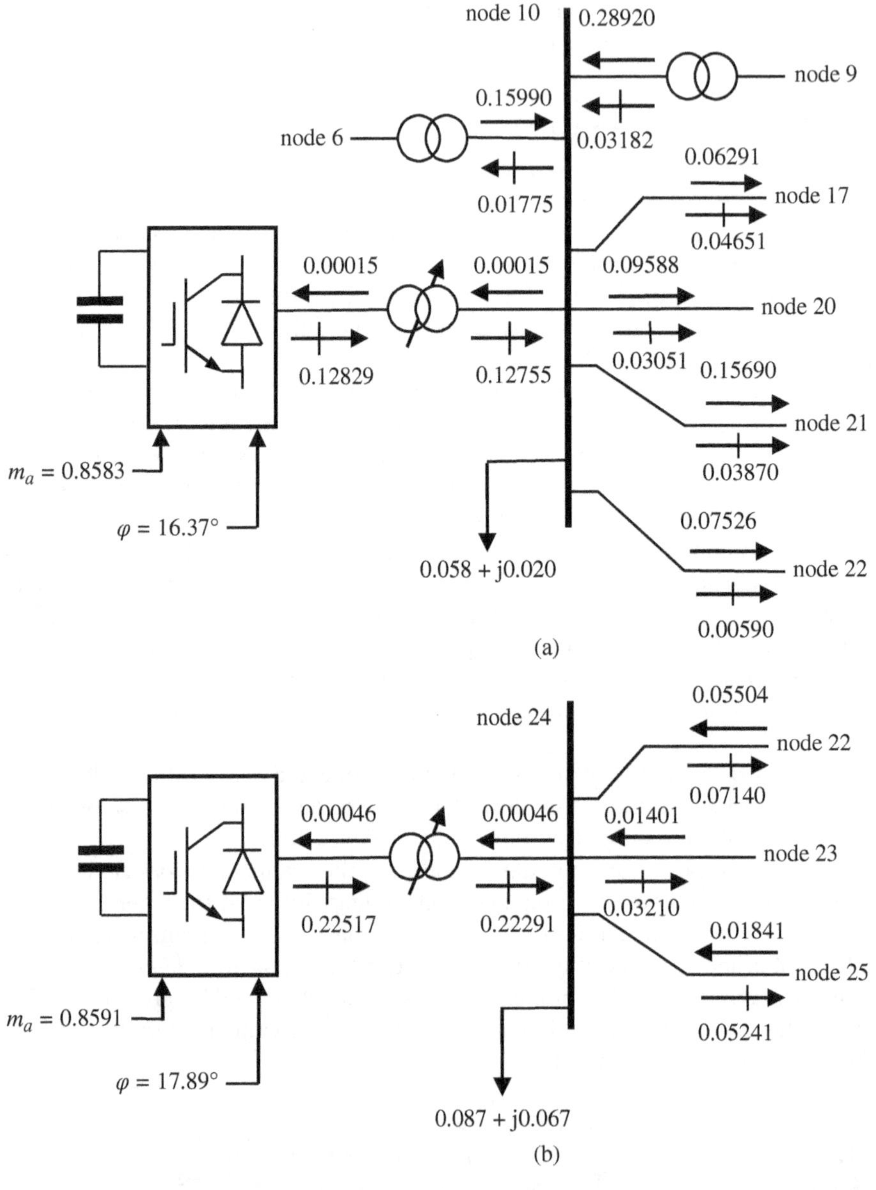

Figure 3.11 Two STATCOM supplying reactive power at buses 10 and 24 of the modified IEEE 30-bus system.

The two STATCOMs use identical parameters. The LTC transformers are assumed to contain no resistance and their reactances are $X_{TR} = 0.05$ p.u. The VSCs series and shunt parameters, in per-unit, are $R_1 = 0.002$, $X_1 = 0.010$ and $G_{sw} = 0.002$, respectively.

The susceptance values used for the case with fixed compensation at buses 10 and 24 are 0.19 and 0.043 p.u., which are the values given in the standard IEEE 30-node system. For the STATCOM case, the voltages at buses 10 and 24 are kept at 1.05 p.u. using the transformers' taps, which stand at 1.0058 and 1.0101, respectively. Also, in

Table 3.1 Voltage magnitudes at the compensated buses in the 30-bus system for two compensation options.

Compensation case	Voltage magnitude (p.u.)	
	Bus 10	Bus 24
None	1.0195	0.9958
Fixed	1.0452	1.0212
STATCOMs	1.05	1.05

Table 3.2 Power loss at the compensated buses in the 30-bus system for two compensation options.

Compensation case	Active power loss (%)	
	Network	STATCOMs
None	5.95	—
Fixed	5.86	—
STATCOMs	5.88	0.06

each case the voltage at the connection node between the VSC and the LTC is kept at 1.05 p.u. using the VSC's amplitude modulation index. As expected, one benefit of shunt compensation is to reduce the system power losses due to an improved voltage profile and this trend is shown in the power loss figures presented in Table 3.2. The STATCOM-type compensation introduces an additional kind of power loss which is associated with the high-frequency switching of the PWM control used by the VSC technology and ohmic losses. The STATCOM losses are quite low in this case because the currents drawn by the two STATCOMs are low compared with the 1 p.u. rated currents, i.e. $0.122\,18\angle{+73.58^\circ}$ and $0.214\,44\angle{+72.01^\circ}$.

3.8 VSC-HVDC Systems Modelling

Power transmission using VSC-HVDC is a relatively recent progression of the HVDC technology, which was originally based on the use of mercury arc valves and replaced in the mid-1970s by solid-state valves of the thyristor type [11]. As stated in Sections 1.2. and 2.4, the two most basic VSC-HVDC configurations are the monopolar, back-to-back and point-to-point schemes, which are shown schematically in Figure 3.12.

Each converter station comprises a VSC and an interfacing transformer. The transformer's primary and secondary windings are connected to the high-voltage power grid and to the AC side of the VSC, respectively – this makes each VSC to be shunt-connected with the AC system, just as if they were two STATCOMs. However, the two VSCs are series connected on their DC sides, sharing a capacitor in the case of the back-to-back configuration and through a DC cable in the case of the point-to-point configuration.

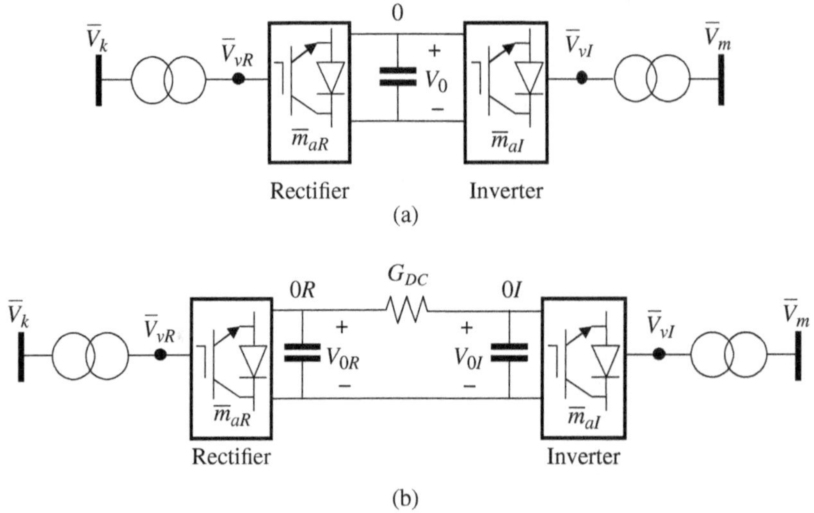

Figure 3.12 Monopolar VSC-HVDC schematic representation: (a) back-to-back circuit; (b) point-to-point circuit. Source: ©IEEE, 2013.

It is likely that in practice the interfacing transformers in most VSC-HVDC installations will not be of the LTC type but rather conventional transformers, owing to the slow action of their electromechanical tap compared with the very fast-acting nature of the VSC's amplitude modulation index. Nevertheless, aiming at maximizing the modelling flexibility at this point, the VSC-HVDC models will be taken to have LTC interfacing transformers.

The fundamental frequency operation of the VSC-HVDC schemes shown in Figure 3.12 may be modelled by taking each VSC to be a variable voltage source behind a coupling impedance, linked by a mismatch active power-constraining equation and solved in a unified manner using Newton-type methods [12–15]. These contributions address key VSC-HVDC modelling issues such as back-to-back and point-to-point schemes [12], multi-terminal schemes [13], state estimation [14] and extensions to optimal power flows [15]. In these contributions the emphasis is on the AC side of the VSC-HVDC links and no DC representation is explicitly available. Hence, incorporation of the switching losses in the DC bus or a DC load is not straightforward to represent in this model owing to its equivalent voltage source nature. An alternative approach to solve the multi-terminal VSC-HVDC power flow problem is put forward in [16], where a sequential numerical approach is used. The VSC-HVDC converters are represented as variable voltage sources to solve the AC part of the network whose calculated values are then injected into a DC conductance matrix representing the multi-node DC network. This is a full VSC-HVDC power flow solution, but the strong convergence characteristics of the Newton-Raphson method are sacrificed owing to the sequential iterative solution adopted.

In contrast, the VSC power flow model presented above may be expanded quite naturally to model VSC-HVDC power flows. The ensuing models will take into account, in aggregated form, the phase-shifting and scaling nature of the PWM control. They also take into account the VSC inductive and capacitive reactive power design limits, switching losses and ohmic losses. Furthermore, the numerical power flow solution is a simultaneous one where the AC and DC circuits are solved together using the Newton-Raphson method, keeping its strong convergence characteristics.

3.8.1 VSC-HVDC Nodal Power Equations

The linearized equation corresponding to the power flow solution of the VSC-HVDC using the Newton-Raphson method is presented in this section. In principle, the complex power model of the VSC, Eqs. (3.41) and (3.42), corresponds to the model of both the rectifier and the inverter. However, extensions may be required to incorporate one or more DC cable resistances, depending on the application.

For instance, the fundamental frequency operation of a combined VSC and cable impedance (resistance) may be represented by combining the cable susceptance model in (3.57) and the VSC model given by Eq. (3.39). The DC cable is assumed to be connected between nodes $0R$ and $0I$ and the VSC is assumed to be connected between nodes $0I$ and vI, as shown in Figure 3.13.

The nodal admittance matrix (3.39) representing the VSC is used here with modified subscripts to indicate that the VSC is playing the role of inverter. Note that $V_{0I} = E_{DCI}$ and it follows that $V_{0R} = E_{DCR}$. The nodal transfer admittance matrix equation of the DC cable is:

$$\begin{pmatrix} \overline{I}_{0R} \\ \overline{I}_{0I} \end{pmatrix} = \begin{pmatrix} G_{DC} & -G_{DC} \\ -G_{DC} & G_{DC} \end{pmatrix} \begin{pmatrix} E_{DCR} \\ E_{DCI} \end{pmatrix} \tag{3.57}$$

A straightforward addition of both individual models yields the compound model representing the VSC-DC cable model.

$$\begin{pmatrix} \overline{I}_{0R} \\ \overline{I}_{0I} \\ \overline{I}_{vI} \end{pmatrix} = \begin{pmatrix} G_{DC} & -G_{DC} & 0 \\ -G_{DC} & (G_{swI} + jk_1^2 m_{aI}^2 B_{eqI}) + \overline{Y}_{00} + G_{DC} & -\overline{Y}_{vI0}\angle -\varphi_I \\ 0 & -\overline{Y}_{vI0}\angle\varphi_I & \overline{Y}_{vIvI} \end{pmatrix} \begin{pmatrix} E_{DCR} \\ E_{DCI} \\ \overline{V}_{vI} \end{pmatrix} \tag{3.58}$$

Let us not forget that $m'_a = k_1 m_a$, as defined in Section 3.6.1.

By extension of the partial result in (3.58), the addition of the nodal transfer admittance matrix of the VSC playing the role of rectifier, connected between nodes vR and $0R$, yields the following expression for the point-to-point VSC-HVDC link:

$$\begin{pmatrix} \overline{I}_{vR} \\ \overline{I}_{0R} \\ \overline{I}_{0I} \\ \overline{I}_{vI} \end{pmatrix} = \begin{pmatrix} \overline{Y}_{vRvR} & -\overline{Y}_{vR0}\angle\varphi_R & 0 & 0 \\ -\overline{Y}_{vR0}\angle -\varphi_R & \begin{matrix}(G_{swR} + jk_1^2 m_{aR}^2 B_{eqR}) \\ +\overline{Y}_{00R} + G_{DC}\end{matrix} & -G_{DC} & 0 \\ 0 & -G_{DC} & \begin{matrix}(G_{swI} + jk_1^2 m_{aI}^2 B_{eqI}) \\ +\overline{Y}_{00I} + G_{DC}\end{matrix} & -\overline{Y}_{vI0}\angle -\varphi_I \\ 0 & 0 & -\overline{Y}_{vI0}\angle\varphi_I & \overline{Y}_{vIvI} \end{pmatrix} \begin{pmatrix} \overline{V}_{vR} \\ E_{DCR} \\ E_{DCI} \\ \overline{V}_{vI} \end{pmatrix} \tag{3.59}$$

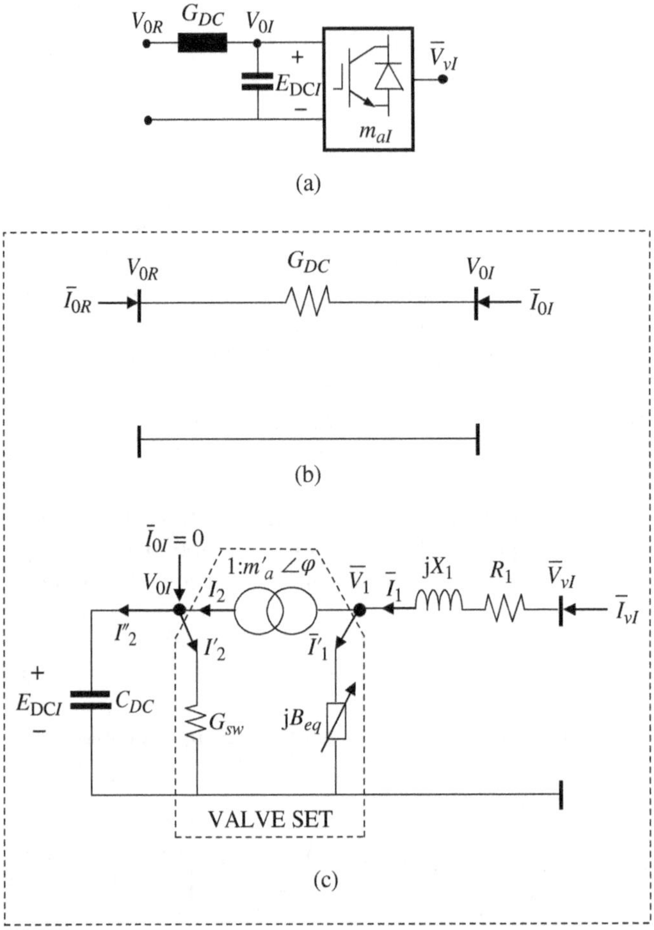

Figure 3.13 VSC-DC cable schematic representation. Source: ©IEEE, 2013.

The nodal complex power equations are derived by multiplying the nodal voltages by the conjugate of the nodal currents:

$$\begin{pmatrix} \overline{S}_{vR} \\ \overline{S}_{0R} \\ \overline{S}_{0I} \\ \overline{S}_{vI} \end{pmatrix} = \begin{pmatrix} \overline{V}_{vR} & 0 & 0 & 0 \\ 0 & E_{DCR} & 0 & 0 \\ 0 & 0 & E_{DCI} & 0 \\ 0 & 0 & 0 & \overline{V}_{vI} \end{pmatrix} \begin{pmatrix} \overline{I}^*_{vR} \\ \overline{I}^*_{0R} \\ \overline{I}^*_{0I} \\ \overline{I}^*_{vI} \end{pmatrix}$$

$$= \begin{pmatrix} \overline{Y}^*_{vRvR}V_{vR}^2 - \overline{Y}^*_{vR0}\angle -\varphi_R \overline{V}_{vR}E_{DCR} \\ -\overline{Y}^*_{vR0}\angle\varphi_R \overline{V}^*_{vR}E_{DCR} + (G_{swR} - jk_1^2 m_{aR}^2 B_{eqR} + \overline{Y}^*_{00R} + G_{DC}) \\ \times E_{DCR}^2 - G_{DC}E_{DCR}E_{DCI} \\ -G_{DC}E_{DCR}E_{DCI} + (G_{swI} - jk_1^2 m_{aI}^2 B_{eqI} + \overline{Y}^*_{00I} + G_{DC}) \\ \times E_{DCI}^2 - \overline{Y}^*_{vI0}\angle\varphi_I \overline{V}^*_{vI}E_{DCI} \\ -\overline{Y}^*_{vI0}\angle -\varphi_I \overline{V}_{vI}E_{DCI} + \overline{Y}^*_{vIvI}V_{vI}^2 \end{pmatrix} \tag{3.60}$$

Following some complex number algebra and separating into real and imaginary quantities, the following equations are arrived at:

$$
\begin{aligned}
P_{vR} &= G_{vRvR}(e_{vR}^2 + f_{vR}^2) - E_{DCR}(e_{vR}[G_{vR0}\cos\varphi_R + B_{vR0}\sin\varphi_R] \\
&\quad - f_{vR}[G_{vR0}\sin\varphi_R - B_{vR0}\cos\varphi_R]) \\
Q_{vR} &= -B_{vRvR}(e_{vR}^2 + f_{vR}^2) - E_{DCR}(e_{vR}[G_{vR0}\sin\varphi_R - B_{vR0}\cos\varphi_R] \\
&\quad + f_{vR}[G_{vR0}\cos\varphi_R + B_{vR0}\sin\varphi_R]) \\
P_{0R} &= (G_{swR} + G_{00R} + G_{DC})E_{DCR}^2 - E_{DCR}(e_{vR}[G_{vR0}\cos\varphi_R + B_{vR0}\sin\varphi_R] \\
&\quad + f_{vR}[G_{vR0}\sin\varphi_R - B_{vR0}\cos\varphi_R]) - G_{DC}E_{DCR}E_{DCI} \\
Q_{0R} &= -(k_1^2 m_{aR}^2 B_{eqR} + B_{00R})E_{DCR}^2 - E_{DCR}(e_{vR}[G_{vR0}\sin\varphi_R - B_{vR0}\cos\varphi_R] \\
&\quad - f_{vR}[G_{vR0}\cos\varphi_R + B_{vR0}\sin\varphi_R]) \\
P_{0I} &= (G_{swI} + G_{00I} + G_{DC})E_{DCI}^2 - E_{DCI}(e_{vI}[G_{vI0}\cos\varphi_I + B_{vI0}\sin\varphi_I] \\
&\quad + f_{vI}[G_{vI0}\sin\varphi_I - B_{vI0}\cos\varphi_I]) - G_{DC}E_{DCI}E_{DCR} \\
Q_{0I} &= -(k_1^2 m_{aI}^2 B_{eqI} + B_{00I})E_{DCI}^2 - E_{DCI}(e_{vI}[G_{vI0}\sin\varphi_I - B_{vI0}\cos\varphi_I] \\
&\quad - f_{vI}[G_{vI0}\cos\varphi_I + B_{vI0}\sin\varphi_I]) \\
P_{vI} &= G_{vIvI}(e_{vI}^2 + f_{vI}^2) - E_{DCI}(e_{vI}[G_{vI0}\cos\varphi_I - B_{vI0}\sin\varphi_I] \\
&\quad + f_{vI}[G_{vI0}\sin\varphi_I + B_{vI0}\cos\varphi_I]) \\
Q_{vI} &= -B_{vIvI}(e_{vI}^2 + f_{vI}^2) - E_{DCI}(-e_{vI}[G_{vI0}\sin\varphi_I + B_{vI0}\cos\varphi_I] \\
&\quad + f_{vI}[G_{vI0}\cos\varphi_I - B_{vI0}\sin\varphi_I])
\end{aligned}
\tag{3.61}
$$

Note that G_{vR0}, B_{vR0}, G_{00R} and B_{00R} are functions of m_{aR} and that G_{vI0}, B_{vI0}, G_{00I} and B_{00I} are functions of m_{aI}.

As with the VSC's formulation, the following constraining equations may be useful in the VSC-HVDC applications:

$$
\begin{aligned}
\Delta P_{0R-vR} &= -P_{0R-vR}^{sch} + P_{0R} \\
\Delta Q_{0R-vR} &= 0 + Q_{0R} \\
\Delta P_{0I-vI} &= P_{0I-vI}^{sch} + P_{0I} \\
\Delta Q_{0I-vI} &= 0 + Q_{0I}
\end{aligned}
\tag{3.62}
$$

3.8.2 VSC-HVDC Linearized Equations

The nodal power equations (3.61) and (3.62) are nonlinear and their solution, for a predefined set of generation and load patterns, may be carried out using the Newton-Raphson method. It should be noted that the VSC making up the HVDC links in Figure 3.12 are connected in series with their interfacing LTC transformers. It follows that a rather large number of parameter regulation options are available for the VSC-HVDC link by making use of the voltage- and power-regulating capabilities of the two VSCs and the voltage-regulating capabilities of the two LTCs. A common practice is to use the rectifier to regulate power on its DC side and to use the inverter to regulate the voltage on its DC side. If available, the taps of the two LTCs are used to regulate voltage magnitudes at their high-voltage sides.

Linearization of (3.61) and (3.62) around the base operating point ($e_{vR}^{(0)}, f_{vR}^{(0)}, \varphi_R^{(0)}, B_{eqR}^{(0)},$) $m_{vR}^{(0)}, e_{vI}^{(0)}, f_{vI}^{(0)}, \varphi_I^{(0)}, B_{eqI}^{(0)}, m_{vI}^{(0)}$ is suitable to regulate power on the DC bus of the rectifier and to regulate the voltage magnitude at both the rectifier and the inverter AC buses. The relevant system of equations is arranged, using compact notation, in the structure shown in Eq. (3.63).

$$\begin{bmatrix} \mathbf{F}_{\mathbf{VSC,R}} \\ \mathbf{F}_{\mathbf{VSC,I}} \\ \mathbf{F}_{\mathbf{DC}} \end{bmatrix} = -\begin{bmatrix} \mathbf{J}_{\mathbf{RR}} & \mathbf{0} & \mathbf{J}_{\mathbf{R0}} \\ \mathbf{0} & \mathbf{J}_{\mathbf{II}} & \mathbf{J}_{\mathbf{I0}} \\ \mathbf{J}_{\mathbf{0R}} & \mathbf{J}_{\mathbf{0I}} & \mathbf{J}_{\mathbf{00}} \end{bmatrix} \begin{bmatrix} \mathbf{\Delta\Phi}_{\mathbf{VSC,R}} \\ \mathbf{\Delta\Phi}_{\mathbf{VSC,I}} \\ \mathbf{\Delta E}_{\mathbf{DC}} \end{bmatrix} \tag{3.63}$$

In addition to the matrix entries corresponding to the two VSCs, making up the HVDC system, there are matrix entries corresponding to the DC cable and mutual matrix terms between the DC nodes and their respective AC nodes. The matrix contributions from the two VSCs are:

$$\mathbf{J}_{\mathbf{RR}} = \begin{bmatrix} \partial\Delta P_{vR}/\partial e_{vR} & \partial\Delta P_{vR}/\partial f_{vR} & \partial\Delta P_{vR}/\partial \varphi_{vR} & \partial\Delta P_{vR}/\partial B_{eqvR} & \partial\Delta P_{vR}/\partial m_{aR} \\ \partial E_{vR}/\partial e_{vR} & \partial E_{vR}/\partial f_{vR} & 0 & 0 & 0 \\ \partial\Delta P_{0R-vR}/\partial e_{vR} & \partial\Delta P_{0R-vR}/\partial f_{vR} & \partial\Delta P_{0R-vR}/\partial \varphi_{vR} & \partial\Delta P_{0R-vR}/\partial B_{eqvR} & \partial\Delta P_{0R-vR}/\partial m_{aR} \\ \partial\Delta Q_{0R-vR}/\partial e_{vR} & \partial\Delta Q_{0R-vR}/\partial f_{vR} & \partial\Delta Q_{0R-vR}/\partial \varphi_{vR} & \partial\Delta Q_{0R-vR}/\partial B_{eqvR} & \partial\Delta Q_{0R-vR}/\partial m_{aR} \\ \partial\Delta Q_{vR}/\partial e_{vR} & \partial\Delta Q_{vR}/\partial f_{vR} & \partial\Delta Q_{vR}/\partial \varphi_{vR} & \partial\Delta Q_{vR}/\partial B_{eqvR} & \partial\Delta Q_{vR}/\partial m_{aR} \end{bmatrix} \tag{3.64}$$

$$\mathbf{J}_{\mathbf{II}} = \begin{bmatrix} \partial\Delta P_{vI}/\partial e_{vI} & \partial\Delta P_{vI}/\partial f_{vI} & \partial\Delta P_{vI}/\partial \varphi_{vI} & \partial\Delta P_{vI}/\partial B_{eqvI} & \partial\Delta P_{vI}/\partial m_{aI} \\ \partial E_{vI}/\partial e_{vI} & \partial E_{vI}/\partial f_{vI} & 0 & 0 & 0 \\ \partial\Delta P_{0I-vI}/\partial e_{vI} & \partial\Delta P_{0I-vI}/\partial f_{vI} & \partial\Delta P_{0I-vI}/\partial \varphi_{vI} & \partial\Delta P_{0I-vI}/\partial B_{eqvI} & \partial\Delta P_{0I-vI}/\partial m_{aI} \\ \partial\Delta Q_{0I-vI}/\partial e_{vI} & \partial\Delta Q_{0I-vI}/\partial f_{vI} & \partial\Delta Q_{0I-vI}/\partial \varphi_{vI} & \partial\Delta Q_{0I-vI}/\partial B_{eqvI} & \partial\Delta Q_{0I-vR}/\partial m_{aI} \\ \partial\Delta_{vI}/\partial e_{vI} & \partial\Delta Q_{vI}/\partial f_{vI} & \partial\Delta Q_{vI}/\partial \varphi_{vI} & \partial\Delta Q_{vI}/\partial B_{eqvI} & \partial\Delta Q_{vI}/\partial m_{aI} \end{bmatrix} \tag{3.65}$$

$$\mathbf{\Delta\Phi}_{\mathbf{VSC,R}} = \begin{bmatrix} \Delta e_{vR} & \Delta f_{vR} & \Delta\varphi_R & \Delta B_{eqR} & \Delta m_{aR} \end{bmatrix}^t \tag{3.66}$$

$$\mathbf{\Delta\Phi}_{\mathbf{VSC,I}} = \begin{bmatrix} \Delta e_{vI} & \Delta f_{vI} & \Delta\varphi_I & \Delta B_{eqI} & \Delta m_{aI} \end{bmatrix}^t \tag{3.67}$$

$$\mathbf{F}_{\mathbf{VSC,R}} = \begin{bmatrix} \Delta P_{vR} & \Delta E_{vR} & \Delta P_{0R-vR} & \Delta Q_{0R-vR} & \Delta Q_{vR} \end{bmatrix}^t \tag{3.68}$$

$$\mathbf{F}_{\mathbf{VSC,I}} = \begin{bmatrix} \Delta P_{vI} & \Delta E_{vI} & \Delta P_{0I-vI} & \Delta Q_{0I-vI} & \Delta Q_{vI} \end{bmatrix}^t \tag{3.69}$$

The mutual terms between the DC nodes and their corresponding AC nodes are:

$$\mathbf{J}_{\mathbf{R0}} = \begin{bmatrix} \partial\Delta P_{vR}/\partial E_{0R} & 0 \\ 0 & 0 \\ \partial\Delta P_{0R-vR}/\partial E_{0R} & 0 \\ \partial\Delta Q_{0R-vR}/\partial E_{0R} & 0 \\ \partial\Delta Q_{vR}/\partial E_{0R} & 0 \end{bmatrix} \tag{3.70}$$

$$\mathbf{J_{I0}} = \begin{bmatrix} 0 & \partial \Delta P_{vI}/\partial E_{0I} \\ 0 & 0 \\ 0 & \partial \Delta P_{0I-vI}/\partial E_{0I} \\ 0 & \partial \Delta Q_{0I-vI}/\partial E_{0I} \\ 0 & \partial \Delta Q_{vI}/\partial E_{0I} \end{bmatrix} \tag{3.71}$$

$$\mathbf{J_{0R}} = \begin{bmatrix} \partial \Delta P_{0R}/\partial e_{vR} & \partial \Delta P_{0R}/\partial f_{vR} & \partial \Delta P_{0R}/\partial \varphi_R & 0 & \partial \Delta P_{0R}/\partial m_{aR} \\ 0 & 0 & 0 & 0 & 0 \end{bmatrix} \tag{3.72}$$

$$\mathbf{J_{0I}} = \begin{bmatrix} 0 & 0 & 0 & 0 & 0 \\ \partial \Delta P_{0I}/\partial e_{vI} & \partial \Delta P_{0I}/\partial f_{vI} & \partial \Delta P_{0I}/\partial \varphi_I & 0 & \partial \Delta P_{0I}/\partial m_{aI} \end{bmatrix} \tag{3.73}$$

The terms corresponding to the DC part of the system are:

$$\mathbf{J_{00}} = \begin{bmatrix} \partial \Delta P_{0R}/\partial E_{0R} & \partial \Delta P_{0R}/\partial E_{0I} \\ \partial \Delta P_{0I}/\partial E_{0R} & \partial \Delta P_{0I}/\partial E_{0I} \end{bmatrix} \tag{3.74}$$

$$\mathbf{\Delta\Phi_{DC}} = \left[\Delta E_{0R} \;\; \Delta E_{0I}\right]^t \tag{3.75}$$

$$\mathbf{F_{DC}} = \left[\Delta P_{0R} \;\; \Delta P_{0I}\right]^t \tag{3.76}$$

If no voltage regulation is exerted at the AC bus of either the rectifier or the inverter, suitable changes take place in Eqs. (3.64)–(3.73), similar to those which were implemented in the VSC with no AC voltage regulation and that led to Eq. (3.46).

Normally, voltage regulation takes place at the DC side of the inverter and the corresponding row and column associated with $0I$ are deleted from Eqs. (3.70)–(3.76). For any practical case, the nodal injected powers at the DC nodes $0R$ and $0I$ will be zero.

3.8.3 Back-to-Back VSC-HVDC Systems Modelling

The back-to-back VSC-HVDC system, also known as zero-distance, may be modelled as a particular case of the point-to-point VSC-HVDC formulation developed in the previous section and comprising Eqs. (3.59)–(3.76), when $G_{DC} = 0$. In such a case, Eq. (3.63) reduces to the following one:

$$\begin{bmatrix} \mathbf{F_{VSC,R}} \\ \mathbf{F_{VSC,I}} \end{bmatrix} = - \begin{bmatrix} \mathbf{J_{RR}} & \mathbf{0} \\ \mathbf{0} & \mathbf{J_{II}} \end{bmatrix} \begin{bmatrix} \mathbf{\Delta_{VSC,R}} \\ \mathbf{\Delta_{VSC,I}} \end{bmatrix} \tag{3.77}$$

In this application only Eqs. (3.64)–(3.69) are relevant. Notice that the Jacobian matrix is decoupled but that the overall system of equations remains tightly coupled through the constrained branch powers (3.62) which make up $\mathbf{F_{VSC,R}}$ and $\mathbf{F_{VSC,I}}$.

3.8.4 VSC-HVDC Numerical Examples

Test Case 6. The test case in this section relates to a simple system where the VSC-HVDC link is used to interconnect two otherwise independent AC systems, as shown in Figure 3.14. The relevant data for this test system is given in Table 3.3.

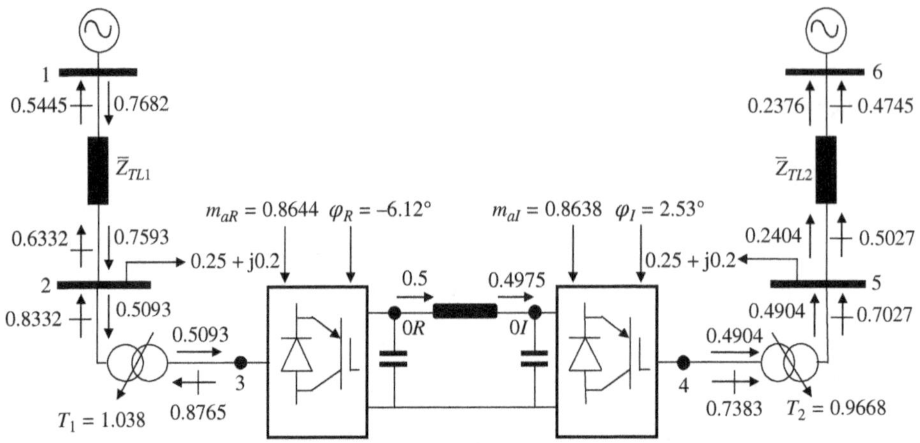

Figure 3.14 Point-to-point VSC-HVDC linking two equivalent AC subsystems.

Table 3.3 Parameter data for the VSC-HVDC system of Test Case 6.

Compensation case	Active power loss (%)	
	Resistance	Reactance
Transmission lines' impedances	0.01	0.10
VSCs' series impedances	0.002	0.01
VSCs' initial shunt admittances	0.002	1
LTCs' series impedances	0	0.05
DC cable resistance	0.01	—

The active and reactive power loads at buses 2 and 5 are 0.25 and 0.20 p.u., respectively. The rectifier is connected between buses 3 and $0R$, the inverter is connected between buses 4 and $0I$, and the DC cable is connected between DC buses $0R$ and $0I$.

The power leaving the rectifier is set at 0.5 p.u. The inverter and rectifier are set to regulate voltage magnitudes at buses 3 and 4 at 1.05 p.u. using m_{aR} and m_{aI}, respectively. Similarly, the voltage magnitudes at buses 2 and 5 are both regulated at 1.05 p.u. using LTC 1 and LTC 2, respectively.

Buses 1 and 6 are taken to be the two slack buses of this asynchronous interconnection. The phase angle at bus 1 provides a reference for the phase angles at buses 2 and 3 whereas the phase angle at bus 6 provides a reference for the phase angles at buses 4 and 5. The full voltage solution is given in Table 3.4.

The solution converges in five iterations to a mismatch tolerance of 10^{-12}. The active and reactive power flows are given in Figure 3.14 where the equivalent generator connected to buses 1 contributes 0.7682 p.u. active power and the equivalent generator connected at bus 6 absorbs 0.2376 p.u. active power. The amplitude modulation ratios and taps for the two VSCs and the two LTCs, respectively, are given in Table 3.5.

The active power loss incurred in the VSC connected to bus 3 stands at 0.93%, with 0.7438% due to switching losses and 0.1962% due to conduction losses. The power loss

Table 3.4 Power flow voltage solution.

Nodes	1	2	3	0*R*	0*I*	4	5	6
V (p.u.)	1.0	1.05	1.05	2.002	2	1.05	1.05	1.0
θ (deg)	0	−4.49	−5.77	—	—	2.36	1.04	0

Table 3.5 Tap values for the two VSCs and the two LTCs.

VSC	1	2	LTC	1	2
m_a	$0.8644\angle{-6.12°}$	$0.8638\angle{2.53°}$	Tap	1.038	0.9668

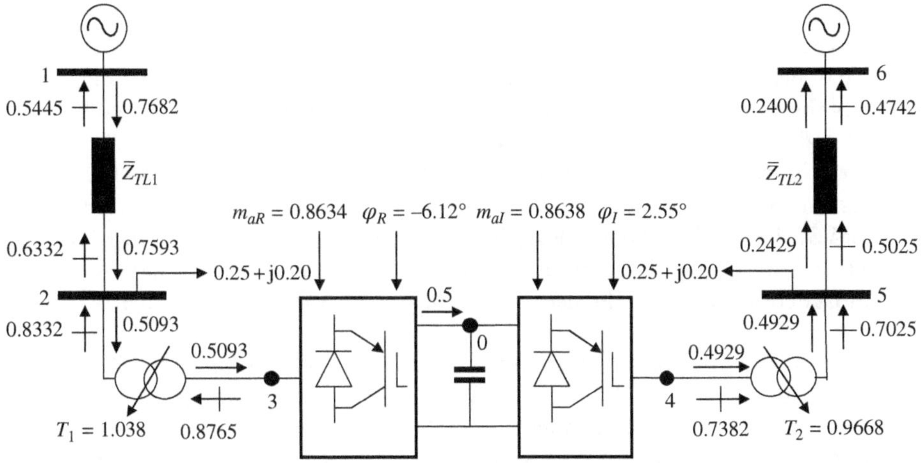

Figure 3.15 Back-to-back VSC-HVDC linking two equivalent AC subsystems.

in the DC cable is 0.25%. The VSC 1 delivers 0.8765 p.u. of reactive power to supply the reactive power load of 0.2 p.u. connected at bus 2. It also caters for the reactive power loss of LTC 1 and together with reactive power absorbed by the transmission line connected between nodes 1 and 2, injects 0.5445 p.u. of reactive power into the slack generator at bus 1. A similar analysis may be carried out for the VSC connected to node 4.

Test Case 7. This test case relates to a system where the VSC-HVDC link is used to interconnect two otherwise independent AC systems represented in a rather simplified form, as shown in Figure 3.15. This test case is a simplified case of Test Case 6, where the DC cable resistance takes a value of zero.

The active and reactive power flows are given in Figure 3.15 where it is shown that the equivalent generator connected at bus 1 contributes 0.7682 p.u. active power and the equivalent generator connected to bus 6 absorbs 0.2400 p.u. active power. The voltage magnitudes at all seven buses are treated as voltage-controlled nodes by appealing to the voltage-regulating capabilities of the two generators, the two LTC transformers and the

Table 3.6 Power flow voltage solution.

Nodes	1	2	3	0	4	5	6
V(p.u)	1.02	1.05	1.05	2	1.05	1.05	1.02
θ (deg)	0	−4.49	−5.77	–	2.38	1.05	0

two VSCs. The phase angles are initialized at 0 in all buses. Buses 1 and 6 are designated to be the two slack buses of this asynchronous interconnection and bus 0 is a DC bus. The phase angle voltage at bus 1 provides a reference for the phase angle voltages at buses 2 and 3 whereas the phase angle voltage at bus 6 is the reference for the phase angle voltages at buses 4 and 5. The voltage solution is given in Table 3.6.

The active power loss incurred in the VSC connected to bus 3 stands at 0.93%, with 0.75% due to switching losses and the rest due to conduction losses – the switching loss is represented by an initial equivalent conductance, G_0, of 0.002 p.u. It delivers 0.8765 p.u. of reactive power to supply the reactive power load of 0.2 p.u. connected at bus 2, with 0.5445 p.u. being absorbed by the slack generator at node 1 and the remaining going to satisfy the reactive power loss incurred by the transmission line connected between nodes 1 and 2 and LTC 1, which stands at 0.332 p.u.

The active power loss incurred in the VSC connected to bus 4 stands at 0.71%. Switching losses are 57% and conduction losses are 0.14%. These power losses are smaller than those of VSC 1 since there is less energy in this part of the network. It delivers 0.7382 p.u. of reactive power to supply the reactive power load of 0.2 p.u. at bus 5 and the reactive power loss of transformer 2. The slack generator at node 6 absorbs 0.4742 p.u. of reactive power.

The complex and real taps corresponding to the two VSCs and the two LTCs, respectively, are given in Table 3.7. VSC 1 and VSC 2 are used to regulate voltage magnitudes at buses 3 and 4 at 1.05 p.u. with actual amplitude modulation indexes m_{aR} and m_{aI} of 0.8634 and 0.8638, respectively. Likewise, LTCs 1 and 2 are used to regulate voltage magnitudes at buses 2 and 5 at 1.05 p.u. with resulting taps $T_1 = 1.038$ and $T_2 = 0.9668$, respectively.

The solution converges in five iterations to a mismatch tolerance of 10^{-12}.

Test Case 8. One point-to-point VSC-HVDC link is used to replace the transmission line connected between nodes 4 and 6 in the IEEE 30-node system. The two VSCs and their interfacing transformers use the same parameters as those in Test Cases 7 and 8. The resistance of the original transmission line (i.e. 0.0119 p.u.) is assumed to be the resistance of the DC cable in the VSC-HVDC link. The rectifier is set to control DC power at 71.58 MW, which is the same amount of active power that would exist in the

Table 3.7 Tap values for the two VSCs and the two LTCs.

VSC	1	2	LTC	1	2
m_a	$0.8634\angle{-6.12°}$	$0.8638\angle 2.55°$	Tap	1.038	0.9668

Table 3.8 Power flow voltage solution.

Nodes	4	*vR*	*R0*	*I0*	*vI*	6
V(p.u)	1.05	1.05	2.0042	2	1.05	1.05
θ (deg)	−10.15	−11.98	—	—	−10.69	−12.34

Table 3.9 Tap values for the two VSCs and the two LTCs.

VSC	Rectifier	Inverter	LTC	1	2
m_a	$0.8634\angle{-12.43°}$	$0.8715\angle{-10.51°}$	Tap	1.0298	1.0711

original AC power flow solution. Furthermore, the DC voltage at the inverter is set at 2 p.u. and the AC voltages at node 4, node 6 and the two connecting buses between the two VSCs and their interfacing transformers are set at 1.05 p.u. The relevant portions of the modified 30-bus system with its power flows is shown in Figure 3.16 and the voltage solution is given in Table 3.8.

The total power losses in the HVDC system stand at 4.33 MW, where 0.427 MW corresponds to ohmic losses in the DC cable, 0.902 MW in the rectifier station and 3 MW in the inverter station. In contrast, the AC transmission line connected between nodes 4 and 6 in the original system incurs only 0.618 MW loss. As expected, the power losses in the DC cable are smaller than in the AC transmission line, but the power losses in the rectifier station are quite close to 1% (1 MW) because the current drawn by the inverter is close to the nominal, i.e. $0.9518\angle{+31.49°}$. On the other hand, the power losses in the inverter station climb to 3% (3 MW) because the current drawn by the inverter goes well above the nominal, i.e. $1.7336\angle{+101.3°}$. This is clearly an unacceptable operational state which calls for an optimization of the VSC-HVDC state variables, a case which will be addressed in Chapter 5 where the topic of optimal power flows is introduced.

The complex and real taps corresponding to the two VSCs and the two LTCs, respectively, are given in Table 3.9.

The equivalent susceptances of VSC 1 and 2, which stand at 0.6252 and 1.5079 p.u., produce 0.6961 and 1.7179 p.u. of reactive power, respectively. The solution converges in six iterations to a mismatch tolerance of 10^{-12}.

3.9 Three-Terminal VSC-HVDC System Model

The nodal power expressions in (3.61) corresponding to the point-to-point VSC-HVDC system may be extended with ease to represent a VSC-HVDC system with three terminals [17]. Without loss of generality, one VSC is taken to act as a rectifier station and the other two VSCs as inverter stations. They are linked on their DC sides by a meshed network of three cables, as shown in Figure 3.17.

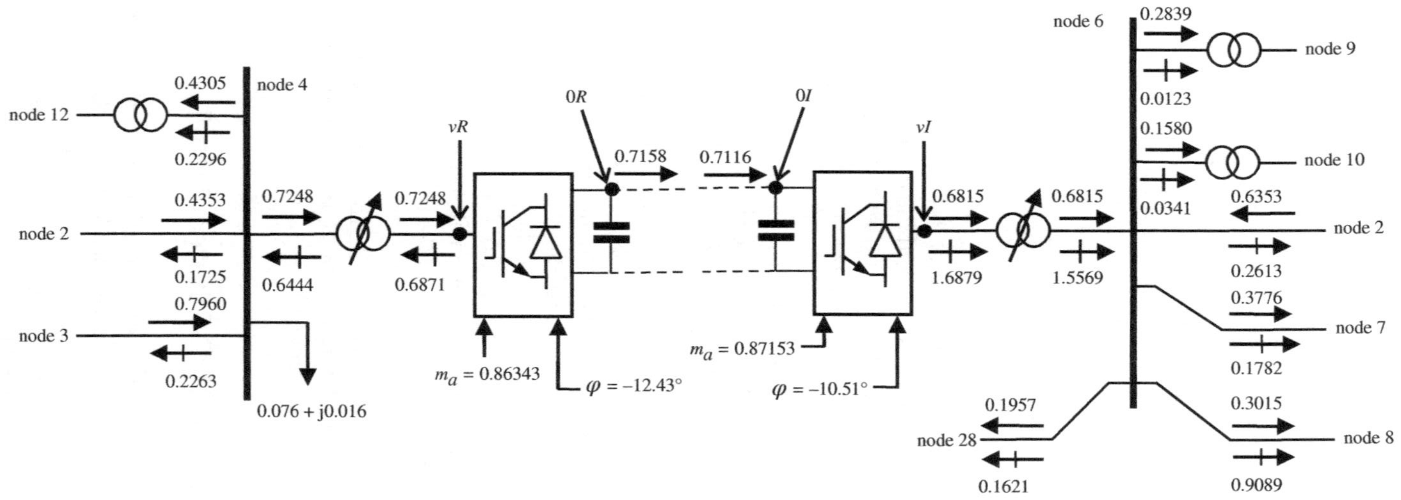

Figure 3.16 One point-to-point VSC-HVDC system linking buses 4 and 6 in the modified IEEE 30-bus system.

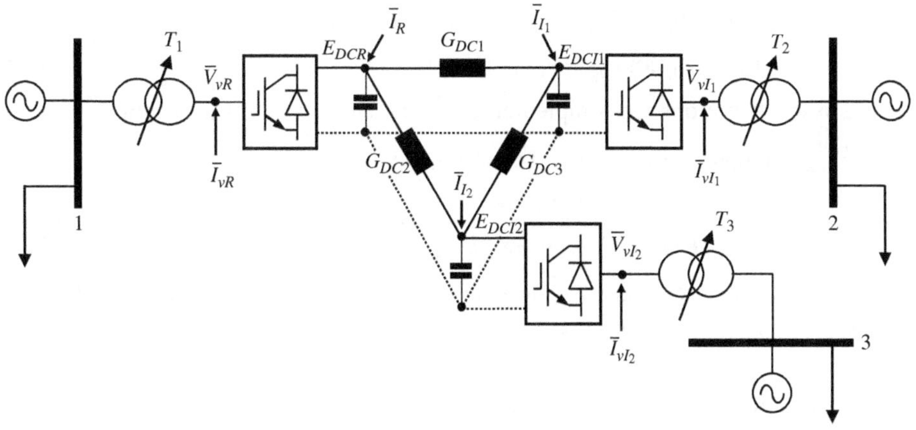

Figure 3.17 A three-terminal VSC-HVDC system. Source: ©ELSEVIER, 2016.

The relevant power expressions are the following:

$$\begin{aligned}
P_{vR} &= G_{vRvR}(e_{vR}^2 + f_{vR}^2) - E_{DCR}(e_{vR}[G_{vR0}\cos\varphi_R + B_{vR0}\sin\varphi_R] \\
&\quad - f_{vR}[G_{vR0}\sin\varphi_R - B_{vR0}\cos\varphi_R]) \\
Q_{vR} &= -B_{vRvR}(e_{vR}^2 + f_{vR}^2) - E_{DCR}(e_{vR}[G_{vR0}\sin\varphi_R - B_{vR0}\cos\varphi_R] \\
&\quad + f_{vR}[G_{vR0}\cos\varphi_R + B_{vR0}\sin\varphi_R]) \\
P_{0R} &= (G_{swR} + G_{00R})E_{DCR}^2 - E_{DCR}(e_{vR}[G_{vR0}\cos\varphi_R + B_{vR0}\sin\varphi_R] \\
&\quad + f_{vR}[G_{vR0}\sin\varphi_R - B_{vR0}\cos\varphi_R]) \\
Q_{0R} &= -(k_1^2 m_{aR}^2 B_{eqR} + B_{00R})E_{DCR}^2 - E_{DCR}(e_{vR}[G_{vR0}\sin\varphi_R - B_{vR0}\cos\varphi_R] \\
&\quad - f_{vR}[G_{vR0}\cos\varphi_R + B_{vR0}\sin\varphi_R])
\end{aligned} \tag{3.78}$$

$$\begin{aligned}
P_{vI_1} &= G_{vI_1vI_1}(e_{vI_1}^2 + f_{vI_1}^2) - E_{DCI_1}(e_{vI_1}[G_{vI_10_1}\cos\varphi_{I_1} - B_{vI_10_1}\sin\varphi_{I_1}] \\
&\quad + f_{vI_1}[G_{vI_10_1}\sin\varphi_{I_1} + B_{vI_10_1}\cos\varphi_{I_1}]) \\
Q_{vI_1} &= -B_{vI_1vI_1}(e_{vI_1}^2 + f_{vI_1}^2) - E_{DCI_1}(-e_{vI_1}[G_{vI_10_1}\sin\varphi_{I_1} + B_{vI_10_1}\cos\varphi_{I_1}] \\
&\quad + f_{vI_1}[G_{vI_10_1}\cos\varphi_{I_1} - B_{vI_10_1}\sin\varphi_{I_1}]) \\
P_{0I_1} &= (G_{swI_1} + G_{00I_1})E_{DCI_1}^2 - E_{DCI_1}(e_{vI_1}[G_{vI_10_1}\cos\varphi_{I_1} + B_{vI_10_1}\sin\varphi_{I_1}] \\
&\quad + f_{vI_1}[G_{vI_10_1}\sin\varphi_{I_1} - B_{vI_10_1}\cos\varphi_{I_1}]) \\
Q_{0I_1} &= -(k_1^2 m_{aI_1}^2 B_{eI_1} + B_{00I_1})E_{DCI_1}^2 - E_{DCI_1}(e_{vI_1}[G_{vI_10_1}\sin\varphi_{I_1} - B_{vI_10_1}\cos\varphi_{I_1}] \\
&\quad - f_{vI_1}[G_{vI_10_1}\cos\varphi_{I_1} + B_{vI_10_1}\sin\varphi_{I_1}])
\end{aligned} \tag{3.79}$$

$$\begin{aligned}
P_{vI_2} &= G_{vI_2vI_2}(e_{vI_2}^2 + f_{vI_2}^2) - E_{DCI_2}(e_{vI_2}[G_{vI_20_2}\cos\varphi_{I_2} - B_{vI_20_2}\sin\varphi_{I_2}]) \\
&\quad + f_{vI_2}[G_{vI_20_2}\sin\varphi_{I_2} + B_{vI_20_2}\cos\varphi_{I_2}]) \\
Q_{vI_2} &= -B_{vI_2vI_2}(e_{vI_2}^2 + f_{vI_2}^2) - E_{DCI_2}(-e_{vI_2}[G_{vI_20_2}\sin\varphi_{I_2} + B_{vI_20_2}\cos\varphi_{I_2}]) \\
&\quad + f_{vI_2}[G_{vI_20_2}\cos\varphi_{I_2} - B_{vI_20_2}\sin\varphi_{I_2}]) \\
P_{0I_2} &= (G_{swI_2} + G_{00I_2})E_{DCI_2}^2 - E_{DCI_2}(e_{vI_2}[G_{vI_20_2}\cos\varphi_{I_2} + B_{vI_20_2}\sin\varphi_{I_2}] \\
&\quad + f_{vI_2}[G_{vI_20_2}\sin\varphi_{I_2} - B_{vI_20_2}\cos\varphi_{I_2}])
\end{aligned}$$

$$Q_{0I_2} = -(k_1^2 m_{aI_2}^2 B_{eI_2} + B_{00I_2})E_{DCI_2}^2 - E_{DCI_2}(e_{vI_2}[G_{vI_20_2} \sin\varphi_{I_2} - B_{vI_20_2}\cos\varphi_{I_2}]$$
$$-f_{vI_2}[G_{vI_20_2}\cos\varphi_{I_2} + B_{vI_20_2}\sin\varphi_{I_2}]) \quad (3.80)$$

These power equations are complemented by the DC power contributions:

$$P_{DCR}^{cal} = (G_{DC1} + G_{DC2})E_{DCR}^2 - G_{DC1}E_{DCI_1} - G_{DC2}E_{DCI_2}$$
$$P_{DCI_1}^{cal} = (G_{DC1} + G_{DC3})E_{DCI_1}^2 - G_{DC1}E_{DCR} - G_{DC3}E_{DCI_2}$$
$$P_{DCI_2}^{cal} = (G_{DC2} + G_{DC3})E_{DCI_2}^2 - G_{DC2}E_{DCR} - G_{DC3}E_{DCI_1} \quad (3.81)$$

Note that G_{vR0}, B_{vR0}, G_{00R} and B_{00R} are functions of m_{aR}; G_{vI_10}, B_{vI_10}, G_{00I_1} and B_{00I_1} are functions of m_{aI_1}; and G_{vI_20}, B_{vI_20}, G_{00I_2} and B_{00I_2} are functions of m_{aI_2}.

3.9.1 VSC Types

Similarly to the standard classification of types of nodes available in conventional AC power flow theory, Table 3.10 introduces the types of VSC stations that are required to solve the generic DC power grid concept put forward in this section. This classification is based on principles of parameter control and application: (i) the slack converter VSC_{Slack} provides voltage control at its DC terminal and it is linked on its AC side to a network which contains synchronous generation; (ii) the converter of type VSC_{Psch} serves the purpose of injecting scheduled power into the DC grid and it is also linked on its AC side to a network with synchronous generation; (iii) the third type of VSC station is the passive converter VSC_{Pass} which is used to interconnect the DC grid with an AC network which contains no synchronous generation of its own.

In the passive AC power grids, the VSC's internal angle, φ, provides the angular reference for the network. The specified and calculated VSC variables are given in Table 3.10 for each type of VSC. In addition to the VSC's control variables, if available, the taps of the associated VSC transformers may enter as control variables in this problem.

3.9.2 Power Mismatches

The power-constraining equations of the type (3.62) are also used in this application. In connection with the three-terminal VSC-HVDC system, the three reactive power-constraining equations are:

$$\Delta Q_{0R-vR} = 0 - Q_{0R}$$
$$\Delta Q_{0I_1-vI_1} = 0 - Q_{0I_1}$$
$$\Delta Q_{0I_2-vI_2} = 0 - Q_{0I_2} \quad (3.82)$$

Table 3.10 Type of VSCs and their control variables.

Type	Known variables	Unknown variables
VSC_{Slack}	E_{DC}, V_v^{spec}	m_a, φ, B_{eq}, e_v, f_v
VSC_{Psch}	P^{sch}, V_v^{spec}	E_{DC}, m_a, φ, B_{eq}, e_v, f_v
VSC_{Pass}	φ, V_v^{spec}	E_{DC}, m_a, B_{eq}, e_v, f_v

Moreover, converters of type VSC_{Psch} require an active power-constraining equation which for the three-terminal example in Figure 3.17, and with no loss of generality, takes the following form:

$$\Delta P_{0I_1-vI_1} = P_{0I_1-vI_1}^{sch} - P_{I_1} \tag{3.83}$$

where $P_{0I_1-vI_1}^{sch}$ is the amount of DC power entering inverter 1 at its DC bus.

3.9.3 Linearized System of Equations

Linearization of the relevant set of mismatch equations around the following base operating point $(e_{vR}^{(0)}, f_{vR}^{(0)}, \varphi_R^{(0)}, B_{eqR}^{(0)}, m_{aR}^{(0)}, e_{vI_1}^{(0)}, f_{vI_1}^{(0)}, \varphi_{I_1}^{(0)}, B_{eqI_1}^{(0)}, m_{aI_1}^{(0)}, E_{DCI_1}^{(0)}, e_{vI_2}^{(0)}, f_{vI_2}^{(0)}, B_{eqI_2}^{(0)}, m_{aI_2}^{(0)}, E_{DCI_2}^{(0)})$ is suitable to regulate power on the DC bus of the rectifier and to regulate the voltage magnitude at both the rectifier AC bus and the two inverters' AC buses. The system of equations is arranged, using compact notation, in the structure shown in (3.84)

$$\begin{bmatrix} \mathbf{F_{VSC,R}} \\ \mathbf{F_{VSC,I_1}} \\ \mathbf{F_{VSC,I_2}} \\ \mathbf{F_{DC}} \end{bmatrix} = -\begin{bmatrix} \mathbf{J_{RR}} & \mathbf{0} & \mathbf{0} & \mathbf{J_R} \\ \mathbf{0} & \mathbf{J_{II_1}} & \mathbf{0} & \mathbf{J_{I_1}} \\ \mathbf{0} & \mathbf{0} & \mathbf{J_{II_2}} & \mathbf{J_{I_2}} \\ \mathbf{0} & \mathbf{J_{{}_1I}} & \mathbf{J_{{}_2I}} & \mathbf{J_{DC}} \end{bmatrix} \begin{bmatrix} \mathbf{\Delta_{VSC,R}} \\ \mathbf{\Delta_{VSC,I_1}} \\ \mathbf{\Delta_{VSC,I_2}} \\ \mathbf{\Delta E_{DC}} \end{bmatrix} \tag{3.84}$$

where the **0** entries are zero-padded matrices of suitable orders.

It is noted that in addition to the matrix entries corresponding to the three VSCs, making up the three-terminal HVDC system, there are matrix entries corresponding to the DC cables and mutual matrix terms between the DC nodes and their respective AC nodes. The matrix contributions from the three VSCs are:

$$\mathbf{J_{RR}} = \begin{bmatrix} \partial\Delta P_{vR}/\partial e_{vR} & \partial\Delta P_{vR}/\partial m_{aR} & \partial\Delta P_{vR}/\partial f_{vR} & 0 & \partial\Delta P_{vR}/\partial\varphi_R \\ \partial\Delta Q_{vR}/\partial e_{vR} & \partial\Delta Q_{vR}/\partial m_{aR} & \partial\Delta Q_{vR}/\partial f_{vR} & 0 & \partial\Delta Q_{vR}/\partial\varphi_R \\ \partial\Delta U_{vR}/\partial e_{vR} & 0 & \partial\Delta U_{vR}/\partial f_{vR} & 0 & 0 \\ \partial\Delta Q_{R-vR}/\partial e_{vR} & \partial\Delta Q_{R-vR}/\partial m_{aR} & \partial\Delta Q_{R-vR}/\partial f_{vR} & \partial\Delta Q_{R-vR}/\partial B_{eqR} & \partial\Delta Q_{R-vR}/\partial\varphi_R \\ \partial\Delta P_R/\partial e_{vR} & \partial\Delta P_R/\partial m_{aR} & \partial\Delta P_R/\partial f_{vR} & 0 & \partial\Delta P_R/\partial\varphi_R \end{bmatrix} \tag{3.85}$$

$$\mathbf{J_{II_1}} = \begin{bmatrix} \partial\Delta P_{vI_1}/\partial e_{vI_1} & \partial\Delta P_{vI_1}/\partial m_{aI_1} & \partial\Delta P_{vI_1}/\partial f_{vI_1} & 0 & \partial\Delta P_{vI1}/\partial\varphi_{I_1} \\ \partial\Delta Q_{vI_1}/\partial e_{vI_1} & \partial\Delta Q_{vI_1}/\partial m_{aI_1} & \partial\Delta Q_{vI_1}/\partial f_{vI_1} & 0 & \partial\Delta Q_{vI1}/\partial\varphi_{I_1} \\ \partial\Delta U_{vI_1}/\partial e_{vI_1} & 0 & \partial\Delta U_{vI_1}/\partial f_{vI_1} & 0 & 0 \\ \partial\Delta Q_{I_1-vI_1}/\partial e_{vI_1} & \partial\Delta Q_{I_1-vI_1}/\partial m_{aI_1} & \partial\Delta Q_{I_1-vI_1}/\partial f_{vI_1} & \partial\Delta Q_{I_1-vI_1}/\partial B_{eq,I_1} & \partial\Delta Q_{I_1-vI_1}/\partial\varphi_{I_1} \\ \partial\Delta P_{I_1-vI_1}/\partial e_{vI_1} & \partial\Delta P_{I_1-vI_1}/\partial m_{aI_1} & \partial\Delta P_{I_1-vI_1}/\partial f_{vI_1} & 0 & \partial\Delta P_{I_1-vI_1}/\partial\varphi_{I_1} \end{bmatrix} \tag{3.86}$$

$$\mathbf{J_{II_2}} = \begin{bmatrix} \partial\Delta P_{vI_2}/\partial e_{vI_2} & \partial\Delta P_{vI_2}/\partial m_{aI_2} & \partial\Delta P_{vI_2}/\partial f_{vI_2} & 0 \\ \partial\Delta Q_{vI_2}/\partial e_{vI_2} & \partial\Delta Q_{vI_2}/\partial m_{aI_2} & \partial\Delta Q_{vI_2}/\partial f_{vI_2} & 0 \\ \partial\Delta E_{vI_2}/\partial e_{vI_2} & 0 & \partial\Delta E_{vI_2}/\partial f_{vI_2} & 0 \\ \partial\Delta Q_{I_2-vI_2}/\partial e_{vI_2} & \partial\Delta Q_{I_2-vI_2}/\partial m_{aI_2} & \partial\Delta Q_{I_2-vI_2}/\partial f_{vI_2} & \partial\Delta Q_{I_2-vI_2}/\partial B_{eqI_2} \end{bmatrix} \tag{3.87}$$

The power mismatch vectors and the state variables increments for the rectifier and inverter stations are:

$$\mathbf{F_{VSC,R}} = \left[\Delta P_{vR}\ \Delta Q_{vR}\ \Delta U_{vR}\ \Delta Q_{R-vR}\ \Delta P_{R}\right]^t \tag{3.88}$$

$$\mathbf{F_{VSC,I_1}} = \left[\Delta P_{vI_1}\ \Delta Q_{vI_1}\ \Delta U_{vI_1}\ \Delta Q_{I_1-vI_1}\ \Delta P_{I_1-vI_1}\right]^t \tag{3.89}$$

$$\mathbf{F_{VSC,I_2}} = \left[\Delta P_{vI_2}\ \Delta Q_{vI_2}\ \Delta U_{vI_2}\ \Delta Q_{I_2-vI_2}\right]^t \tag{3.90}$$

$$\mathbf{\Delta\Phi_{VSC,R}} = \left[\Delta e_{vR}\ \Delta m_{aR}\ \Delta f_{vR}\ \Delta B_{eqR}\ \Delta\varphi_{R}\right]^t \tag{3.91}$$

$$\mathbf{\Delta\Phi_{VSC,I_1}} = \left[\Delta e_{vI_1}\ \Delta m_{aI_1}\ \Delta f_{vI_1}\ \Delta B_{eqI_1}\ \Delta\varphi_{I_1}\right]^t \tag{3.92}$$

$$\mathbf{\Delta\Phi_{VSC,I_2}} = \left[\Delta e_{vI_2}\ \Delta m_{aI_2}\ \Delta f_{vI_2}\ \Delta B_{eqI_2}\right]^t \tag{3.93}$$

The mutual terms between the DC nodes and their corresponding AC nodes are:

$$\mathbf{J_R} = \begin{bmatrix} 0 & 0 \\ 0 & 0 \\ 0 & 0 \\ 0 & 0 \\ \partial\Delta P_R/\partial E_{DCI_1} & \partial\Delta P_R/\partial E_{DCI_2} \end{bmatrix} \tag{3.94}$$

$$\mathbf{J_{I_1}} = \begin{bmatrix} \partial\Delta P_{vI_1}/\partial E_{DCI_1} & 0 \\ \partial\Delta Q_{vI_1}/\partial E_{DCI_1} & 0 \\ 0 & 0 \\ \partial\Delta Q_{I_1-vI_1}/\partial E_{DCI_1} & 0 \\ \partial\Delta P_{I_1-vI_1}/\partial E_{DCI_1} & 0 \end{bmatrix} \tag{3.95}$$

$$\mathbf{J_{I_2}} = \begin{bmatrix} 0 & \partial\Delta P_{vI_2}/\partial E_{DCI_2} \\ 0 & \partial\Delta Q_{vI_2}/\partial E_{DCI_2} \\ 0 & 0 \\ 0 & \partial\Delta Q_{I_2-vI_2}/\partial E_{DCI_2} \end{bmatrix} \tag{3.96}$$

$$\mathbf{J_{{}_1I}} = \begin{bmatrix} \partial\Delta P_{I_1}/\partial e_{vI_1} & \partial\Delta P_{I_1}/\partial m_{aI_1} & \partial\Delta P_{I_1}/\partial f_{vI_1} & 0 & \partial\Delta P_{I_1}/\partial\varphi_{I_1} \\ 0 & 0 & 0 & 0 & 0 \end{bmatrix} \tag{3.97}$$

$$\mathbf{J_{{}_2I}} = \begin{bmatrix} 0 & 0 & 0 & 0 \\ \partial\Delta P_{I_2}/\partial e_{vI_2} & \partial\Delta P_{I_2}/\partial m_{aI_2} & \partial\Delta P_{I_2}/\partial f_{vI_2} & 0 \end{bmatrix} \tag{3.98}$$

The terms corresponding to the DC part of the system are:

$$\mathbf{J_{DC}} = \begin{bmatrix} \partial\Delta P_{I_1}/\partial E_{DCI_1} & \partial\Delta P_{I_1}/\partial E_{DCI_2} \\ \partial\Delta P_{I_2}/\partial E_{DCI_2} & \partial\Delta P_{I_2}/\partial E_{DCI_2} \end{bmatrix} \tag{3.99}$$

$$\mathbf{F_{DC}} = \left[\Delta P_{I_1}\ \Delta P_{I_2}\right]^t \tag{3.100}$$

$$\mathbf{\Delta E_{DC}} = \left[\Delta E_{DCI_1}\ \Delta E_{DCI_2}\right]^t \tag{3.101}$$

Notice that $\mathbf{\Delta E_{DC}}$ is a vector that contains only DC voltages as state variables.

If no voltage regulation is exerted at the AC bus of either the rectifier or any of the two inverters, then suitable changes take place in (3.85)–(3.96). For instance, if no voltage regulation takes place in the AC bus of the rectifier, then (3.85), (3.88), (3.91) and (3.94) are modified as shown below:

$$\mathbf{J_{RR}} = \begin{bmatrix} \partial \Delta P_{vR}/\partial e_{vR} & \partial \Delta P_{vR}/\partial f_{vR} & 0 & \partial \Delta P_{vR}/\partial \varphi_R \\ \partial \Delta Q_{vR}/\partial e_{vR} & \partial \Delta Q_{vR}/\partial f_{vR} & 0 & \partial \Delta Q_{vR}/\partial \varphi_R \\ \partial \Delta Q_{R-vR}/\partial e_{vR} & \partial \Delta Q_{R-vR}/\partial f_{vR} & \partial \Delta Q_{R-vR}/\partial B_{eqR} & \partial \Delta Q_{R-vR}/\partial \varphi_R \\ \partial \Delta P_R/\partial e_{vR} & \partial \Delta P_R/\partial f_{vR} & 0 & \partial \Delta P_R/\partial \varphi_R \end{bmatrix} \tag{3.102}$$

$$\mathbf{J_R} = \begin{bmatrix} 0 & 0 \\ 0 & 0 \\ 0 & 0 \\ \partial \Delta P_R/\partial E_{DCI_1} & \partial \Delta P_R/\partial E_{DCI_2} \end{bmatrix} \tag{3.103}$$

$$\mathbf{F_{VSC,R}} = \left[\Delta P_{vR} \ \Delta Q_{vR} \ \Delta Q_{R-vR} \ \Delta P_R\right]^t \tag{3.104}$$

$$\mathbf{\Delta_{VSC,R}} = \left[\Delta e_{vR} \ \Delta f_{vR} \ \Delta B_{eqR} \ \Delta \varphi_R\right]^t \tag{3.105}$$

Notice that the state variable m_{aR} charged with regulating the AC voltage at node vR becomes a constant parameter.

If voltage regulation takes place at any of the DC buses, the corresponding row and column are deleted from (3.94)–(3.101). It should be remarked that the voltage must be specified in at least one of the buses of the DC network. Such a node plays the role of reference node in the DC network and in this three-terminal VSC-HVDC example this role has been assigned to the rectifier station which is VSC_{slack} type.

The increments of the state variables in vector (3.91), calculated at iteration (r), are used to update the state variables, as follows:

$$\begin{aligned} e_{vR}^{(r)} &= e_{vR}^{(r-1)} + \Delta e_{vR}^{(r)} \\ f_{vR}^{(r)} &= f_{vR}^{(r-1)} + \Delta f_{vR}^{(r-1)} \\ m_{aR}^{(r)} &= m_{aR}^{(r-1)} + \Delta m_{aR}^{(r-1)} \\ B_{eqR}^{(r)} &= B_{eqR}^{(r-1)} + \Delta B_{eqR}^{(r)} \\ \varphi_R^{(r)} &= \varphi_R^{(r-1)} + \Delta \varphi_R^{(r)} \end{aligned} \tag{3.106}$$

Similar expressions exist for updating the state variables of the inverters contained in vectors (3.92) and (3.93).

The updating of the DC voltages using vector (3.101) is carried out as follows:

$$\begin{aligned} E_{DCI_1}^{(r)} &= E_{DCI_1}^{(r-1)} + \Delta E_{DCI_1}^{(r)} \\ E_{DCI_2}^{(r)} &= E_{DCI_2}^{(r-1)} + \Delta E_{DCI_2}^{(r)} \end{aligned} \tag{3.107}$$

It should be noted that when all entries relating to the second inverter (entries with subscripts I_2) are removed in (3.78)–(3.107), the three-terminal VSC-HVDC model reduces neatly to the more particular case of the point-to-point VSC-HVDC link model.

3.10 Multi-Terminal VSC-HVDC System Model

Moving into the multi-terminal direction, expression (3.84) may be extended to represent a VSC-HVDC system with multiple terminals consisting of n rectifying stations, m inverting stations and a complex DC network of cables. Such a general expression takes the following form:

$$\begin{bmatrix} \mathbf{F}_{\mathbf{VSC},\mathbf{R}_1} \\ \vdots \\ \mathbf{F}_{\mathbf{VSC},\mathbf{R}_n} \\ \mathbf{F}_{\mathbf{VSC},\mathbf{I}_1} \\ \vdots \\ \mathbf{F}_{\mathbf{VSC},\mathbf{I}_m} \\ [\mathbf{F}_{\mathbf{DC}}] \end{bmatrix} = - \begin{bmatrix} \mathbf{J}_{\mathbf{RR}_1} & \cdots & \mathbf{0} & \mathbf{0} \cdots \mathbf{0} & & & \\ \vdots & \ddots & \vdots & \vdots \ \ddots \ \vdots & & & [\mathbf{J}_{\mathbf{R0}}] \\ \mathbf{0} & \cdots & \mathbf{J}_{\mathbf{RR}_n} & \mathbf{0} \cdots \mathbf{0} & & & \\ \mathbf{0} \cdots \mathbf{0} & & & \mathbf{J}_{\mathbf{II}_1} & \cdots & \mathbf{0} & \\ \vdots \ \ddots \ \vdots & & & \vdots & \ddots & \vdots & [\mathbf{J}_{\mathbf{I0}}] \\ \mathbf{0} \cdots \mathbf{0} & & & \mathbf{0} & \cdots & \mathbf{J}_{\mathbf{II}_m} & \\ & [\mathbf{J}_{\mathbf{0R}}] & & & [\mathbf{J}_{\mathbf{0I}}] & & [\mathbf{J}_{\mathbf{00}}] \end{bmatrix} \begin{bmatrix} \mathbf{\Delta}_{\mathbf{VSC},\mathbf{R}_1} \\ \vdots \\ \mathbf{\Delta}_{\mathbf{VSC},\mathbf{R}_n} \\ \mathbf{\Delta}_{\mathbf{VSC},\mathbf{I}_1} \\ \vdots \\ \mathbf{\Delta}_{\mathbf{VSC},\mathbf{I}_m} \\ [\mathbf{\Delta E}_{\mathbf{DC}}] \end{bmatrix} \tag{3.108}$$

In this expression, each one of the n mismatch terms $\mathbf{F}$ with subscript VSC,R and VSC,I takes the form of the vector expressions (3.88)–(3.90) depending on the type of converter it is, namely VSC_{Slack}, VSC_{Psch} or VSC_{Pass}. Likewise, the vectors of incremental state variables $\mathbf{\Delta\Phi}$ take the form of vectors (3.91)–(3.93). By the same token, matrices $\mathbf{J}$ with subscripts RR and II take the form of matrices (3.85)–(3.87).

The vectors and matrices of higher dimensionality that appear in (3.108) have the following structure:

$$[\mathbf{J}_{\mathbf{R0}}] = \begin{bmatrix} \mathbf{J}_{\mathbf{R0}_1} & \cdots & \mathbf{0} & \mathbf{0} \cdots \mathbf{0} \\ \vdots & \ddots & \vdots & \vdots \ \ddots \ \vdots \\ \mathbf{0} & \cdots & \mathbf{J}_{\mathbf{R0}_n} & \mathbf{0} \cdots \mathbf{0} \end{bmatrix} \tag{3.109}$$

$$[\mathbf{J}_{\mathbf{I0}}] = \begin{bmatrix} \mathbf{0} \cdots \mathbf{0} & \mathbf{J}_{\mathbf{I0}_1} & \cdots & \mathbf{0} \\ \vdots \ \ddots \ \vdots & \vdots & \ddots & \vdots \\ \mathbf{0} \cdots \mathbf{0} & \mathbf{0} & \cdots & \mathbf{J}_{\mathbf{I0}_m} \end{bmatrix} \tag{3.110}$$

$$[\mathbf{J}_{\mathbf{0R}}] = \begin{bmatrix} \mathbf{J}_{\mathbf{0R}_1} & \cdots & \mathbf{0} & \mathbf{0} \cdots \mathbf{0} \\ \vdots & \ddots & \vdots & \vdots \ \ddots \ \vdots \\ \mathbf{0} & \cdots & \mathbf{J}_{\mathbf{0R}_n} & \mathbf{0} \cdots \mathbf{0} \end{bmatrix} \tag{3.111}$$

$$[\mathbf{J}_{\mathbf{0I}}] = \begin{bmatrix} \mathbf{0} \cdots \mathbf{0} & \mathbf{J}_{\mathbf{0I}_1} & \cdots & \mathbf{0} \\ \vdots \ \ddots \ \vdots & \vdots & \ddots & \vdots \\ \mathbf{0} \cdots \mathbf{0} & \mathbf{0} & \cdots & \mathbf{J}_{\mathbf{0I}_m} \end{bmatrix} \tag{3.112}$$

The various entries in (3.109) are vectors with the generic i-th terms given below, depending on whether the converter i is type VSC_{Psch} or type VSC_{Pass}:

$$\mathbf{J}_{\mathbf{RDC}_i} = \left[\partial \Delta P_{vR_i}/\partial E_{DCR_i} \ \ \partial \Delta Q_{vR_i}/\partial E_{DCR_i} \ \ 0 \ \ \partial \Delta Q_{R_i-vR_i}/\partial E_{DCR_i} \ \ \partial \Delta P_{R_i-vR_i}/\partial E_{DCR_i}\right]^t \tag{3.113}$$

$$\mathbf{J}_{\mathbf{RDC}_i} = \left[\partial \Delta P_{vR_i}/\partial E_{DCR_i} \ \ \partial \Delta Q_{vR_i}/\partial E_{DCR_i} \ \ 0 \ \ \partial \Delta Q_{R_i-vR_i}/\partial E_{DCR_i}\right]^t \tag{3.114}$$

In cases when the converter type is VSC_{Slack} (fixed DC voltage), the corresponding vector term would be a zero-padded vector of suitable dimensions and it is removed from (3.109). As implied from this equation, there will be as many of these zero-padded vectors as the number of VSC converters there are in the multi-terminal VSC-HVDC system. However, caution has to be exercised because vectors having a location of the VSC_{Slack}-type converter and DC nodes with a connection to this converter will contain non-zero entries – this situation is well exemplified by (3.94) of the three-terminal example in Section 3.9. Equations (3.113) and (3.114) and the preceding discussion apply to the vector entries in matrix (3.110) by exchanging the subscript **R** by **I**.

The various entries in (3.111) are vectors with the generic i-th terms given below, depending on whether the converter i is type VSC_{Psch} or type VSC_{Pass}:

$$\mathbf{J}_{\mathbf{DCR}_i} = \left[\partial \Delta P_{R_i}/\partial e_{vR_i} \;\; \partial \Delta P_{R_i}/\partial m_{aR_i} \;\; \partial \Delta P_{R_i}/\partial f_{vR_i} \;\; 0 \;\; \partial \Delta P_{R_i}/\partial \varphi_{R_i}\right] \tag{3.115}$$

$$\mathbf{J}_{\mathbf{DCR}_i} = \left[\partial \Delta P_{R_i}/\partial e_{vR_i} \;\; \partial \Delta P_{R_i}/\partial m_{aR_i} \;\; \partial \Delta P_{R_i}/\partial f_{vR_i} \;\; 0\right] \tag{3.116}$$

If the converter type is VSC_{Slack} then the corresponding vector term would be a zero-padded vector of suitable dimensions and it is removed from (3.111). There will be as many of these zero-padded vectors as the number of VSC converters there are in the multi-terminal VSC-HVDC system. Equations (3.115) and (3.116) and the preceding discussion apply to the vector entries in matrix (3.112) by exchanging the subscript **R** for **I**.

The Jacobian matrix of the DC network takes the following general form:

$$[\mathbf{J}_{\mathbf{DC}}] = \begin{bmatrix} J_{1,1} & \cdots & J_{,1n} & J_{1,n+1} & \cdots & J_{1,n+m} \\ \vdots & \ddots & \vdots & \vdots & \ddots & \vdots \\ J_{n,1} & \cdots & J_{n,n} & J_{n,n+1} & \cdots & J_{n,n+m} \\ J_{n+1,1} & \cdots & J_{n+1,n} & J_{n+1,n+1} & \cdots & J_{n+1,n+m} \\ \vdots & \ddots & \vdots & \vdots & \ddots & \vdots \\ J_{n+m,1} & \cdots & J_{n+m,n} & J_{n+m,n+1} & \cdots & J_{n+m,n+m} \end{bmatrix} \tag{3.117}$$

with the row and column corresponding to the VSC_{Slack}-type converter being removed from (3.117).

The remaining terms, corresponding to the DC side of the network, in the general expression (3.108) are:

$$[\mathbf{F}_{\mathbf{DC}}] = \left[\Delta P_{0,1} \;\cdots\; \Delta P_{0,n} \;\; \Delta P_{0,n+1} \;\cdots\; \Delta P_{0,n+m}\right]^t \tag{3.118}$$

$$[\mathbf{\Delta\Phi}_{\mathbf{DC}}] = \left[E_{0,1} \;\cdots\; E_{0,n} \;\; E_{0,n+1} \;\cdots\; E_{0,n+m}\right]^t \tag{3.119}$$

with the entries corresponding to the VSC_{Slack}-type converter being removed from (3.118) and (3.119).

Note that the generalized reference frame is highly sparse and the use of sparsity techniques plays a useful role in actual software implementations.

3.10.1 Multi-Terminal VSC-HVDC System with Common DC Bus Model

In the particular case when there is no DC network but all the converter stations in the multi-terminal VSC-HVDC system are connected back-to-back (i.e. share a common

DC bus), a simplification of (3.108) reduces to the following:

$$\begin{bmatrix} \mathbf{F}_{\mathrm{VSC,R}_1} \\ \vdots \\ \mathbf{F}_{\mathrm{VSC,R}_n} \\ \mathbf{F}_{\mathrm{VSC,I}_1} \\ \vdots \\ \mathbf{F}_{\mathrm{VSC,I}_m} \end{bmatrix} = -\begin{bmatrix} \mathbf{J}_{\mathrm{RR}_1} & \cdots & \mathbf{0} & \mathbf{0} & \cdots & \mathbf{0} \\ \vdots & \ddots & \vdots & \vdots & \ddots & \vdots \\ \mathbf{0} & \cdots & \mathbf{J}_{\mathrm{RR}_n} & \mathbf{0} & \cdots & \mathbf{0} \\ \mathbf{0} & \cdots & \mathbf{0} & \mathbf{J}_{\mathrm{II}_1} & \cdots & \mathbf{0} \\ \vdots & \ddots & \vdots & \vdots & \ddots & \vdots \\ \mathbf{0} & \cdots & \mathbf{0} & \mathbf{0} & \cdots & \mathbf{J}_{\mathrm{II}_m} \end{bmatrix} \begin{bmatrix} \mathbf{\Delta}_{\mathrm{VSC,R}_1} \\ \vdots \\ \mathbf{\Delta}_{\mathrm{VSC,R}_n} \\ \mathbf{\Delta}_{\mathrm{VSC,I}_1} \\ \vdots \\ \mathbf{\Delta}_{\mathrm{VSC,I}_m} \end{bmatrix} \tag{3.120}$$

Notice that the Jacobian matrix is decoupled but that the overall system of equations remains tightly coupled through the constrained branch powers (3.83) which make up the n rectifiers and m inverters in (3.120), namely, $\mathbf{F}_{\mathrm{VSC,R}_i}$ and $\mathbf{F}_{\mathrm{VSC,I}_j}$, with $i = 1 \ldots n$ and $j = 1 \ldots m$.

3.10.2 Unified Solutions of AC-DC Networks

Since the multi-terminal VSC-HVDC system is used to interconnect a number of otherwise independent networks, the linearized form of the overall electrical power system's equations at a given iteration (r) is:

$$\underbrace{\begin{bmatrix} \mathbf{F}_{\mathrm{AC}} \\ \mathbf{F}_{\mathrm{VSC,R}} \\ \vdots \\ \mathbf{F}_{\mathrm{VSC,I}} \\ \vdots \\ \mathbf{F}_{\mathrm{DC}} \end{bmatrix}^{(r)}}_{\mathbf{F}^{(r)}_{\mathrm{AC/DC}}} = -\begin{bmatrix} \\ & \mathbf{J}_{\mathrm{AC/DC}} & \\ \\ \end{bmatrix}^{(r)} \underbrace{\begin{bmatrix} \Delta\mathbf{\Psi}_{\mathrm{AC}} \\ \Delta\mathbf{\Phi}_{\mathrm{VSC,R}} \\ \vdots \\ \Delta\mathbf{\Phi}_{\mathrm{VSC,I}} \\ \vdots \\ \Delta\mathbf{E}_{\mathrm{DC}} \end{bmatrix}^{(r)}}_{\Delta\mathbf{\Psi}^{(r)}_{\mathrm{AC/DC}}} \tag{3.121}$$

where $\mathbf{J}_{\mathrm{AC/DC}}$ accommodates the first-order partial derivatives of all the mismatches of the AC networks $\mathbf{F}_{\mathrm{AC}}$, the VSCs $\mathbf{F}_{\mathrm{VSC,R}}$ and $\mathbf{F}_{\mathrm{VSC,I}}$ and the DC network $\mathbf{F}_{\mathrm{DC}}$, with respect to all the state variables involved. The mismatches relating to any FACTS device that may exist in any of the AC power grids [4] would be included in $\mathbf{F}_{\mathrm{AC}}$. Correspondingly, $\Delta\mathbf{\Psi}_{AC}$ is a vector that accommodates all the AC networks' state variables, including those pertaining to any FACTS device.

It should be remarked that the manner in which the VSCs of the multi-terminal VSC-HVDC system are being modelled and assembled in an all-encompassing frame-of-reference, gives rise to a simultaneous, true unified solution of the AC-DC system, as illustrated in the flow diagram in Figure 3.18.

3.10.3 Unified vs Quasi-Unified Power Flow Solutions

The flow diagram in Figure 3.18 gives the overall solution procedure for a generic AC-DC power grid. This is in connection with Eq. (1.121); it corresponds to a unified power flow solution using the Newton-Raphson method exhibiting quadratic convergence. A typical solution converges in four and five iterations to satisfy convergence tolerances of $\varepsilon = 1e{-}6$ and $\varepsilon = 1e{-}12$, respectively. This is regardless of the number and size of AC and DC subsystems present in the overall AC-DC power grid, provided the iterative solution is properly initialized.

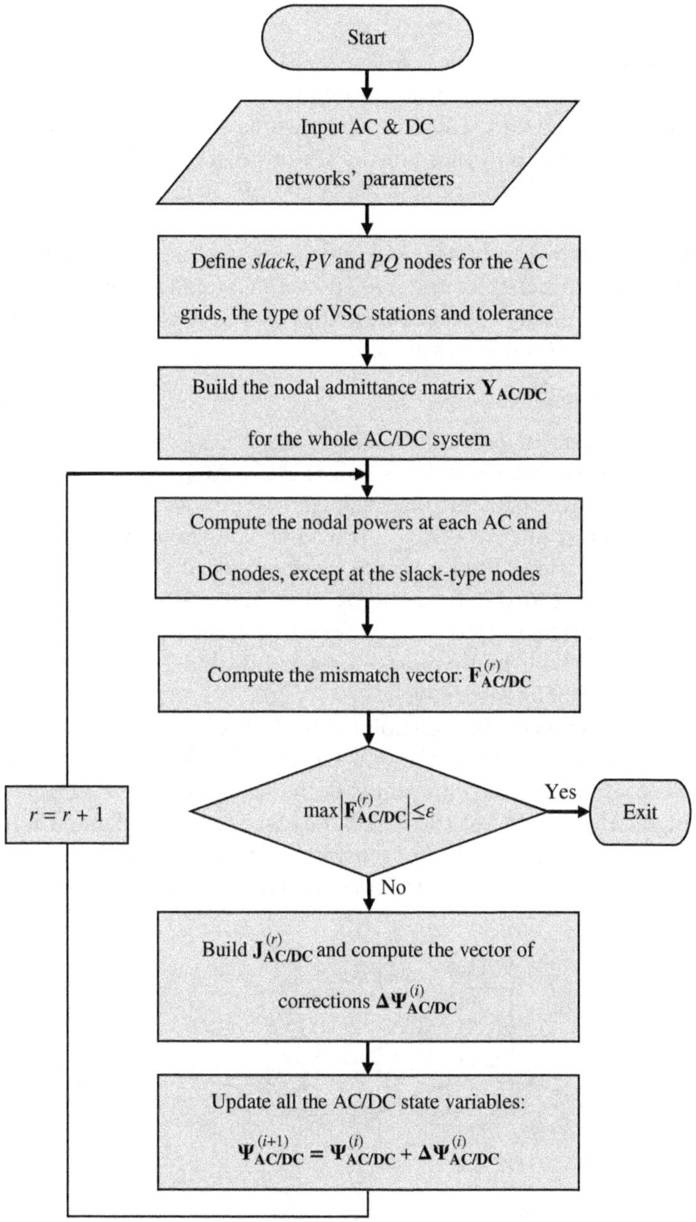

Figure 3.18 Flow diagram of a true unified solution of the multi-terminal VSC-HVDC system. Source: ©ELSEVIER, 2016.

As expected, the quadratic convergence characteristics of this method outperform the convergence characteristics of the sequential and quasi-unified power flow solutions of multi-terminal VSC-HVDC systems, which have been published in the open literature in recent years [10, 16].

3.10.4 Test Case 9

The six-terminal VSC-HVDC network shown in Figure 3.19 is used to illustrate the applicability of the frame-of-reference. It is meant to signify a stage of development of the undersea power grid in the North Sea [18]. The AC networks are taken to represent the UK grid, the NORDEL grid and the Union for the Co-ordination of Transmission of Electricity (UCTE) grid, with total power demands of 60 GW, 30 GW and 450 GW, respectively.

For the purpose of this illustrative test case, and with no loss of generality for the frame-of-reference presented in this chapter, these AC power networks are represented by their Thevenin equivalents coupled to reactive ties. The equivalent is determined by Eq. (3.122), where a power factor $pf = 0.95$ is selected.

$$R_{th} + jX_{th} = [P(1 + j\tan(\cos^{-1}(pf)))]^{-1} \tag{3.122}$$

Table 3.11 gives the parameters in per-unit values of the VSCs, DC ring cables, Thevenin equivalents and the reactive ties for systems AC_1, AC_3 and AC_5. Their respective slack generators control their terminal voltage magnitudes at 1.0 p.u. The VSC_a is selected to be a converter of type VSC_{Slack} providing voltage regulation at its DC bus at $E_{DCnom} = 2$ p.u. The stations VSC_c and VSC_e are modelled as converters of type VSC_{Psch} and set to draw each 250 MW from the DC ring. The stations VSC_b, VSC_d and VSC_f are modelled as converters of type VSC_{Pass}. The LTC transformers are set to uphold the voltage magnitude at their high-voltage buses at 1.05 and 1 p.u. for (VSC_a, VSC_c, VSC_e) and (VSC_b, VSC_d, VSC_f), respectively.

The bus of network AC_2 represents the in-feed point of the Valhall oil platform, which lies 294 km off the Norwegian coast and has a power demand of 78 MW. The buses of networks AC_4 and AC_6 represent the collector points of the wind parks at the Dogger Bank and the German Bight. The former lies 100 km off the east coast of England, where permission has been granted for a 7.2 GW of wind generation development; for the purpose of this test case, a power injection of 400 MW is assumed. The latter is a wind park

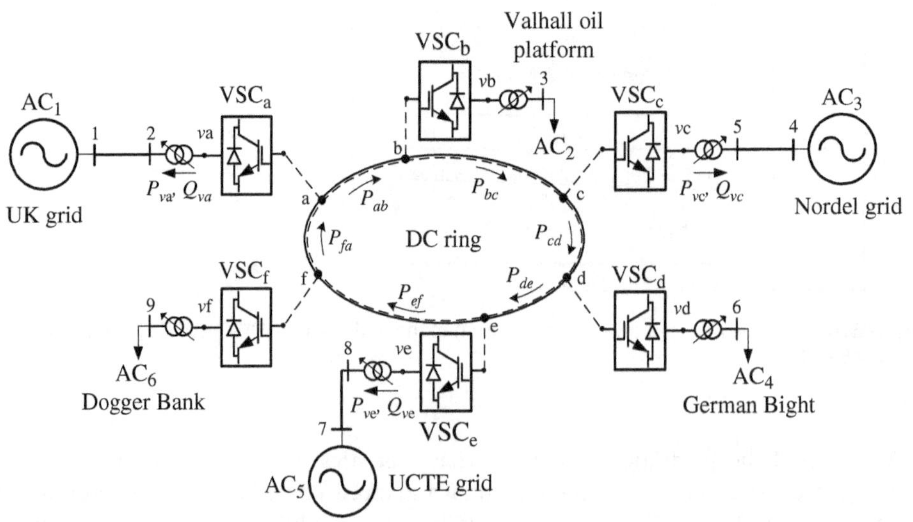

Figure 3.19 Multi-terminal VSC-HVDC system with a DC ring. Source: ©IEEE, 2016.

Table 3.11 Parameters of the VSCs, AC_1, AC_3, AC_5 and DC networks, with electrical parameters in p.u. and distances in km.

VSCs and LTCs						
G_0	Z_1	Y_{filter}		Z_{reac}	Z_T	
0.01	1e-4 + j0.001	0 + j0.20		0.0015 + j0.15	0.006 + j0.06	
AC networks						
Network	V_{Th}	R_{Th}		X_{Th}	*Tie line*	
AC_1	1.00	1.504e-3		4.943e-4	$X_s = 0.06$	
AC_2	1.00	3.0e-3		9.887e-4	$X_s = 0.06$	
AC_3	1.00	2.005e-4		6.591e-5	$X_s = 0.06$	
Length (km) and resistance of the DC ring sections						
	a-b	*b-c*	*c-d*	*d-e*	*e-f*	*f-a*
Length	300	300	250	200	300	250
R_{DC}	0.0058	0.0058	0.0048	0.0039	0.0058	0.0048

that lies 100 km off the German coast and comprises 80 5 MW wind turbines, thus with a total power injection of 400 MW. Hence, given that the networks AC_2, AC_4 and AC_6 possess no synchronous generation of their own, the stations VSC_b, VSC_d and VSC_f are modelled as converters of type VSC_{Pass}. The 1600 km length submarine cables making up the DC ring are taken to be rated at ±160 kV and to have a resistance value of 0.02 Ω km^{-1}. The nominal powers of the converters are 500 MVA.

The power flow simulation results are provided in Tables 3.12–3.14. As expected, the convergence to tolerances of 10^{-6} and 10^{-12} were reached in four and five iterations, respectively, this being the hallmark of a true unified iterative solution using the power flow Newton-Raphson method [19].

Table 3.12 gives a summary of the values of the VSCs' state variables. Power is being injected into the DC ring through converters VSC_d and VSC_f and it is observed that their respective DC voltages are higher than the reference voltage provided by VSC_a. The angle φ of the converters VSC_b, VSC_d and VSC_f has been kept to zero since these

Table 3.12 State variables solution per VSC.

VSC No	VSC Type	E_{DC} (p.u.)	m_a	φ (deg)	LTC tap	P_{loss} (MW)
a	VSC_{Slack}	2.0000	0.8567	11.3806	1.0127	0.7641
b	VSC_{Pass}	1.9996	0.7927	0.0	1.0009	0.3029
c	VSC_{Psch}	1.9997	0.8573	13.4952	1.0129	1.0699
d	VSC_{Pass}	2.0010	0.8012	0.0	0.9963	3.1377
e	VSC_{Psch}	2.0005	0.8600	13.4811	1.0143	1.0684
f	VSC_{Pass}	2.0013	0.8011	0.0	0.9963	3.1383

Convergence: $\varepsilon = 10^{-6}$ takes four iterations and $\varepsilon = 10^{-12}$ takes five iterations.

Table 3.13 Voltages and powers injected at the AC networks.

Network	V (p.u.)	θ	P_{inj} (MW)	Q_{inj} (MVAR)
AC_1	1.05	6.8625	209.7336	94.0067
AC_2	1.00	−1.9359	78	0.0
AC_3	1.05	8.1370	248.1699	91.2079
AC_4	1.00	9.7335	−400	0.0
AC_5	1.05	8.1492	248.1436	104.2285
AC_6	1.00	9.7335	−400	0.0

Table 3.14 Power flows in the DC ring.

DC cables	P_{DC} (MW)		P_{loss} (MW)
	send-rec	rec-send	
a–b	59.0374	−59.0272	0.0101
b–c	−19.3486	19.3497	0.0010
c–d	−269.3497	269.5239	0.1741
d–e	125.4183	−125.3877	0.0306
e–f	−124.6122	124.6572	0.0450
f–a	270.2843	−270.1092	0.1751

converters provide the angular references for networks AC_2, AC_4 and AC_6. Notice that there is a marked difference between the modulation ratios of the group of converters (VSC_a, VSC_c, VSC_e) and the group of converters (VSC_b, VSC_d, VSC_f). This is due to the different voltages selected for each group.

The power loss incurred by each VSC is also reported in Table 3.12. Converters VSC_d and VSC_f are the stations that incur more power loss whereas VSC_b is the unit that incurs less power loss since this converter is the one that draws less power from the DC ring. The nodal voltages, active and reactive powers injected at each AC network terminals are reported in Table 3.13. It should be noted that more than 90 MVAR are injected by the LTCs of VSC_a, VSC_c, VSC_e to uphold the target values of 1.05 p.u. at AC_1, AC_3 and AC_5, respectively. Given that both VSC_c and VSC_e are set to draw 250 MW from the DC ring, the remaining amount of power that flows through VSC_a towards AC_1 stands at 209.7336 MW.

The scheduled power control of a converter type VSC_{Psch} applies at its DC bus; hence, the power delivered at its AC terminal is expected to be smaller than its corresponding DC power P^{sch} owing to the power loss incurred in the converter. In this test case, the powers delivered to AC_3 and AC_5 stand at 248.1699 and 248.1436 MW, respectively. Also, it should be noticed that the power injected by the LTCs at AC_4 and AC_6 carry a negative sign, which correctly accounts for the fact that the powers generated by the wind parks are being injected into the DC ring. The opposite occurs for the case of the oil platform, which draws power from the DC grid at the terminals of the network AC_2.

The power flows in the DC ring are listed in Table 3.14. It is noted that the ring sectors that are the most heavily loaded are the cables connected between nodes c–d and f–a, each carrying approximately 269 MW and 270 MW, respectively. Conversely, the cable sections of the DC ring that carry less power are the sections a–b and b–c since they link to converter VSC_b, which, in turn, is the converter that draws less power from the DC ring. The total power loss in the DC ring stands at 0.4361 MW.

3.11 Conclusions

The power flow Newton-Raphson method is one of the most widely used power systems application tools in today's electrical power industry. An in-depth study of the power flow Newton-Raphson method using rectangular coordinates has been presented in this chapter, starting from first principles.

The STATCOM is a key element of the FACTS technology. It is the modern counterpart of the well-established SVC and forms the basic building block with which other more advanced FACTS equipment may be built, such as the UPFC and the various forms of VSC-HVDC links. Indeed, the latter application has blurred the line between the FACTS and HVDC transmission options. Accordingly, STATCOM models aimed at power flow solutions using the Newton-Raphson method have been introduced. The use of two STATCOM models linked on their DC sides through a cable has yielded a versatile and comprehensive model of the point-to-point VSC-HVDC configuration. When the cable's length is zero, the model becomes that of a back-to-back VSC-HVDC link. This model represents a paradigm shift in the way the fundamental frequency, positive sequence modelling of VSC-HVDC links is carried out. Extension of the point-to-point VSC-HVDC link model to include one more VSC station has enabled the representation of the three-terminal VSC-HVDC system, where three types of VSC converters have been defined, according to their operation. This has elucidated the theoretical framework on which the multi-terminal VSC-HVDC system stands. The frame-of-reference corresponds to the Jacobian equation of a generic AC-DC power flow solution using the Newton-Raphson method, which exhibits quadratic convergence.

References

1 Stagg, G.W. and El-Abiad, A.H. (1968). *Computer Methods in Power Systems Analysis*. McGraw-Hill.

2 Arrillaga, J. and Watson, N.R. (2004). *Computer Modelling of Electrical Power Systems*. John Wiley & Sons.

3 Tinney, W.F. and Hart, C.E. (1967). Power flow solution by Newton's method. *IEEE Transactions on Power Apparatus and Systems* PAS-86 (11): 1449–1460.

4 Acha, E., Fuerte-Esquivel, C.R., Ambriz-Perez, H., and Angeles-Camacho, C. (2004). *FACTS Modelling and Simulation in Power Networks*. John Wiley & Sons.

5 Acha, E., Kazemtabrizi, B., and Castro, L.M. (2013). A new VSC-HVDC model for power flows using the Newton-Raphson method. *IEEE Transactions on Power Systems* 28 (3): 2602–2612.

6 Phadke, A.G. and Thorp, J.S. (2008). *Synchronized Phasor Measurements and Their Applications*. Springer.

7 Acha, E. and Kazemtabrizi, B. (2013). A new STATCOM model for power flows using the Newton-Raphson method. *IEEE Transactions on Power Systems* 28 (3): 2455–2465.

8 Hingorani, N.G. and Gyugyi, N. (1999). *Understanding FACTS: Concepts and Technology of Flexible AC Transmission Systems*. Wiley–IEEE Press.

9 Nilsson, J.W. and Riedel, S. (2010). *Electric Circuits*. Prentice Hall.

10 Baradar, M. and Ghandhari, M. (2013). A multi-option unified power flow approach for hybrid AC/DC grids incorporating multi-terminal VSC-HVDC. *IEEE Transactions on Power Systems* 28 (3): 2376–2383.

11 Arrillaga, J. (1998). *High Voltage Direct Current Transmission*. IET.

12 Angeles-Camacho, C., Tortelli, O., Acha, E., and Fuerte-Esquivel, C.R. (2003). Inclusion of a high voltage DC-voltage source converter model in a Newton-Raphson power flow algorithm. *IEE Proceedings: Generation Transmission and Distribution* 150: 691–696.

13 Zhang, X.P. (2004). Multi-terminal voltage-sourced converter-based HVDC models for power flow analysis. *IEEE Transactions Power Systems* 19 (4): 1877–1884.

14 de la Villa, A., Acha, E., and Gomez Exposito, A. (2008). Voltage source converter modelling for power system state estimation: STATCOM and VSC-HVDC. *IEEE Transactions Power Systems* 23 (4): 1552–1559.

15 Pizano-Martinez, A., Fuerte-Esquivel, C.R., Ambriz-Perez, H., and Acha, E. (2007). Modelling of VSC-HVDC systems for Newton-Raphson OPF algorithms. *IEEE Transactions Power Systems* 22 (4): 1794–1803.

16 J. Beerten, S. Cole and R. Belmans (2010) "A Sequential AC/DC power flow algorithm for networks containing multi-terminal VSC HVDC systems", in Proc. Conf. IEEE PES Winter Meeting 2010, Minneapolis, 25–29 July 2010, pp. 1–7.

17 Acha, E. and Castro, L.M. (2016). A generalized frame of reference for the incorporation of multi-terminal VSC-HVDC systems in power flow solutions. *Electric Power Systems Research* 136: 415–424.

18 T.K. Vrana, R.E. Torres-Olguin, B. Liu and T.M. Haileselassie, "The North Sea Super Grid – A Technical Perspective", 9th IET Int. Conf. AC and DC Power Trans., 19–21 October 2010, London, pp. 1–5.

19 Fuerte-Esquivel, C.R. and Acha, E. (1997). A Newton-type algorithm for the control of power flow in electrical power networks. *IEEE Transactions on Power Systems* 12 (4): 1474–1480.

3.A Appendix

Using the nodal admittance matrix of the VSC in (3.39), expressions for the real and imaginary parts of the currents at nodes vR and 0 may be determined and given below:

$$\begin{bmatrix} CR_{vR} \\ CI_{vR} \\ CR_0 \\ CI_0 \end{bmatrix} = \begin{bmatrix} (G_1 e_{vR} - B_1 f_{vR}) - k_1 m_a \{G_1 E_{DC} \cos\varphi - B_1 E_{DC} \sin\varphi\} \\ (G_1 e_{vR} + B_1 f_{vR}) - k_1 m_a \{G_1 E_{DC} \sin\varphi + B_1 E_{DC} \cos\varphi\} \\ (k_2 [CR_{vR}^2 + CI_{vR}^2] + k_1^2 m_a^2 G_1) E_{DC} - k_1 m_a \{(G_1 e_{vR} - B_1 f_{vR}) \cos\varphi + (G_1 f_{vR} + B_1 e_{vR}) \sin\varphi\} \\ k_1^2 m_a^2 (B_1 + B_{eq}) E_{DC} - k_1 m_a \{-(G_1 e_{vR} - B_1 f_{vR}) \sin\varphi + (G_1 f_{vR} + B_1 e_{vR}) \cos\varphi\} \end{bmatrix} \tag{3.A.1}$$

From which the partial derivatives of the four currents with respect to all the relevant state variables are obtained:

$$\partial CR_{vR}/\partial e_{vR} = G_2$$
$$\partial CR_{vR}/\partial f_{vR} = -B_2$$
$$\partial CR_{vR}/\partial E_{DC} = -k_1 m_a\{G_2\cos\varphi - B_2\sin\varphi\}$$
$$\text{where } \partial CR_{vR}/\partial E_{DC} = 0 \text{ if } E_{DC} \text{ is fixed}$$
$$\partial CR_{vR}/\partial m_a = -k_1 E_{DC}\{G_2\cos\varphi - B_2\sin\varphi\}$$
$$\partial CR_{vR}/\partial\varphi = k_1 m_a E_{DC}\{G_2\sin\varphi + B_2\cos\varphi\}$$
$$\partial CR_{vR}/\partial B_{eq} = 0$$

$$\partial CI_{vR}/\partial e_{vR} = B_2$$
$$\partial CI_{vR}/\partial f_{vR} = G_2$$
$$\partial CI_{vR}/\partial E_{DC} = -k_1 m_a\{G_2\sin\varphi + B_2\cos\varphi\}$$
$$\text{where } \partial CI_{vR}/\partial E_{DC} = 0 \text{ if } E_{DC} \text{ is fixed}$$
$$\partial CI_{vR}/\partial m_a = -k_1 E_{DC}\{G_2\sin\varphi + B_2\cos\varphi\}$$
$$\partial CI_{vR}/\partial\varphi = -k_1 m_a E_{DC}\{G_2\cos\varphi - B_2\sin\varphi\}$$
$$\partial CI_{vR}/\partial B_{eq} = 0$$

$$\partial CR_0/\partial e_{vR} = -k_1 m_a\{G_2\cos\varphi + B_2\sin\varphi\} + 2k_2E_{DC}(CR_{vR}\cdot\partial CR_{vR}/\partial e_{vR} + CI_{vR}\cdot\partial CI_{vR}/\partial e_{vR})$$
$$\partial CR_0/\partial f_{vR} = k_1 m_a\{B_2\cos\varphi - G_2\sin\varphi\} + 2k_2E_{DC}(CR_{vR}\cdot\partial CR_{vR}/\partial f_{vR} + CI_{vR}\cdot\partial CI_{vR}/\partial f_{vR})$$
$$\partial CR_0/\partial E_{DC} = k_2[CR_2^2 + CI_2^2] + k_1^2 m_a^2 G_2 + 2k_2E_{DC}(CR_{vR}\cdot\partial CR_{vR}/\partial E_{DC} + CI_{vR}\cdot\partial CI_{vR}/\partial E_{DC})$$
$$\text{where } \partial CR_0/\partial E_{DC} = 0 \text{ if } E_{DC} \text{ is fixed}$$
$$\partial CR_0/\partial m_a = 2k_1^2 m_a G_2 E_{DC} - k_1\{(G_2e_{vR} - B_2f_{vR})\cos\varphi + (G_2f_{vR} + B_2e_{vR})\sin\varphi\} + 2k_2E_{DC}(CR_{vR}\cdot\partial CR_{vR}/\partial m_a + CI_{vR}\cdot\partial CI_{vR}/\partial m_a)$$
$$\partial CR_0/\partial\varphi = k_1 m_a\{(G_2e_{vR} - B_2f_{vR})\sin\varphi - (G_2f_{vR} + B_2e_{vR})\cos\varphi\} + 2k_2E_{DC}(CR_{vR}\cdot\partial CR_{vR}/\partial\varphi + CI_{vR}\cdot\partial CI_{vR}/\partial\varphi)$$
$$\partial CR_0/\partial B_{eq} = 0$$

$$\partial CI_0/\partial e_{vR} = k_1 m_a\{G_1\sin\varphi - B_1\cos\varphi\}$$
$$\partial CI_0/\partial f_{vR} = -k_1 m_a\{B_1\sin\varphi + G_1\cos\varphi\}$$
$$\partial CI_0/\partial E_{DC} = k_1^2 m_a^2(B_1 + B_{eq}) \text{ where } \partial CI_0/\partial E_{DC} = 0 \text{ if } E_{DC} \text{ is fixed}$$
$$\partial CI_0/\partial m_a = 2k_1^2 m_a(B_1 + B_{eq})E_{DC} + k_1\{(G_1e_{vR} - B_1f_{vR})\sin\varphi - (G_1f_{vR} + B_1e_{vR})\cos\varphi\}$$
$$\partial CI_0/\partial\varphi = k_1 m_a\{(G_1e_{vR} - B_1f_{vR})\cos\varphi + (G_1f_{vR} + B_1e_{vR})\sin\varphi\}$$
$$\partial CI_0/\partial B_{eq} = k_1^2 m_a^2 E_{DC}$$

3.B Appendix

With reference to the nodal admittance matrix of the LTC in (3.50), expressions for the real and imaginary parts of the currents at nodes k and vR may be determined and given below:

$$\begin{bmatrix} CR_k + jCI_k \\ CR_{vR} + jCI_{vR} \end{bmatrix} = \begin{bmatrix} (G_l + jB_l) & -T(G_l + jB_l) \\ -T(G_l + jB_l) & T^2(G_l + jB_l) \end{bmatrix} \begin{bmatrix} e_k + jf_k \\ e_{vR} + jf_{vR} \end{bmatrix} \tag{3.B.1}$$

$$\begin{bmatrix} CR_k \\ CI_k \\ CR_{vR} \\ CI_{vR} \end{bmatrix} = \begin{bmatrix} (G_l e_k - B_l f_k) - T(G_l e_{vR} - B_l f_{vR}) \\ (G_l f_k + B_l e_k) - T(G_l f_{vR} + B_l e_{vR}) \\ -T(G_l e_k - B_l f_k) + T^2(G_l e_{vR} - B_l f_{vR}) \\ -T(G_l f_k + B_l e_k) + T^2(G_l f_{vR} + B_l e_{vR}) \end{bmatrix} \tag{3.B.2}$$

From which expressions for the partial derivatives of the real and imaginary parts of the nodal currents with respect the real and imaginary parts of the nodal voltages are as follows:

$\partial CR_k/\partial e_k = G_l$

$\partial CR_k/\partial f_k = -B_l$

$\partial CR_k/\partial e_{vR} = -TG_l$

$\partial CR_k/\partial f_{vR} = TB_l$

$\partial CR_k/\partial T = -(G_l e_{vR} - B_l f_{vR})$

$\partial CI_k/\partial e_k = B_l$

$\partial CI_k/\partial f_k = G_l$

$\partial CI_k/\partial e_{vR} = -TB_l$

$\partial CI_k/\partial f_{vR} = -TG_l$

$\partial CI_k/\partial T = -(G_l f_{vR} + B_l e_{vR})$

$\partial CR_{vR}/\partial e_k = -TG_l$

$\partial CR_{vR}/\partial f_k = TB_l$

$\partial CR_{vR}/\partial e_{vR} = T^2 G_l$

$\partial CR_{vR}/\partial f_{vR} = -T^2 B_l$

$\partial CR_{vR}/\partial T = 2T(G_l e_{vR} - B_l f_{vR}) - (G_l e_k - B_l f_k)$

$\partial CI_{vR}/\partial e_k = -TB_l$

$\partial CI_{vR}/\partial f_k = -TG_l$

$\partial CI_{vR}/\partial e_{vR} = T^2 B_l$

$\partial CI_{vR}/\partial f_{vR} = T^2 G_l$

$\partial CI_{vR}/\partial T = 2T\{G_l f_{vR} + B_l e_{vR}\} - (G_l f_2 + B_l e_2)$

The nodal active and reactive powers at the sending and receiving ends of the LTC transformer, are given below:

$$\begin{bmatrix} P_2 + jQ_2 \\ P_3 + jQ_3 \end{bmatrix} = \begin{bmatrix} e_2 + jf_2 & 0 \\ 0 & e_3 + jf_3 \end{bmatrix} \begin{bmatrix} CR_2 - jCI_2 \\ CR_3 - jCI_3 \end{bmatrix} \Rightarrow \begin{bmatrix} P_2 \\ Q_2 \\ P_3 \\ Q_3 \end{bmatrix} = \begin{bmatrix} e_2 CR_2 + f_2 CI_2 \\ f_2 CR_2 - e_2 CI_2 \\ e_3 CR_3 + f_3 CI_3 \\ f_3 CR_3 - e_3 CI_3 \end{bmatrix} \tag{3.B.3}$$

These power expressions, in mismatch nodal power form, similar to expressions (3.18) and (3.19), are linearized around a base operating point to yield:

$$\begin{bmatrix} \Delta P_k \\ \Delta Q_k \\ \Delta P_{vR} \\ \Delta Q_{vR} \\ \Delta E_k \end{bmatrix} = \begin{bmatrix} \partial P_k/\partial e_k & \partial P_k/\partial f_k & \partial P_k/\partial e_{vR} & \partial P_k/\partial f_{vR} & \partial P_k/\partial T \\ \partial Q_k/\partial e_k & \partial Q_k/\partial f_k & \partial Q_k/\partial e_{vR} & \partial Q_k/\partial f_{vR} & \partial Q_k/\partial T \\ \partial P_{vR}/\partial e_k & \partial P_{vR}/\partial f_k & \partial P_{vR}/\partial e_{vR} & \partial P_{vR}/\partial f_{vR} & \partial P_{vR}/\partial T \\ \partial Q_{vR}/\partial e_k & \partial Q_{vR}/\partial f_k & \partial Q_{vR}/\partial e_{vR} & \partial Q_{vR}/\partial f_{vR} & \partial Q_{vR}/\partial T \\ \partial E_k/\partial e_k & \partial E_k/\partial f_k & 0 & 0 & 0 \end{bmatrix} \begin{bmatrix} \Delta e_k \\ \Delta f_k \\ \Delta e_{vR} \\ \Delta f_{vR} \\ \Delta T \end{bmatrix} \tag{3.B.4}$$

Notice that the voltage expression (3.26) provides the fifth mismatch equation in (3.B.4). The individual Jacobian entries are given below:

$$\frac{\partial P_k}{\partial e_k} = CR_k + e_k\frac{\partial}{\partial e_k}CR_k + f_k\frac{\partial}{\partial e_k}CI_k$$
$$\frac{\partial P_k}{\partial f_k} = CI_k + e_k\frac{\partial}{\partial f_k}CR_k + f_k\frac{\partial}{\partial f_k}CI_k$$
$$\frac{\partial P_k}{\partial e_{vR}} = e_k\frac{\partial}{\partial e_{vR}}CR_k + f_k\frac{\partial}{\partial e_{vR}}CI_k$$
$$\frac{\partial P_k}{\partial f_{vR}} = e_k\frac{\partial}{\partial f_{vR}}CR_k + f_k\frac{\partial}{\partial f_{vR}}CI_k$$

$$\frac{\partial Q_k}{\partial e_k} = -CI_k + f_k\frac{\partial}{\partial e_k}CR_k - e_k\frac{\partial}{\partial e_k}CI_k$$
$$\frac{\partial Q_k}{\partial f_k} = CR_k + f_k\frac{\partial}{\partial f_k}CR_k - e_k\frac{\partial}{\partial f_k}CI_k$$
$$\frac{\partial Q_k}{\partial e_{vR}} = f_k\frac{\partial}{\partial e_{vR}}CR_k - e_k\frac{\partial}{\partial e_{vR}}CI_k$$
$$\frac{\partial Q_k}{\partial f_{vR}} = f_k\frac{\partial}{\partial f_{vR}}CR_k - e_k\frac{\partial}{\partial f_{vR}}CI_k$$

$$\frac{\partial P_{vR}}{\partial e_k} = e_{vR}\frac{\partial}{\partial e_k}CR_{vR} + f_{vR}\frac{\partial}{\partial e_k}CI_{vR}$$
$$\frac{\partial P_{vR}}{\partial f_k} = e_{vR}\frac{\partial}{\partial f_k}CR_{vR} + f_{vR}\frac{\partial}{\partial f_k}CI_{vR}$$
$$\frac{\partial P_{vR}}{\partial e_{vR}} = CR_{vR} + e_{vR}\frac{\partial}{\partial e_{vR}}CR_{vR} + f_{vR}\frac{\partial}{\partial e_{vR}}CI_{vR}$$
$$\frac{\partial P_{vR}}{\partial f_{vR}} = CI_{vR} + e_{vR}\frac{\partial}{\partial f_{vR}}CR_{vR} + f_{vR}\frac{\partial}{\partial f_{vR}}CI_{vR}$$

$$\frac{\partial Q_{vR}}{\partial e_k} = f_{vR}\frac{\partial}{\partial e_k}CR_{vR} - e_{vR}\frac{\partial}{\partial e_k}CI_{vR}$$
$$\frac{\partial Q_{vR}}{\partial f_k} = f_{vR}\frac{\partial}{\partial f_k}CR_{vR} - e_{vR}\frac{\partial}{\partial f_k}CI_{vR}$$
$$\frac{\partial Q_{vR}}{\partial e_{vR}} = -CI_{vR} + f_{vR}\frac{\partial}{\partial e_{vR}}CR_{vR} - e_{vR}\frac{\partial}{\partial e_{vR}}CI_{vR}$$
$$\frac{\partial Q_{vR}}{\partial f_{vR}} = CR_{vR} + f_{vR}\frac{\partial}{\partial f_{vR}}CR_{vR} - e_{vR}\frac{\partial}{\partial f_{vR}}CI_{vR}$$

$$\frac{\partial P_k}{\partial T} = e_k\frac{\partial}{\partial T}CR_k + f_k\frac{\partial}{\partial T}CI_k$$
$$\frac{\partial Q_k}{\partial T} = f_k\frac{\partial}{\partial T}CR_k - e_k\frac{\partial}{\partial T}CI_k$$
$$\frac{\partial P_{vR}}{\partial T} = e_{vR}\frac{\partial}{\partial T}CR_{vR} + f_{vR}\frac{\partial}{\partial T}CI_{vR}$$
$$\frac{\partial Q_{vR}}{\partial f_3} = f_{vR}\frac{\partial}{\partial T}CR_{vR} - e_{vR}\frac{\partial}{\partial T}CI_{vR}$$

4

Optimal Power Flows

4.1 Introduction

The power flow application presented in Chapter 3 and the optimal power flow (OPF) application addressed in this chapter share one common objective: to assess the steady-state operating conditions of an electrical power network operating under a fixed set of generation and load pattern, at a given point in time.

However, the objectives and applicability of the OPF application are far wider than those of conventional power flows. Normally, the OPF is used by transmission system operators (TSOs) for planning the short-term operation of the power network ahead of real-time operation subject to various technical, economic and grid security constraints [1, 2]. In this context, the TSO solves successive instances of OPF scenarios to generate the most economic power dispatch – ahead of real-time operation – in a power system undergoing energy price variations in real-time markets while adhering to transmission network constraints [2]. The application of OPF as a forward operational planning tool for near-real-time planning is shown in Figure 4.1, within the context of an overall framework for transmission system planning. It spans from several hours to minutes, ahead of real-time operation, depending on system operational requirements/conditions.

Today's power grids are highly complex systems, which if not properly planned and operated are very prone to instabilities due to the rather stressing conditions to which they are subjected and a wide range of unforeseen events. In this context, the OPF is used as a reliable power system analysis tool for determining the most suitable scenarios that ensure the continued safe, reliable and economic operation of the system. Problems such as frequency events due to imbalances between generation and demand may be prevented by running a series of security-constraint OPF problems, which ensure that all the undesirable network events are properly accounted for, with necessary security measures in place to mitigate their ensuing effects [1].

Arguably, the conventional power flows when constrained by system security limits will produce, if applicable, feasible solutions. However, such solutions are not necessarily optimal since the conventional power flows do not incorporate an objective function. Objective functions, such as generators' cost of active power dispatch, or transmission losses will vary under different operating conditions and therefore, unlike conventional power flow solutions, OPF produce solutions that are both technically feasible and *optimal*. As a result of this, a seemingly comparable power flow solution and an optimal power solution may not necessarily agree with each other [3].

VSC-FACTS-HVDC: Analysis, Modelling and Simulation in Power Grids, First Edition.
Enrique Acha, Pedro Roncero-Sánchez, Antonio de la Villa Jaén, Luis M. Castro and Behzad Kazemtabrizi.

Companion website: www.wiley.com/go/acha_vsc_facts

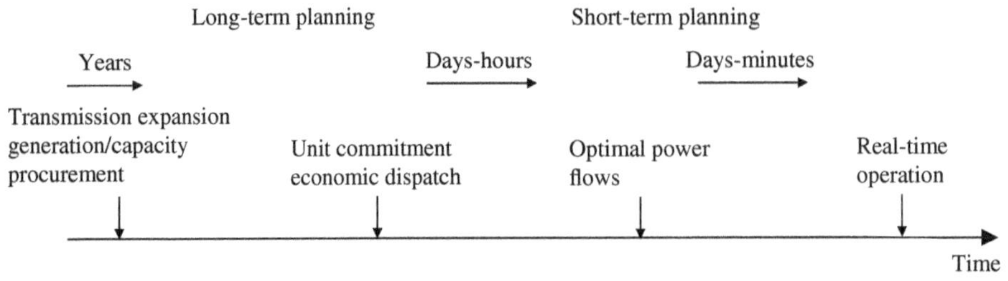

Figure 4.1 Transmission system planning framework.

It would be fair to say that such an enlargement in modelling and analysis capabilities of the OPF solution compared with the conventional power flow solution comes at a price, in terms of a significant increase in formulation complexity and, in most cases, a significant increase in the computational burden.

In this chapter, the general nonlinear formulation of the OPF problem for application in power systems operational planning is put forward. Furthermore, all the models presented for conventional AC and DC power flows in Chapter 3, expressed here in polar form, are extended for incorporation into the nonlinear formulation of the OPF for solving optimal assessments of AC and DC power systems.

4.2 Power Flows in Polar Coordinates

The conventional power flow formulation using the Newton-Raphson method in rectangular coordinates was presented in Chapter 3. The OPF formulation given in this chapter may be seen as an extension of the conventional power flow formulation in polar coordinates, given that the overall power systems optimization problem, as well as the associated constraints, is more amenable to a treatment in polar coordinates. For instance, using polar coordinates, voltage limits are directly enforced on nodal voltage magnitudes [1]. Moreover, the use of polar coordinates leads to a simpler model description for HVDC and FACTS equipment for purposes of enforcing control, e.g. voltage control may be defined by simply adding a constraint on the nodal voltage magnitudes within the optimization formulation.

Hence, the power flow problem in polar coordinates is succinctly introduced in this section as a preamble before proceeding with the detailed treatment of the OPF AC formulation using Newton's method. In order to formulate the power flow in polar coordinates, the generic bus l in Figure 3.1 is referred to. In this node, the complex power into bus l is defined by Eq. (4.1), corresponding to the product of the complex nodal voltage and the conjugate of the complex current injected into the node,

$$\bar{S}_l = \bar{V}_l \bar{I}_l^* \tag{4.1}$$

Replacing the total nodal current injection, $\bar{I}_l$, with the expression given in Eq. (3.13) gives an expression for the total complex power injection at bus l, as follows:

$$\bar{S}_l = \bar{V}_l \cdot \sum_{m=1}^{n} \bar{Y}_{lm}^* \bar{V}_m^* \tag{4.2}$$

In Eq. (4.2), the term $\overline{Y}_{lm}$ represents the individual $lm-th$ element in the nodal admittance matrix of the system given in expression (3.10).

In order to formulate the nodal power injections in polar coordinates, the complex nodal voltages and admittance terms in Eq. (4.2) are expressed in terms of magnitudes and angles:

$$\overline{S}_l = \sum_{m=1}^{n} V_l Y_{lm} V_m e^{j(\theta_l - \theta_m - \gamma_{lm})} \tag{4.3}$$

Equation (4.3) may be resolved into its real and imaginary components to give the expressions for nodal active and reactive power injections into bus l:

$$P_l = \sum_{m=1}^{n} V_l Y_{lm} V_m \cos(\theta_l - \theta_m - \gamma_{lm}) \tag{4.4}$$

$$Q_l = \sum_{m=1}^{n} V_l Y_{lm} V_m \sin(\theta_l - \theta_m - \gamma_{lm}) \tag{4.5}$$

From the application of Kirchhoff's current law at bus l, it follows that there must be a power balance at this node at steady-state operation. Therefore, the power balance at node l may be written as:

$$h_l(\theta_l, V_l) = \begin{pmatrix} h_{P_l}(\theta_l, V_l) \\ h_{Q_l}(\theta_l, V_l) \end{pmatrix} = \begin{pmatrix} P_l^{gen} - P_l^{load} - P_l^{calc} \\ Q_l^{gen} - Q_l^{load} - Q_l^{calc} \end{pmatrix} = 0 \tag{4.6}$$

The aim of the power flow is to find the operating state $\mathbf{x} = (\mathbf{V}, \boldsymbol{\theta})^t$ that upholds the network constraint in (4.6). To such an end, the nonlinear power flow equations in (4.4)–(4.5) are linearized (around an initial operating point) and solved iteratively using the Newton-Raphson method.

The linearized system of equations for the whole system is presented below, in polar coordinates, relating the nodal active and reactive power mismatches and the increments of state variables through the relevant Jacobian matrix:

$$\begin{bmatrix} \Delta\mathbf{P} \\ \Delta\mathbf{Q} \end{bmatrix} = - \begin{bmatrix} \partial\Delta\mathbf{P}/\partial\boldsymbol{\theta} & \partial\Delta\mathbf{P}/\partial\mathbf{V} \\ \partial\Delta\mathbf{Q}/\partial\boldsymbol{\theta} & \partial\Delta\mathbf{Q}/\partial\mathbf{V} \end{bmatrix} \begin{bmatrix} \Delta\boldsymbol{\theta} \\ \Delta\mathbf{V} \end{bmatrix} \tag{4.7}$$

In the following sections the AC OPF is presented, beginning with an overview of the OPF problem, its various formulations and solution methods.

4.3 Optimal Power Flow Formulation

In its most general form, the OPF problem is formulated as a *constrained* nonlinear, non-convex optimization problem, which may be expressed in the form of Eq. (4.8). In this expression, **x** and **u** are the vectors of *state* and *control* variables, respectively. These variables may differ at different pre-contingency and post-contingency timeframes, with $k \in K = \{0, \ldots, n_c\}$ representing $n_c + 1$ operating states, i.e. one normal state and n_c abnormal 'contingency' states. In (4.8), f is the *objective function,* i.e. a function of system operational state variables, which is to be optimized – for conventional plants the objective function is normally the generators' fuel cost function. The functions h

and g are functions for *equality* and *inequality* constraints, respectively, associated with the system model and any relevant operational constraints (e.g. thermal limits of transmission lines, voltage limits, transformer tap limits, generators' reactive power limits and any other specific operational limits for power flow controllers).

$$\min_{\mathbf{x}_0,\mathbf{u}_0} \quad f_0(\mathbf{x}_0,\mathbf{u}_0)$$

$$s.t. \quad \begin{cases} \mathbf{h}_k(\mathbf{x}_k,\mathbf{u}_k) = 0 \\ \mathbf{g}_k(\mathbf{x}_k,\mathbf{u}_k) \leq 0 \\ \mathbf{m}_k(\mathbf{u}_0) = \mathbf{u}_k \end{cases} \quad \forall k \in K \tag{4.8}$$

Function m is generally a linear mapping, which describes the degree of *allowable* change in post-contingency remedial actions with respect to pre-contingency state and control variables [4]. For example, some TSOs allow a degree of post-contingency violation of line thermal limits within a pre-specified timeframe and before any other post-contingency control functions are deemed to take place.

A candidate (pre-contingency) operating state that remains secure without any post-contingency control function, other than the automatic controls that will kick in anyway following a contingency, is termed a *preventively secure* system state. With the widespread use of power electronic controls in power systems and the rising levels of renewable generation integration, most power networks cannot continue to operate under such conservative margins and some degree of post-contingency remedial action is normally required by system operators following a contingency in order to alleviate post-contingency violations. Such system states are termed *correctively secure* operating states [1]. A full OPF problem may contain both preventive and corrective secure states at the same time.

The main purpose of solving an OPF problem is to obtain an 'optimum' operating point for a specific objective function, e.g. fuel cost for conventional generators, while adhering to the system's operational limits. Operational limits may include, but not exclusively, generators' reactive power limits, line thermal and static stability limits, voltage limits, and any limits on associated control equipment such as transformer tap changers, shunt compensators and, depending on the system configuration, any other power flow controller. The constraints imposed on the system's state variables and on nodal power flow equations essentially form the boundaries of the OPF solution space [3].

4.4 The Lagrangian Methods

The OPF problem in the form of Eq. (4.8) is essentially a nonlinear mathematical programming problem where an objective function is to be minimized subject to a set of technical and economic constraints that describe the operation of the power system. Nonlinear mathematical programs are the group of optimization problems where the objective function or the constraints are nonlinear [5, 6].

There is a large number of ways to formulate and solve nonlinear OPF problems. In this chapter, the focus is on the well-known class of *Lagrangian methods* [5]. The Lagrangian methods solve a constrained nonlinear optimization problem by formulating an unconstrained *Lagrangian function*. By linearizing the Lagrangian and solving it through an iterative process – such as Newton's method – a solution may be found which satisfies the necessary optimality conditions for the problem at hand [5]. Therefore, it may be said that such a solution corresponds to a local optimum of the original problem for which a Lagrangian was formed.

For the OPF problem given in Eq. (4.8), the Lagrangian function is formulated as shown in (4.9), representing the weighted sum of equality and inequality constraints added to the original objective function. The Lagrangian function applies to any operating condition and in order to simplify the formulation, the indices relating to operating conditions (i.e. $k \in K$) are dropped from the equation:

$$L(\mathbf{z}, \boldsymbol{\lambda}, \boldsymbol{\mu}) = f(\mathbf{z}) + \sum_{i \in E} \lambda_i h_i(\mathbf{z}) + \sum_{j \in A \subset I} \mu_j g_j(\mathbf{z}) \tag{4.9}$$

In (4.9) the sets E and I contain all equality and inequality constraints, respectively, and $\boldsymbol{\lambda}$ and $\boldsymbol{\mu}$ – often referred to as *dual variables* – are their associated vectors of *Lagrange multipliers* – which are used to enforce the equality and *active* inequality constraints, respectively. The set A is termed the *active set* and is made up of the inequality constraints that have become active (i.e. have been violated). Active inequality constraints are therefore enforced to their limits throughout the solution process. The system variables vector $\mathbf{z} = (\mathbf{x}, \mathbf{u})^t$ is defined as the vector containing all the state and control variables.

Once the problem is formulated, the ensuing unconstrained problem in Eq. (4.9) is solved for an optimum operating point $(\mathbf{z}^{opt}, \boldsymbol{\lambda}^{opt}, \boldsymbol{\mu}^{opt})^t$, using a suitable numerical method such as Newton's method. The necessary optimality conditions introduced in the next section are essential for checking whether or not a solution of the Lagrangian function in (4.9) is also an optimum for the original problem.

4.4.1 Necessary Optimality Conditions (Karush-Kuhn-Tucker Conditions)

The Karush-Kuhn-Tucker (KKT) conditions define the set of necessary optimality conditions that must be satisfied if an optimum operating point $(\mathbf{z}^{opt}, \boldsymbol{\lambda}^{opt}, \boldsymbol{\mu}^{opt})^t$ is to be reached [5]. Mathematically, the KKT conditions may be defined as the first-order optimality conditions listed in (4.10).

$$\begin{cases} \nabla_{\mathbf{z}} L(\mathbf{z}^{opt}, \boldsymbol{\lambda}^{opt}, \boldsymbol{\mu}^{opt}) = 0 \\ \nabla_{\boldsymbol{\lambda}} L(\mathbf{z}^{opt}, \boldsymbol{\lambda}^{opt}, \boldsymbol{\mu}^{opt}) = \mathbf{h}(\mathbf{z}^{opt}) = 0 \\ \boldsymbol{\mu}^{opt^t} . g(\mathbf{z}^{opt}) = 0 \end{cases} \tag{4.10}$$

The first two conditions in (4.10) define a system of nonlinear equations, which may be solved through an appropriate linearization using Newton's method at its core. The first condition essentially states that at the optimum, the gradient of the Lagrangian should be zero. The second and third conditions in (4.10) state that all equality and active inequality constraints have to be satisfied at the optimum for the solution to have met the criterion of optimality. The third condition specifically ensures that at the optimum, any constraint violations have been removed by means of the Lagrange multipliers. This is carried out mathematically by activating/deactivating the relevant inequality constraint, as shown in Eq. (4.11), which enforces the inequality constraint to its limits. If the inequality constraint remains within bounds $\mu_j^{opt} = 0$, and $\mu_j^{opt} > 0$ otherwise [5].

$$\mu_j^{opt} g(\mathbf{z}^{opt}) = 0 \rightarrow \begin{cases} \mu_j^{opt} > 0 \;\Rightarrow\; g(\mathbf{z}^{opt}) = 0 \\ \mu_j^{opt} = 0 \;\Rightarrow\; g(\mathbf{z}^{opt}) \leq 0 \end{cases} \qquad \forall j \in I \tag{4.11}$$

It should be noted that the KKT conditions above guarantee a local optimum for general non-convex functions only. However, if the problem is convex, the KKT conditions will guarantee a global optimum. From an engineering perspective and specifically referring to the OPF problem, the set of KKT conditions is used to test whether or not a technically feasible solution is also a local optimum for the problem in (4.8). This means that for an actual power system, the optimum operating point, i.e. the set of complex nodal voltages and control variables set-points, satisfies the set of equality constraints, i.e. nodal active and reactive power balance equations, and all equipment operates within design limits while maintaining a given objective function at its optimum value.

4.5 AC OPF Formulation

In its most general form, the OPF is inherently a nonlinear and non-convex formulation resulting from the nonlinear nature of the power flow equations given in (4.1). The full AC OPF formulation in polar coordinates is given in Eq. (4.12), which is, in turn, a derivation of the general optimization form given in (4.8). For simplicity, only the pre-contingency operating state, i.e. $k = 0$, is shown. It is assumed here that no controls, other than the ones associated with the conventional synchronous generators, are active.

$$\begin{aligned} &\min_{\theta,V} \quad F(\boldsymbol{\theta}, \mathbf{V}, \mathbf{P}^{gen}) \\ &s.t. \quad \begin{cases} \mathbf{h}(\boldsymbol{\theta}, \mathbf{V}) = 0 \\ \mathbf{g}(\boldsymbol{\theta}, \mathbf{V}) \leq \mathbf{g}^{\max} \\ \boldsymbol{\theta}^{\min} \leq \boldsymbol{\theta} \leq \boldsymbol{\theta}^{\max} \\ \mathbf{V}^{\min} \leq \mathbf{V} \leq \mathbf{V}^{\max} \end{cases} \end{aligned} \tag{4.12}$$

In Eq. (4.12), the vector function $\mathbf{h}(\theta, \mathbf{V})$ is defined explicitly as a function of the system nodal power balance equations:

$$\mathbf{h}(\boldsymbol{\theta}, \mathbf{V}) = \begin{pmatrix} \mathbf{h}_{\mathbf{P}}(\boldsymbol{\theta}, \mathbf{V}) \\ \mathbf{h}_{\mathbf{Q}}(\boldsymbol{\theta}, \mathbf{V}) \end{pmatrix} = \begin{pmatrix} P_{g,l}^{gen} - P_{d,l}^{load} - P_l^{calc} \\ Q_{g,l}^{gen} - Q_{d,l}^{load} - Q_l^{calc} \end{pmatrix} = 0 \quad \forall l \in L,\; g \in G,\; d \in D \tag{4.13}$$

In Eq. (4.13), sets L, G, and D are defined as the sets of all the network buses, generators and loads, respectively. Equation (4.13) arises from an application of Kirchhoff's current law, which holds true for a feasible solution. These equations find applicability in conventional AC power systems as well as in hybrid AC-DC systems. There may be other *control* equality constraints associated with power flow controllers or FACTS equipment, which may be suitably added to the set of equality constraints in Eq. (4.12).

Similarly, the function $\mathbf{g}(\theta, \mathbf{V})$ in Eq. (4.14) is defined as a complex-valued function over the set of state variables defining the inequality constraints. For an actual system, the inequality constraints generally refer to the transmission line thermal limits, but depending on the nature of the problem, e.g. when a DC approximation is used, they could include active power limits and/or current limits. In general, the set of inequality constraints include the following:

$$\mathbf{g}(\boldsymbol{\theta}, \mathbf{V}) = \begin{cases} \left\|\overline{S}_l\right\|^2 \leq S_l^{\max} \quad \forall l \in L \\ P_g^{gen,min} \leq P_g^{gen} \leq P_g^{gen,max} \quad \forall g \in G \\ Q_g^{gen,min} \leq Q_g^{gen} \leq Q_g^{gen,max} \quad \forall g \in G \end{cases} \tag{4.14}$$

Normally, the angle constraints apply to the slack bus only. On the other hand, all buses must satisfy the voltage magnitude constraints. The inequality constraint violations are enforced through an appropriate constraint-handling method, which is explained in Sections 4.5.5 and 4.5.6.

4.5.1 Objective Function

More often than not, the OPF objective function $F(\theta, \mathbf{V}, \mathbf{P}^{gen})$ is the fuel cost–output power function of the thermal synchronous generators normally employed in economic dispatch assessments [3], which may be assumed to have the following general quadratic form [7, 8]:

$$F(\boldsymbol{\theta}, \mathbf{V}, \mathbf{P}^{gen}) = \sum_{g \in G} (a_g + b_g P_g^{gen} + c_g P_g^{gen2}) \tag{4.15}$$

In Eq. (4.15) the coefficients a, b and c define the generator's quadratic cost curve.

The generator's active power dispatch may be initialized first by solving a *lossless economic dispatch* problem, ignoring the transmission constraints in Eq. (4.12) and assuming an equal marginal cost for the generators' outputs. This criterion of equal marginal costs provides a suitable starting point for the OPF algorithm [3].

$$\frac{\partial}{\partial P_g} F(\boldsymbol{\theta}, \mathbf{V}, \mathbf{P}^{gen}) = \lambda_{ini} \quad \forall g \in G \tag{4.16}$$

However, the OPF solution determines an optimum feasible dispatch problem taking into account all the transmission constraints and optimality conditions. Alternative objective functions may be used which largely depend on the nature of the problem being solved. For operational planning, the aim of the TSO is normally to obtain the most economic power dispatch for all the generators and in a free market environment this will depend on the market design and market rules. The objective could also be the minimization of active power transmission losses, reactive power losses and even equipment placing.

4.5.2 Linearized System of Equations

The Lagrangian function for the AC OPF problem given in (4.12) may be expressed as follows:

$$L(\boldsymbol{\theta}, \mathbf{V}, \boldsymbol{\lambda}, \boldsymbol{\mu}) = F(\boldsymbol{\theta}, \mathbf{V}, \mathbf{P}^{gen}) + \sum_{i \in E} \lambda_i h_i(\boldsymbol{\theta}, \mathbf{V}) + \sum_{j \in A \subset I} \mu_j g_j(\boldsymbol{\theta}, \mathbf{V}) \tag{4.17}$$

To solve the function in (4.17) towards the optimum, the Lagrangian may be linearized using Newton's method, resulting in a system of linearized equations, given by Eq. (4.18).

$$\nabla^2 L(\mathbf{z}, \boldsymbol{\omega}) \times \begin{pmatrix} \Delta\mathbf{z} \\ \Delta\boldsymbol{\omega} \end{pmatrix} = -\nabla L(\mathbf{z}, \boldsymbol{\omega}) \tag{4.18}$$

To simplify the formulation we may assume that $\mathbf{z} = (\boldsymbol{\theta}, \mathbf{V}, \mathbf{P}^{gen})^t$ and $\boldsymbol{\omega} = (\boldsymbol{\lambda}, \boldsymbol{\mu})^t$ are vectors of system variables and multipliers, respectively. If the set of active inequality constraints is empty, i.e. $A = 0$, then Eq. (4.18) does not include the multipliers μ and may be written down as follows:

$$\begin{bmatrix} \nabla^2_{\mathbf{zz}} L & \nabla^2_{\mathbf{z}\boldsymbol{\lambda}} L \\ \nabla^2_{\boldsymbol{\lambda}\mathbf{z}} L & \nabla^2_{\boldsymbol{\lambda\lambda}} L \end{bmatrix} \begin{bmatrix} \Delta\mathbf{z} \\ \Delta\boldsymbol{\lambda} \end{bmatrix} = - \begin{bmatrix} \partial L/\partial\mathbf{z} \\ \partial L/\partial\boldsymbol{\lambda} \end{bmatrix} \tag{4.19}$$

In Eq. (4.19), the Hessian and Jacobian terms of the Lagrangian Eq. (4.17), i.e. matrices of second- and first-order partial differentials, respectively, as well as the gradient terms, are defined as follows:

$$\begin{cases} \nabla^2_{\mathbf{zz}} L = \nabla^2_{\mathbf{zz}} F + \boldsymbol{\lambda}^t \nabla^2_{\mathbf{zz}} \mathbf{h}(\mathbf{z}) \\ \nabla^2_{\mathbf{z}\boldsymbol{\lambda}} L = \nabla^2_{\boldsymbol{\lambda}\mathbf{z}} L = \nabla_{\mathbf{z}} \mathbf{h}(\mathbf{z}) \\ \nabla^2_{\boldsymbol{\lambda\lambda}} L = 0 \\ \partial L/\partial\mathbf{z} = \nabla_{\mathbf{z}} L(\mathbf{z}) \\ \partial L/\partial\boldsymbol{\lambda} = \mathbf{h}(\mathbf{z}) \end{cases} \tag{4.20}$$

A full iterative solution of Eq. (4.19) requires the convergence of an *inner loop* to a specified tolerance. In principle at least, convergence of this inner loop may be achieved in a true quadratic convergence fashion. Accurate values of state variables and multipliers are obtained upon convergence of the inner loop. However, it is likely that one or more variables may have violated their operational limits. Hence, all inequality constraints are checked at this point for limits violations and the active set is updated by enforcing the violated inequalities using their corresponding multipliers, i.e. $\boldsymbol{\mu}$. In other words, if the active set is no longer empty then their corresponding multipliers are non-zero, i.e. $\boldsymbol{\mu} \neq 0$. This means that the Lagrangian function now takes the form of Eq. (4.17), including all the equality constraints as well as the inequalities constraints which have been violated, becoming part of the active set. This would complete one global iteration. Note that the updating of the active set takes place following a Gauss-type procedure.

The iterative process is ready to start the next global iteration, which calls for the solution of the linearized Lagrangian expression in Eq. (4.18). It involves a recalculation of the state variables and multipliers using the inner iterative loop governed by the Newton-Raphson method. Upon convergence, all inequality constraints are checked again for any change in their status concerning limits violations, i.e. the active set is updated. If one or more inequality constraints have changed their status, then a new global iteration is started; conversely, if no further change in the active set is detected, then the overall iterative process, including global and internal iterations, comes to an end.

It should be remarked that the active set is updated outside the inner loop, a procedure that impairs the overall convergence of the OPF solution using Newton's method. Furthermore, and although in theory the inner loop may be executed in a true convergent fashion, a decelerating factor ($0 < \beta < 1$) is used at the point of updating the state variables and Lagrange multipliers at the end of each local iteration. This is particularly the case during the local iterations of the first two global iterations [9]. Use of a deceleration factors impairs the quadratic convergence characteristics of the Newton-Raphson procedure, i.e. the number of local iterations will increase. However, experience has shown that this is a very powerful resource owing to the highly nonlinear nature of the problem at hand.

4.5.3 Augmented Lagrangian Function

An alternative to the original Lagrangian function given in (4.17) is to augment it by a *penalty function* for active inequality constraints [3]. The *augmented Lagrangian function* is then formed as shown in Eq. (4.21):

$$L(\boldsymbol{\theta}, \mathbf{V}, \boldsymbol{\lambda}, \boldsymbol{\mu}) = F(\boldsymbol{\theta}, \mathbf{V}, \mathbf{P}^{gen}) + \sum_{i\in E} \lambda_i h_i(\boldsymbol{\theta}, \mathbf{V}) + \sum_{j\in A\subset I} \psi_j(\mu_j, g_j(\boldsymbol{\theta}, \mathbf{V})) \tag{4.21}$$

The penalty function for the active set in (4.21), in more explicit form, is given in Eq. (4.22):

$$\psi_j = \begin{cases} \mu_j(g_j(\boldsymbol{\theta}, \mathbf{V}) - g_j^{\max}) + \frac{\alpha}{2}\|g_j(\boldsymbol{\theta}, \mathbf{V}) - g_j^{\max}\|^2 & \forall j \in A \subset I \\ zero \quad \textbf{otherwise} \end{cases} \tag{4.22}$$

This equation is, essentially, an augmentation of the term $\boldsymbol{\mu}^t\mathbf{g}$ in the Lagrangian in Eq. (4.17) by the penalty term: $\frac{\alpha}{2}\|g - g^{\max}\|^2$. This is the reason why the *augmented* Lagrangian term is used.

The penalty function ψ combines both the method of multipliers and penalty functions for enforcing the active inequality constraints. The non-zero penalty term α is started with a rather large value. However, some care needs to be exercised in its assignation since too large a value may turn the numerical problem into an ill-conditioned one [3]. Nevertheless, the effect of the penalty function ψ is that it discourages the solution search from moving in a direction that increases the violation; hence, upon convergence, all the active inequality constraints are enforced to their respective limits – this means that at this point the third KKT constraint should hold, i.e. $\mu_j(g_j(\boldsymbol{\theta}, \mathbf{V}) - g_j^{\max}) = 0$.

It should be noted that for the sake of simplicity, only the upper limits are shown explicitly. However, the above explanation applies with no change when inequality constraints violate their lower limits, the only thing required is to substitute the upper limit by the lower limit in expression (4.22).

In summary, the iterative algorithm comprises an inner Newton-Raphson loop and an outer global loop where the algorithm checks for any active inequality constraint violations. The multipliers associated with the active inequalities in the $i-th$ iteration are updated in the outer loop and take the following general form:

$$\mu_j^{(i)} = \begin{cases} \mu_j^{(i-1)} + \alpha^{(i-1)}\|g_j(\boldsymbol{\theta}, \mathbf{V}) - g_j^{\max}\|^2 & \text{if}\, \mu_j^{(i-1)} \geq \alpha^{(i-1)}(g_j(\boldsymbol{\theta}, \mathbf{V}) - g_j^{\max}) \quad \forall j \in A \subset I \\ \text{zero} \quad \text{otherwise} \end{cases} \tag{4.23}$$

It should be remarked that only violations for the upper limits are shown explicitly. This would conclude the first global (outer) iteration. The second loop now includes both equality and inequality constraints shown in Eq. (4.17) [3].

4.5.4 Selecting the OPF Solution Algorithm

It should be said that there is a wide range of algorithms for solving a nonlinear optimization problem such as the OPF. However, the method of the augmented Lagrangian function is a popular choice in the electrical power industry and it has been chosen here as the solution method.

The steps for solving the OPF are given below for two key processes: (i) for forming the augmented Lagrangian function; (ii) for linearizing the constraints and solving towards the optimum solution using Newton's method. Nonetheless, the reader is encouraged to study alternative methods for solving nonlinear problems, such as interior point methods (IPMs) [6] and gradient methods [10]. These alternative solutions are beyond the subject matter of this book.

Step One. Initialize the OPF procedure using a *lossless* economic dispatch.

Step Two. Set global iterations to 1.

Step Three. Form the Lagrangian function system $L(\mathbf{z}, \boldsymbol{\omega})$.

Step Four. Form the linearized Lagrangian function and solve for $\Delta\mathbf{z}$ and $\Delta\boldsymbol{\omega}$.

Step Five. If $\Delta\mathbf{z}$ is within the tolerance limits, stop and go to Step Six. Otherwise, repeat Steps Three and Four, successively, until the tolerance criterion fulfils. If the maximum number of inner iterations is exceeded, terminate the iterative process.

Step Six. Check for binding inequalities. If the active set is empty, the OPF solution has been reached and the overall process is stopped. Otherwise, go to Step Seven.

Step Seven. Update the active set and go to Step Three. If the maximum of global iterations permitted has been exceeded, terminate the process.

4.5.5 Control Enforcement in the OPF Algorithm

In OPF solutions using Newton's method, the equality constraint set is expanded to include *control constraints*. The control constraints are any constraints explicitly used by the control equipment in the power grid. A good example of this is the automatic voltage regulator (AVR) controllers used to regulate nodal voltages at buses where synchronous generators are connected, as long as they operate within their reactive power limits [3].

Another example of this is the nodal voltage regulation that transformers with tap-changing capabilities may exert when operating within their maximum/minimum control range. Other controllers may include power electronic mechanisms, such as FACTS and HVDC converters, and energy storage systems. In any case, suitable control variables are defined and the OPF solution ensures that such control variables attain values that establish an optimal operation of the power grid subject to all the economic, operational and security constraints that have been defined a priori.

The control constraints may be defined, in general form, by the following function:

$$C(\mathbf{x}, \mathbf{u}) = 0 \tag{4.24}$$

The Lagrangian for the function (4.24) is given in (4.25). This is used to enforce the control actions defined in (4.24) throughout the solution process:

$$L_C = \boldsymbol{\lambda}_C C(\mathbf{x}, \mathbf{u}) \tag{4.25}$$

Of particular interest in this book is the voltage source converter's operational design limits for both active and reactive powers, which are enforced by means of a set of equality constraints of the type (4.25). These are expanded to include any voltage control and power flow control in STATCOMs and VSC-HVDC systems.

4.5.6 Handling Limits of State Variables

Pure penalty functions may be used to enforce violated limits on state variables, such as nodal voltage magnitudes for voltage-controlled buses using AVRs, load tap changers (LTCs) or any other kind of voltage regulator.

A suitable penalty function may be a quadratic equation of the form given in (4.26) [9]:

$$\Phi(x_1, \dots, x_n) = \frac{1}{2}(S^W)(\Delta \mathbf{x})^2 \tag{4.26}$$

The term S^W is a penalty factor which is a large non-zero integer values. The vector $\Delta\mathbf{x} = (\Delta x_1, \dots, \Delta x_n)^t$ is defined as the vector of all violated state variables, which must be enforced to their respective limits. The penalty function in (4.26) has the Jacobian and Hessian terms given in Eqs. (4.27) and (4.28), respectively.

$$\nabla_x \Phi(x_1, \dots, x_n) = \frac{1}{2} S^W \left(\frac{\partial}{\partial \Delta x_1} \Delta x_1^2, \dots, \frac{\partial}{\partial \Delta x_n} \Delta x_n^2 \right)^t \tag{4.27}$$

$$\nabla_{xx}^2 \Phi(x_1, \dots, x_n) = \frac{1}{2} S^W \left(\frac{\partial^2}{\partial \Delta x_1^2} \Delta x_1^2, \dots, \frac{\partial^2}{\partial \Delta x_n^2} \Delta x_n^2 \right)^t \tag{4.28}$$

It is observed that by choosing a penalty function with a quadratic form, it yields a Hessian matrix which is positive definite (i.e. $\nabla_{\mathbf{xx}}^2 \Phi > 0$). This enables application of convexification around the violated limit, $\Delta\mathbf{x}$, which in turn helps to improve the numerical efficiency of the solution algorithm. The reader should be aware that selecting a good initial value for the penalty term is a highly empirical exercise, which is rooted in experience and trial and error [3]. Choosing too large a value may lead to ill-conditioning of the solution whereas small values may lead to a poor rate of convergence and possible stagnation. Experience shows that a value of 10^{10} is a good penalty factor in most cases.

4.5.7 Handling Limits of Functions

Sometimes it becomes necessary to enforce a control action throughout the OPF solution process to represent the action of control equipment in the power system. One classic example is the synchronous generator's reactive power control. Under normal circumstances, the generator keeps constant nodal voltage at the high-voltage bus of its connecting transformer by way of its AVR system, using field excitation control [8]. However, synchronous generators operating under stressful operating conditions may hit their reactive power limit and the AVR is no longer able to maintain nodal voltage at the specified magnitude. To model such an operating condition in the OPF solution, it becomes necessary to enforce the reactive power limit for the violated generator in question by defining an explicit Lagrangian function in the form of Eq. (4.24) [3].

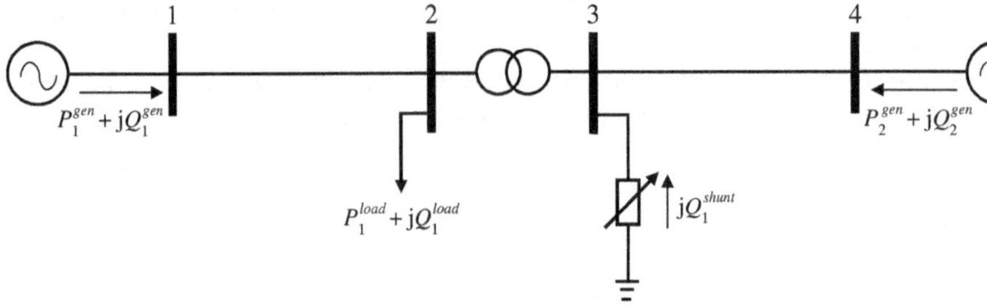

Figure 4.2 A contrived power system diagram.

4.5.8 A Simple Network Model

In this section, the process of formulating the full AC OPF problem given in Eq. (4.12) is described in step-by-step fashion, applied to the contrived four-node power system given in Figure 4.2.

This example is designed to show how the various components in the power system are modelled and formulated as an optimization problem. The process involves (i) identifying the state variables, (ii) identifying necessary equality and inequality constraints, and (iii) choosing an appropriate solution method – for this particular example, the augmented Lagrangian method has been chosen.

In the following steps the sets of buses, generators and shunt compensators are defined as $L = \{1, .., 4\}$, $G = \{1, 2\}$, $S = \{1\}$ respectively. All the appropriate state variables are indexed over their corresponding sets. For example, all nodal voltages are indexed over the set of nodes, L, all generators over the set G, and the shunt compensator over the set S.

4.5.8.1 Step One – Identifying State and Control Variables

The state variables applied to the fictitious four-node power system in Figure 4.2 are given in Eq. (4.29):

$$\mathbf{z} = (\theta_2, .., \theta_4, V_1, \dots, V_4, B_1^{shunt}, P_1^{gen}, P_2^{gen})^t \tag{4.29}$$

Appropriate upper and lower limits apply to all variables given in (4.28). As stated earlier, the angle on the slack bus (system reference bus) is fixed to zero. All other nodal voltage magnitudes as well as angles are included in the system state variables vector. The vector of control variables in this case includes the power output of both generators as well as the reactive shunt susceptance of the shunt compensator.

4.5.8.2 Step Two – Identifying Constraints

The equality constraints include the nodal power balance equations at each node in the general form given in (4.13). For this reason, it is necessary to first calculate the nodal power injections at each bus using Eq. (4.30).

$$\overline{S}_l = \sum_{m=1}^{4} V_l.Y_{lm}.V_n e^{j(\theta_l - \theta_m - \gamma_{lm})} \quad \forall l \in L = \{1, .., 4\} \tag{4.30}$$

In (4.30) the term Y_{lm} is the corresponding admittance element in the system admittance matrix for the branch connecting nodes l and m, where m goes from 1 to 4. It should be noted that this also includes the transformer's reactance element connecting nodes 2 and 3.

The reactive power constraint for the shunt compensator in node 3 is given in (4.31).

$$Q_{s,l}^{shunt} = -jV_l^2 B_s^{shunt} \quad \forall l = \{3\}, s = \{1\} \tag{4.31}$$

The nodal power balance equations in the form of equality constraints may be formed as shown in Eq. (4.32), which follows the convention set given in Eq. (4.13). Moreover, the set of inequality constraints is formed following the convention set in Eq. (4.14). The ensuing inequality constraint set is shown in Eq. (4.33).

$$\begin{cases} h_1 = \begin{pmatrix} h_{P_1} \\ h_{Q_1} \end{pmatrix} = \begin{pmatrix} P_1 - P_{1,1}^{gen} \\ Q_1 - Q_{1,1}^{gen} \end{pmatrix} = 0 \\ h_2 = \begin{pmatrix} h_{P_2} \\ h_{Q_2} \end{pmatrix} = \begin{pmatrix} P_2 + P_{1,2}^{load} \\ Q_1 + Q_{1,2}^{load} \end{pmatrix} = 0 \\ h_3 = h_{Q_3} = Q_3 - Q_{1,3}^{shunt} = 0 \\ h_4 = \begin{pmatrix} h_{P_4} \\ h_{Q_4} \end{pmatrix} = \begin{pmatrix} P_4 - P_{2,4}^{gen} \\ Q_4 - Q_{2,4}^{gen} \end{pmatrix} = 0 \end{cases} \tag{4.32}$$

$$\begin{aligned} \mathbf{g}(\boldsymbol{\theta}, \mathbf{V}) = \|(P_g^{gen,\min} + jQ_g^{gen,\min})\|^2 \le \|(P_g^{gen} + jQ_g^{gen})\|^2 \\ \le \|(P_g^{gen,\max} + jQ_g^{gen,\max})\|^2 \quad \forall g \in G = \{1, 2\} \end{aligned} \tag{4.33}$$

In this example, the thermal limits of transmission lines have been neglected but in actual systems these limits are also enforced as part of the overall OPF formulation.

4.5.8.3 Step Three – Forming the Lagrangian Function

Equation (4.34) shows the Lagrangian function at the start of the solution process where the set of active inequality constraints is empty.

$$L(\boldsymbol{\theta}, \mathbf{V}, \boldsymbol{\lambda}) = F(\boldsymbol{\theta}, \mathbf{V}, \mathbf{P}^{gen}) + \sum_{i=1}^{4} \lambda_i h_i(\boldsymbol{\theta}, \mathbf{V}) \tag{4.34}$$

Notice that in (4.34), the objective function is defined as the summation of the individual generators' quadratic cost functions as shown in (4.35):

$$F(\boldsymbol{\theta}, \mathbf{V}, \mathbf{P}^{gen}) = \sum_{g=1}^{2} (a_g + b_g P_g^{gen} + c_g P_g^{gen^2}) \tag{4.35}$$

It should be noted that the initial dispatch of the generators to enable the calculation of the starting point of the objective function is carried out through a lossless economic dispatch, as stated in Section 4.4.1.

4.5.8.4 Step Four – Linearized System of Equations

The linearized system of equations for this test network is shown in (4.36). It contains the Hessian and the Jacobian terms of the Lagrangian (4.35).

$$\begin{pmatrix} \nabla^2_{\theta^2}L & \nabla^2_{\theta\mathbf{V}}L & 0 & \nabla_{\theta}\mathbf{h} & 0 \\ \nabla^2_{\mathbf{V}\theta}L & \nabla^2_{\mathbf{V}^2}L & \nabla^2_{\mathbf{V}B}L & \nabla_{\mathbf{V}}\mathbf{h} & 0 \\ 0 & \nabla^2_{B\mathbf{V}}L & \nabla^2_{B^2}L & \nabla_{B}\mathbf{h} & 0 \\ \nabla_{\theta}\mathbf{h} & \nabla_{\mathbf{V}}\mathbf{h} & \nabla_{B}\mathbf{h} & 0 & \nabla_{\mathbf{P}^{gen}}\mathbf{h} \\ 0 & 0 & 0 & \nabla_{\mathbf{P}^{gen}}\mathbf{h} & \nabla^2_{\mathbf{P}^{gen2}}F \end{pmatrix} \begin{pmatrix} \Delta\theta \\ \Delta\mathbf{V} \\ \Delta B_s \\ \Delta\lambda \\ \Delta\mathbf{P}^{gen} \end{pmatrix} = - \begin{pmatrix} \nabla_{\theta}L \\ \nabla_{\mathbf{V}}L \\ \nabla_{B}L \\ \Delta\mathbf{h} \\ \nabla_{\mathbf{P}^{gen}}L \end{pmatrix} \quad (4.36)$$

It should be noted that each element in the matrix of coefficients in (4.36) is itself a submatrix, containing Hessian (second-order) terms and Jacobian (first-order) terms of the Lagrangian, as defined by Eq. (4.21).

4.5.8.5 Step Five – Implementation of the Augmented Lagrangian

If and when the augmented Lagrangian function is required in order to enforce the active set, throughout the course of the solution the active constraints are enforced to their respective limits using the penalty functions in the form given in (4.22). This yields the following augmented Lagrangian function:

$$L(\theta, \mathbf{V}, \lambda, \mu) = F(\theta, \mathbf{V}, \mathbf{P}^{gen}) + \sum_{i=1}^{4} \lambda_i h_i(\theta, \mathbf{V}) + \sum_{j \in A \subset I} \psi_j(\mu_j, g_j) \quad (4.37)$$

In (4.37), the general complex function g is defined by (4.33), corresponding to the set of functions for active inequality constraints. For example, take the case when the active set includes a voltage magnitude violation in node 2 (i.e. $V_2^{(2)} > V_2^{\max}$) at the end of the first global iteration. This involves the formation of the following augmented Lagrangian function for the next iteration:

$$L^{(2)}(\theta, \mathbf{V}, \lambda, \mu) = F(\theta, \mathbf{V}, \mathbf{P}^{gen}) + \sum_{i=1}^{4} \lambda_i h_i(\theta, \mathbf{V}) + \mu_1(V_2 - V_2^{\max}) + \frac{\alpha}{2}(V_2^{(2)} - V_2^{\max})^2 \quad (4.38)$$

Hence, the linearized system of equations in (4.36) is augmented by the respective Hessian and Jacobian terms corresponding to the new augmented Lagrangian function.

In the solution process, the Hessian term becomes a rather large positive value (i.e. $\nabla^2_{V_2^2}L >> 0$), which precludes the search direction from increasing the violation:

$$\frac{\partial^2}{\partial V_2^2} L^{(2)}(\theta, \mathbf{V}, \lambda, \mu) = \sum_{i=1}^{4} \lambda_i \frac{\partial^2}{\partial V_2^2} h_i(\theta, \mathbf{V}) + \alpha \quad (4.39)$$

The Jacobian term, on the other hand, penalizes the gradient vector and prevents it from deviating any further from the limit boundaries of the solution space:

$$\frac{\partial}{\partial V_2} L^{(2)}(\theta, \mathbf{V}, \lambda, \mu) = \sum_{i=1}^{4} \lambda_i \frac{\partial^2}{\partial V_2^2} h_i(\theta, \mathbf{V}) + \mu_2 + \alpha(V_2^{(2)} - V_2^{\max}) \quad (4.40)$$

This will also ensure the third KKT condition is satisfied at the solution. It should be noted that a pure quadratic penalty function in the form of Eq. (4.26) may be used for enforcing the control in, for example, voltage-regulated buses.

4.5.9 Recent Extensions in the OPF Problem

Over the past decade, advances in computation technology, coupled with rapid developments in power systems theory and practice, have enabled both system operators and the wider power systems research community to extend considerably the basic OPF formulation given in (4.1). Such extensions include approximation methods, such as *convex relaxation*, which approximates the non-convex OPF problem to a convex problem, reducing the associated computational burden when solving large systems. The OPF can also be approximated to be a linear mathematical program by setting all voltage magnitudes to constant, essentially solving a DC-like problem instead of a full AC power flow problem [11]. It should be noted that this is only in the sense that the formulation uses only real numbers as opposed to complex numbers.

There are other forms of formulating the OPF problem. For instance, the well-known IPM has been used to formulate the inequality-constrained OPF problem by imposing logarithmic barrier functions on the objective function to create equality-constrained problems, which could then be solved using the method of Lagrange multipliers [2, 6, 12]. The IPM is essentially a Lagrangian method which creates a dual objective function to be solved using the method of Lagrange multipliers described in Section 4.4. Moreover, the deterministic OPF may be extended to be a non-deterministic formulation to account for uncertainties in input data such as renewable power generation outputs, flexible load patterns (e.g. electric vehicles and demand-side management integration) and power systems contingencies [4]. These extensions are beyond the scope of this book and the reader is encouraged to refer to the relevant literature for more in-depth discussions on these topics.

4.5.10 Test Case: IEEE 30-Bus System

In this section, the IEEE 30-bus system[1] is used as test system. The OPF solution for this system is formulated and solved for a range of operating conditions using general-purpose, nonlinear mathematical programming solvers. The process of formulating and solving the OPF problem follows the approach described in Sections 4.4.1–4.5.7.

4.5.10.1 Test System

The IEEE 30-bus test system comprises 6 generators, 34 transmission lines, 4 transformers, 30 load points and 2 shunt compensators, which are placed at nodes 10 and 24. The system voltage levels are 132/33 kV and the base power is taken to be 100 MVA.

4.5.10.2 Problem Formulation

Following the general expression in (4.12), the benchmark test system is formulated as a full AC OPF problem. To begin with, no controls other than the generators' power outputs apply. It follows that the vector of state/control variables is defined as $\mathbf{z} = (\boldsymbol{\theta}, \mathbf{V}, \mathbf{P}^{gen})^t$ and the problem is bound to the equality and inequality constraints given in Eqs. (4.13) and (4.14), respectively.

The objective of the OPF solution is to find a feasible minimum-cost solution for the synchronous generators' power dispatch subject to constraints. The fuel costs–output powers characteristics of the synchronous generators are modelled as polynomial

1 System data may be found from the following link: http://labs.ece.uw.edu/pstca/pf30/pg_tca30bus.htm.

Table 4.1 Generator cost function coefficients.

Generator number	Bus number	*a* ($ MW^{-2})	*b* ($ MW^{-1})	*c* ($)
1	1	0.0384	20	0
2	2	0.25	20	0
3	5	0.01	40	0
4	8	0.01	40	0
5	11	0.01	40	0
6	13	0.01	40	0

Table 4.2 Transformer data (load tap changers).

Branch number	Sending bus	Receiving bus	Transformer tap	Upper limit	Lower limit
11	6	9	0.978	1.3	0.7
12	6	10	0.969	1.3	0.7
15	4	12	0.932	1.3	0.7
36	28	27	0.968	1.3	0.7

Table 4.3 Shunt devices.

Bus number	Initial susceptance (MVAR)
10	19
24	4.3

quadratic cost functions, as in Eq. (4.15). The cost functions are calculated in units of US $ h^{-1}. The synchronous generators data is given in Table 4.1. Transformers and shunt compensators data are given in Tables 4.2 and 4.3, respectively. Table 4.4 shows the synchronous generators limits.

4.5.10.3 OPF Test Cases

Further to the base case (benchmark OPF solution), four other test cases are solved to investigate various aspects of the OPF formulation, using the IEEE 30-bus system. The benchmark test case solution verifies the model objects created in the advanced interactive multidimensional modelling system (AIMMS) environment[2] and compares it against the solution furnished by a well-known Matlab-based power system simulation package, i.e. MATPOWER.[3] The model object created in AIMMS is a fully customizable

2 AIMMS is a multi-purpose optimization model-based language. A free academic license may be obtained from: https://aimms.com.

3 MATPOWER is a Matlab-based open source software available to download from: http://www.pserc.cornell.edu/matpower.

Table 4.4 Generator limits.

Generator number	Active power (upper limit) (MW)	Reactive power (upper limit) (MVAR)	Reactive power (lower limit) (MVAR)
1	360.2	10	0
2	140	50	−40
3	100	40	−40
4	100	40	−40
5	100	24	−24
6	100	24	−24

model of the IEEE 30-bus system, which follows the problem formulation given in Eq. (4.12). This model can be modified with ease to accommodate further linear and nonlinear constraints for exerting control actions at various points of the power system.

Three test cases with voltage regulation are carried out using the four tap-changing transformers available in the IEEE 30-bus system. The transformers' taps are modelled as additional constraints using the model given in Chapter 3, with the taps treated as continuous variables. If required, additional control constraints in the form of Eq. (4.26) may be added to exert nodal voltage control.

The three OPF solutions with voltage regulation are solved using the augmented Lagrangian method embedded within the MINOS solver. Furthermore, the OPF solutions have been carried out with an interior-point solver (IPOPT) and with an additional solver called CONOPT, which uses a generalized gradient method and has been developed for solving highly nonlinear problems. In all cases solved as part of this exercise, the OPF solutions converge to exactly the same results and within the same bounds defined by the constraints.

4.5.10.4 Benchmark Test Case (With No Voltage Control)

There is no nodal voltage regulation in the base test case and the transformers' taps are taken to be fixed parameters. A summary of results is presented in Tables 4.5 and 4.6 as given by the solvers AIMMS and MATPOWER, respectively. It should be noted

Table 4.5 Optimal power flow solution: real powers (benchmark test).

Generator number	Final solution (AIMMS) (MW)	Final solution (MATPOWER) (MW)
1	212.23	212.23
2	36.23	36.23
3	29.35	29.35
4	12.93	12.93
5	4.4	4.4
6	0	0

Table 4.6 Optimal power flow solution: reactive powers (benchmark test).

Generator number	Final solution (AIMMS) (MVAR)	Final solution (MATPOWER) (MVAR)
1	0.00	0.00
2	27.17	27.17
3	29.94	29.94
4	40.00	40.00
5	9.10	9.10
6	7.72	7.72

that their respective solutions converge to exactly the same optimal values. Given the nonlinearity of the considered constraints – nodal power balance equations – the KKT conditions only guarantee a local optimum for the primal problem, which is defined by Eq. (4.12).

4.5.10.5 Test Case with Voltage Control Using Variable Transformers Taps (Case I)

In this test case, the vector of state/control variables is extended to include the variable tap **T** for the on-load tap changer transformers. Hence, the vector takes the following form: $\mathbf{z}_{cont} = (\boldsymbol{\theta}, \mathbf{V}, \mathbf{P}^{gen}, \mathbf{T})^t$. No further constraints are added to the constraints set used in the benchmark test case. However, during the course of the OPF solution, the tap is now treated as a variable as opposed to a constant parameter.

4.5.10.6 Test Case with Nodal Voltage Regulation (Case II)

In this test case, an additional constraint is added to regulate the nodal voltages of the transformer buses to keep the values obtained in the benchmark test case. The nodal voltages are kept fixed by adding a constraint to the constraints set (4.26).

The target nodal voltage magnitudes and the associated transformers' taps values, which are now part of the state variables vector, are given in Table 4.7. It should be noticed from this table that the OPF solution has converged to the same tap ratios as in the benchmark test case, which in that particular solution were fixed values. This result further validates the accuracy response of the OPF model using the AIMMS solver.

Table 4.7 Voltage regulation at transformer buses – controlled test II.

Transformer number	Regulated bus	Target voltage	Transformer final tap value
1	9	1.0422	0.978
2	10	1.0385	0.969
3	12	1.0498	0.932
4	27	1.0215	0.968

Table 4.8 Voltage regulation at transformer buses – controlled test III.

Transformer number	Regulated bus	Target voltage	Transformer final tap value
1	9	1.05	0.901
2	10	1.01	1.145
3	12	1.00	1.041
4	27	1.00	0.977

As expected, there are no changes in the results compared to the nodal voltage magnitudes in the benchmark test case – both test cases converge to the same solution.

4.5.10.7 Test Case with Nodal Voltage Regulation (Case III)

The OPF model now considers a case of variable operating conditions. The test considers that the nodal voltage magnitudes at the transformers' receiving end buses are controlled to predetermined values. The same process applies as in the previous test case, i.e. nodal voltage regulation I, where the transformers' taps are taken to be state variables. Table 4.8 shows the transformers' taps values at the optimum solution, corresponding to the target voltage values shown in Table 4.8.

Constraint Handling

1. There is a voltage constraint violation at bus 11, which is handled with the use of the augmented Lagrangian function method. The voltage is enforced to its corresponding upper limit of 1.06 p.u.
2. Likewise, the synchronous generators connected at nodes 8 and 13 are bound to their respective upper limits of 40 and 24 MVAR, respectively. This contrasts with the reactive power behaviour of the slack generator, which is kept to its lower limit, i.e. zero.

4.5.10.8 A Summary of Results

AIMMS uses the nonlinear solver CONOPT whereas MATPOWER uses Matlab's interior point solver MIPS. A summary of the OPF results for both benchmark and voltage-controlled tests is given in Table 4.9.

It can be observed that the best optimal solution is obtained when the transformers' taps are free to vary within their limits and no additional constraints are imposed on the system. In this case the total value of the objective function is \$8902.69 h^{-1} for the six generators. Conversely, the worst solution among the five tests carried out corresponds to the final case where nodal voltage magnitudes are controlled at predetermined values, signifying that this is a more constrained operating point. In this case the total value of the objective function gives a cost of \$8909.3 h^{-1}, which increases the cost by \$6.7 h^{-1}.

The nodal voltage profiles for the five test cases are shown in Figure 4.3 for the 30 nodes of the test system, where the changes in nodal voltage magnitudes are shown when comparing the tests with voltage control and the benchmark tests. In all instances, the OPF algorithms converge to local optimal solutions, with all the constraints satisfied at their limits.

Table 4.9 OPF summary results – IEEE 30-bus system.

Model object	OPF solver	Test	Convergence time	Optimality of primal	Final objective solution
MATPOWER	MIPS – Interior point	Benchmark	3.9 s	Locally optimal	$8906.14 h^{-1}
AIMMS	MINOS – Augmented Lagrangian	Benchmark	0.05 s – 11 Global iterations	Locally optimal	$8906.14 h^{-1}
AIMMS	MINOS – Augmented Lagrangian	Controlled I – Variable tap changer	0.05 sec – 11 Global iterations	Locally optimal	$8902.69 h^{-1}
AIMMS	MINOS – Augmented Lagrangian	Controlled II – Voltage regulation I	0.05 sec – 7 Global iterations	Locally pptimal	$8906.18 h^{-1}
AIMMS	MINOS – Augmented Lagrangian	Controlled III – Voltage regulation II	0.03 sec – 6 Global iterations	Locally optimal	$8909.30 h^{-1}

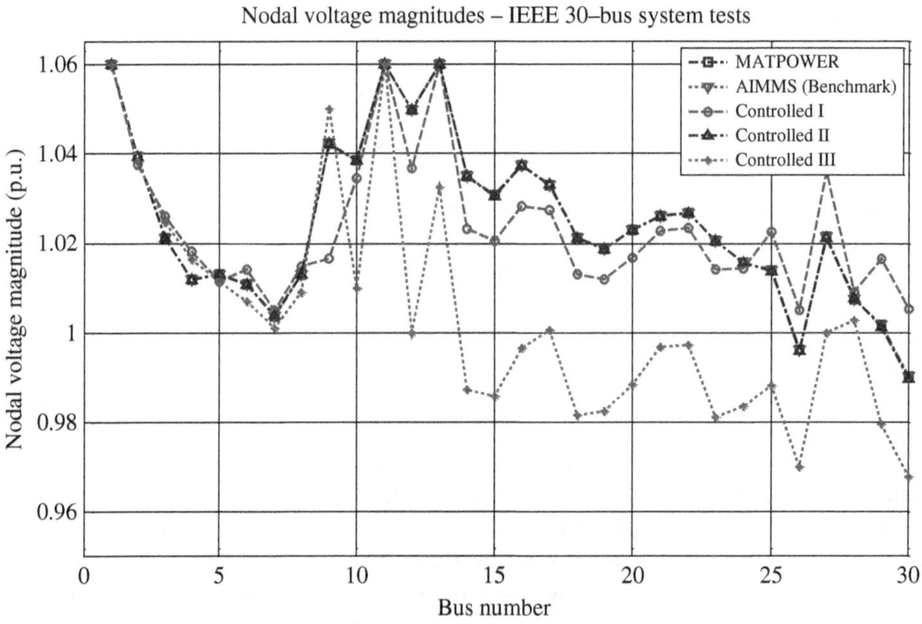

Figure 4.3 IEEE 30-bus test results: nodal voltage magnitudes.

The synchronous generators' power dispatches for both active and reactive powers are shown in Figure 4.4 for the benchmark and the third test case, i.e. nodal voltage regulation to a priori set values. Notice the changes in the reactive power dispatch brought about by changes in the nodal voltage magnitude settings. The change in the optimum active power dispatch is minimum.

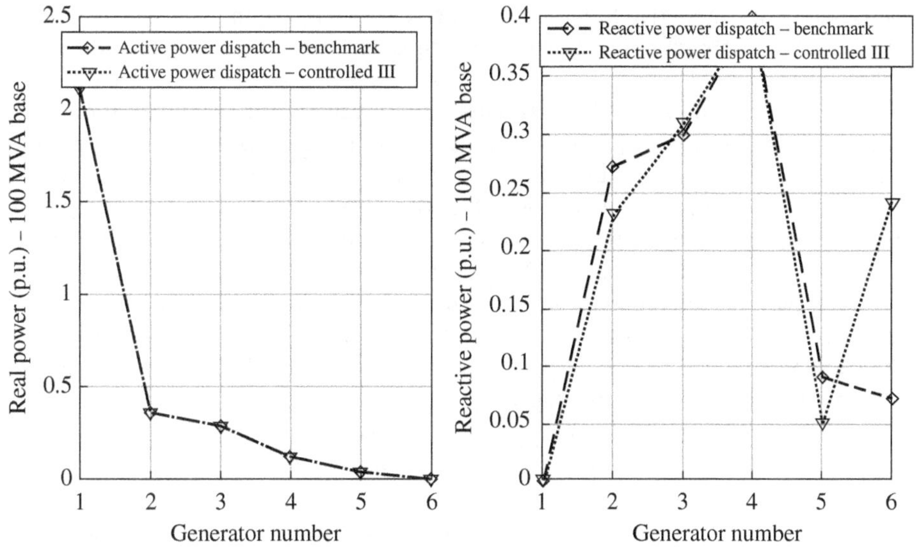

Figure 4.4 Generators' power outputs at optimum.

4.6 Generalization of the OPF Formulation for AC-DC Networks

The original OPF formulation shown in (4.12) is extended in this section to incorporate explicitly all the control constraints necessary to model combined AC-DC network.

If vector $\mathbf{z} = (\theta, \mathbf{V}, \mathbf{P}^{gen})^t$ contains the original vector of state variables for the AC network, then this is expanded to encompass the state variables corresponding to the DC network and DC equipment, as described in Eq. (4.41).

$$\min_{\mathbf{z},\mathbf{x}_{dc},\mathbf{u}_{dc}} \quad F(\mathbf{z}, \mathbf{x}_{dc}, \mathbf{u}_{dc})$$
$$s.t. \quad \begin{cases} \mathbf{h}^{acdc}(\mathbf{z}, \mathbf{x}_{dc}, \mathbf{u}_{dc}) = 0 \\ \mathbf{C}^{dc}(\mathbf{x}_{dc}, \mathbf{u}_{dc}) = 0 \\ \mathbf{g}^{acdc}(\mathbf{z}, \mathbf{x}_{dc}, \mathbf{u}_{dc}) \leq \mathbf{g}^{acdc\max} \\ \mathbf{g}^{dc}(\mathbf{x}_{dc}, \mathbf{u}_{dc}) \leq \mathbf{g}^{dc\max} \\ \mathbf{z}^{\min} \leq \mathbf{z} \leq \mathbf{z}^{\max} \\ \mathbf{x}_{dc}^{\min} \leq \mathbf{x}_{dc} \leq \mathbf{x}_{dc}^{\max} \\ \mathbf{u}_{dc}^{\min} \leq \mathbf{u}_{dc} \leq \mathbf{u}_{dc}^{\max} \end{cases} \tag{4.41}$$

The objective function is taken to be the generators' cost functions and for the sake of simplicity in the formulation, any post-contingency operating states are neglected to begin with. These will be incorporated as the iterative solution progresses and there are constraints violations.

In Eq. (4.41) the vector $\mathbf{z}_{dc} = (\mathbf{x}_{dc}, \mathbf{u}_{dc})^t$ is a vector containing DC state and control variables, to be defined at a later stage. Moreover, the explicit equality constraints associated with DC equipment are contained in the complex function set $\mathbf{C}^{dc}(\mathbf{x}_{dc}, \mathbf{u}_{dc})$, which possesses a set of Lagrangian functions. Each piece of DC equipment has an associated inequality constraints set defined by $\mathbf{g}^{dc}$, as well as limits on the DC state and control variables.

The OPF formulation in (4.41) represents a more generalized formulation than the one given in (4.12). It is now possible to solve the power flow equations of a combined AC-DC system using a unified reference frame. However, the assumption of the existence of a unified frame-of-reference for solving power flow equations hinges on defining the appropriate set of nodal power balance equations – h^{acdc}, one that makes no notional difference between the DC and AC networks. In other words, and from the perspective of a mathematical programming solver, there should be no distinctions between the AC and DC sides of the network. This assumption is illustrated in Figure 4.5.

The Lagrangian function of the equation set in (4.41) is given by Eq. (4.42), assuming that the active set is empty at this stage.

$$L^{acdc}(\mathbf{z}, \mathbf{z}_{dc}) = F(\mathbf{z}, \mathbf{z}_{dc}) + \sum_{i \in E} \lambda_i h_i^{acdc}(\mathbf{z}, \mathbf{z}_{dc}) + \sum_{j \in E^{dc}} \lambda_j^{dc} C_j^{dc}(\mathbf{z}_{dc}) \tag{4.42}$$

In (4.42) the vector $\boldsymbol{\lambda}^{dc} = (\lambda_1^{dc}, \dots, \lambda_N^{dc})^t$ is the vector containing Lagrange multipliers pertaining to DC equipment or DC systems. The set E^{dc} is the set of all DC control constraints defined by the complex function $C(\mathbf{z}) = 0$.

The linearized system of equations given in (4.18) and (4.19) for the AC OPF problem may be expanded to include the compound AC-DC problem, as follows:

$$\begin{bmatrix} \nabla^2_{\mathbf{zz}} L^{acdc} & \nabla^2_{\mathbf{zz}_{dc}} L^{acdc} & \nabla_{\mathbf{z}}(\mathbf{h}^{acdc}) & \nabla_{\mathbf{z}}(\mathbf{C}^{dc}) \\ \nabla^2_{\mathbf{z}_{dc}\mathbf{z}} L^{acdc} & \nabla^2_{\mathbf{z}_{dc}\mathbf{z}_{dc}} L^{acdc} & \nabla_{\mathbf{z}_{dc}}(\mathbf{h}^{acdc}) & \nabla_{\mathbf{z}_{dc}}(\mathbf{C}^{dc}) \\ \nabla_{\mathbf{z}}(\mathbf{h}^{acdc}) & \nabla_{\mathbf{z}_{dc}}(\mathbf{h}^{acdc}) & 0 & 0 \\ \nabla_{\mathbf{z}}(\mathbf{C}^{dc}) & \nabla_{\mathbf{z}_{dc}}(\mathbf{C}^{dc}) & 0 & 0 \end{bmatrix} \begin{bmatrix} \Delta\mathbf{z} \\ \Delta\mathbf{z}_{dc} \\ \Delta\boldsymbol{\lambda} \\ \Delta\boldsymbol{\lambda}_{dc} \end{bmatrix} = - \begin{bmatrix} \nabla_{\mathbf{z}} L^{acdc} \\ \nabla_{\mathbf{z}_{dc}} L^{acdc} \\ \Delta\mathbf{h}^{acdc} \\ \Delta\mathbf{C}^{dc} \end{bmatrix} \tag{4.43}$$

This system of linear equations may be solved by iteration towards a candidate optimum, using Newton's method. The necessary optimality conditions introduced in Section 4.3 guarantee that a candidate solution for the system of equations in (4.43) is a local optimum. It should be remarked that the key to integrating AC-DC power flows in a single frame of reference rests on the ability to define a suitable nodal power balance equation set – $\mathbf{h}^{acdc}$.

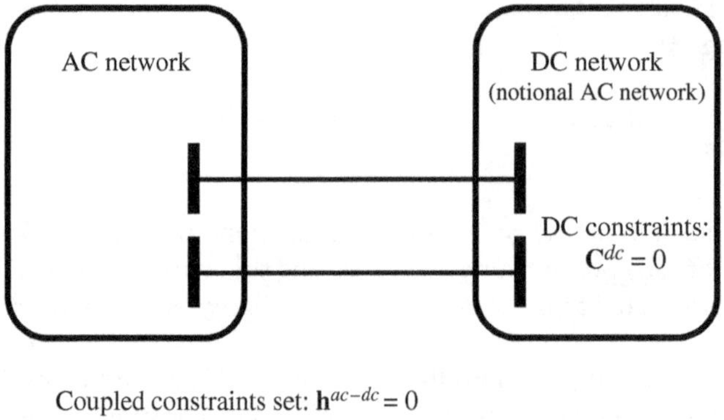

Figure 4.5 Network model for AC-DC unified formulations.

4.7 Inclusion of the VSC Model in OPF

The VSC equivalent circuit, for which the full model was developed in Chapter 3 in rectangular coordinates, is shown schematically in Figure 4.6. It should be brought to the reader's attention that there are alternative VSC models for both power flows and OPFs, which can be used instead of the VSC model shown in Figure 4.6. For example, [13] presents a converter model where the VSC is described by a controllable voltage source. However, such a model is more contrived than the one presented here. For instance, it does not explicitly contain a DC bus. Nonetheless, the lack of a DC bus in the voltage source model does not prevent it from being used in the STATCOM and VSC-HVDC power flow applications. but it is amenable only to iterative, numerical solutions of the sequential kind. That is, the AC and DC subsystems are solved separately, in sequences, exchanging partial results between the AC and DC networks at each iteration until there are no discernible changes in results in both networks. This contrasts with the numerical solutions presented in Chapters 3–6 of this book, where iterative unified solutions of the AC and DC subsystems are carried out. This is possible owing to the use of the model shown in Figure 4.6, which is essentially the model of a STATCOM and represents the VSC stations of the many forms of VSC-HVDC systems.

4.7.1 VSC Power Balance Equations

Following the generalized OPF formulation introduced in Section 4.6, the VSC nodal power balance equations may be written down, in compact form, as follows:

$$h_c^{vsc}(\boldsymbol{\theta}_{vR}, \mathbf{V}_{vR}, \mathbf{E}_{dc}, \mathbf{B}_{eq}, \mathbf{m_a}, \boldsymbol{\Phi}) = 0 \quad \forall c \in Cv \tag{4.44}$$

The function in (4.44) represents the coupled AC-DC nodal power flows at the sending node vR and at the receiving node 0 of the converter station. The set Cv is defined to be the set comprising all the VSC converter stations in the system.

Figure 4.6 VSC converter station with its unified equivalent circuit.

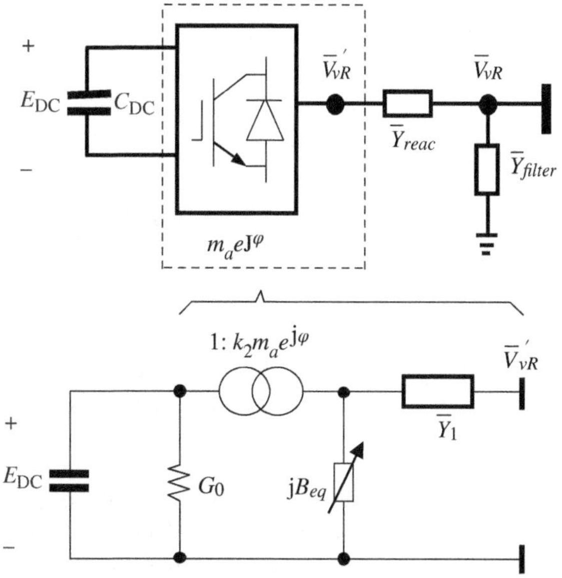

Building on the model presented in Chapter 3, it is noted that the AC and DC sides of the converter are coupled together by way of the AC voltage equation given in (4.45), which is valid for any VSC converter station – for simplicity, the converter index c is being dropped:

$$\overline{V}_1 = k_2 \overline{m}_a E_{dc} = k_2 m_a E_{dc} e^{j\phi} \tag{4.45}$$

It should be noted that this equation is in fact equation (3.32) used in the previous chapter and it is reproduced here to enable a smooth development of the VSC-OPF formulation. The factor k_2 allows for application of Eq. (4.45) to any kind of AC-DC converter, including modular multi-level converters. For a two-level, three-phase VSC the constant is: $k_2 = \sqrt{\frac{3}{8}}$.

For the VSC station given in Figure 4.6, the nodal power equations can be calculated using Eq. (4.46):

$$\overline{S}_c^{vsc} = \text{diag}(\overline{V}_c^{vsc}).\overline{I}_c^{vsc*} = \text{diag}(\overline{V}_c^{vsc}).\{Y_{vsc_c}.\overline{V}_c^{vsc}\}^* \quad \forall c \in Cv \tag{4.46}$$

Expanding Eq. (4.46) results in an equation similar to the one derived in the previous chapter for the full VSC model, but all the variables are expressed using polar coordinates:

$$\begin{pmatrix} \overline{S}_{vR_c} \\ \overline{S}_{0_c} \end{pmatrix} = \begin{pmatrix} V_{vR_c} e^{j\theta_{vR}} & 0 \\ 0 & E_{dc_c} \end{pmatrix} \cdot \left\{ \begin{pmatrix} \overline{Y}^*_{vRvR_c} & -\overline{Y}^*_{vR0_c} e^{-j\varphi_c} \\ -\overline{Y}^*_{vR0_c} e^{j\varphi_c} & (G_{sw} - jk_2^2 m_a^2 B_{eq}) + \overline{Y}^*_{0_c} \end{pmatrix} \begin{pmatrix} V_{vR_c} e^{-j\theta_{vR_c}} \\ E_{dc_c} \end{pmatrix} \right\}$$

$$\forall c \in Cv \tag{4.47}$$

where

$$\begin{aligned} \overline{Y}_{vRvR} &= \overline{Y}_{filter} + \Delta\overline{Y} = Y_{vRvR} e^{j\gamma_{vRvR}} \\ \overline{Y}_{vR0} &= k_2 m_a \Delta\overline{Y} = Y_{vR0} e^{j\gamma_{vR0}} \\ \overline{Y}_{00} &= k_2^2 m_a^2 \Delta\overline{Y} = Y_{00} e^{j\gamma_{00}} \end{aligned} \tag{4.48}$$

$$\Delta\overline{Y} = \frac{\overline{Y}_1 \overline{Y}_{reac}}{(\overline{Y}_1 + \overline{Y}_{reac})} \tag{4.49}$$

Note that the index c has been dropped.

Expressions for the nodal power equations at the sending and receiving ends of the VSC may be derived by developing the complex voltage–current relationship in (4.47). The ensuing equations are given in (4.50).

$$\begin{cases} \overline{S}_{vR} = (P_{vR} + jQ_{vR}) = V_{vR}^2 Y_{vRvR} e^{-j\gamma_{vRvR}} - V_{vR} E_{dc} Y_{vR0} e^{j(\theta_{vR} - \gamma_{vR0} - \phi)} \\ \overline{S}_0 = (P_0 + jQ_0) = E_{dc}^2 [(G_{sw} - jk_2^2 m_a^2 B_{eq}) + Y_{00} e^{-j\gamma_{00}}] - V_{vR} E_{dc} Y_{vR0} e^{j(\phi - \theta_{vR} - \gamma_{vR0})} \end{cases} \tag{4.50}$$

When integrating this model into the OPF formulation, one reactive power constraint needs to be included in the form of a control constraint on the DC side of the VSC to enforce reactive power flow to zero at the DC side. It should be noted that there is no reactive power in DC circuits. To force zero reactive power flow on the DC circuit, a notional variable susceptance, B_{eq}, is employed in the AC side of the converter. The zero reactive power constraint is:

$$Q_0 = 0 \tag{4.51}$$

The VSC formulation in OPF requires the equality constraints given in Eqs. (4.50) and (4.51), which are added to the system Lagrangian function. The vector of state and control variables for the VSC is given in Eq. (4.52):

$$\mathbf{z}_{vsc} = (\theta_{vR}, V_{vR}, E_{dc}, B_{eq}, m_a, \varphi)^t \tag{4.52}$$

Moreover, the vector of Lagrange multipliers for the equality constraints in Eqs. (4.50) and (4.51) is defined as follows:

$$\boldsymbol{\lambda}_{vsc} = (\boldsymbol{\lambda}_{vR}, \lambda_{P_0}, \lambda_{Q_0})^t \tag{4.53}$$

where the term $\lambda_{vR} = (\lambda_{vRP}, \lambda_{vRQ})^t$ is the vector of Lagrange multipliers associated with the active and reactive power balance equations at the VSC's AC terminal, respectively.

Once the vector of variables/multipliers is identified for the VSC station, the power balance equations for the VSC station, in the form of equality constraints, given in Eq. (4.44), may be written down as follows:

$$L^{vsc}(\mathbf{z}_{vsc}, \boldsymbol{\lambda}_{vsc}) = \boldsymbol{\lambda}_{vR}^t \mathbf{h}^{vR} + \boldsymbol{\lambda}_{P_0}^t C_P^{dc} + \boldsymbol{\lambda}_{Q_0}^t C_Q^{dc} \tag{4.54}$$

Note that in Eq. (4.54), the following nodal power balance equations are defined as equality constraints for both AC and DC sides of the converter station:

$$\begin{cases} \mathbf{h}_c^{vR} = \begin{pmatrix} h_{P_c}^{vR} \\ h_{Q_c}^{vR} \end{pmatrix} = \begin{pmatrix} P_{vR_c} - P_{g,c}^{gen} + P_{d,c}^{load} \\ Q_{vR_c} - Q_{g,c}^{gen} + Q_{d,c}^{load} \end{pmatrix} = 0 \quad \forall c \in Cv, g \in G, d \in D \\ C_P^{dc} = P_{0_c} - P_{0_c}^{sch} = 0 \qquad \forall c \in Cv \\ C_Q^{dc} = Q_{0_c} - Q_{0_c}^{sch} = 0 \qquad \forall c \in Cv \end{cases} \tag{4.55}$$

In Eq. (4.55), two control constraints are defined explicitly for the DC side of the VSC station to enable: (i) control/schedule of DC power flow to a predetermined value, P^{sch}, and (ii) zero reactive power output at the DC side of the converter station to be maintained. The latter point means that within the OPF formulation, the reactive power at the DC side of the converter station is always set to zero, $Q_0^{sch} = 0$.

4.7.2 VSC Control Considerations

The VSC control strategies, described in Chapter 3 for power flows, apply with little change to the case of OPF. The VSC can be used to control rather effectively the nodal voltage magnitude at the VSC's AC side, i.e. node vR, while keeping the DC side voltage, E_{dc}, constant.

To exert voltage control within the OPF solution, we use penalty functions similar to the one described in Section 4.5.3. In the simulation scenarios given in this chapter, the controls are automatically handled by the solver. The VSC variables are initialized in a similar manner as for the conventional power flow solution. All the VSC stations in the simulations presented in this chapter operate within their linear regions of m_a operation, assuming purely sinusoidal voltage and current waveforms. The limits on the equivalent susceptance, B_{eq}, define the boundaries of VSC station operation in terms of reactive power operation. Similarly, limits on the active power operation of the VSC may also be enforced by the introduction of suitable constraints.

4.7.3 VSC Linearized System of Equations

The linearized system of equations, given in this section in compact form, possesses the same structure as the linearized system of equations given in Eq. (4.43) but for the VSC with a Lagrangian function given in Eq. (4.54).

$$\begin{pmatrix} \mathbf{H}_{vsc} & \mathbf{J}_{vsc} \\ \mathbf{J}_{vsc}^t & 0 \end{pmatrix} \begin{pmatrix} \Delta\mathbf{z}_{vsc} \\ \Delta\boldsymbol{\lambda}_{vsc} \end{pmatrix} = -\begin{pmatrix} \partial L^{vsc}/\partial\mathbf{z}_{vsc} \\ \partial L^{vsc}/\partial\boldsymbol{\lambda}_{vsc} \end{pmatrix} \tag{4.56}$$

In Eq. (4.56) the following Hessian and Jacobian terms are defined as follows:

$$\begin{cases} \mathbf{H}_{vsc} = \boldsymbol{\lambda}_{vR}^t \nabla_{\mathbf{z}_{vsc}^2}^2 \mathbf{h}^{vR} + \boldsymbol{\lambda}_{P_0}^t \nabla_{\mathbf{z}_{vsc}^2}^2 C_P^{dc} + \boldsymbol{\lambda}_{Q_0}^t \nabla_{\mathbf{z}_{vsc}^2}^2 C_Q^{dc} \\ \mathbf{J}_{vsc} = \nabla_{\mathbf{z}_{vsc}} \mathbf{h}^{vR} + \nabla_{\mathbf{z}_{vsc}} C_P^{dc} + \nabla_{\mathbf{z}_{vsc}} C_Q^{dc} \end{cases} \tag{4.57}$$

Similarly, the gradient terms are defined as follows:

$$\begin{cases} \dfrac{\partial L^{vsc}}{\partial\mathbf{z}_{vsc}} = \boldsymbol{\lambda}_{vR}^t \dfrac{\partial\mathbf{h}^{vR}}{\partial\mathbf{z}_{vsc}} + \boldsymbol{\lambda}_{P_0}^t \dfrac{\partial C_P^{dc}}{\partial\mathbf{z}_{vsc}} + \boldsymbol{\lambda}_{Q_0}^t \dfrac{\partial C_Q^{dc}}{\partial\mathbf{z}_{vsc}} \\ \dfrac{\partial L^{vsc}}{\partial\boldsymbol{\lambda}_{vsc}} = \Delta\mathbf{h}^{vR} + \Delta C_P^{dc} + \Delta C_Q^{dc} \end{cases} \tag{4.58}$$

4.8 The Point-to-Point and Back-to-Back VSC-HVDC Links Models in OPF

The VSC OPF formulation is extended to model the point-to-point VSC-HVDC link and, as a particular case, the back-to-back VSC-HVDC link. In this respect, a general frame-of-reference to solve OPF problems afforded by Eq. (4.41) becomes a rather useful device for modelling these kinds of VSC-HVDC links, with the variables/multipliers explicitly defined for the two VSC stations together with their corresponding Lagrangian functions.

The power balances that exist in the point-to-point and back-to-back VSC-HVDC links, useful for the OPF solution, are shown in Figure 4.7a,b, respectively.

Each converter station is connected in shunt with its corresponding AC system using a step-up transformer, which may be taken to be an LTC transformer for the sake of modelling convenience, although in practice they may well be conventional transformers. Nonetheless, the VSC interfacing transformer, whether a conventional transformer or an LTC one, is handled as separate components within the OPF formulation. The VSC and the transformer each will have their own Lagrangian function and equality constraints, i.e. nodal power balance equations at the VSC and transformer interfacing nodes.

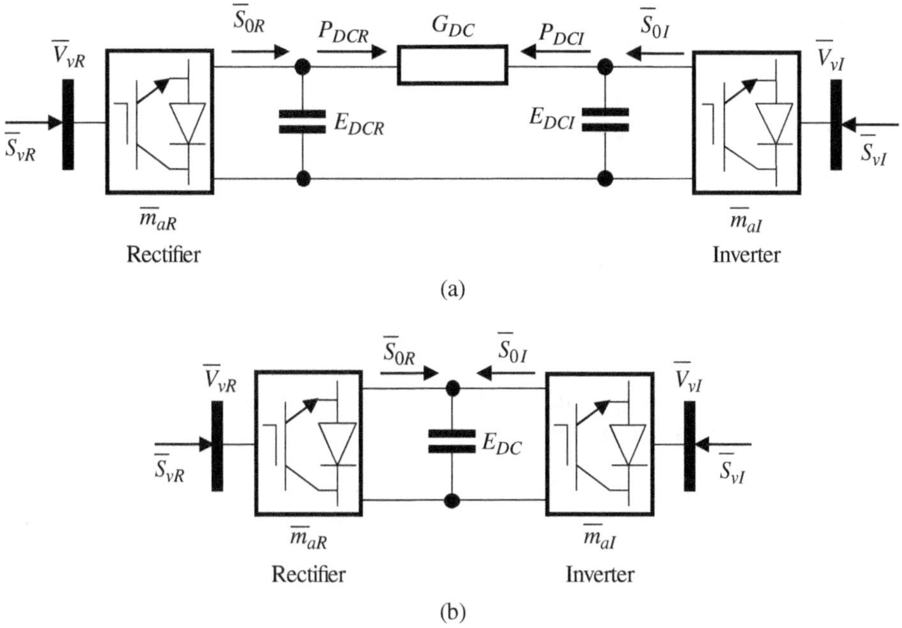

Figure 4.7 Power balances in VSC-HVDC links: (a) point-to-point, (b) back-to-back (interfacing transformers not shown).

4.8.1 VSC-HVDC Link Power Balance Formulation

From the modelling viewpoint, there will be two sets of nodal power balance equations for each VSC station (i.e. rectifier and inverter), similarly to Eq. (4.50), with the proviso that the DC power loss, i.e. $R_{DC}I_{DC}^2$, should be taken into account in point-to-point VSC-HVDC topologies. Naturally, this power loss is nil in back-to-back VSC-HVDC topologies.

In order to identify the power balance equations for the DC link, the circuit in Figure 4.8 proves rather useful.

The terms $\bar{S}_{0R}^0$ and $\bar{S}_{0I}^0$ correspond to the VSC nodal power equations at the DC sides of the inverter and the rectifier converters, as given by Eq. (4.50). To derive an expression for the nodal power injections at both DC nodes, the active power losses in the cable

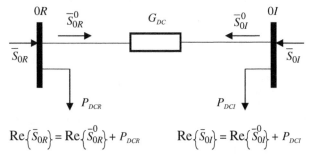

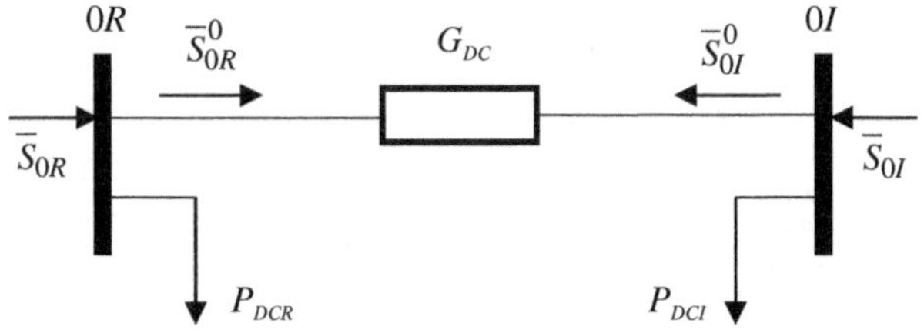

$$\mathrm{Re}\{\bar{S}_{0R}\} = \mathrm{Re}\{\bar{S}_{0R}^0\} + P_{DCR} \qquad \mathrm{Re}\{\bar{S}_{0I}\} = \mathrm{Re}\{\bar{S}_{0I}^0\} + P_{DCI}$$

Figure 4.8 The VSC-HVDC link for a point-to-point configuration with the DC link nodal power balance equations.

require consideration. Since there is no reactive power flow in a DC cable, the reactive power loss is nil and the following equation set represents the nodal power balance of the DC link in Figure 4.8:

$$\begin{cases} \mathrm{Re}\{\overline{S}_{0R}\} = \mathrm{Re}\{\overline{S}_{0R}^{0}\} + P_{DCR} \\ \mathrm{Im}\{\overline{S}_{0R}\} = \mathrm{Im}\{\overline{S}_{0R}^{0}\} + 0 \\ \mathrm{Re}\{\overline{S}_{0I}\} = \mathrm{Re}\{\overline{S}_{0I}^{0}\} + P_{DCI} \\ \mathrm{Im}\{\overline{S}_{0I}\} = \mathrm{Im}\{\overline{S}_{0I}^{0}\} + 0 \end{cases} \tag{4.59}$$

where $\left(\overline{S}_{0R}^{0}\ \overline{S}_{0I}^{0}\right)^{t}$ is the vector of nodal power injections at the DC side, which were derived using Eqs. (4.46) and (4.47) and given by Eq. (4.50).

The expression (4.60) encapsulates the DC active power losses in the cable, which is derived following a similar procedure as in (4.46) and (4.47).

$$\begin{aligned} \begin{pmatrix} P_{DCR} \\ P_{DCI} \end{pmatrix} &= \begin{pmatrix} E_{DCR} & \\ & E_{DCI} \end{pmatrix} \left\{ \begin{pmatrix} G_{DC} & -G_{DC} \\ -G_{DC} & G_{DC} \end{pmatrix} \begin{pmatrix} E_{DCR} \\ E_{DCI} \end{pmatrix} \right\} \\ &= \begin{pmatrix} E_{DCR}^2 G_{DC} - G_{DC} E_{DCI} E_{DCR} \\ E_{DCI}^2 G_{DC} - G_{DC} E_{DCI} E_{DCR} \end{pmatrix} \end{aligned} \tag{4.60}$$

Substituting Eqs. (4.50) and (4.60) into (4.59) yields the full set of nodal power injections for the rectifiers and the inverter stations:

$$\begin{aligned} \overline{S}_{vR} &= (P_{vR} + \mathrm{j}Q_{vR}) = V_{vR}^2 Y_{vRvR} e^{-\mathrm{j}\gamma_{vRvR}} - V_{vR} E_{DCR} Y_{vR0} e^{\mathrm{j}(\theta_{vR} - \gamma_{vR0} - \phi_R)} \\ \overline{S}_{0R} &= (P_{0R} + \mathrm{j}Q_{0R}) = E_{DCR}^2 [(G_{swR} - \mathrm{j}k_2^2 m_{aR}^2 B_{eqR}) + Y_{00R} e^{-\mathrm{j}\gamma_{00R}}] \\ &\quad - V_{vR} E_{DCR} Y_{vR0} e^{\mathrm{j}(\phi_R - \theta_{vR} - \gamma_{vR0})} + P_{DCR} \\ \overline{S}_{vI} &= (P_{vI} + \mathrm{j}Q_{vI}) = V_{vI}^2 Y_{vIvI} e^{-\mathrm{j}\gamma_{vIvI}} - V_{vI} E_{DCI} Y_{vI0} e^{\mathrm{j}(\theta_{vI} - \gamma_{vI0} - \phi_I)} \\ \overline{S}_{0I} &= (P_{0I} + \mathrm{j}Q_{0I}) = E_{DCI}^2 [(G_{swI} - \mathrm{j}k_2^2 m_{aI}^2 B_{eqI}) + Y_{00I} e^{-\mathrm{j}\gamma_{00I}}] \\ &\quad - V_{vI} E_{DCI} Y_{vI0} e^{\mathrm{j}(\phi_I - \theta_{vI} - \gamma_{vI0})} + P_{DCI} \end{aligned} \tag{4.61}$$

It should be noted that the terms $\overline{Y}_{vR0}$ and $\overline{Y}_{vI0}$ are functions of m_{aR} and m_{aI} – refer to the VSC equivalent admittance definition given in Eq. (4.48).

From Eqs. (4.59) and (4.60), and taking into account the equality constraints given in Eq. (4.55), the complete set of equality constraints for the point-to-point VSC-HVDC link is given in Eq. (4.62). This includes any binding equality constraint for enforcing active power control on its DC side.

$$\begin{cases} h_c^{vi} = \begin{pmatrix} h_{P_c}^{vi} \\ h_{Q_c}^{vi} \end{pmatrix} = \begin{pmatrix} P_{vi_c} - P_{g,c}^{gen} + P_{d,c}^{load} \\ Q_{vi_c} - Q_{g,c}^{gen} + Q_{d,c}^{load} \end{pmatrix} = 0 \quad \forall c \in Cv, g \in G, d \in D \\ h_c^{0i} = \mathrm{Re}\{\overline{S}_{0i_c} - \overline{S}_{0i_c}^{0}\} - P_{DCi_c} = 0 \quad \forall c \in Cv \\ C_{Pi}^{dc} = P_{0i_c} - P_{0i_c}^{sch} = 0 \qquad \forall c \in Cv \\ C_{Qi}^{dc} = Q_{0i_c} - Q_{0i_c}^{sch} = 0 \qquad \forall c \in Cv \end{cases} \quad \text{and } i = \{R, I\} \tag{4.62}$$

It would be appropriate at this point to bring to the reader's attention that for the back-to-back VSC-HVDC link, the DC power loss is set to zero in the formulation. Hence, Eq. (4.59) simplifies to the following:

$$\begin{cases} \mathrm{Re}\{\overline{S}_{0R}\} = \mathrm{Re}\{\overline{S}_{0R}^{0}\} + 0 \\ \mathrm{Im}\{\overline{S}_{0R}\} = \mathrm{Im}\{\overline{S}_{0R}^{0}\} + 0 \\ \mathrm{Re}\{\overline{S}_{0I}\} = \mathrm{Re}\{\overline{S}_{0I}^{0}\} + 0 \\ \mathrm{Im}\{\overline{S}_{0I}\} = \mathrm{Im}\{\overline{S}_{0I}^{0}\} + 0 \end{cases} \tag{4.63}$$

4.8.2 VSC-HVDC Link Control

There are several control options available for the VSC rectifier and inverter stations. It was stated in Chapter 3 that in a point-to-point VSC-HVDC configuration, the rectifier station is normally used to regulate power at its DC side whereas the inverter station is used to regulate its DC voltage. These control laws may be added to the general OPF formulation by simply defining additional control constraints in the form of Eq. (4.24) for power flow control at either side of the converter stations. Alternatively, pure penalty functions may be used for enforcing nodal voltage magnitudes at either side of the converter, i.e. AC or DC. The DC power flow control is already included as an equality power constraint on the DC side, given by the set of constraints in Eq. (4.62). More specifically, when the rectifier station exerts DC power flow control at its DC, we have:

$$C_{P_R}^{dc} = P_{0R} - P_{0R}^{sch} = 0 \tag{4.64}$$

For the case of the point-to-point configuration shown in Figure 4.7a, and assuming that the rectifier regulates DC power flow and that the inverter regulates DC nodal voltage, the Lagrangian function may be written down as follows:

$$\begin{aligned} L^{dc\ link}(\mathbf{z}_{vscR}, \mathbf{z}_{vscI}, \boldsymbol{\lambda}_R, \boldsymbol{\lambda}_I) = {} & \lambda_{vR}^t \mathbf{h}^{vR} + \lambda_{vI}^t \mathbf{h}^{vI} + \lambda_{0R} h^{0R} + \lambda_{0I} h^{0I} + \\ & + \lambda_{P_{0R}} C_{P_R}^{dc} + \lambda_{Q_{0R}} C_{Q_R}^{dc} + \lambda_{Q_{0I}} C_{Q_I}^{dc} \end{aligned} \tag{4.65}$$

In Eq. (4.65), the following vectors of state/controls variables and Lagrange multipliers are defined for the DC link:

$$\begin{cases} \mathbf{z}_{vscR} = (\mathbf{z}_{vR}, \mathbf{z}_{0R})^t \\ \mathbf{z}_{vscI} = (\mathbf{z}_{vI}, \mathbf{z}_{0I})^t \end{cases} \tag{4.66}$$

$$\begin{cases} \boldsymbol{\lambda}_R = (\lambda_{vR}, \lambda_{0R}, \lambda_{PQ_R})^t \\ \boldsymbol{\lambda}_I = (\lambda_{vI}, \lambda_{0I}, \lambda_{PQ_I})^t \end{cases} \tag{4.67}$$

Furthermore, in Eqs. (4.66) and (4.67) the following subvectors are defined:

$$\begin{cases} \mathbf{z}_{vR} = (\theta_{vR}, V_{vR}, B_{eqR}, m_{aR}, \phi_R)^t \\ \mathbf{z}_{vI} = (\theta_{vI}, V_{vI}, B_{eqI}, m_{aI}, \phi_I)^t \end{cases} \tag{4.68}$$

$$\begin{cases} \mathbf{z}_{0R} = (E_{dcR})^t \\ \mathbf{z}_{0I} = (E_{dcI})^t \end{cases} \tag{4.69}$$

$$\begin{cases} \boldsymbol{\lambda}_{PQR} = (\lambda_{P_{0R}}, \lambda_{Q_{0R}})^t \\ \boldsymbol{\lambda}_{PQI} = (\lambda_{P_{0I}}, \lambda_{Q_{0I}})^t \end{cases} \tag{4.70}$$

4.8.3 VSC-HVDC Full Set of Equality Constraints

The schematic diagram of a point-to-point VSC-HVDC link and its associated constraints is shown in Figure 4.9. However, note that not all the constraints are active at the same time and that their inclusion in the active set depends on which converter control modes have been selected. The only exceptions to the above are the zero reactive power constraints, i.e. $C^{dc}_{Q_R} = 0$ and $C^{dc}_{Q_I} = 0$, which are always binding given that they correspond to physical constraints.

The vector of state/control variables for the AC-DC system is given in Eqs. (4.71)–(4.74), including subvectors for the AC and DC networks and the DC link.

$$\mathbf{Z}_{ACDC} = (\mathbf{Z}_{AC}, \mathbf{z}_{vscR}, \mathbf{z}_{vscI})^t \tag{4.71}$$

$$\mathbf{Z}_{AC} = (\mathbf{P}^{gen}, \boldsymbol{\theta}_{AC}, \mathbf{V}_{AC}, \mathbf{T})^t \tag{4.72}$$

$$\mathbf{z}_{vscR} = (\mathbf{z}_{vR}, \mathbf{z}_{0R})^t \tag{4.73}$$

$$\mathbf{z}_{vscI} = (\mathbf{z}_{vI}, \mathbf{z}_{0I})^t \tag{4.74}$$

The following vectors are defined for the Lagrange multipliers associated to the nodal power balances (and other constraints) for both the AC system and the DC link.

$$\boldsymbol{\Lambda}_{ACDC} = (\boldsymbol{\Lambda}_{AC}, \boldsymbol{\lambda}_R, \boldsymbol{\lambda}_I)^t \tag{4.75}$$

where

$$\boldsymbol{\Lambda}_{AC} = (\boldsymbol{\lambda}_{AC})^t \tag{4.76}$$

$$\boldsymbol{\lambda}_R = (\lambda_{vR}, \lambda_{0R}, \lambda_{PQR})^t \tag{4.77}$$

$$\boldsymbol{\lambda}_I = (\lambda_{vI}, \lambda_{0I}, \lambda_{PQI})^t \tag{4.78}$$

As shown in Figure 4.9, three sets of constraints are required for the entire AC-DC system: the set of AC constraints, the set of DC constraints and the set of *coupled* AC-DC

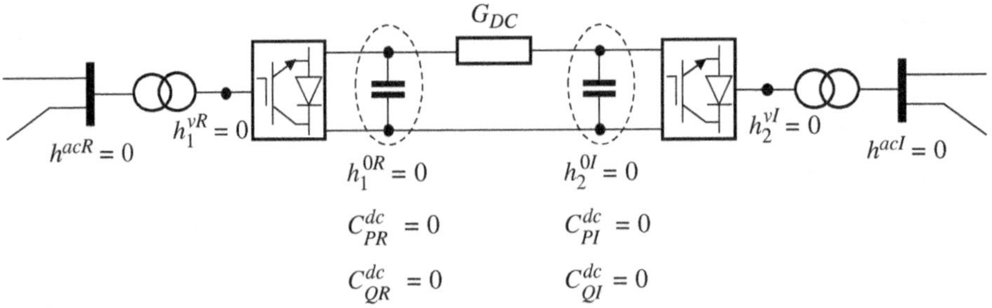

Figure 4.9 Hybrid AC-DC network model with the full set of coupled AC-DC constraints.

constraints. They are defined as follows:

$$\begin{cases} \mathbf{h}^{ac} = \begin{pmatrix} \mathbf{h_P} \\ \mathbf{h_Q} \end{pmatrix} = 0 \\ h_c^{vi} = \begin{pmatrix} h_{P_c}^{vi} \\ h_{Q_c}^{vi} \end{pmatrix} = \begin{pmatrix} P_{vi_c} - P_{g,c}^{gen} + P_{d,c}^{load} \\ Q_{vi_c} - Q_{g,c}^{gen} + Q_{d,c}^{load} \end{pmatrix} = 0 \quad \forall c \in Cv, g \in G, d \in D \quad \text{and } i = \{R, I\} \\ h_c^{0i} = \mathrm{Re}\{\overline{S}_{0i_c} - \overline{S}_{0i_c}^{0}\} - P_{DCi_c} = 0 \quad \forall c \in Cv \\ C_{Pi}^{dc} = P_{0i_c} - P_{0i_c}^{sch} = 0 \quad \forall c \in Cv \\ C_{Qi}^{dc} = Q_{0i_c} - Q_{0i_c}^{sch} = 0 \quad \forall c \in Cv \end{cases} \tag{4.79}$$

The Lagrangian function for the overall AC-DC system is:

$$L^{acdc}(\mathbf{Z}_{ACDC}, \mathbf{\Lambda}_{ACDC}) = F(\mathbf{Z}_{AC}) + L^{ac}(\mathbf{Z}_{AC}, \mathbf{\Lambda}_{AC}) + L^{dc}(\mathbf{z}_{vscR}, \mathbf{z}_{vscI}, \boldsymbol{\lambda}_R, \boldsymbol{\lambda}_I) \tag{4.80}$$

Note that no active inequality constraints are assumed to exist at this point of the solution; when the inequality limits become violated, as the solution proceeds they will be enforced using either penalty functions or the augmented Lagrangian method.

The Lagrangian function of the AC system takes the following form:

$$L^{ac}(\mathbf{Z}_{AC}, \mathbf{\Lambda}_{AC}) = \mathbf{\Lambda}_{\mathbf{AC}}^{\mathbf{t}} \mathbf{h}^{ac}(\mathbf{Z}_{AC}) \tag{4.81}$$

The function $\mathbf{h}^{ac}(\mathbf{Z}_{AC})$ refers to the vector of nodal power balance equations at the AC side of the system, excluding the AC nodes of the DC link. Similarly, the Lagrangian term pertaining to the DC link is defined in Eq. (4.65), with the constraint set defined in Eq. (4.62).

The Lagrangian function for the DC link is:

$$\begin{aligned} L^{dc\ link}(\mathbf{z}_{vscR}, \mathbf{z}_{vscI}, \boldsymbol{\lambda}_R, \boldsymbol{\lambda}_I) &= \boldsymbol{\lambda}^t{}_{vR} \mathbf{h}^{vR} + \boldsymbol{\lambda}^t{}_{vI} \mathbf{h}^{vI} + \lambda_{0R} h^{0R} + \lambda_{0I} h^{0I} + \\ &+ \lambda_{P_{0R}} C_{P_R}^{dc} + \lambda_{Q_{0R}} C_{Q_R}^{dc} + \lambda_{Q_{0I}} C_{Q_I}^{dc} \end{aligned} \tag{4.82}$$

4.8.4 Linearized System of Equations

With reference to Eqs. (4.71)–(4.82), the linearized system of equations for the entire AC-DC system may be written as follows:

$$\begin{pmatrix} \mathbf{H}_{AC} & \mathbf{H}_{ACR} & \mathbf{H}_{ACI} & \mathbf{J}_{AC} & \mathbf{J}_{ACR} & \mathbf{J}_{ACI} \\ \mathbf{H}_{ACR}^t & \mathbf{H}_R & \mathbf{H}_{RI} & \mathbf{J}_{RAC} & \mathbf{J}_{VRR} & \mathbf{J}_{VRI} \\ \mathbf{H}_{ACI}^t & \mathbf{H}_{RI}^t & \mathbf{H}_I & \mathbf{J}_{IAC} & \mathbf{J}_{VIR} & \mathbf{J}_{VII} \\ \mathbf{J}_{AC}^t & \mathbf{J}_{ACR}^t & \mathbf{J}_{ACI}^t & 0 & 0 & 0 \\ \mathbf{J}_{RAC}^t & \mathbf{J}_{VRR}^t & \mathbf{J}_{VRI}^t & 0 & 0 & 0 \\ \mathbf{J}_{IAC}^t & \mathbf{J}_{VIR}^t & \mathbf{J}_{VII}^t & 0 & 0 & 0 \end{pmatrix} \begin{pmatrix} \mathbf{Z}_{AC} \\ \mathbf{z}_{vscR} \\ \mathbf{z}_{vscI} \\ \mathbf{\Lambda}_{AC} \\ \boldsymbol{\lambda}_R \\ \boldsymbol{\lambda}_I \end{pmatrix} = - \begin{pmatrix} \mathbf{g}_{AC} \\ \mathbf{g}_R \\ \mathbf{g}_I \\ \Delta\mathbf{h}^{ac} \\ \Delta\mathbf{h}_R^{vsc} \\ \Delta\mathbf{h}_I^{vsc} \end{pmatrix} \tag{4.83}$$

In Eq. (4.83), the Hessian terms corresponding to the original AC Hessian, defined in Eqs. (4.19) and (4.20), are:

$$\mathbf{H}_{AC} = \nabla^2_{\mathbf{Z}^2_{AC}} L^{acdc}(\mathbf{Z}_{AC}) = \nabla^2_{\mathbf{Z}^2_{AC}} F(\mathbf{Z}_{AC}) + \boldsymbol{\Lambda}^t_{AC} \nabla^2_{\mathbf{Z}^2_{AC}} \mathbf{h}^{ac}(\mathbf{Z}_{AC}) \tag{4.84}$$

The additional terms in Eq. (4.83), arising from the inclusion of the DC link, in the Hessian submatrix are:

$$\mathbf{H}_{ACR} = \nabla^2_{\mathbf{Z}_{AC}\mathbf{z}_{vscR}} L^{dc\,link} = \boldsymbol{\lambda}^t{}_{vR} \nabla^2_{\mathbf{Z}_{AC}\mathbf{z}_{vscR}} \mathbf{h}^{vR} \tag{4.85}$$

$$\mathbf{H}_{ACI} = \nabla^2_{\mathbf{Z}_{AC}\mathbf{z}_{vscI}} L^{dc\,link} = \boldsymbol{\lambda}^t{}_{vI} \nabla^2_{\mathbf{Z}_{AC}\mathbf{z}_{vscI}} \mathbf{h}^{vI} \tag{4.86}$$

$$\mathbf{H}_R = \nabla^2_{\mathbf{z}^2_{vscR}} L^{dc\,link} = \boldsymbol{\lambda}^t{}_{vR} \nabla^2_{\mathbf{z}^2_{vscR}} \mathbf{h}^{vR} + \lambda_{0R} \nabla^2_{\mathbf{z}^2_{vscR}} h^{0R} + \lambda_{P_{0R}} \nabla^2_{\mathbf{z}^2_{vscR}} P_{0R} + \lambda_{Q_{0R}} \nabla^2_{\mathbf{z}^2_{vscR}} Q_{0R} \tag{4.87}$$

$$\mathbf{H}_I = \nabla^2_{z^2_{vscI}} L^{dc\,link} = \boldsymbol{\lambda}^t{}_{vI} \nabla^2_{\mathbf{z}^2_{vscI}} \mathbf{h}^{vI} + \lambda_{0I} \nabla^2_{\mathbf{z}^2_{vscI}} h^{0I} + \lambda_{Q_{0I}} \nabla^2_{\mathbf{z}^2_{vscI}} Q_{0I} \tag{4.88}$$

$$\mathbf{H}_{RI} = \nabla^2_{\mathbf{z}_{vscR}\mathbf{z}_{vscI}} L^{dc\,link} = 0 \tag{4.89}$$

The Jacobian term associated with the AC network corresponds to Eq. (4.20) and it is reproduced here for the sake of completeness:

$$\mathbf{J}_{AC} = \nabla^2_{\mathbf{Z}_{AC}\boldsymbol{\Lambda}_{AC}} L^{acdc} = \nabla_{\mathbf{Z}_{AC}} h^{ac}(\mathbf{Z}_{AC}) \tag{4.90}$$

The additional Jacobian terms in Eq. (4.83) are the following:

$$\mathbf{J}_{ACR} = \nabla^2_{\mathbf{Z}_{AC}\boldsymbol{\lambda}_R} L^{acdc} = \nabla_{\mathbf{Z}_{AC}} \mathbf{h}^{vR} \tag{4.91}$$

$$\mathbf{J}_{ACI} = \nabla^2_{\mathbf{Z}_{AC}\boldsymbol{\lambda}_I} L^{acdc} = \nabla_{\mathbf{Z}_{AC}} \mathbf{h}^{vI} \tag{4.92}$$

$$\mathbf{J}_{RAC} = \nabla^2_{\mathbf{z}_{vscR}\boldsymbol{\Lambda}_{AC}} L^{acdc} = \nabla_{\mathbf{z}_{vscR}} \mathbf{h}^{ac} \tag{4.93}$$

$$\mathbf{J}_{IAC} = \nabla^2_{\mathbf{z}_{vscI}\boldsymbol{\Lambda}_{AC}} L^{acdc} = \nabla_{\mathbf{z}_{vscI}} \mathbf{h}^{ac} \tag{4.94}$$

$$\mathbf{J}_{VRR} = \nabla^2_{\mathbf{z}_{vscR}\boldsymbol{\lambda}_R} L = \nabla_{\mathbf{z}_{vscR}} (\mathbf{h}^{vR} + h^{0R} + P_{0R} + Q_{0R}) \tag{4.95}$$

$$\mathbf{J}_{VII} = \nabla^2_{\mathbf{z}_{vscI}\boldsymbol{\lambda}_I} L = \nabla_{\mathbf{z}_{vscI}} (\mathbf{h}^{vI} + h^{0I} + Q_{0I}) \tag{4.96}$$

$$\mathbf{J}_{VRI} = \nabla^2_{\mathbf{z}_{vscR}\boldsymbol{\lambda}_I} L = 0 \tag{4.97}$$

$$\mathbf{J}_{VIR} = \nabla^2_{\mathbf{z}_{vscI}\boldsymbol{\lambda}_R} L = 0 \tag{4.98}$$

The gradient terms are:

$$g_{AC} = \frac{\partial}{\partial \mathbf{Z}_{AC}} L^{acdc} \tag{4.99}$$

$$g_R = \frac{\partial}{\partial \mathbf{z}_{vscR}} L^{acdc} \tag{4.100}$$

$$g_I = \frac{\partial}{\partial \mathbf{z}_{vscR}} L^{acdc} \tag{4.101}$$

4.9 Multi-Terminal VSC-HVDC Systems in OPF

The point-to-point VSC-HVDC link model presented in Section 4.8 is expanded in this section to model any number of VSC converters making up a multi-terminal VSC-HVDC system of an arbitrary configuration. To illustrate this point, refer to the schematic circuit shown in Figure 4.10, comprising n VSC stations, i.e. n terminals, where $n = n_R + n_I$, with n_R and n_I being the number of stations working as rectifiers and inverters, respectively.

The nodal power constraints at each DC bus of the multi-terminal VSC-HVDC system are shown on the figure. These constraints are an extension of the point-to-point VSC-HVDC formulation, representing the nodal power balance at the DC node, and active and reactive power flow control constraints. It should be remarked that the reactive power flow constraint is always set to zero to maintain zero reactive power flow in the DC link – refer to Eq. (4.51).

In the OPF multi-terminal VSC-HVDC solution, one VSC is selected to maintain constant DC voltage at its DC bus, acting essentially as a slack node and becoming a voltage reference for the other nodal voltages in the DC grid. Notice that this concept is not different from that employed in Chapter 3 for the conventional power flow solution of multi-terminal VSC-HVDC systems. However, in the OPF solution the implementation is carried out quite simply by enforcing a DC nodal voltage constraint, as shown in Section 4.5.6. It will become apparent in Section 4.9.2 – in the test case – that the nature of the unified formulation makes it possible to treat this as simply another constraint on the state variables and the enforcement is carried out by the chosen method in the underlying optimization solver.

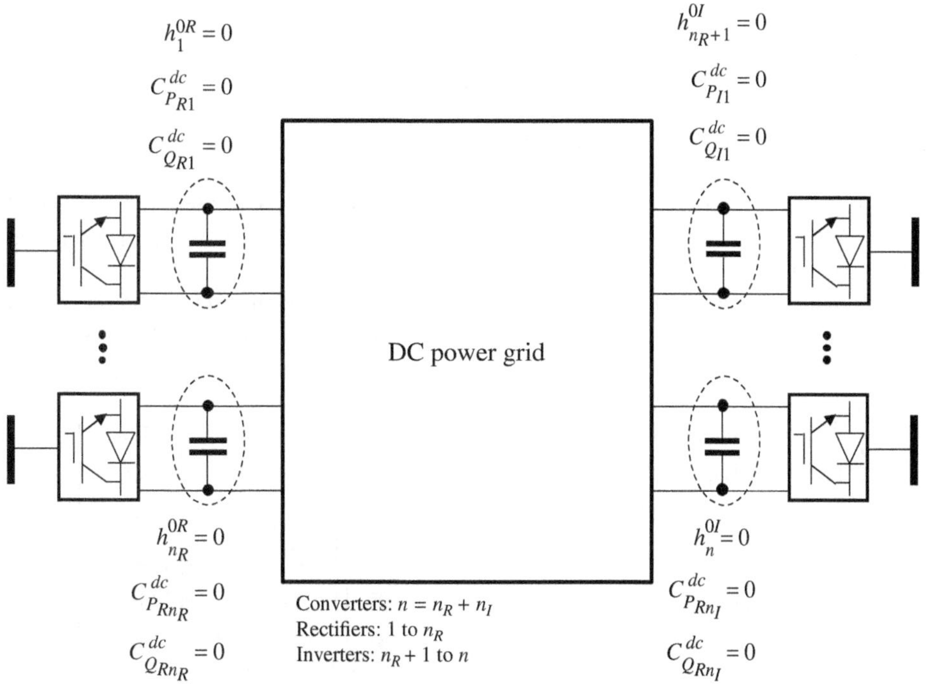

Figure 4.10 Multi-terminal VSC-HVDC system with n-VSC stations.

4.9.1 The Expanded, General Formulation

The number of nodal power balance equations used in (4.79) for the case of the point-to-point VSC-HVDC link is expanded to encompass the n VSC terminal making up the multi-terminal VSC-HVDC system – one for each VSC station. In the generic Figure 4.10 is assumed that n_R stations are operating as rectifiers and that n_I stations are operating as inverters.

Hence, the following vectors of state/control variables are set up:

$$\mathbf{Z}_{AC-DC} = (\mathbf{Z}_{AC}, \mathbf{Z}_{vscR}, \mathbf{Z}_{vscI})^t \tag{4.102}$$

$$\mathbf{Z}_{AC} = (\mathbf{P}^{gen}, \mathbf{\theta}_{AC}, \mathbf{V}_{AC}, \mathbf{T})^t \tag{4.103}$$

$$\mathbf{Z}_{vscR} = (\mathbf{Z}_{vR}, \mathbf{Z}_{0R})^t \tag{4.104}$$

$$\mathbf{Z}_{vscI} = (\mathbf{Z}_{vI}, \mathbf{Z}_{0I})^t \tag{4.105}$$

In Eqs. (4.102)–(4.105) the vector $\mathbf{Z}_{AC}$ is for the AC power network, the vector $\mathbf{Z}_{vscR} = (z_{vR1}, z_{vR2}, \ldots. z_{vRn_R}, z_{0R1}, z_{0R2}, \ldots. z_{0Rn_R})^t$ is for the n_R-terminals in the system acting as rectifiers and $\mathbf{Z}_{vscI} = (z_{vI1}, z_{vI2}, \ldots. z_{vIn_I}, z_{0I1}, z_{0I2}, \ldots. z_{0In_I})^t$ is for the n_I-terminals in the system acting as inverters.

Similarly, the vector of Lagrange multipliers for the multi-terminal system can be defined as follows:

$$\mathbf{\Lambda}_{AC-DC} = (\mathbf{\Lambda}_{AC}, \mathbf{\Lambda}_R, \mathbf{\Lambda}_I)^t \tag{4.106}$$

$$\mathbf{\Lambda}_{AC} = (\boldsymbol{\lambda}_{AC})^t \tag{4.107}$$

$$\mathbf{\Lambda}_R = (\boldsymbol{\lambda}_{vR}, \boldsymbol{\lambda}_{0R}, \boldsymbol{\lambda}_{PQR})^t \tag{4.108}$$

$$\mathbf{\Lambda}_I = (\boldsymbol{\lambda}_{vI}, \boldsymbol{\lambda}_{0I}, \boldsymbol{\lambda}_{PQI})^t \tag{4.109}$$

In Eqs. (4.106)–(4.109) the vector Λ_{AC} is for the AC power network, the vector $\boldsymbol{\lambda}_{vR} = (\lambda_{vR1}, \lambda_{vR2}, \ldots. \lambda_{vRn_R})^t$ and $\boldsymbol{\lambda}_{0R} = (\lambda_{0R1}, \lambda_{0R2}, \ldots. \lambda_{0Rn_R})^t$ is for the n_R-terminals in the system acting as rectifiers and $\boldsymbol{\lambda}_{vI} = (\lambda_{vI1}, \lambda_{vI2}, \ldots. \lambda_{vIn_R})^t$ and $\boldsymbol{\lambda}_{0I} = (\lambda_{0I1}, \lambda_{0I2}, \ldots. \lambda_{0In_I})^t$ is for the n_I-terminals in the system acting as inverters. The terms λ_{PQR} and λ_{PQI} are Lagrange multipliers, which are used to enforce the active and reactive power constraints at the DC bus of the corresponding receiving and sending end converters.

The linearized system of Eq. (4.83) may be expanded to accommodate the linearized representation of the AC system and that of the multi-terminal VSC-HVDC system, as shown below:

$$\begin{pmatrix} [\mathbf{H}]_{AC} & [\mathbf{H}]_{ACR} & [\mathbf{H}]_{ACI} & [\mathbf{J}]_{AC} & [\mathbf{J}]_{ACR} & [\mathbf{J}]_{ACI} \\ [\mathbf{H}]^t_{ACR} & [\mathbf{H}]_{R} & [\mathbf{H}]_{RI} & [\mathbf{J}]_{RAC} & [\mathbf{J}]_{VR} & [\mathbf{J}]_{VRI} \\ [\mathbf{H}]^t_{ACI} & [\mathbf{H}]^t_{RI} & [\mathbf{H}]_{I} & [\mathbf{J}]_{IAC} & [\mathbf{J}]_{VIR} & [\mathbf{J}]_{VI} \\ [\mathbf{J}]^t_{AC} & [\mathbf{J}]^t_{ACR} & [\mathbf{J}]^t_{ACI} & [\mathbf{0}] & [\mathbf{0}] & [\mathbf{0}] \\ [\mathbf{J}]^t_{RAC} & [\mathbf{J}]^t_{VR} & [\mathbf{J}]^t_{VRI} & [\mathbf{0}] & [\mathbf{0}] & [\mathbf{0}] \\ [\mathbf{J}]^t_{IAC} & [\mathbf{J}]^t_{VIR} & [\mathbf{J}]^t_{VI} & [\mathbf{0}] & [\mathbf{0}] & [\mathbf{0}] \end{pmatrix} \begin{pmatrix} [\mathbf{Z}]_{AC} \\ [\mathbf{Z}]_{vscR} \\ [\mathbf{Z}]_{vscI} \\ [\mathbf{\Lambda}]_{AC} \\ [\mathbf{\Lambda}]_{R} \\ [\mathbf{\Lambda}]_{I} \end{pmatrix} = - \begin{pmatrix} [\mathbf{g}]_{AC} \\ [\mathbf{g}]_{R} \\ [\mathbf{g}]_{I} \\ [\Delta\mathbf{h}]^{ac} \\ [\Delta\mathbf{h}]^{vsc}_{R} \\ [\Delta\mathbf{h}]^{vsc}_{I} \end{pmatrix} \tag{4.110}$$

The Hessian and Jacobian entries in (4.110) have the same patterns as those given in Eqs. (4.84)–(4.98) for the case of the point-to-point VSC-HVDC link, but for the multi-terminal system there will be n terms as opposed to two terms. This higher dimensionality of the multi-terminal VSC-HVDC system is reflected in (4.110) using brackets.

4.9.2 Multi-Terminal VSC-HVDC Test Case

In this section, the five-bus AC test system in [3] is modified to embed in it a multi-terminal VSC-HVDC transmission system, as shown in Figure 4.11. This is now a hybrid AC-DC test system comprising two synchronous generators and three VSC terminals, effectively forming an interconnected AC-DC transmission system. The data for the modified hybrid test system is given in Tables 4.9–4.14.

This example serves the purpose of illustrating the applicability of the unified OPF formulation presented in this chapter for solving AC-DC systems and to assess the impact that converter losses have on the overall system performance. The system has been modelled in the AIMMS solution environment. The optimization solution uses the augmented Lagrangian algorithm already available in AIMMS.

4.9.2.1 DC Network

The data for the three converter stations, including interfacing transformers and filters, is given in Table 4.10. The data for the three DC lines is given in Table 4.11. The DC voltage information for the three buses is given in Table 4.12.

The relevant VSC constraints are given in Table 4.13.

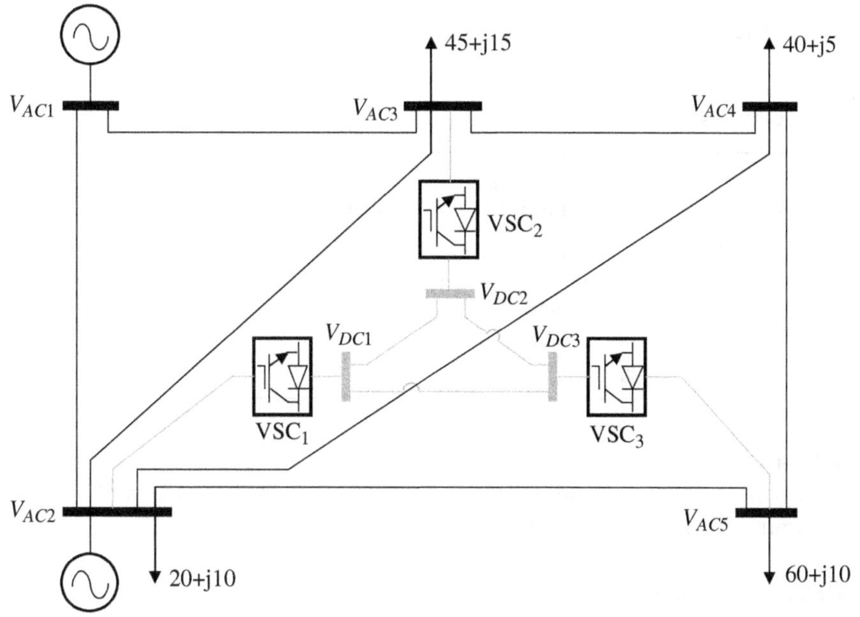

Figure 4.11 Modified five-bus power system [3] to include a tree-terminal VSC-HVDC system.

Table 4.10 VSC data.

VSC No	AC bus	DC bus	Rated power (MVA)	Base DC voltage (kV)	Transformer resistance (p.u.)	Transformer inductance (p.u.)	Shunt filter susceptance (p.u.)	Switching losses (nominal current)
1	VAC2	VDC1	100	345	0.0015	0.1121	0	1%
2	VAC3	VDC2	100	345	0.0015	0.1121	0	1%
3	VAC5	VDC3	100	345	0.0015	0.1121	0	1%

Table 4.11 DC lines data.

Line	Line resistance (p.u.)	DC voltage (kV)	DC power (MW)
1	0.0260	345	100
2	0.0365	345	100
3	0.0260	345	100

Table 4.12 DC bus data.

DC bus	Max voltage (p.u.)	Min voltage (p.u.)
VDC 1	1.10	0.9
VDC 2	1.01	1.01
VDC 3	1.10	0.9

Table 4.13 VSC constraints.

Power constraints	AC active power (MW)		AC reactive power (MW)		AC output voltage (p.u.)	
VSC	Max	Min	Max	Min	Max	Min
1	500	−500	500	−500	1.20	0.8
2	500	−500	500	−500	1.20	0.8
3	500	−500	500	−500	1.20	0.8

4.9.2.2 AC Network

The AC voltage magnitude constraints at the nodes are given in Table 4.14 and the generator constraints are given in Table 4.15.

4.9.2.3 Objective Function

The objective is to produce a minimum-cost solution by reducing the generators' fuel costs. A simple cost function is chosen for the two generators, given by the following

Table 4.14 AC voltage magnitude constraints.

AC node	Max voltage (p.u.)	Min voltage (p.u.)
AC 1 (slack)	1.02	1.0
VAC 2	1.02	1.0
VAC 3	1.10	0.9
VAC 4	1.10	0.9
VAC 5	1.10	0.9

Table 4.15 Generator constraints.

Generator	Max P (MW)	Min P (MW)	Max Q (MVAR)	Min Q (MVAR)	Voltage reference
1	250	10	500	−500	1.02
2	40	0	40	−40	1.02

linear function.

$$f(P_g^{gen}) = cP_g^{gen} \quad \forall g \in G \tag{4.111}$$

4.9.2.4 Summary of OPF Results

Two OPF solutions are carried out: (i) with no converter's internal losses, and (ii) with converter internal losses included. In the latter case, the power loss is taken to be a function of the actual VSC's output current, such as described by the multi-terminal VSC-HVDC model in Chapter 3. In both cases the DC voltage at converter VSC2 is kept at a reference value of 1.01 p.u. The test system has been implemented in the solution environment afforded by AIMMS.

For the case when the converters are taken to incur no switching power loss, two different solutions tools were used to find out the optimum operating point. Both application tools furnished the same solution. One OPF algorithm was coded using Matlab scripting and the other uses AIMMS. The case when the VSC's switching power losses are taken into account was only carried out using AIMMS. Initially, values of 1% are assigned to the three VSC converters, as seen in Table 4.10. A summary of the results is given in Table 4.16.

It is observed from these results that the converters contribute a small portion of the overall system power loss. It is in fact about 5.5% of the total power loss, with the rest corresponding to power loss in the seven AC transmission lines and three DC transmission lines. The converter results for both solution cases are given in Tables 4.17 and 4.20, respectively.

In both solution cases the nodal voltage magnitudes on the AC and DC sides of the network have been kept within their limits and no constraint violations have been observed during the solution process. Note that since the problem was formulated as a nonlinear program, the results are guaranteed only to be locally optimal, owing to its nonconvex

Table 4.16 A summary of the hybrid test system results.

Model object	OPF solver	Test	Convergence time	Optimality of primal	Total converter losses	Total system losses	Final objective solution
AIMMS	MINOS – Augmented Lagrangian	Lossless converters	0.02 s – 7 Global iterations	Locally optimal	0 MW	3.79 MW	$8102 h^{-1}
AIMMS	MINOS – Augmented Lagrangian	Lossy converters	0.05 s – 11 Global iterations	Locally optimal	0.23 MW	4.14 MW	$8119 h^{-1}

nature. In both solutions of the test case, the converters are free to set their AC voltages within the allowable limits, except VSC2, which sets the DC link's reference voltage – it acts as slack bus for the DC network.

A selection of the most relevant VSC converter, DC network and AC network operational parameters is shown below. For the sake of clarity and comprehensiveness, the results are separated into DC and AC network results (Tables 4.17–4.22).

DC Network

4.9.2.5 Converter Outputs – No Converter Losses

Table 4.17 Converters outputs summary.

Converter No.	AC bus voltage (magnitude)	AC bus voltage (phase angle)	DC bus voltage	Output active power (MW)	Output reactive power (MVAR)	DC bus power (MW)	Converter losses (MW)
VSC1	1.004	−3.201	1.017	54.78	8.22	−54.73	0
VSC2	0.996	−4.625	1.010	−20.49	13.66	20.50	0
VSC3	0.994	−5.023	1.008	−33.74	10.08	33.76	0

Table 4.18 DC power lines.

DC line	Sending end power (MW)	Receiving end power (MW)
VDC1 – VDC2	28.64	−28.43
VDC1 – VDC3	26.09	−25.85
VDC2 – VDC3	7.93	−7.91

Table 4.19 Converters PWM performance.

Converter No.	Amplitude modulation ratio	Phase angle (°)
VSC1	0.8143	−11.74
VSC2	0.8338	−1.475
VSC3	0.8243	−0.261

4.9.2.6 Converter Outputs – With Converter Losses

Table 4.20 Converters outputs summary.

Converter No.	AC bus voltage (magnitude)	AC bus voltage (phase angle)	DC bus voltage	Output active power (MW)	Output reactive power (MVAR)	DC bus power (MW)	Converter losses (MW)
VSC1	1.006	−3.147	1.015	37.55	3.85	−37.52	0.14
VSC2	0.992	−4.922	1.010	−12.34	9.68	12.35	0.02
VSC3	0.991	−5.486	1.008	−24.71	7.98	24.72	0.07

Table 4.21 DC lines.

DC line	Sending end power (MW)	Receiving end power (MW)
VDC1 – VDC2	19.07	−18.98
VDC1 – VDC3	18.31	−18.20
VDC2 – VDC3	6.661	−6.660

Table 4.22 Converters PWM performance.

Converter No.	Amplitude modulation ratio	Phase angle (°)
VSC1	0.8127	−9.010
VSC2	0.8232	−2.996
VSC3	0.8183	−1.580

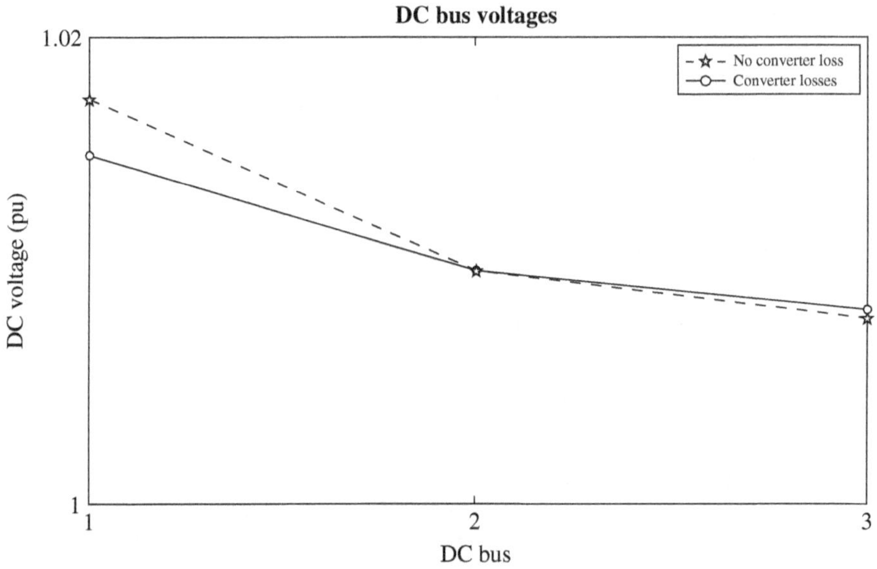

Figure 4.12 DC network voltage profile.

It is apparent from these results that in the first test case, power flow is entering into the DC meshed system through VSC1: importing 54.78 MW. In turn, 20.50 MW and 33.75 MW are put back into the AC system through converters VSC2 and VSC3, respectively. In the second test case, the power import into the DC mesh system is also through VSC1 but with a reduced value of 37.55 MW and power is put back into the AC network through converters VSC2 and VSC3, at values of 12.34 MW and 24.72 MW, respectively.

The different imports of power flow through VSC1 are due to the different DC voltages that exist at the bus DC1 for both test cases. The DC voltages are shown in Figure 4.12 for the three DC buses. It is seen from the figure that VSC1 keeps a higher voltage in the case when power losses are not included. This difference in voltage magnitude at VSC1 makes clear the importance of having a realistic converter loss model included in the OPF formulation. Note that in both cases, VSC2 provides the reference voltage for the whole of the DC network, acting as slack bus for the DC network. It should also be noted that apart from the voltage regulation in bus DC2 by VSC2, there is no other active control.

The PWM's amplitude modulation ratios and phase angles of the VSC converters are given in Tables 4.19 and 4.22 for both test cases. The results show that the three converters operate within their linear regions, i.e. below 1. It is also seen from the VSC's phase angles that converter VSC1 carries the largest load in both cases, which reflects in their largest phase angles, with values of −11.74°and −9.01°, respectively.

AC Network

Table 4.23 shows the generator outputs for both test cases.

Table 4.23 Generators outputs.

Generator Bus	No converter loss		With converter loss	
	Active power (MW)	Reactive power (MVAR)	Active power (MW)	Reactive power (MVAR)
VAC1	128.79	−7.72	129.14	−8.38
VAC2	40	9.50	40	15

4.9.2.7 Power Flows in AC Transmission Lines – With No Converter Losses

Table 4.24 Active power flows in the AC network.

AC Lines	Sending end active power (MW)	Receiving end active power (MW)
VAC1 – VAC2	94.590 8	−92.869 9
VAC1 – VAC3	34.198 7	−33.293 3
VAC2 – VAC3	13.884 6	−13.769 9
VAC2 – VAC4	17.676 3	−17.490 4
VAC2 – VAC5	26.529 7	−26.250 6
VAC3 – VAC4	22.565 3	−22.514 0
VAC4 – VAC5	0.004 41	−0.004 39

Table 4.25 Reactive power flow in the AC network.

AC Lines	Sending end reactive power (MVAR)	Receiving end reactive power (MVAR)
VAC1 – VAC2	−5.3197	4.3353
VAC1 – VAC3	−2.4048	0.0220
VAC2 – VAC3	−1.8131	−1.8440
VAC2 – VAC4	−1.7356	−1.6990
VAC2 – VAC5	−1.2903	−0.8673
VAC3 – VAC4	−1.0922	−0.7332
VAC4 – VAC5	−2.5678	−2.3709

4.9.2.8 Power Flows in AC Transmission Lines – With Converter Losses

Table 4.26 Active power flows in the AC network.

AC lines	Sending end active power (MW)	Receiving end active power (MW)
VAC1 – VAC2	92.591 4	−90.940 6
VAC1 – VAC3	36.552 4	−35.524 1
VAC2 – VAC3	17.740 1	−17.551 6
VAC2 – VAC4	21.193 8	−20.925 7
VAC2 – VAC5	34.455 8	−20.925 8
VAC3 – VAC4	20.416 8	−33.985 0
VAC4 – VAC5	1.300 11	−1.298 61

Table 4.27 Reactive power flows in the AC network.

AC lines	Sending end reactive power (MVAR)	Receiving end reactive power (MVAR)
VAC1 – VAC2	−6.874 13	5.671 7
VAC1 – VAC3	−1.504 30	−0.474 0
VAC2 – VAC3	−0.318 76	−3.107 9
VAC2 – VAC4	−0.449 91	−2.731 8
VAC2 – VAC5	0.190 78	−1.767 2
VAC3 – VAC4	−2.394 57	0.555 5
VAC4 – VAC5	−2.823 80	−2.079 4

The active and reactive power flows in the AC transmission lines are shown in Tables 4.24–4.27. They correspond to the cases with converter losses and no converter losses and complement the power flows results in the DC transmission lines given in Tables 4.18 and 4.21, respectively.

4.10 Conclusion

The basic principles of the OPF problem as a nonlinear optimization problem have been explained in this chapter. TSOs use the OPF application tool for the operational planning and near real-time dispatch of utility-size power grids.

The theoretical requirements for developing a unified OPF formulation of hybrid AC-DC systems have been laid out and applied to develop a comprehensive frame-of-reference suitable for the optimal solution of modern power systems. This involved developing the OPF descriptor of the AC-DC kernel, namely the VSC. This was followed by expanding the kernel model to build up models of VSC-HVDC links and multi-terminal VSC-HVDC systems of an arbitrary configuration.

It has been shown that in order to achieve the more complex VSC-HVDC solutions, the use of a set of appropriately defined constraints is mandatory in order to model the DC grid. This yields a unified formulation for achieving OPF solutions of combined AC-DC systems using Newton's method, exhibiting true quadratic convergence for the inner loop. The equipment and operational constraints are enforced in the outer loop and quadratic convergence is not possible here. Nevertheless, the unified formulation makes it simple for power system modellers to quickly formulate and solve combined AC-DC networks with no additional modifications to the solver structure, as demonstrated in this chapter using AIMMS.

A simple test system has been solved using this solution environment. The example illustrates one possible role that a multi-terminal DC system may play in future power systems, namely that a multi-terminal DC system may be used to re-route power, thus alleviating congestion, all while adhering to security and economic constraints. It should be remarked that the OPF formulation presented in this chapter is capable of solving the full network, AC and DC networks, in a unified manner, with its ensuing strong convergence characteristics.

References

1 B. Stott and O. Alsaç, (2012) "Optimal power flow – basic requirements for real-life problems and their solutions", White paper, Arizona, USA, 1 June 2012. https://pdfs.semanticscholar.org/673e/416fa38c04611e90b12cbdb1a74ca08367db.pdf

2 Vaahedi, E. (2014.). *Practical Power System Operation*. IEEE Press.

3 Acha, E., Fuerte-Esquivel, C.R., Ambriz-Perez, H., and Angeles-Camacho, C. (2004). *FACTS Modelling and Simulation in Power Networks*. John Wiley & Sons.

4 Capitanescu, F. (2018). Critical review of recent advanced and further developments needed in AC optimal power flow. *Electric Power Systems Research* vol. 136: 57–68.

5 Bertsekas, D.P. (1982). *Constrained Optimization and Lagrange Multiplier Methods*. Academic Press.

6 Boyd, S. and Vandenberghe, L. (2009). *Convex Optimization*. Cambridge University Press.

7 Saccomano, F. (2003). *Electric Power Systems Analysis and Control*. IEEE Press.

8 Stevenson, W.D. and Grainger, J. (1994). *Power System Analysis*. McGraw-Hill.

9 Kazemtabrizi, B. and Acha, E. (2014). An advanced STATCOM model for optimal power flows using Newton's method. *IEEE Transactions on Power Systems* vol. 29 (2): 514–525.

10 Dommel, H.W. and Tinney, W.F. (1968). Optimal power flow solutions. *IEEE Transactions on Power Apparatus & Systems* PAS-97: 37–47.

11 Zimmerman, R.D., Murillo-Sanchez, C.E., and Thomas, R.J. (2011). MATPOWER: steady-state operations, planning and analysis tools for power systems research and education. *IEEE Transactions on Power Systems* 26 (1): 12–19.

12 Pardalos, P. and Resende, M.G.C. (2002). *Handbook of Applied Optimization*. Oxford University Press.

13 B. Kazemtabrizi and E. Acha, (2012), "A Comparison Study Between Mathematical Models of Static VAR Compensators Aimed at Optimal Power Flow Solutions", 16th IEEE Mediterranean Electrotechnical Conference, pp. 1125–1128.

5

State Estimation

5.1 Introduction

The wide-area blackout that affected the northeast region of the United States of America in 1965 prompted the electrical utilities to embark on new procedures that ensured a higher level of service reliability. From that point onwards, the electricity supply industry started making use of concepts such as safety analysis, safety indexes, stability analysis and optimization, all of which nowadays are standard features in energy management system (EMS) technology.

The power network's measurement and control systems were very basic at the time when the wide-area blackout occurred; they could only monitor, in real time, the status of switches, the system frequency and the voltage magnitudes and powers measured by current transformers and voltage transformers at the substation level. A major drawback was that the readings arriving at the control centre were contaminated with the noise inherent to the measurement chain and calibration errors.

The early efforts to develop a more reliable measurement system were aimed at acquiring, at few seconds' intervals, the network's topology (switches' status) and analogue (measurement equipment) information in order to configure a real-time database. This process of data acquisition, detection and signalling of the power network, together with the use of graphical displays and storage of all the events, comprise the supervisory control and data acquisition (SCADA) system.

Once the database is configured, all the applications that reside in the EMS take as data input the operating state of the power network corresponding to that point in time. Such a snapshot of the system is used as input information to a large number of power systems applications; hence, the fidelity and accuracy of this information are reflected in the final, overall results provided by the EMS and in the ensuing analysis and decisions taken.

There are three fundamental issues in real-time power systems applications, which are not satisfactorily addressed by the type of power flow algorithms described in Chapters 3 and 4 of this book:

(i) The number of measurements available is usually much larger than the minimum amount of information required to determine the system state (i.e. there is redundancy).

VSC-FACTS-HVDC: Analysis, Modelling and Simulation in Power Grids, First Edition.
Enrique Acha, Pedro Roncero-Sánchez, Antonio de la Villa Jaén, Luis M. Castro and Behzad Kazemtabrizi.

Companion website: www.wiley.com/go/acha_vsc_facts

(ii) Owing to inherent errors in the instrumentation, there is always a degree of inconsistency between the values of the available measurements.
(iii) The loss of certain measurements, for periods of time, is something to be expected; hence, the number and the distribution of measurements will change from one instant to the next.

Owing to the basic assumptions under which the classic power flow formulation is developed, its use in real-time applications becomes impractical. A paradigm change in which the power systems equations are solved, which fully took advantage of the rich source of real-time data that is available in a control centre, is credited to F. Schweppe and his team at the Massachusetts Institute of Technology (MIT), who in 1970 published a method to carry out state estimation in electrical power networks [1]. The state estimation process allows the determination of the state variables of a power system using statistical criteria to the entire set of available measurements, in real time. Thus, the estimated state computed at a given point in time yields the statistically optimal state that can be obtained from the available information and models that describe the power system's behaviour. The state estimators have evolved to become crucial elements in the control of today's electrical power networks [2, 3].

Over the years, a number of HVDC and FACTS equipment models have been developed and incorporated into state estimation formulations. The early publications [4–7] addressed the combined solution of AC and DC power networks interconnected by thyristor-based HVDC links. The work reported in [8] describes a sequential solution method with improved rate of convergence by using a decoupled method. In [9], simplified state estimation models of the thyristor-controlled series compensator (TCSC) and the unified power flow controller (UPFC) were put forward and in [10] a more elaborated model for the UPFC is reported where the controller's operating constraints are taken into account. New models for the TCSC, UPFC and static var compensator (SVC) are proposed in [11, 12], where additional control variables are incorporated as state variables, with their initial conditions specified to enhance the algorithm's rate of convergence. Additional models of FACTS equipment are presented in [13, 14].

In reference [15], a basic model of the voltage source converter is developed using first principles and then extensions are made to include the representation of the STATCOM and VSC-HVDC links. This model incorporates an explicit representation of the VSC's DC circuit and takes into account the converter power losses. In [16] the model of a multi-terminal VSC-HVDC system is introduced. The outputs of phasor measurement units (PMUs) were identified, from the outset, to be measurements capable of improving the estimation process of a power network [17, 18].

5.2 State Estimation of Electrical Networks

The measurements that are used to form the real-time database are carried out at the substations using current and potential transformers (i.e. transducers). The amplitudes of the instantaneous voltage and current waveforms measured by the transducers are used to calculate the RMS values of voltage and current together with the corresponding values of active power and reactive power. This information is concentrated in remote terminal units (RTUs) together with information of the contact positions of the switchgear equipment and passed onto the SCADA systems located in the control centres.

In modern substations where the standard IEC 61850 is employed, the measurements and the status of the switchgear equipment are directly processed by intelligent electronic devices (IEDs). The captured information is shared with other IEDs through a local area network (LAN) and from there onto the SCADA systems.

As with any kind of measurement carried out in the electrical power system, there is a degree of built-in noise. The noise results from errors introduced during the various stages of the measurement chain, comprising the transducers, the SCADA system and the communication system. For instance, if several voltage measurements were available at a given bus, it is very unlikely that the measured values would be exactly the same. Likewise, the power balance at a given bus may differ from zero.

In the control centre, after receiving information from the various substations, the priority is to obtain a reliable assessment of the operational state of the electrical power system, corresponding to the point in time when the measurements were captured. The power system state estimator carries out this function in various steps, as indicated in

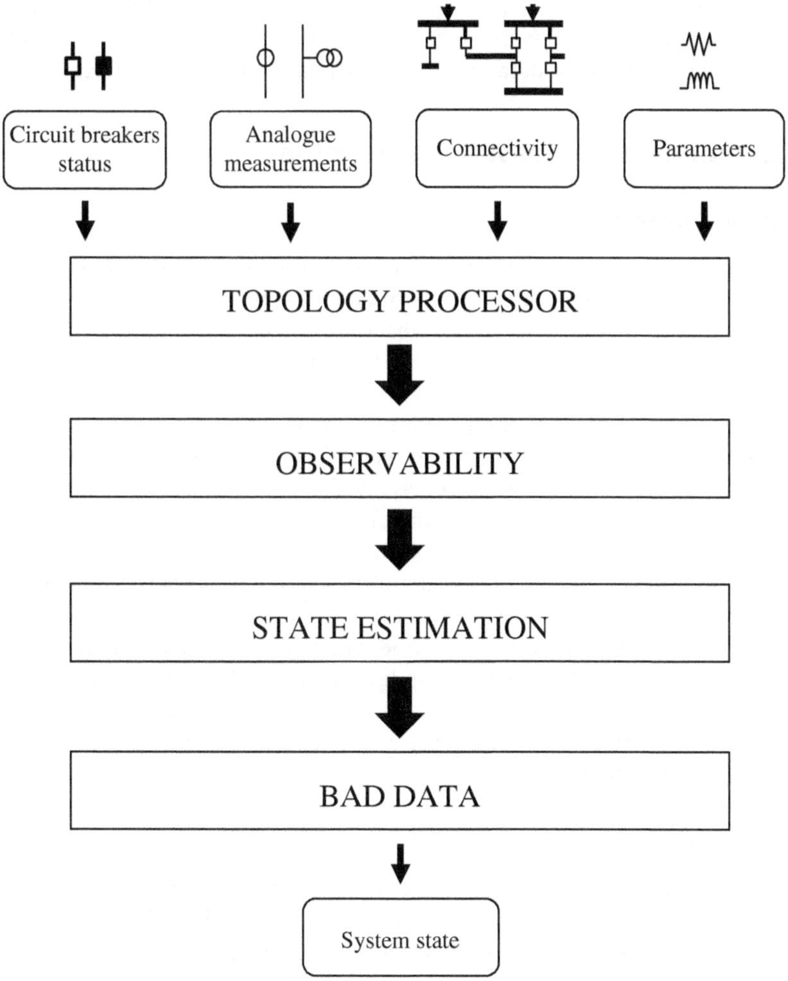

Figure 5.1 State estimation process.

Figure 5.1. It uses the dynamic information received from the substations and the static information corresponding to the electrical parameters of the equipment that make up the power network and the network's connectivity.

The various steps involved in the state estimation process are succinctly described below:

- *Topology processor*. All the information sent from the substations to the EMS is collected in this module; a model of the power network is built and the available measurements are assigned to the newly established model.
- *Observability*. In this module, an analysis of the distribution of the network measurements is carried out. It checks whether or not the available information is sufficient to calculate the estimated system state and determines the observable areas of the network.
- *Estimation*. The state estimation algorithm resides in this module; it furnishes an estimate of the system state.
- *Bad data identification*. The process of detection and identification of bad data (measurements that are grossly in error) is carried out in this module. In the absence of bad data, the estimated state is given a clean bill of health. However, if bad data is detected, this is deleted from the data set and the estimation process is repeated.

5.3 Network Model and Measurement System

5.3.1 Topological Processing

It is quite normal for a power system network to change its topology as a result of control actions, maintenance operations and power network contingencies. The state estimator is a power systems application tool that runs in real time and uses the topology that the power network exhibits at the point in time when the state is estimated. The topology processor module carries out such a task. It uses a static database that contains the connection status of the various pieces of equipment making up the network: transmission lines, transformers, switches, measuring devices, etc. Moreover, a dynamic database is used – it comprises the real-time information received at the control centre from the substations. This information contains the measurements received from the measuring equipment installed at the various substations, such as voltage, current, power, discrete information corresponding to the operational status (open or closed) of the switchgear equipment (e.g. circuit breakers and disconnectors).

This information is used to build a network model known as the bus-branch model. The branches are associated to the transmission lines and transformers that are energized. The branches are joined at their respective ends through electrical nodes. Hence, the number of nodes that are part of the energized power network at a given point in time is determined in this module.

5.3.2 Network Model

In general, the various models used to determine the estimation of the network's state assume that the power system operates under steady-state, balanced conditions – this is similar to the simplified assumptions adopted in the power flow formulation presented in Chapters 3, 4 and 6. Hence, an AC power network with n nodes requires a state vector $\mathbf{x}$ with $2n$ variables to fully describe its operational state.

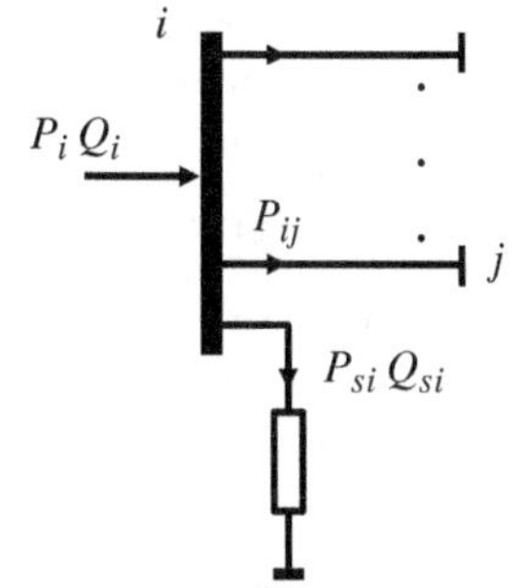

Figure 5.2 Nodal power balances at a generic node *i*.

One available option is to use the polar representation, i.e. voltage magnitude and phase angle:

$$\mathbf{x} = [V_1, \dots, V_n, \theta_1, \dots, \theta_n]^T \tag{5.1}$$

where V_i and θ_i are the voltage magnitude and phase associated with a generic node i, where $i = 1 \dots n$. This basic state vector may be extended to incorporate additional variables aiming at achieving a more convenient description of a given piece of equipment and to estimate the value of the new state variable based on available measurements. A load tap changer (LTC) transformer is a good example of this, where the position of the transformer tap may be estimated. To this end, the off-nominal turns ratio of the transformer tap position, which is a non-voltage variable, is incorporated as an additional variable in the state vector. Likewise, HVDC and FACTS equipment may also contribute non-voltage state variables to the state vector – this issue will be discussed later in the chapter.

A generic node i is depicted in Figure 5.2; it emphasizes the relationship between the branch and the node elements, amenable to the construction of the branch-node model used in power systems state estimation.

The bus-branch model is amenable to a straightforward balance of active and reactive powers at each node of the power network, say at the generic node i:

$$P_i = P_{si} + \sum_{j \subset \Omega} P_{ij} \tag{5.2}$$

$$Q_i = Q_{si} + \sum_{j \subset \Omega} Q_{ij} \tag{5.3}$$

where P_i and Q_i are the net active and reactive powers injected at node i; P_{ij} and Q_{ij} are power flows of active and reactive power through the branch $i-j$; Ω is the set of all nodes connecting to node i; and the variables P_{si} and Q_{si} represent the active and reactive powers of the shunt element connected to node i.

Both branches and shunt components will be described with reference to the particular model employed to represent a given piece of equipment. In this application, the equipment model should be formulated in terms of only electrical parameters (i.e. voltage, current and power) as a function of the state variables and their own electrical parameters.

Having determined the values of the state variables and using readily available information of the parameters of the equipment making up the network, it follows that the power flows throughout the branches and shunt elements of the power network are determined with ease. In turn, using the power balance equations (5.2) and (5.3), the power injections at each node of the network are determined.

In a conventional state estimator, the following premise is taken at the outset: given an electrical magnitude z, it is always possible to determine a function $h(\cdot)$, such that the magnitude z may be expressed as a function of the state vector $\mathbf{x}$, i.e.

$$z = h(\mathbf{x}) \tag{5.4}$$

The models of equipment used in conventional state estimation are similar to those used in the conventional power flow problem presented in Chapter 3. However, in this chapter the nodal voltages used as state variables are expressed in polar form. The nodal power expressions of the conventional power system equipment are given below:

- Transmission lines are described by

$$P_{ij} = V_i^2(G_{si} + G_{ij}) - V_iV_j(G_{ij}\cos\theta_{ij} + B_{ij}\sin\theta_{ij}) \tag{5.5}$$

$$Q_{ij} = -V_i^2(B_{si} + B_{ij}) - V_iV_j(G_{ij}\sin\theta_{ij} - B_{ij}\cos\theta_{ij}) \tag{5.6}$$

  where G_{ij} and B_{ij} are the series conductance and susceptance of the branch, G_{si} and B_{si} are the shunt conductance and susceptance at node i and $\theta_{ij} = \theta_i - \theta_j$.
- LTC transformers are described by

$$\begin{aligned} P_{ij} &= V_i^2G_{ij} - TV_iV_j(G_{ij}\cos\theta_{ij} + B_{ij}\sin\theta_{ij}); \\ P_{ji} &= T^2V_j^2G_{ij} - TV_iV_j(G_{ij}\cos\theta_{ji} + B_{ij}\sin\theta_{ji}) \end{aligned} \tag{5.7}$$

$$\begin{aligned} Q_{ij} &= -V_i^2B_{ij} - TV_iV_j(G_{ij}\sin\theta_{ij} - B_{ij}\cos\theta_{ij}); \\ Q_{ji} &= -T^2V_j^2B_{ij} - TV_iV_j(G_{ij}\sin\theta_{ji} - B_{ij}\cos\theta_{ji}) \end{aligned} \tag{5.8}$$

  where T is the off-nominal tap ratio of the LTC transformer.

These equations are the starting point to develop power equations with which to calculate the active and reactive powers at both ends of the branches.

In nodes with shunt-connected devices, the nodal powers may be expressed in terms of their conductance and susceptance, say at node i: $P_{si} = V_i^2G_{shi}$ and $Q_{si} = -V_i^2B_{shi}$. Other measurements of electrical quantities such as the nodal voltage magnitudes, phase angles and LTC transformer taps may be directly expressed since these are variables already included in the state vector.

The measurements associated to the injections of active and reactive powers at a given node may be formulated using the power balance equations given by (5.2) and (5.3) and the power equations (5.5)–(5.8).

5.3.3 The Measurements System Model

This section addresses the incorporation of the available measurements in the power system into the estimation process. As already pointed out, the measurements distribution in a power grid is not always the same. In fact, the number, distribution and type of measurements available in a power network are closely related to the voltage level at which the measurements are carried out. The high-voltage transmission system often has the largest number of measurements relating to voltage, active power and reactive power. In these networks the number of measurements is usually high, exhibiting high redundancy. In low-voltage distribution networks the number of measurements tends to be low, with current and voltage magnitude being the most popular types of measurements.

The raw measurement set used in the estimation process may change from one estimation process to the next, this being a function of the information that is available in the SCADA system at the point in time when the estimation is carried out. This feature represents a major difference between the state estimation formulation and the power flow formulation. In the latter, an n-node network will have $2n$ specified nodal quantities and $2n$ state variables. From the outset, the type of node is intrinsically linked, in a semi-fixed format, to the type of specified variables, e.g. voltage magnitude, phase angle voltage, active power and reactive power.

In general, in the state estimation problem, the existing redundancy in the measurement vector with m measurements will be much greater than the number of $2n$ state variables. This redundancy in the measurements enables the state estimation process to obtain an operational filtered status of the power system and to develop functions for detection and identification of bad data.

In the topology processor, once the bus-branch model has been determined, the available measurements are assigned to the network model. These measurements may be classified according to the element associated with the measurement:

a) Nodal measurements such as nodal voltage magnitude, injections of net active and reactive powers at a node or active and reactive powers drawn by a shunt element.
b) Measurements in branches such as the active and reactive power flows in transmission lines and transformers.
c) Measurements of parameters associated with the model representation of a piece of equipment. A case in point is the tap position of an LTC transformer singled out for estimation, with the transformer's tap position being an available measurement.

The raw value of the measurements may be expressed as a function of the system's state variables using (5.9). Moreover, the measurements model in the state estimation adds an error term to the exact value of the measurement.

$$\tilde{\mathbf{z}} = \begin{bmatrix} \tilde{z}_1 \\ \tilde{z}_2 \\ \vdots \\ \tilde{z}_m \end{bmatrix} = \begin{bmatrix} h_1(x_1, x_2, \dots, x_{2n}) \\ h_2(x_1, x_2, \dots, x_{2n}) \\ \vdots \\ h_m(x_1, x_2, \dots, x_{2n}) \end{bmatrix} + \begin{bmatrix} e_1 \\ e_2 \\ \vdots \\ e_m \end{bmatrix} \tag{5.9}$$

where $\tilde{\mathbf{z}}$ is a column vector containing the measured values and e_i (with $i = 1, \dots, m$) is an error term associated to measurement i. The error term e_i associated to a given measurement $\tilde{z}_i$ is a random variable with a Gaussian distribution of zero mean and standard deviation σ_i. Hence, the higher the quality of a measuring instrument, the smaller its standard deviation will be.

By way of example, the measured voltage magnitude at a given node i is carried out directly on a state variable and its formulation may be expressed as follows:

$$\tilde{V}_i = V_i + e_{Vi} \tag{5.10}$$

The active and reactive power flows at one end of a transmission line are formulated using Eqs. (5.5) and (5.6), as follows:

$$\tilde{P}_{ij} = V_i^2(G_{si} + G_{ij}) - V_i V_j(G_{ij} \cos\theta_{ij} + B_{ij} \sin\theta_{ij}) + e_{pij} \tag{5.11}$$

$$\tilde{Q}_{ij} = -V_i^2(B_{si} + B_{ij}) - V_i V_j(G_{ij} \sin\theta_{ij} - B_{ij} \cos\theta_{ij}) + e_{qij} \tag{5.12}$$

The measurements errors are taken to be random variables independent from each other. Hence, the covariance matrix of errors is a diagonal matrix of size $m \times m$, where m

is the total number of measurements available. The entries in this matrix are the square of the standard deviations:

$$\mathbf{R_z} = \begin{bmatrix} \ddots & & \\ & \sigma_i^2 & \\ & & \ddots \end{bmatrix} \tag{5.13}$$

The matrix of weighted measurements is calculated as the inverse of the covariance matrix:

$$\mathbf{W} = \mathbf{R_z}^{-1} = \begin{bmatrix} \ddots & & \\ & \frac{1}{\sigma_i^2} & \\ & & \ddots \end{bmatrix} \tag{5.14}$$

The diagonal entries may be interpreted as the relative weight of a measurement bearing in mind the quality of the measuring equipment used to acquire it – the higher its quality, the higher the weight of its associated measurement [19].

A special case in the state estimation formulation is the transit node, where the balance of active and reactive power injected at the node is zero. To implement this feature in the state estimation model it is necessary to incorporate a *virtual measurement* comprising injections of active and reactive power of zero values at the node. Since the power balances (5.2) and (5.3) in the transit node are exact, the error terms are always zero, i.e. not random variables.

One way to combine the virtual and the conventional measurements is to give the former an error term with a standard deviation of a very low value. It follows that the virtual measurements possess a higher relative weight than the conventional measurements and the estimated values of the virtual injections is almost zero. This approach simplifies the use of such measurements but it may introduce ill-conditioning problems in the state estimation formulation. To circumvent the problem, the virtual measurements are introduced as equality constraints during the optimization process of the state estimation solution. This method is presented in Section 5.4.2.

As a corollary to this discussion, most FACTS devices models in state estimation incorporate power balances which must be fulfilled in an exact manner. In such cases the power balances are included as virtual measurements in the state estimation process. This issue is addressed in Section 5.6.

5.4 Calculation of the Estimated State

5.4.1 Solution by the Normal Equations

In the previous section, the network model with the measurements in terms of the state variables was formulated. In compact form, Eq. (5.9) may be expressed as follows:

$$\tilde{\mathbf{z}} = \mathbf{h}(\mathbf{x}) + \mathbf{e} \tag{5.15}$$

It should be emphasized that the main objective of the state estimation process is to determine, using the measurements available at a given point in time, the value of the state variables that best describe the operating state of the power grid. This requires establishing a methodology to solve the system of equations (5.15). This system comprises m non-linear equations, each associated with a measurement, and $2n$ unknowns associated with $2n$ state variables.

Under certain conditions, which are discussed in the observability section, the system of equations (5.15) is usually over-determined. The most popular solution method in power systems state estimation is the weighted least-squares estimation (WLS).

In the WLS method, the estimated value of the state variables is obtained by solving an optimization problem, which consists of minimizing the following objective function:

$$J(x) = \sum_{i=1}^{m} w_{ii} r_i^2 \tag{5.16}$$

where the variable r_i, termed residual of a measurement, is obtained as follows:

$$r_i = \tilde{z}_i - h_i(\mathbf{x}) \tag{5.17}$$

This quantifies the difference between the measured value and the estimated value of the measurement. The variable w_{ii} is the relative weight associated to the measurement $\tilde{z}_i$ and corresponding to the diagonal element in the weights matrix defined in (5.14).

The objective function (5.16) may then be formulated in matrix form as follows:

$$J(\mathbf{x}) = [\tilde{\mathbf{z}} - \mathbf{h}(\mathbf{x})]^T \mathbf{W} [\tilde{\mathbf{z}} - \mathbf{h}(\mathbf{x})] \tag{5.18}$$

Hence, the estimated value of the state vector $\hat{x}$ by means of the WLS estimator corresponds to the solution of the following optimization problem:

$$\min\, [\tilde{\mathbf{z}} - \mathbf{h}(\mathbf{x})]^T \mathbf{W} [\tilde{\mathbf{z}} - \mathbf{h}(\mathbf{x})] \tag{5.19}$$

The solution process involves the checking of the first-order optimality conditions:

$$g(\mathbf{x}) = \frac{\partial J(\mathbf{x})}{\partial \mathbf{x}} = 0 \tag{5.20}$$

which yields

$$-\mathbf{H}^T(\mathbf{x}) \mathbf{W} [\tilde{\mathbf{z}} - \mathbf{h}(\mathbf{x})] = 0 \tag{5.21}$$

where

$$\mathbf{H}(\mathbf{x}) = \frac{\partial \mathbf{h}(\mathbf{x})}{\partial \mathbf{x}} \tag{5.22}$$

is the Jacobian matrix.

Expanding the function $g(\mathbf{x})$ into its Taylor series around the state vector $\mathbf{x}^k$, an approximated function may be established by neglecting the higher-order terms:

$$\mathbf{g}(\mathbf{x}) \approx \mathbf{g}(\mathbf{x}^k) + \mathbf{G}(\mathbf{x}^k)\Delta\mathbf{x} \tag{5.23}$$

where

$$\mathbf{G}(\mathbf{x}^k) = \frac{\partial \mathbf{g}(\mathbf{x}^k)}{\partial \mathbf{x}} \tag{5.24}$$

$\mathbf{G}(\mathbf{x}^k)$ is called the gain matrix and can be computed by

$$\mathbf{G}(\mathbf{x}^k) = \mathbf{H}^T(\mathrm{x}^k) \mathbf{W} \mathbf{H}(\mathrm{x}^k) \tag{5.25}$$

The solution of the non-linear equation (5.25) may be obtained by iteration using the system of equations known as the normal equation:

$$\mathbf{G}(\mathbf{x}^k)\Delta\mathbf{x}^{k+1} = \mathbf{H}^T(\mathbf{x}^k) \mathbf{W} [\tilde{\mathbf{z}} - \mathbf{h}(\mathbf{x}^k)] \tag{5.26}$$

where $\Delta\mathbf{x}^{k+1} = \mathbf{x}^{k+1} - \mathbf{x}^k$.

The steps to determine the estimated values of the state vector using the WLS algorithm are:

1) Set the iteration counter $k = 0$ and voltage magnitudes and transformer turns ratio at unity values, and voltage phase angles at zero values, i.e. flat start.
2) Determine the Jacobian matrix $\mathbf{H}^T(\mathbf{x}^k)$ and compute the estimated value of the measurements $\mathbf{h}(\mathbf{x}^k)$.
3) Set up and solve the system of normal equations defined in (5.26) to obtain $\Delta\mathbf{x}^{k+1}$.
4) Test for convergence: $\max(\Delta\mathbf{x}^{k+1}) < \delta$, where δ is the convergence threshold specified a priori. If the convergence criterion has been satisfied, stop.
5) Update the state vector $\mathbf{x}^{k+1} = \mathbf{x}^k + \Delta\mathbf{x}^{k+1}$. Go to step 2.

The implementation of this algorithm should be carried out in an efficient manner if the aim is the solution of utility-size power networks. These networks may comprise thousands of nodes and several thousands of available measurements. The various matrices involved in the problem are huge and their solution requires the use of sparse matrix techniques. On the other hand, in solving the normal equations (5.26), the inverse of the gain matrix is calculated not directly but using factorization techniques such as Cholesky's factorization or orthogonal factorization [20]. Once the network state $\hat{x}$ has been estimated, it is then possible to obtain the values of the estimated measurements using Eq. (5.27):

$$\hat{\mathbf{z}} = \mathbf{h}(\hat{\mathbf{x}}) \tag{5.27}$$

where $\hat{\mathbf{z}}$ is a column vector containing the estimated values of the measurements. In turn, the covariance matrix of the estimated measurements $\mathbf{R}_{\hat{\mathbf{z}}}$ can be calculated using the following expression:

$$\mathbf{R}_{\hat{\mathbf{z}}} = \mathbf{H}\mathbf{G}^{-1}\mathbf{H}^{\mathrm{T}} \tag{5.28}$$

The diagonal elements of the covariance matrix are the variances associated to each one of the estimated measurements. It may be argued that, in general, thanks to the redundancy presented by the measurement system and the estimation process itself, the estimated measurements have greater accuracy than the values of the raw measurements obtained at the substation level. It follows that the standard deviations of the estimated measurements have comparable or lower standard deviations than those associated with the errors of the raw measurements. Hence, the estimated status would be, on the whole, more accurate.

5.4.2 Equality-Constrained WLS

As discussed, the transit nodes must incorporate virtual measurements with zero active and reactive power injections and these measurements could be incorporated as equality constraints into the process of calculating the estimated state. The optimization problem to be solved is formulated as follows:

$$\begin{aligned} &\min \quad J = [\tilde{\mathbf{z}} - \mathbf{h}(\mathbf{x})]^T \mathbf{W} [\tilde{\mathbf{z}} - \mathbf{h}(\mathbf{x})] \\ &\text{subject to} \quad \mathbf{c}(\mathbf{x}) = 0 \end{aligned} \tag{5.29}$$

where $\mathrm{c}(\mathbf{x})$ contains the virtual measurements.

The estimated state may be calculated using the following Lagrangian function:

$$L = [\tilde{\mathbf{z}} - \mathbf{h}(\mathbf{x})]^T \mathbf{W}[\tilde{\mathbf{z}} - \mathbf{h}(\mathbf{x})] - \lambda^T \mathbf{c}(\mathbf{x}) \tag{5.30}$$

Applying optimality conditions principles to Eq. (5.30) yields the following nonlinear system of equations, which can be solved by iteration using a Gauss-Newton's method:

$$\begin{bmatrix} \mathbf{H}^T\mathbf{W}\mathbf{H} & \mathbf{C}^T \\ \mathbf{C} & \mathbf{0} \end{bmatrix} \begin{bmatrix} \Delta\mathbf{x} \\ -\boldsymbol{\lambda} \end{bmatrix} = \begin{bmatrix} \mathbf{H}^T\mathbf{W}\Delta\mathbf{z} \\ -\mathbf{c}(\mathbf{x}) \end{bmatrix} \tag{5.31}$$

It should be noted that the calculation of the system state using this approach bears a high degree of resemblance to the power system optimization method presented in Chapter 4.

5.4.3 Observability Analysis and Reference Phase

When parts of the network lack sufficient measurements, it is not possible to develop the estimation process. To proceed with the state estimation process, it becomes necessary to include a minimum number of *pseudo-measurements*. These pseudo-measurements are incorporated into the estimation process by resorting to historical databases. This process is developed in the observability module.

Another key aspect which is taken care of in the observability module is the allocation of the phase angle reference. Each observable island present in the electrical system is assigned a phase angle reference. However, it should be noted that the node with the assigned phase angle has no influence on the estimated values of voltage magnitudes or powers throughout the network.

This point is further elaborated by means of the example network shown in Figure 5.3, where all branches have a series impedance value of 0.05 + j0.10 p.u. Notice that the operational status is the same in both representations of the network, with the only difference being the selection of reference node and the assigned value of phase angle. In Case (a), node 1 is assigned to be the reference and is given a zero value whereas in Case (b), node 2 is assigned to be the reference and is given a value of 30°.

It may be seen that the power flows and nodal power injections are the same in both cases. The only discernible difference is in the phase angle voltages. In Case (b), all the phases are rotated by 33° with respect to the phases in Case (a), with 3° resulting from the change in the selection of reference node and 30° coming from the adopted reference value. Nonetheless, the phase angle differences across each one of the branches are identical in both cases. Since the power flows in a given branch depend on the voltage magnitudes and phase angle difference across the branch, both network states yield identical power flows.

It follows that the operating state of the power network is independent of the value assigned to the reference phase angle; although it is usually taken to be zero, it can in fact take any other value. The value assigned to the reference phase angle does not change during the estimation process; the estimated and the assigned values coincide. Hence, it is normal practice to assign a zero value to the reference phase angle and to remove this variable from the set of state variables. This reduces the number of state variables to $2n-1$.

Alternatively, all the phase angles are incorporated in the state vector. This enables the straightforward assignment of any reference phase angle to the measurements

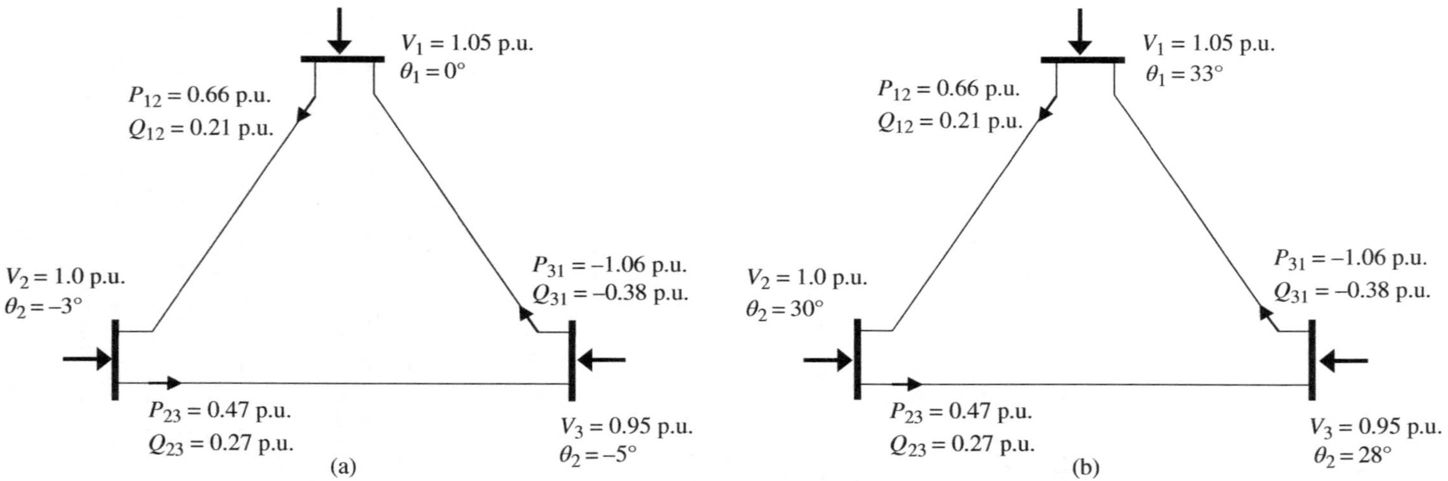

Figure 5.3 Phase reference selection.

vector; it involves adding a phase angle pseudo-measurement (which is usually zero) corresponding to the node which has been selected to be the reference phase angle. This is the approach implemented in the program weighted least squares state estimator (WLS-SE), which will be introduced in the section below.

When the synchrophasor measurements provided by the PMUs are incorporated into the state estimation process, all the phase angles may be given explicit representation in the state vector. As discussed in Chapter 1, synchrophasors already have a common reference time, which is why it is not necessary to include a phase reference pseudo-measurement; this issue is addressed in Section 5.7.

5.4.4 Weighted Least Squares State Estimator (WLS-SE) Using Matlab Code

A computer program with which to calculate the estimated state of power systems of small size has been developed using the Matlab environment. The algorithm employed is the equality-constrained WLS described in the previous sections.

The main program comprises four steps:

(i) All the necessary data with which to carry out the estimation process is read.
(ii) The estimated state is calculated.
(iii) An analysis of bad data is carried out.
(iv) Results are displayed.

Appendix 5.A gives additional information concerning the characteristics of the state estimation application and the input data format required by the WLS-SE computer program.

To illustrate the application of the state estimation process, a small test network is used [21]. As shown in Figure 5.4, this power network comprises five buses and seven transmission lines. The branch parameters are given in Table 5.1.

Two test cases are included in this section. The measurements sets included in the WLS-SE program have been carefully selected to ensure that some of the key concepts are highlighted. However, it should be noted that the measurements used in all test cases are editable (see Appendix 5.A). In the first test case, a state estimation process is developed with all the measurements taken to have exact values, i.e. with no noise. The measurements set comprises the voltage magnitudes at all nodes; the phase angle

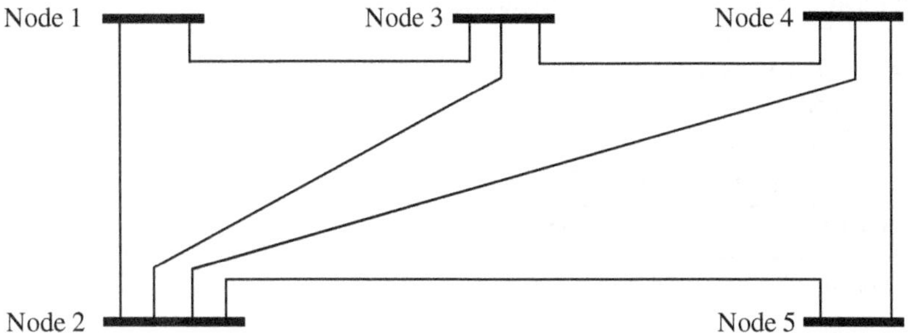

Figure 5.4 Tutorial network for WLS-SE.

Table 5.1 WLS-SE tutorial network parameters.

Branch		Resistance (p.u)	Reactance (p.u)	Total susceptance (p.u)
1	2	0.02	0.06	0.06
1	3	0.08	0.24	0.05
2	3	0.06	0.18	0.04
2	4	0.06	0.18	0.04
2	5	0.04	0.12	0.03
3	4	0.01	0.03	0.02
4	5	0.08	0.24	0.05

reference assigned to node 1; the injections of active and reactive powers at the nodes 1, 2 and 5; the active and reactive power flows at both ends of the transmission lines linking nodes 1–2 and 1–5; and active and reactive power flows at only one end of the remaining transmission lines. The measurements of voltage magnitude, power injection and power flow are given typical values of standard deviations: 0.004, 0.01 and 0.008 p.u., respectively.

The first test case uses the following data files:

Network file:	Network_tutorial
Measurements file:	Measure_tutorial_network
SE configuration file:	Config_1

Running the main program WLS_SE.p yields the estimated state shown in Table 5.2.

The estimation process takes four iterations to converge to a specified tolerance of 1e-5 for all the variables involved. As expected, since exact measurements were used, the final value of the objective function is zero. Also, notice that the furnished estimated state should coincide with a power flow solution that uses the same input data as in this example, bearing in mind that the input data for the power flow application is more restricted than for the state estimation application.

Table 5.2 WLS-SE results for the tutorial network with exact measurements.

Node	Nodal voltages	
	Magnitude	Phase (deg)
1	1.0600	0
2	1.0450	−4.9790
3	1.0100	−12.7197
4	1.0188	−10.3247
5	1.0204	−8.7834

A second test case assesses the situation when the exact measurements contain noise. This uses a new estimator configuration (file_configuration = 'Config_2') which adds random noise to each one of the exact measurements.

Network file:	Network_tutorial
Measurements file:	Measure_tutorial_network
SE configuration file:	Config_2

It can be observed that due to the presence of noise in the measurements, the estimate differs from the exact state obtained previously. It can also be seen that in this case the value of the objective function may not be considered to be zero. It is interesting to see that the standard deviations of the estimated measurements have values that are, in general, lower than the corresponding values of the raw measurements. For instance, at Node 1 the standard deviation associated with the estimated voltage magnitude takes a value of 0.0018 p.u., which contrasts with the standard deviation of the measurement which stands at 0.0040 p.u. This represents a 55% reduction in the value of standard deviation. This result illustrates one of the benefits of having redundancy in the measurements, with the estimation process significantly increasing the accuracy of the estimated operating status of the network.

It should be mentioned that owing to the random nature of the noise incorporated in the measurements, each run will yield different results. This may be further corroborated by carrying out new simulations including new measurements to the estimation process.

5.5 Bad Data Identification

5.5.1 Bad Data

In addition to improving the accuracy of the estimated state of the power network, a measurement set with in-built redundancy may be used to detect and identify bad data. A measurement is considered bad data when the difference between the measured value and the estimated value is larger than what can be considered to be noise introduced by the measurement chain. The presence of bad data is undesirable because it distorts the estimated state.

Prior to the estimation process, the measurements are filtered and it is possible to identify at this stage some bad data directly from the measurements values. Consider, for example, the case of measurements of voltage magnitude whose values are much lower than the nominal value, some of them even being zero. These values are discarded and not included in the vector of raw measurements. However, pre-filtering may not remove all the erroneous measurements and the actual estimation process incorporates a module to identify and eliminate the remaining bad data. This is carried out at the end of the estimation process when a new and more comprehensive test is done to eliminate the remaining erroneous measurements; it takes advantage of the statistic nature of the estimated state. This test, which is described below, searches for the largest normalized residual.

In realistic environments it is not possible to know the exact value of the measurements, hence statistical criteria may be applied to find out the wrong measurements with a certain level of confidence. The reference value of the measurement is taken to be an estimated state. In order to develop this method it becomes necessary to have a previously estimated state.

Any measurement identified as bad data must be removed from the available set of measurements. The process comes to an end when a new estimation of the state is carried out and no more bad data is found.

5.5.2 The Largest Normalized Residual Test

The most popular test to detect and identify bad data in the state estimation process is the largest normalized residual. This test is based on the statistical properties of the estimated residuals $\hat{r}_i$ associated to the measurements, which are calculated after having obtained the estimated state using:

$$\hat{r}_i = \tilde{z}_i - h_i(\hat{\mathbf{x}}) = \tilde{z}_i - \hat{z}_i \tag{5.32}$$

The covariance matrix associated to the estimated residuals is given as:

$$\mathbf{R}_{\hat{\mathbf{r}}} = \mathbf{R}_{\mathbf{z}} - \mathbf{H}\mathbf{G}^{-1}\mathbf{H}^{\mathrm{T}} \tag{5.33}$$

where the diagonal elements of this matrix are the variances of the residuals. The normalized residuals are:

$$r_i^{\mathrm{N}} = \frac{\hat{r}_i}{\sqrt{\mathbf{R}_{\hat{\mathbf{r}}}(i,i)}} \tag{5.34}$$

where $\sqrt{\mathbf{R}_{\hat{\mathbf{r}}}(i,i)}$ is the standard deviation of the residual estimate $\hat{r}_i$. The normalized residual of a measurement, calculated in this manner, follows a normal distribution of zero mean and unity standard deviation. This enables verification of whether or not a measurement is bad data by simply comparing the normalized residual value and a pre-specified threshold value. A popular threshold value in power systems state estimation is 3; a measurement with a normalized residual higher than 3 will have a high probability of being bad data.

Bad data and any other measurement associated to it will exhibit normalized residuals which may also surpass the threshold value of 3. It may be shown that in the case of a single bad data, the normalized residual with the highest value corresponds to the erroneous measurement. The test for detection and identification of bad data may be implemented using the following algorithm:

1) Obtain an estimated state $\hat{\mathbf{x}}$.
2) Calculate the diagonal elements of the covariance matrix of the residuals $\mathrm{diag}(\mathrm{R}_{\hat{\mathrm{r}}})$.
3) Calculate the normalized residuals of the measurements r_i^{N} $(i = 1, \ldots, m)$.
4) If the largest normalized residual is smaller than 3, then stop.
5) Delete from the estimation process the measurement with the highest normalized residual.
6) Go to step 1.

Notice that only one bad data is eliminated per cycle. Upon completion of this process the estimated state of the power grid becomes known together with the subset of measurements classified as bad data. This subset does not participate in the estimation process.

A special type of measurement is the critical measurement. If one or more of them are eliminated from the set of measurements, the entire system becomes unobservable. The estimator has no redundancy associated with these critical measurements, so their estimated values coincide with their measured values. Furthermore, their residual values are zero. It should be noted that in this situation a zero value of the residual bears no relationship whatsoever to a higher quality of the estimated value.

The estimated variance associated to the residual of a critical measurement is zero and this prevents the critical measurements from being incorporated into the bad data identification test, since they would present singularity problems at the point of calculating the normalized residuals. It may be concluded that the errors in critical measurements can be neither detected nor identified.

In cases of reduced redundancy, the test of the largest normalized residual enables the presence of an error to be detected but not its identification. It shows as a number of measurements with the same values of normalized residuals. Hence, the error is detected but the measurement that causes the error cannot be identified.

5.5.3 Bad Data Identification Using WLS-SE

The estimator WLS-SE enables an analysis of bad data using the largest normalized residual test. Consider the second test case presented in Section 5.4.4, which includes the addition of Gaussian noise to the exact measurements. It may be seen that, in general, the value of the normalized residuals associated to the noisy measurements is lower than the threshold value of 3. Owing to the random nature of the incorporated noise, each execution of the WLS-SE program will yield different values of measurements and associated normalized residuals. It is to be expected that in some cases the random error assigned to a measurement exhibits a high value and that the normalized residual can exceed the threshold value of 3.

To run the third test case, the WLS-SE program uses the files:

Network file:	Network_tutorial
Measurements file:	Measure_tutorial_network_bad_data
SE configuration file:	Config_1

It can be seen how in the estimation process of the largest residual, a value as large as 16.3 is present. It corresponds to the measurement of active power flowing between nodes 1 and 2. It is quite clear that such a large value of normalized residual indicates the presence of bad data. To corroborate this, it is sufficient to eliminate the measurement from the file containing the raw measurements. Running the WLS-SE program again, it is verified that all the normalized residuals have zero values, i.e. the rest of the measurements have their exact values.

It should be remarked how a single measurement with a wrong value may yield several measurements with non-zero normalized residuals, some of them going above the threshold value of 3. This is due to the adverse effect that the bad data inflicts on the outcome of the estimate process, with its impact being more severe in the vicinity of the wrong measurement.

New simulations can be carried out by simply changing the erroneous measurement, or by modifying the number of available measurements in the vicinity of the erroneous measurement and with it the local redundancy.

5.6 FACTS Device State Estimation Modelling in Electrical Power Grids

5.6.1 Incorporation of New Models in State Estimation

The inclusion of models of new equipment into the state estimation processor should enable a more detailed estimation of the operating state of the electrical power grid. This would be the case because of the incorporation of new measurements, which increases redundancy, thus improving the accuracy and identification of bad data.

The incorporation of new equipment representation in the network model requires:

a) enlargement of the state vector to incorporate all the new basic variables
b) addition of the set of constraints characterizing the new devices. Such constraints usually involve new parameters pertaining to the devices being modelled
c) augmentation of the measurements set so that the new devices can be monitored.

In general, the incorporation of a FACTS device representation into the power system model would increase the node count. A VSC-based FACTS device comprises an AC circuit and a DC circuit; normally, the AC circuit will connect to a point of the AC power grid through a coupling transformer. Hence, the AC point of connection between the two devices introduces a new node, as illustrated in Figure 5.5. The VSC's coupling transformer is suitably represented by a conventional branch, which enables the incorporation of measurements at each end of the branch.

In addition to the newly created AC nodes, the inclusion of VSCs in the circuit introduces DC nodes. These nodes may be isolated, such as in the case of STATCOMs, or may be interconnected, as is the case when two or more VSCs are used to form HVDC links, forming meshed networks in the case of multi-terminal VSC-HVDC systems. It will be shown below, for each of the devices, that each DC node available would enable the incorporation of suitable DC measurements.

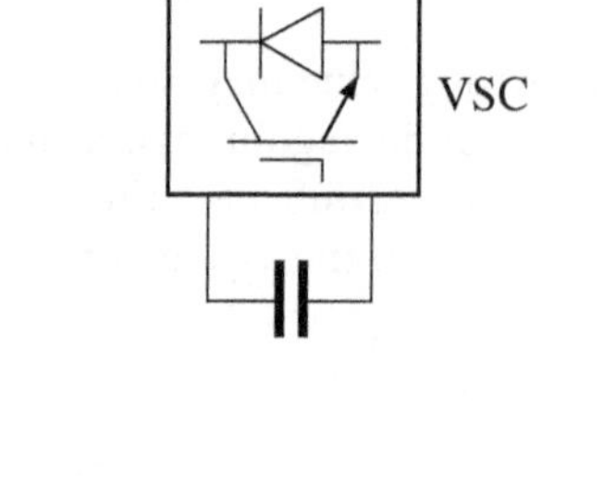

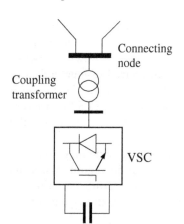

Figure 5.5 Schematic representation of the VSC and its coupling transformer.

5.6.2 Voltage Source Converters

A VSC on its own may not be considered to be a FACTS device, let alone a VSC-HVDC link. However, the VSC constitutes the kernel used by a number of compound devices, some of which are addressed in the remainder of this chapter. Hence, a suitable state estimation model of the VSC is developed below which will then be used to form the state estimation models of the STATCOM, UPFC, VSC-HVDC and multi-terminal VSC-HVDC.

As elucidated previously, the VSC model introduces two new nodes into the overall state estimation model, one associated with its AC side and the other with its DC side, as shown in Figure 5.6.

Node v on the AC side connects to the low-voltage side of the coupling transformer, which may be represented in a conventional manner. The state variables associated with this node are voltage magnitude, V_v, and phase angle, θ_v.

The DC node enables the connection of the VSC to a connecting DC system, which may be as simple as a DC bus, with a capacitor connected to it, or as complex as a meshed DC power grid. In order to add generality to the model description, the state variables making up the DC part of the state vector are not explicitly stated. Instead, the DC state variables are given, specifically, for each of the VSC-based devices.

In this application, it is desirable to take account of the VSC power losses and to include them in the active power balance set. Moreover, these power losses may be separated into two terms [15], one for switching losses and the other for conduction losses:

$$P_{loss} = P^c_{loss} + P^s_{loss} \tag{5.35}$$

As indicated in Chapter 3, the conduction losses, P^c_{loss}, also termed ohmic losses, are well represented by means of a series resistor, R_1, connected to the AC node, whereas the switching losses, P^s_{loss}, may be represented by means of a shunt conductance, G_{sw}, connected to the DC node. Hence, the VSC power losses are defined by the

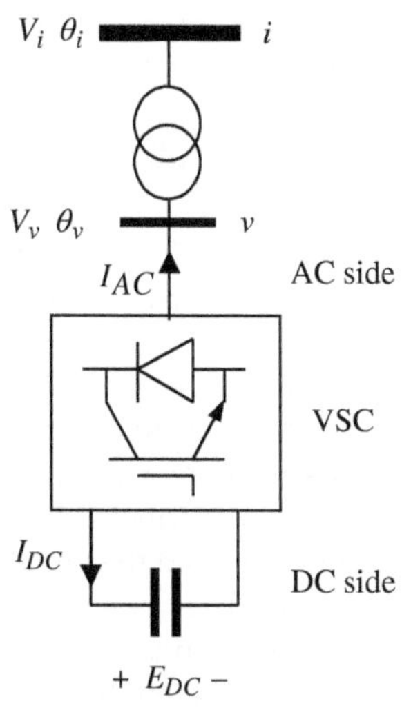

Figure 5.6 Representation of the VSC model in state estimation.

following relationship:

$$P_{loss} = R_1 I_{AC}^2 + G_{sw} E_{DC}^2 \tag{5.36}$$

where I_{AC} is the AC current delivered by the VSC and E_{DC} is the DC voltage of the VSC station.

The active power balance in the VSC may be expressed as follows:

$$P_{AC} + P_{DC} + P_{loss} = 0 \tag{5.37}$$

where P_{AC} represents the active power delivered by the convertor at its AC node whereas P_{DC} is the power delivered by the converter at its DC node.

The power P_{DC} may be calculated using available information of voltage and current:

$$P_{DC} = E_{DC} I_{DC} \tag{5.38}$$

where I_{DC} is the DC current of the VSC station.

Substituting (5.38) into (5.37) and combining with the VSC's power loss equation (5.36), the active power balance is reformulated as follows:

$$P_{AC} + E_{DC} I_{DC} + R_1 I_{AC}^2 + G_{sw} E_{DC}^2 = 0 \tag{5.39}$$

In order to include this equation into the VSC model, it becomes necessary to express all the variables involved in terms of state variables. In general, the DC voltage, E_{DC}, is taken to be a state variable. Concerning the DC current I_{DC}, it may also be taken directly to be a state variable or, alternatively, it may be expressed to be a function of other DC state variables. The variables P_{AC}, and the square of the AC current I_{AC}, delivered by the VSC at its AC node, are elaborated below. The powers contributed by the VSC at its AC node may be derived by resorting to power balances, active and reactive, at node v:

$$P_{AC} = P_{vi} - P_v \tag{5.40}$$

$$Q_{AC} = Q_{vi} - Q_v \tag{5.41}$$

where P_{vi} and Q_{vi} are the active and reactive power flows through the coupling transformer; P_v and Q_v are the nodal active and reactive power injections at node v. In turn, the square of the current through the VSC, I_{AC}, may be expressed in terms of Eqs. (5.40) and (5.41), as follows:

$$I_{AC}^2 = \frac{P_{AC}^2 + Q_{AC}^2}{V_v^2} \tag{5.42}$$

The power flows, P_{vi} and Q_{vi}, may be expressed in terms of the AC state variables (nodal voltages), using the conventional model of the coupling transformer (5.7) and (5.8).

In cases when the power injections P_v and Q_v are different from zero, owing to an element connected to node v, the associated power equations are expressed as a function of state variables. By way of example, let us assume that the element connected to node v is a shunt element. The associated injected power may be calculated using the shunt admittance and the nodal voltage magnitude at node v.

Notice that when it is not possible, or not desirable, to include in the formulation the associated model of the power injections P_v and Q_v, then the power variables must be incorporated into the state vector as new state variables.

It is often the case that the VSC is directly connected to the coupling transformer, leading to zero power injections, P_v and Q_v, at this node. In this case, the VSC powers at the AC node are calculated using the power equations (5.7) and (5.8):

$$P_{AC} = P_{vi} = G_{sc}V_v^2 - V_v V_i(G_{sc}\cos\gamma + B_{sc}\sin\gamma) \tag{5.43}$$

$$Q_{AC} = Q_{vi} = -B_{sc}V_v^2 - V_v V_i(G_{sc}\sin\gamma - B_{sc}\cos\gamma) \tag{5.44}$$

where G_{sc} and B_{sc} are the real and imaginary parts of the coupling transformer's series admittance, the tap of the coupling transformer is assumed $T = 1$ and $\gamma = \theta_v - \theta_i$.

On the other hand, the square of the AC current leaving the transformer can be obtained by substituting the Eqs. (5.43) and (5.44) into (5.42), giving the following result:

$$I_{AC}^2 = I_{vi}^2 = (G_{sc}^2 + B_{sc}^2)(V_v^2 + V_i^2 - 2V_v V_i\cos\gamma) \tag{5.45}$$

In this chapter we shall assume that the VSC is directly connected to the coupling transformer, so that expressions (5.43)–(5.45) are applicable.

It should be noted that other transformer models, such as tap-changing or phase-shifting transformers, may be included instead, but slightly different equations will be required – refer to Sections 3.2.3 and 3.2.4.

The VSC model, described above, may incorporate new measurements relating to the actual control of the VSC, such as the modulation index m_a of the PWM control, and the phase-angle control $\gamma = \theta_v - \theta_i$. Notice that this phase-angle control signal γ, representing the phase-angle difference between the terminal buses of the coupling transformer, is directly formulated by means of the associated state variables at the two ends of the coupling transformer.

The modulation index m_a relates the line-to-line RMS voltage magnitude in the AC side of the converter to the DC-side voltage [22]. As explained in Section 3.6.1, this constitutes a coupling between the AC and DC sides of the converter:

$$V_v^{LL} = \sqrt{\frac{3}{8}}m_a E_{DC} \tag{5.46}$$

where V_v^{LL} is the RMS value of the fundamental-frequency, line-to-line voltage and E_{DC} is the DC-side voltage.

The above equation, expressed in per unit, is integrated into the VSC model, considering as base voltages $V_{B,AC}^{LL}$ and $E_{B,DC}$ for the AC side and DC side, respectively,

$$V_v = k_1 m_a E_{DC} \tag{5.47}$$

where

$$k_1 = \frac{E_{B,DC}}{V_{B,AC}^{LL}} \cdot \sqrt{\frac{3}{8}} \tag{5.48}$$

V_v and E_{DC} are the AC-side and DC-side voltages expressed in per unit, the former being a line-line voltage.

In the examples included in this chapter it is considered that the base voltage of the AC and DC side have the same value. In this way, the constant k_1 takes the value 0.6124.

In the remainder of this chapter, the VSC model is used to develop the state estimation models of VSC-FACTS and VSC-HVDC equipment.

5.6.3 STATCOM

In this section, we apply the generic model of the VSC developed above to obtain the state estimation model of a STATCOM. As explained in Section 1.2.2, the combination of a VSC and its connecting transformer is termed, in the parlance of power systems engineers, a STATCOM. As illustrated in Figure 5.6, it uses only one node on its DC side, with a capacitor connected to it.

The STATCOM model incorporated into the state estimator uses as state variables the AC voltage magnitude and its phase angle at node v, namely V_v and θ_v, and the DC voltage, E_{DC}:

$$X_{\text{STATCOM}} = [V_v, \theta_v, E_{DC}] \tag{5.49}$$

It should be noted that this model assumes that the power delivered by the VSC at its DC node is zero; hence, the DC current is also zero. This would correspond to the case when the DC capacitor is fully charged, a fact that happens during steady-state operation.

The model is completed by incorporating the active power balance (5.39) as either virtual measurement or equality constraint. It follows that the only power exchanged between the DC and AC nodes, through the coupling transformer, is the power corresponding to the converter losses, i.e. conduction and switching power losses. In this way, incorporating the converter power losses (5.36) in the active power restriction equation yields:

$$P_{vi} + R_1 I_{vi}^2 + G_{sw} E_{DC}^2 = 0 \tag{5.50}$$

This restriction may be formulated as a function of the state variables on the AC node employing the expressions (5.43)–(5.45), in order to formulate the active power flow, P_{vi}, and the square of the current flow, I_{vi}, through the coupling transformer.

Table 5.3 summarizes the measurements set associated with this STATCOM model.

A special case of the STATCOM operation is when it is possible to have power exchanges between the AC and DC sides of the device, something that is achievable when there is an energy storage device connected to the DC node, such as a battery pack. This type of FACTS equipment, introduced in Section 1.2.4, is known as BESS. From the vantage of the state estimation application, this VSC model is extended to

Table 5.3 STATCOM measurements.

Conventional measurements		Control measurements	
AC voltage	$\tilde{V}_i = V_i + \varepsilon_{Vi}$ $\tilde{V}_v = V_v + \varepsilon_{Vj}$	Modulation index	$\tilde{m}_a = \frac{V_v}{k_1 E_{DC}} + \varepsilon_{m_a}$
AC power flows	$\tilde{P}_{vi} = G_{sc} V_v^2 - V_v V_i (G_{sc} \cos\gamma + B_{sc} \sin\gamma) + \varepsilon_{pvi}$ $\tilde{Q}_{vi} = -B_{sc} V_v^2 - V_v V_i (G_{sc} \sin\gamma - B_{sc} \cos\gamma) + \varepsilon_{qvi}$ The reverse power flows are realized by simply exchanging the subscripts v and i and changing the sign of the phase-angle γ.	Phase-angle control signal	$\tilde{\gamma} = \gamma + \varepsilon_\gamma$
DC voltage	$\tilde{E}_{DC} = E_{DC} + \varepsilon_{E_{DC}}$		

Table 5.4 Extended STATCOM measurements.

Measurements	
DC current	$\tilde{I}_{DC} = I_{DC} + \varepsilon_{I_{DC}}$
DC power	$\tilde{P}_{DC} = E_{DC}I_{DC} + \varepsilon_{P_{DC}}$

incorporate the output DC current, I_{DC}, into the state vector. Hence, the new model of the STATCOM has the following state vector:

$$X_{\text{STATCOM}}^{\text{amp}} = [V_v, \theta_v, E_{DC}, I_{DC}] \tag{5.51}$$

Furthermore, the active power balance is extended to incorporate a new term in order to take account of the power exchanged with the DC side of the STATCOM. The restriction term (5.50) becomes:

$$E_{DC}I_{DC} + P_{vi} + R_1 I_{vi}^2 + G_{sw}E_{DC}^2 = 0 \tag{5.52}$$

With this extended model, it is now possible to incorporate new measurements on the DC side of the STATCOM – shown in Table 5.4.

It should be noted that this extended model is suitable to use in cases when it is not physically possible to exchange power between the AC and DC sides of the STATCOM owing to, say, the lack of an energy storage source. In such a case, it suffices to set the DC current restriction to zero, i.e. $I_{DC} = 0$.

5.6.4 STATCOM Model in WLS-SE

The software WLS-SE includes three test cases, which use the extended STATCOM model, developed in the previous section. The basic network of the tutorial section is expanded to include one STATCOM model. The STATCOM's coupling transformer is connected to Node 3 of the tutorial network and a new node is created, Node 6, to connect the STATCOM. The transformer impedance is 0.01 + j 0.1 p.u. The STATCOM series resistance is 0.02 p.u. and the shunt conductance is 0.002 p.u.

The study case 4, using WLS-SE, employs the following data files:

Network file:	Network_tutorial_STATCOM
Measurements file:	Measure_STATCOM_case_a
SE configuration file:	Config_1

In this example, the STATCOM is set to inject reactive power at Node 3, as shown in Figure 5.7.

The active power taken from the network at Node 3 is exactly the power required to supply the power losses of the VSC and its coupling transformer, since this VSC is unable to contribute active power at all. This test case was solved using the extended model of the STATCOM; hence, it has become necessary to use a zero DC current restriction. This example uses exact measurements, including the measurements associated to the VSC control and DC voltage.

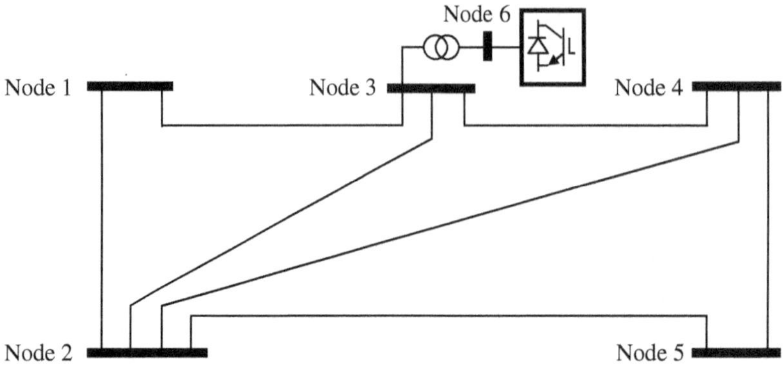

Figure 5.7 Tutorial network including one STATCOM.

It may be of some instructive value to assess the performance of the STATCOM model when the VSC's power losses are assumed to be negligibly small. This requires that both the series resistance and shunt conductance of the VSC be set to zero in the relevant data file. This corresponds to the test case 5 and in WLS-SE. These changes have been implemented in the following data files:

Network file:	Network_tutorial_STATCOM_VSC_no_losses
Measurements file:	Measure_STATCOM_case_a
SE configuration file:	Config_1

The exact measurements are the same as those used in the previous example; however, the VSC power losses have not been included in the model. The results furnished by WLS-SE show that the measurements of active power flows though the coupling transformer 3-6 has a normalized residue value of 1.0, which incorrectly indicates a noisy value of the measurements given that in all fairness, it is the correct value. This example highlights how incomplete/inaccurate models may distort the estimated state, having a negative impact on the normalized residual values of some of the measurements associated with the device.

In test case 6, which also uses the STATCOM connected to Node 3, WLS-SE uses the following data files:

Network file:	Network_tutorial_STATCOM
Measurements file:	Measure_STATCOM_case_b
SE configuration file:	Config_2

In this example, the STATCOM possesses the ability to inject power at the DC node, which flows towards the AC network. It employs the full STATCOM model – without the restriction that nullifies the DC current on the VSC's DC node. Instead, it incorporates this current as a measurement. Notice that this case also includes the power measurement on the DC side of the VSC. In this example, the state estimator configuration incorporates noise in all the measurements.

5.6.5 Unified Power Flow Controller

As introduced in Section 1.2.4, the UPFC comprises two VSCs connected in a back-to-back configuration on their DC sides. At their AC sides, one VSC is connected to the secondary winding of a transformer whose primary winding is connected in shunt to the AC power grid. At the opposite end, a second VSC is connected to the secondary winding of a transformer whose primary winding is connected in series with the AC power grid, as shown in Figure 5.8.

The state estimation model developed in this section includes the power losses associated with both converters. Hence, the generic VSC model introduced in Section 5.6.2, is suitable to represent each one of the VSCs making up the UPFC.

In addition to the nodal voltages at the AC side of both VSCs, namely nodes s and r, the model incorporates the DC voltage and the DC current as state variables. Accordingly, the state variable set associated to the UPFC model is:

$$X_{\text{UPFC}} = [V_s, \theta_s, V_r, \theta_r, E_{DC}, I_{DC}] \tag{5.53}$$

The complete UPFC model requires the active power balance equations for each VSC. The balance in the shunt-connected VSC is the following:

$$E_{DC}I_{DC} + P_{sk} + R_{ssh}I_{sk}^2 + G_{psh}E_{DC}^2 = 0 \tag{5.54}$$

where R_{ssh} and G_{psh} are the parameters associated to the power losses in this converter.

The restriction (5.54) is expressed as a function of the AC state variables given in (5.43)–(5.45), to formulate the active power flow P_{sk} and the square de la current, I_{sk}, flowing through the shunt-connected, coupling transformer, $s-k$.

The following expressions have been used in the formulation of the measurements of power flow through the series-connected transformer.

$$S_{km} = P_{km} + jQ_{km} = V_k I_{km}^* \tag{5.55}$$
$$S_{mk} = P_{mk} + jQ_{mk} = V_m I_{mk}^* \tag{5.56}$$

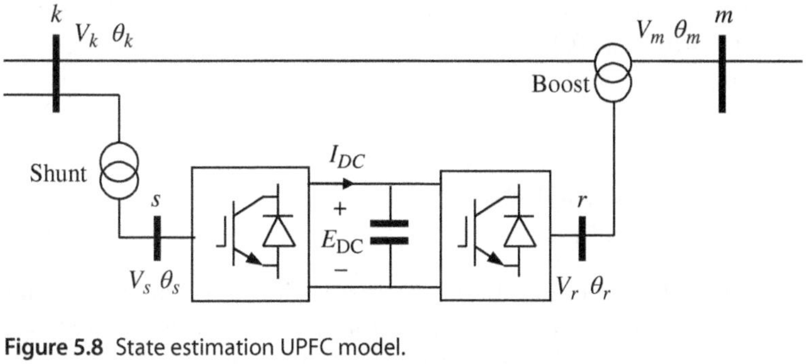

Figure 5.8 State estimation UPFC model.

where the currents can be expressed as:

$$I_{km} = (V_k - V_r - V_m)Y_b \tag{5.57}$$

$$I_{mk} = -I_{km} \tag{5.58}$$

with the parameter Y_b being the series admittance of the shunt transformer. The power flows can be formulated using the state variables as:

$$S_{km} = V_k(V_k^* - V_m^*)Y_b^* - V_k V_r^* Y_b^* \tag{5.59}$$

$$S_{mk} = V_m(V_m^* - V_k^*)Y_b^* + V_m V_r^* Y_b^* \tag{5.60}$$

In the series-connected VSC, the active power balance equation is:

$$-E_{DC}I_{DC} + P_r + R_{ssr}I_r^2 + G_{psr}E_{DC}^2 = 0 \tag{5.61}$$

where R_{ssr} and G_{psr} are the parameters associated to the power losses of this converter.

The variables P_r and I_r correspond to the active power and current delivered by the series-connected converter at node r. In order to express these constraints as a function of the AC state variables, we use the active and reactive power expressions in the series-connected converter:

$$S_r = P_r + jQ_r = V_r I_r^* \tag{5.62}$$

Considering that $I_r = I_{mk}$, the power flows are computed as:

$$S_r = V_r(V_r^* - V_k^*)Y_b^* + V_r V_m^* Y_b^* \tag{5.63}$$

The square of the current, I_r, in the converter may be expressed as follows:

$$I_r^2 = \frac{P_r^2 + Q_r^2}{V_r^2} \tag{5.64}$$

The measurements set of the UPFC is summarized in Table 5.5.

5.6.6 The UPFC Model in WLS-SE

In this section we analyze the results of a case included in WLS-SE, where a UPFC is added to the tutorial's test system. The UPFC is connected to Node 3 in series with the transmission line 8-4 (previously labelled as 3-4). The shunt-connected VSC is added to the grid through the coupling transformer 6-3, as shown in Figure 5.9. The series impedance of both transformers is: 0.01 + j 0.1 p.u.

The files corresponding to this case are:

Network file:	Network_tutorial_UPFC
Measurements file:	Measure_UPFC
SE configuration file:	Config_1

Notice that there is no nodal power injection at node 8, i.e. it is the connection point of the UPFC and line 8-4 (transit node). To represent such condition in WLS-SE, we require equality power constraints with null injected powers at node 8. Execution of this example shows that the measurement of active power flow from node 8 to node 4 is flagged as bad data, with the rest of the measurements being exact measurements.

Table 5.5 UPFC measurements.

Conventional measurements		Control measurements	
AC voltage	$\tilde{V}_r = V_r + \varepsilon_{Vr}$ $\tilde{V}_k = V_k + \varepsilon_{Vk}$ $\tilde{V}_s = V_s + \varepsilon_{Vs}$ $\tilde{V}_m = V_m + \varepsilon_{Vm}$	Modulation index	$\tilde{m}_{as} = \dfrac{V_s}{k_1 E_{DC}} + \varepsilon_{m_{as}}$ $\tilde{m}_{ar} = \dfrac{V_r}{k_1 E_{DC}} + \varepsilon_{m_{ar}}$
AC power flows	$\tilde{P}_{sk} = G_{Ts}V_s^2 - V_sV_k(G_{Ts}\cos\theta_{sk} + B_{Ts}\sin\theta_{sk}) + \varepsilon_{psk}$ $\tilde{Q}_{sk} = -B_{Ts}V_s^2 - V_sV_k(G_{Ts}\sin\theta_{sk} - B_{Ts}\cos\theta_{sk}) + \varepsilon_{qsk}$ The opposite flows can be obtained by interchanging the indexes *s* and *k*. $\tilde{P}_{km} = G_{Tb}V_k^2 - V_kV_m(G_{Tb}\cos\theta_{km} + B_{Tb}\sin\theta_{km}) -$ $-V_kV_r(G_{Tb}\cos\theta_{kr} + B_{Tb}\sin\theta_{kr}) + \varepsilon_{pkm}$ $\tilde{Q}_{km} = -B_{Tb}V_k^2 - V_kV_m(G_{Tb}\sin\theta_{km} - B_{Tb}\cos\theta_{km}) -$ $-V_kV_r(G_{Tb}\sin\theta_{kr} - B_{Tb}\cos\theta_{kr}) + \varepsilon_{qkm}$ The opposite flows can be obtained by interchanging the indexes *k* and *m*. $\tilde{P}_r = G_{Tb}V_r^2 - V_rV_k(G_{Tb}\cos\theta_{rk} + B_{Tb}\sin\theta_{rk}) +$ $+V_rV_m(G_{Tb}\cos\theta_{rm} + B_{Tb}\sin\theta_{rm}) + \varepsilon_{pr}$ $\tilde{Q}_r = -B_{Tb}V_r^2 - V_rV_k(G_{Tb}\sin\theta_{rk} - B_{Tb}\cos\theta_{rk}) +$ $+V_rV_m(G_{Tb}\sin\theta_{rm} - B_{Tb}\cos\theta_{rm}) + \varepsilon_{qr}$	Phase-angle control signal	$\tilde{\theta}_{sk} = \theta_s - \theta_k + \varepsilon_{\theta sk}$ $\tilde{\theta}_{rm} = \theta_r - \theta_m + \varepsilon_{\theta rm}$ $\tilde{\theta}_{rk} = \theta_r - \theta_k + \varepsilon_{\theta rk}$ $\tilde{\theta}_{km} = \theta_k - \theta_m + \varepsilon_{\theta km}$
DC voltage	$\tilde{E}_{DC} = E_{DC} + \varepsilon_{E_{DC}}$		
DC current	$\tilde{I}_{DC} = I_{DC} + \varepsilon_{I_{DC}}$		
DC power	$\tilde{P}_{DC} = E_{DC}I_{DC} + \varepsilon_{P_{DC}}$		

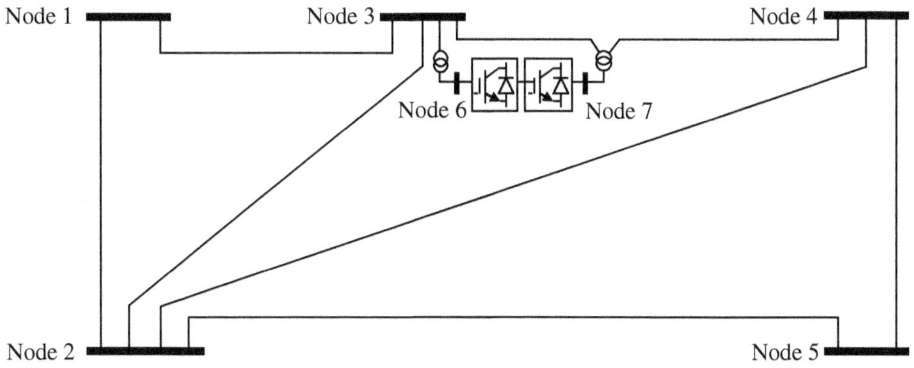

Figure 5.9 Tutorial network including a UPFC.

It is interesting to assess the case that results from removing the restrictions of null power injection at node 8. This situation would present itself if we were to have power injections at the node but no available measurements. The results show normalized residues larger than 4.5. In this case, two of the measurements have large values: the active power flow 8-4 and the active power injection at node 4. The test of the largest normalized residue detects the presence of an error in the measurements set, but it is not possible to identify the measurement that causes the error. This anomalous situation is due to the lack of sufficient measurement redundancy around the bad data, i.e. there is insufficient information to identify the bad data.

5.6.7 High Voltage Direct Current Based on Voltage Source Converters

The VSC generic model is now applied to develop the VSC-HVDC model for state estimation. It incorporates two AC nodes, $AC1$ and $AC2$, which use the following conventional state variables (nodal complex voltages) for their representation: V_{AC1}, θ_{AC1}, V_{AC2}, θ_{AC2}. Moreover, the VSC-HVDC introduces two DC nodes, which are linked through a DC cable of resistance, R_{DC}, as shown in Figure 5.10. The model described below represents well the converters when operating either as a rectifier or as an inverter.

It should be recalled that the DC resistance, R_{DC}, is taken to have a zero value in cases of a back-to-back configuration. The new electrical quantities associated to the DC circuit are the DC voltages and currents at the VSC-1 and VSC-2 nodes, together with the modulation indexes of the two VSCs:

$$[E_{DC1}, E_{DC2}, I_{DC1}, I_{DC2}, m_{a1}, m_{a2}] \tag{5.65}$$

where the subscripts 1 and 2 refer to the DC nodes of the VSC-1 and VSC-2, respectively. In turn, using these variables it is possible to calculate the injected powers at the respective DC nodes. Four out of the six variables associated to the DC circuit may be derived from two basic variables, using the available relationships of modulation indexes (5.47):

$$m_{a1} = \frac{V_{AC1}}{k_1 E_{DC1}} \qquad m_{a2} = \frac{V_{AC2}}{k_1 E_{DC2}} \tag{5.66}$$

and two other relationships which occur naturally in the DC circuit:

$$I_{DC1} = -I_{DC2} \tag{5.67}$$

and

$$I_{DC1} = \frac{E_{DC1} - E_{DC2}}{R_{DC}} \tag{5.68}$$

It is said that the variables associated to the DC circuit possess two-degree freedom. It should be noted that the relationship (5.68) is undefined when applied to the back-to-back configuration. In order to overcome this singularity in the model, we select an alternative set of state variables, independent of the DC cable resistance.

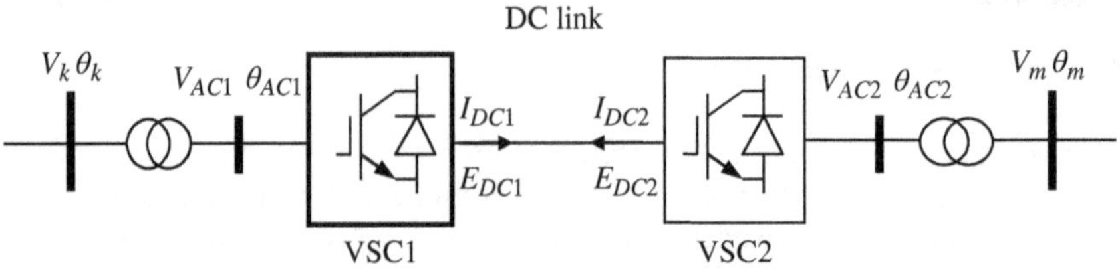

Figure 5.10 VSC-HVDC model.

These are the voltage and current at one of the DC buses. Hence, the alternative state vector contains the following set of variables:

$$X_{VSC-HVDC} = [V_{AC1}, \theta_{AC1}, V_{AC2}, \theta_{AC2}, E_{DC1}, I_{DC1}] \tag{5.69}$$

Using this state variable set, the DC electrical quantities at the VSC-2 are defined by the following relationships:

$$I_{DC2} = -I_{DC1} \tag{5.70}$$

$$E_{DC2} = E_{DC1} - I_{DC1}R_{DC} \tag{5.71}$$

Notice that these relationships are well suited to the null value of the DC resistance, R_{DC}, of the back-to-back HVDC link, where $E_{DC2} = E_{DC1}$.

In order to complete the model, it is necessary to incorporate a power balance restriction equation (5.39) at each VSC. At the VSC-1, where we have taken the state variables, the active power balance is:

$$P_{AC1\,k} + E_{DC1}I_{DC1} + R_{11}I^2_{AC1\,k} + G_{sw1}E^2_{DC1} = 0 \tag{5.72}$$

where R_{11} and G_{sw1} are the parameters associated with the VSC-1's power losses.

This restriction is expressed in terms of the AC state variables used in (5.43)–(5.45) in order to formulate the active power flow, P_{AC1k}, and the square of the current flow, I_{AC1k}, through the coupling transformer $AC1 - k$.

For the VSC-2, the active power balance expression is formulated as follows:

$$P_{AC2\,m} + E_{DC2}I_{DC2} + R_{12}I^2_{AC2\,m} + G_{sw2}E^2_{DC2} = 0 \tag{5.73}$$

where R_{12} and G_{sw2} are the parameters responsible for the power losses in VSC-2.

Employing the relationships (5.70) and (5.71), the power balance equation (5.73) may be expressed as a function of the state variables in the DC circuit:

$$\begin{aligned} P_{AC2\,m} &- (1 + 2R_{DC}G_{sw2})E_{DC1}I_{DC1} + (R_{DC} + R^2_{DC}G_{sw2})I^2_{DC1} \\ &+ G_{sw2}E^2_{DC1} + R_{12}I^2_{AC2m} = 0 \end{aligned} \tag{5.74}$$

The active power, $P_{AC2\,m}$, and the square of the current, $I_{AC2\,m}$, in the coupling transformer are formulated as a function of the state variables applying the expressions (5.43)–(5.45) of the coupling transformer $AC2 - m$.

The measurements of the VSC-HVDC state estimation model developed above are summarized in Table 5.6.

It should be remarked that the VSC-HVDC model just developed for the point-to-point case is also valid for the back-to-back case, where it is only necessary to set the resistance of the DC cable, R_{DC}, to zero.

5.6.8 VSC-HVDC Model in WLS-SE

In this section, we present two examples relating to the VSC-HVDC, which are included in WLS-SE, using the VSC-HVDC models presented in the section above.

In test case 7, the transmission line connected between nodes 3 and 4 in the tutorial network is replaced by a point-to-point VSC-HVDC link, as shown in Figure 5.11.

Coupling to the AC grid is carried out using transformers 3-6 and 4-7, each having a series impedance of 0.01 + j 0.1 p.u. The rectifier and inverter VSCs are connected

Table 5.6 VSC-HVDC measurements.

Conventional measurements		Control measurements	
AC voltage	$\tilde{V}_{AC1} = V_{AC1} + \varepsilon_{V_{AC1}}$ $\tilde{V}_k = V_k + \varepsilon_{V_k}$ $\tilde{V}_{AC2} = V_{AC2} + \varepsilon_{V_{AC2}}$ $\tilde{V}_m = V_m + \varepsilon_{V_m}$	Modulation index	$\tilde{m}_{a1} = \dfrac{V_{AC1}}{k_1 E_{DC1}} + \varepsilon_{m_{a1}}$ $\tilde{m}_{a2} = \dfrac{V_{AC2}}{k_1 E_{DC2}} + \varepsilon_{m_{a2}}$
AC power flows	$\tilde{P}_{AC1k} = G_{AC1k}V_{AC1}^2 - V_{AC1}$ $V_k(G_{AC1k}\cos\gamma_1 + B_{AC1k}\sin\gamma_1) + \varepsilon_{p_{AC1k}}$ $\tilde{Q}_{AC1k} = -B_{AC1k}V_{AC1}^2 - V_{AC1}$ $V_k(G_{AC1k}\sin\gamma_1 - B_{AC1k}\cos\gamma_1) + \varepsilon_{q_{AC1k}}$ The opposite flows can be obtained by interchanging the indexes *AC1* and *k*, and changing the sign of the phase-angle γ_1. The coupling transformer *AC2-m* has a similar formulation.	Phase-angle control signal	$\tilde{\gamma}_1 = \theta_{AC1} - \theta_k + \varepsilon_{\gamma_1}$ $\tilde{\gamma}_2 = \theta_{AC2} - \theta_m + \varepsilon_{\gamma_2}$
DC voltage	$\tilde{E}_{DC1} = E_{DC1} + \varepsilon_{E_{DC1}}$ $\tilde{E}_{DC2} = E_{DC1} - R_{DC}I_{DC1} + \varepsilon_{E_{DC2}}$		
DC current	$\tilde{I}_{DC1} = I_{DC1} + \varepsilon_{I_{DC1}}$ $\tilde{I}_{DC2} = -I_{DC1} + \varepsilon_{I_{DC2}}$		
DC power	$\tilde{P}_{DC1} = E_{DC1}I_{DC1} + \varepsilon_{P_{DC1}}$ $\tilde{P}_{DC2} = R_{DC}I_{DC1}^2 - E_{DC1}I_{DC1} + \varepsilon_{P_{DC2}}$		

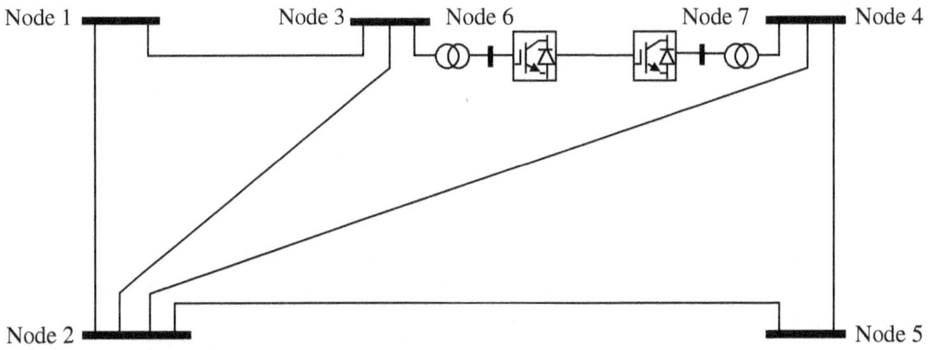

Figure 5.11 Tutorial network including VSC-HVDC.

to nodes 6 and 7, each having a series resistance and a shunt conductance of 0.02 p.u. and 0.002 p.u. It should be borne in mind that these parameters impact directly on the conduction and switching losses, respectively.

The DC cable which links both VSCs has a resistance value of 0.02 p.u. In this test case, the estate variables have been taken to correspond to the VSC at node 6. The data files corresponding to this example are:

Network file:	Network_tutorial_VSC_HVDC_ptp
Measurements file:	Measure_VSC_HVDC_ptp
SE configuration file:	Config_1

Execution of the program reveals the existence of bad data. The error is identified to be a wrong measurement in the DC circuit; more specifically, it is the DC voltage of the VSC at node 7. In order to verify that this is the case, the wrong measurement is eliminated and the program is run again, showing that all the normalized residuals have a null value, i.e. it yields exact measurements.

For test case 8, WLS-SE uses the following files:

Network file:	Network_tutorial_VSC_HVDC_btb
Measurements file:	Measure_VSC_HVDC_btb
SE configuration file:	Config_2

In this example we replace the transmission line 3-4 of the tutorial network with a back-to-back VSC-HVDC. Except for the DC resistance, which is taken to be zero, all other parameters of the VSC-HVDC are kept the same as in the previous test case. When analyzing the results, keep in mind that noise has been added to the measurements to make them more realistic.

5.6.9 Multi-terminal HVDC

So far, all the devices addressed in this chapter have comprised one and two DC nodes. In contrast, a multi-terminal VSC-HVDC system with n VSCs will have one or many DC nodes, up to n nodes. This point is discussed in more detail in Section 1.2.6.

In any case, the n converters are linked on their DC sides through a network of DC cables or, indeed, a single DC bus, such as in the case of a back-to-back MT-VSC-HVDC. In a multi-terminal system, each VSC sets the border between an AC system and the DC system. The coupling of a VSC with the AC grid is carried out through a coupling transformer, as shown in Figure 5.12.

The first step towards establishing the model of the multi-terminal VSC-HVDC is to select its state variables. As with the other VSC-based devices presented in this chapter, the state vector incorporates the AC nodal voltages of each VSC. Concerning the state variables of the DC system, the selection is the voltages at the DC nodes of each VSC. Hence, the state vector of the MT-VSC-HVDC is the following extended state vector:

$$X_{\text{Multi}} = [V_{AC1}, V_{AC2} \dots V_{ACn}, \theta_{AC1}, \theta_{AC2} \dots \theta_{ACn}, E_{DC1}, E_{DC2} \dots E_{DCn}] \tag{5.75}$$

We now look at the available measurements for the multi-terminal system and their formulations. Concerning the measurements of the AC sides of the VSCs, the formulation is the same as the one employed by the model described in Section 5.6.2. Concerning the DC grid, the injected currents at the DC nodes may be derived from the nodal DC conductance matrix, $\mathbf{G}_{DC}$, and the DC nodal voltages:

$$\mathbf{I}_{DC} = \mathbf{G}_{DC}\mathbf{E}_{DC} \tag{5.76}$$

where $\mathbf{E}_{DC}$ and $\mathbf{I}_{DC}$ are vectors of voltages and injected currents, respectively, at all the DC nodes.

Accordingly, the current I_{DCi} injected by the VSC connected to the DC node i may be expressed as a function of the state variables using:

$$I_{DCi} = \mathbf{g}_{DCi}\mathbf{E}_{DC} \tag{5.77}$$

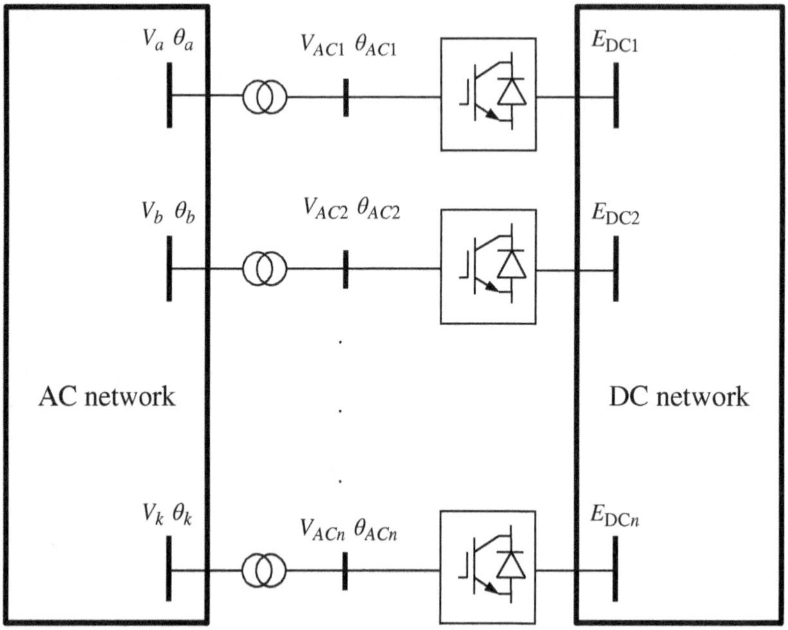

Figure 5.12 Multi-terminal network.

where $\mathbf{g}_{DCi}$ is a row vector extracted from the nodal conductance matrix corresponding to node i.

On the other hand, an inner measurement of current in the DC grid, between nodes i and j, would have the following form:

$$I_{DC\,ij} = \frac{E_{DCi} - E_{DCj}}{R_{DCij}} \tag{5.78}$$

where R_{DCij} is the resistance of the DC link between nodes i and j.

The injected power by this converter to node i of the DC grid is:

$$P_{DCi} = E_{DCi} I_{DCi} = E_{DCi}\, \mathbf{g}_{DCi} \mathbf{E}_{DC} \tag{5.79}$$

To complete the model, it is only necessary to incorporate an active power balance equation for each VSC in the system. Let us consider the case of a VSC which connects to nodes i and r on its DC and AC sides, respectively. Moreover, the VSC connects to the AC grid through a coupling transformer, using nodes r and j, so that the power balance equation is formulated as follows:

$$E_{DCi}\, \mathbf{g}_{DCi} \mathbf{E}_{DC} + P_{rj} + R_{1i} I_{rj}^2 + G_{swi} E_{DCi}^2 = 0 \tag{5.80}$$

where G_{swi} and R_{1i} are the parameters associated to the power losses of this converter. Table 5.7 summarizes the measurements of the MT-VSC-HVDC model developed above.

Table 5.7 Multi-terminal measurements.

Conventional measures		Control measures	
AC voltage	$\tilde{V}_r = V_r + \varepsilon_{Vr}$ $\tilde{V}_j = V_j + \varepsilon_{Vj}$	Modulation index	$\tilde{m}_{ri} = \frac{V_r}{k_1 E_{dci}} + \varepsilon_m$
AC power flows	$\tilde{P}_{rj} = G_{rj}V_r^2 - V_rV_j(G_{rj}\cos\gamma_{rj} + B_{rj}\sin\gamma_{rj}) + \varepsilon_{prj}$ $\tilde{Q}_{rj} = -B_{rj}V_r^2 - V_rV_j(G_{rj}\sin\gamma_{rj} - B_{rj}\cos\gamma_{rj}) + \varepsilon_{qrj}$ The opposite flows can be obtained by exchanging the indexes r and j.	Phase-angle control signal	$\tilde{\gamma}_{rj} = \theta_r - \theta_j + \varepsilon_{\theta rj}$
DC voltage	$\tilde{E}_{DCi} = E_{DCi} + \varepsilon_{E_{DC}}$		
DC current	$\tilde{I}_{DCi} = \mathbf{g}_{DCi}\mathbf{E}_{DC} + \varepsilon_{I_{DCi}}$ (Injected) $\tilde{I}_{DCij} = \dfrac{E_{DCi} - E_{DCj}}{R_{DCij}} + \varepsilon_{I_{DCij}}$ (In DC link i-j)		
DC power	$\tilde{P}_{DCi} = E_{DCi}\,\mathbf{g}_{DCi}\mathbf{E}_{DC} + \varepsilon_{P_{DC}}$ (Injected) $\tilde{P}_{DCij} = \dfrac{E_{DCi}^2 - E_{DCi}E_{DCj}}{R_{DCij}} + \varepsilon_{I_{DCij}}$ (In DC link i-j)		

5.6.10 MT-VSC-HVDC Model in WLS-SE

In this section, the state estimation of an MT-VSC-HVDC system incorporated into WLS-SE is analyzed. The system has been formed by incorporating into the tutorial grid a three-node DC grid. The DC grid connects to the AC grid in nodes 2, 3 and 4, as shown in Figure 5.13. The impedances of the coupling transformers 2-8, 3-6 and 4-7 are 0.02 + j 0.1, 0.01 + j 0.7 and 0.01 + j 0.7, respectively. The resistances of the DC lines 1-2, 2-3 and 1-3 are 0.2, 0.1 and 0.15, respectively.

The files with the relevant information for this case, in WLS-SE, are:

Network file:	Network_tutorial_Multiterminal
Measurements file:	Measure_Multiterminal
SE configuration file:	Config_1

The results furnished by the estimator show some wrong measurements. The largest normalized residue is 11.6, which is associated to the measurement of the AC voltage magnitude at node 4. By eliminating this measurement and executing once more the estimation process, it is found that one more bad data is present. In this case the largest normalized residue is 3.5, which corresponds to the measurement of power in the DC link 1-3. By eliminating this measurement and executing the estimation process again, it shows that all measurements are exact, since all the normalized residues carry null values.

This example illustrates how the method of the largest normalized residue is used to identify one or more bad data. However, a word of caution is needed; this characteristic of the largest normalized residue cannot be generalized. In this example, the correlation

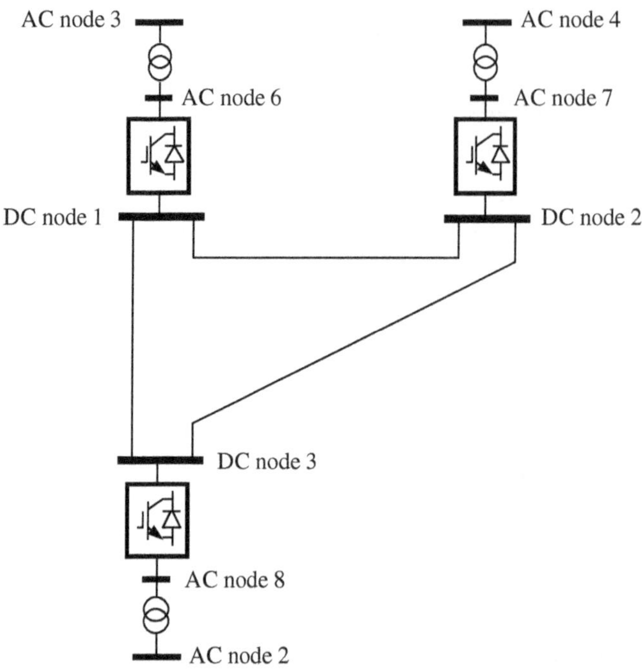

Figure 5.13 Multi-terminal DC network and coupling transformers.

between the two wrong measurements is weak and this is the reason it has been possible to identify them in two stages. When the wrong measurements are strongly correlated, the method of the largest normalized residue may fail. Furthermore, in the presence of unfortunate combinations of errors, some valid measurements may be flagged as bad data by the method of the largest normalized residue.

5.7 Incorporation of Measurements Furnished by PMUs

5.7.1 Incorporation of Synchrophasors in State Estimation

As stated in Section 1.4 of Chapter 1, PMU devices are used to acquire synchronized measurements of voltage and current phasors in power grids, termed synchrophasors. In power systems state estimation, the synchrophasors find quite a natural place in the state estimation processes, enabling a larger amount of information over the operational state of the power grid. In general, a state estimation containing synchrophasor measurements would exhibit higher precision, owing to the higher quality of these types of measurements [23].

The implementation of a conventional state estimator in an electrical power grid enables the operational status of the power grid at a given point in time to be determined. The power grid is taken to remain static throughout one estimation cycle; this involves the time that passes between the measurements' capture at the substations and their transmission to the control centre. Throughout this interval, the measurements are taken to remain stationary. It is unavoidable that some small differences will exist between the various measurements at their time of capture. Nevertheless, if the operating conditions do not change significantly, this does not seem to affect the estimation process unduly.

However, the measured synchrophasors do not allow for any slack at the point in time at which they are taken, given that any delay impacts adversely the measured value. Hence, all the synchrophasor measurements aimed at state estimation applications should be synchronized to correspond to the same sampling time.

One way to incorporate the synchrophasors into the estate estimation process is to carry out the implementation in two stages [24]. In the first stage, we carry out a conventional state estimation process, without using the synchrophasor measurements. In the second stage, we use the synchrophasor measurements together with the results from the first stage to start a new, more refined, state estimated process. It should be noted that the second stage can be formulated in such a way that the estimation process be treated as a linear process and, accordingly, non-iterative.

The advantage of this approach is that the results from the first stage can be obtained with a commercial, conventional state estimator. The disadvantage of using this approach is that it does not exploit fully the great potential that synchrophasors have in improving the grid's state estimation. If the synchrophasor measurements are not incorporated during the first stage, neither the grid's observability nor the treatment of bad data will be improved. Furthermore, the calculated state estimation will not be optimal since all the available measurements, conventional and non-conventional, were not considered simultaneously.

An alternative approach is to incorporate the synchrophasors in the state estimation process directly, as a new type of measurement. In this way, the calculated state estimation will be optimum and the synchrophasors will improve both the grid's observability and the treatment of bad data. The section below is used to describe this approach in more detail.

5.7.2 Synchrophasors Formulations

The PMU measurements, which are involved in the state estimation process, are:

- synchrophasor of nodal voltages
- synchrophasor of branch currents
- synchrophasor of nodal current injections.

In state estimation PMU applications, the voltage and current waveforms are assumed to be purely sinusoidal which, in generic form, may be expressed by the following time expression:

$$x(t) = X_m \cos(\omega t + \phi) \tag{5.81}$$

where X_m is the signal amplitude, ϕ is the signal phase shift with respect to an arbitrary reference and $\omega = 2\pi f$, with f being the frequency.

This time expression has an associated synchrophasor, which may be written down in two different, commutable ways: the polar and rectangular representations. In the former case, the synchrophasor $\overline{X}$ is expressed in terms of the RMS value of X_m and its argument ϕ, i.e.

$$\overline{X} = Xe^{j\phi} \tag{5.82}$$

where $X = X_m/\sqrt{2}$ is the RMS value of X_m.

In case of the rectangular representation, the synchrophasor is expressed in terms of its real and imaginary parts, X_{Re} and X_{Im}, respectively,

$$\overline{X} = X_{\mathrm{Re}} + \mathrm{j}\, X_{\mathrm{Im}} \tag{5.83}$$

where $X_{\mathrm{Re}} = X\cos\phi$ and $X_{\mathrm{Im}} = X\sin\phi$ are the real and imaginary parts of $\overline{X}$, respectively.

The standard IEEE C37.118 2005 'Standard for Synchrophasors for Power Systems' allows both output formats for a PMU synchrophasor, the polar and the rectangular representations. Each representation has distinctive advantages when applied to either voltage or current measurements.

It is customary for state estimators to represent the nodal voltages in polar form. Hence, this provides an incentive for representing the synchrophasors of voltage in polar form, i.e. using magnitude and phase. These will not require a different treatment from those employed for the conventional measurements of voltage magnitude:

$$\tilde{V}_i = V_i + \varepsilon_{Vi} \tag{5.84}$$

Concerning the phase voltages, these show similar characteristics to the magnitudes, in the sense that they are also direct synchrophasor measurements of a state variable. Even though conventional state estimators do not normally incorporate phase voltage measurements, their inclusion in the state estimator presents no difficulty at all, simply being expressed as:

$$\tilde{\theta}_i = \theta_i + \varepsilon_{\theta i} \tag{5.85}$$

On the other hand, the current synchrophasors possess a non-linear representation whether they are expressed in polar form or rectangular form. In the former case, it becomes necessary to use the current magnitude as measurement. This type of measurement is widely used in distribution systems. However, its formulation introduces ill-conditioning terms in the Jacobian matrix, for reduced values of the current magnitude. This anomalous condition shows up at the first iteration when the iterative solution is started from a flat voltage profile, with currents taking a null value, leading to some Jacobian terms having undetermined values.

An alternative to overcome this problem is to use the square of the current as the measurement, leading to well-conditioned Jacobian terms. However, the quality of the measurement worsens as the associated error increases.

Hence, for current synchrophasors it is recommended to use the rectangular form. This way, ill-conditioning problems are avoided and the original precision of the measurement is preserved.

Using the rectangular representation, the current flow, in phasor form, from node i to node j, is:

$$I_{ij} = I_{\mathrm{Re},ij} + \mathrm{j}\, I_{Im,ij} \tag{5.86}$$

Their real and imaginary parts may be expressed as a function of the state variables through the following non-linear equations:

$$I_{\mathrm{Re},ij} = (G_{ij} + G_{si})\, V_i \cos\theta_i - (B_{ij} + B_{si})\, V_i \sin\theta_i - G_{ij} V_j \cos\theta_j + B_{ij} V_j \sin\theta_j \tag{5.87}$$

$$I_{\mathrm{Im},ij} = (G_{ij} + G_{si})\, V_i \sin\theta_i + (B_{ij} + B_{si})\, V_i \cos\theta_i - B_{ij} V_j \cos\theta_j - G_{ij} V_j \sin\theta_j \tag{5.88}$$

where $G_{ij} + \mathrm{j}\, B_{ij}$ is the series admittance of branch $i - j$ and $G_{si} + \mathrm{j}\, B_{si}$ is the shunt admittance at node i.

The synchrophasor of the total current injected into node i may be expressed as the sum of the real and the imaginary parts, respectively, of all the current flowing through the branches which connect to that node as follows:

$$I_{\text{Re},i} = \sum_{j\in\Omega} I_{r,ij} \tag{5.89}$$

$$I_{Im,i} = \sum_{j\in\Omega} I_{i,ij} \tag{5.90}$$

where Ω is the set of nodes connecting to node i.

5.7.3 Phase Reference

As pointed out in Section 5.4, in a conventional state estimation process, i.e. one where no synchrophasor measurements are employed, it is necessary to assign a phase reference to each observable island in the power grid. However, when synchrophasor measurements are incorporated into the estimation process, there is no need to adopt a phase reference. In this case, the adopted phase reference is determined by the synchronizing signal transmitted across the GPS. Only one synchrophasor measurement is sufficient (either voltage or current) for the estimation process to have a phase reference. Accordingly, when synchrophasor measurements are available, no pseudo-measurements of phase reference should be included in the state estimation process. For this reason, as in the conventional estimator, when synchrophasor measurements are included in the state estimation process, all the nodal phases of the power grid are incorporated as state variables in the state vector.

One potential problem that arises from not adopting the customary zero value of phase reference is that the nodal phase angles furnished by PMUs may be very far away from zero. The reason is that at nodes where no synchrophasor measurements are available, the initial values of nodal phase angles state variables are normally taken to be zero, a situation which will impinge on the convergence of the state estimation process. An alternative to ameliorate this convergence problem is to adopt as initial value of nodal phase state variables the measured value, and at nodes where no synchrophasor measurements are available, the average value of all the synchrophasor phase measurements available.

5.7.4 PMU Outputs in WLS-SE

In this section, we show the results of one test case in WLS-SE where synchrophasor measurements of voltage and currents are included. We use the tutorial power grid, which includes a PMU at node 2. The available outputs enable the incorporation of voltage synchrophasors at node 2 and current synchrophasors in the transmission lines that connect at node 2. The files for this example are:

Network file:	Network_tutorial
Measurements file:	Measure_PMU
SE configuration file:	Config_2

It should be noted that two voltage magnitude measurements appear at node 2: one corresponding to the conventional measurements system and the other to the output of the PMU device. Since the measurements incorporate noise, both measurement values differ; however, the quality of the synchrophasor measurement is higher.

It should be remarked that since we have included synchrophasor measurements, no pseudo-measurement of phase reference is required in this case.

5.A Appendix

5.A.1 Input Data and Output Results in WLS-SE

5.A.1.1 Input Data

The input information for the program WLS_SE.p has been grouped in three main blocks. Each block contains an editable m-file.

The first block contains the network data, the second block incorporates the measurements and the third block contains the state estimator configuration. Table 5.A.1 gives the file names used in the examples presented in Chapter 5.

All the information can be modified using the Matlab editor. The variables information is given below.

5.A.1.2 Network Data

The node information is included in the variable 'ac_nodes' and the network branches in the variable 'ac_branch'. The variable 'ac_nodes' is a matrix which contains one row per

Table 5.A.1 Data files.

Case	Network_data	Measurement_data	Config_data
1	Network_tutorial	Measure_tutorial_network	Config_1
2	Network_tutorial	Measure_tutorial_network	Config_2
3	Network_tutorial	Measure_tutorial_network_ bad_data	Config_1
4	Network_tutorial_STATCOM	Measure_STATCOM_case_a	Config_1
5	Network_tutorial_STATCOM_ no_losses	Measure_STATCOM_case_a	Config_1
6	Network_tutorial_STATCOM	Measure_STATCOM_case_b	Config_2
7	Network_tutorial_VSC_HVDC_ptp	Measure_VSC_HVDC_ptp	Config_1
8	Network_tutorial_VSC_HVDC_btb	Measure_VSC_HVDC_btb	Config_2
9	Network_tutorial_UPFC	Measure_UPFC	Config_1
10	Network_tutorial_Multiterminal	Measure_Multiterminal	Config_1
11	Network_tutorial	Measure_PMU	Config_1

Table 5.A.2 Variable 'ac_node' format.

ac_nodes	
1	Number of node
2	Not used
3	Not used
4	VSC type: (0) Without VSC converter (3) VSC STATCOM (4) HVDC sending VSC (5) HVDC receiving VSC (6) UPFC shunt VSC (7) UPFC series VSC (8) Multi-terminal VSC-HVDC
5	VSC series resistance related to switching losses
6	VSC shunt conductance related to conduction losses
7	In type 5 (VSC-HVDC) dc link resistance In type 8 (multiterminal HVDC) related dc-side node
8	In type 4 (VSC-HVDC) receiving node In type 5 (VSC-HVDC) sending node

Table 5.A.3 Variable 'ac_branch' format.

ac_branch	
1	'From' node
2	'To' node
3	Series resistance
4	Series reactance
5	Total shunt susceptance
6	Not used
7	4 in boosting transformer (UPFC) 0 in other case

node. The nodes are assigned the row number that they occupy in the matrix. The information associated to each node is entered column-wise, according to the format indicated in Table 5.A.2

The variable 'ac_branch' contains one row per branch. The branches are assigned the row number that they occupy in the matrix. The information associated to each branch is entered column-wise, according to the format indicated in Table 5.A.3.

In the case of a multi-terminal HVDC grid, the additional variables 'Adc_inc' and 'dc_link' are required. The variable 'Adc_inc' is the node-branch incidence matrix for the dc grid. The variable 'dc_link' is a column vector containing in each row the resistance of the corresponding dc branch.

5.A.1.3 Measurements Data

There are two variables in this block, 'z' and 'c'. The variable 'z' is a matrix containing one row per available measurement. In turn, the information associated with each measurement is entered, column-wise, in the matrix, following the format given in Table 5.A.4.

The variable 'c' is a matrix containing one row per equality restriction included, with the information associated with each restriction entered, column-wise, in the matrix, following the format given in Table 5.A.5.

Table 5.A.4 Variable 'z' format.

z	
1	Type of measurement: 1 Voltage magnitude (Vi) 2 Voltage phase angle (Oi) 3 Active power flow (Pij) 4 Reactive power flow (Qij) 7 Active power injection (Pi) 8 Reactive power injection (Qi) 12 Real part of the current synchrophasor through a branch (Re(Iij)) 13 Imaginary part of the current synchrophasor through a branch (Im(Iij)) 14 Real part of the current injection synchrophasor (Re(Ii)) 15 Imaginary part of the current injection synchrophasor (Im(Ii)) 25 dc voltage in STATCOM, HVDC VSC-sending and UPFC dc-link (VSC s Edc) 26 dc current in STATCOM, HVDC VSC-sending and UPFC dc-link (VSC s Idc) 27 dc power in STATCOM, HVDC VSC-sending and UPFC dc-link (VSC s Pdc) 28 Modulation index STATCOM, HVDC VSC-sending and UPFC VSC-shunt (VSC m) 29 dc voltage in HVDC VSC-receiving (VSC r Edc) 30 dc current in HVDC VSC-receiving (VSC r Idc) 31 HVDC VSC-receiving dc power (VSC r Pdc) 32 Modulation index in HVDC VSC-receiving and UPFC VSC-series (VSC r m) 33 Modulation index in Multiterminal HVDC VSC (m VSC Mul) 34 Nodal phase difference (Oi-Oj) 40 UPFC active power flow at the ac-side node in series-VSC (UPFC psVSC) 41 UPFC reactive power flow at ac-side node in series-VSC (UPFC qsVSC) 45 Multiterminal HVDC dc voltage (Vdc Mul) 46 Multiterminal HVDC dc current injection at VSC (Idc_i Mul) 48 Multiterminal HVDC dc current flow in dc-link (Idc_ij Mul) 49 Multiterminal HVDC dc power flow in dc-link (Pdc_i Mul)
2	For a node measurement number. For a branch flow measurement number. This value is positive if the measurement is taken in the node 'from' of the corresponding branch in variable 'ac_branch'. If the measurement is taken in the opposite end of the branch, i.e. node 'to', the branch number takes a negative value
3	Measured value
4	Quality of measurement (standard deviation)

Table 5.A.5 : Variable 'c' format.

c	
1	Type of constraint: 7 Active power injection (Pi) 8 Reactive power injection (Qi) 26 Current at VSC dc side in STATCOM (VSC s Idc) 50 STATCOM, HVDC VSC-sending, UPFC VSC-shunt active power balance (VSC s bal) 51 HVDC VSC-receiving active power balance (VSC r bal) 52 UPFC VSC-receiving active power balance (UPFC r bal) 53 Multiterminal HVDC VSC active power balance (Multi. bal)
2	Related number of the ac node.
3	Not used. This value must be 0

The elimination of a measurement from the estimation process does not require the corresponding row to be deleted. It is sufficient to comment out the relevant row using the character '%'. In general, for all the cases presented in this chapter, all the possible measurements have been included in the relevant matrices and the measurements that have not been included in the estimation process have simply been commented out. This facilitates the incorporation of new measurements in new study cases.

It has been stated above that it is possible to eliminate measurements; however, some caution needs to be exercised since the WLS-SE program has not been fitted with an observability analysis. It should be remarked that to carry out a successful state estimation calculation, it is necessary to count with a number and distribution of measurements that warrant it.

5.A.1.4 State Estimator Configuration

This block uses three variables: 'noise', 'converg' and 'iter_fin'. The variable 'noise' specifies whether a Gaussian noise has been added to the measurement prior to the state estimation process. If the variable is given a value of 1, then random noise is incorporated to the measurement value. The value of the variable 'converg' is the convergence tolerance used in the iterative process. The variable 'iter_fin' is the maximum number of allowed iterations.

5.A.2 Output Results

Once the estimation process has run, the program shows the results on the screen. The information is presented in three blocks. The first block contains information relating to the general characteristics of the estimation process, including the data files used and the configuration parameters of the state estimator. The second block presents the parameters associated to the power grid just estimated. The third block includes information relating to the estimation process and the estimated values.

In the next section the estimated values of the state variables are presented and then the estimated values of the measurements. In order to identify bad data, the normalized residual value of each measurement is specified. For completeness, all the restrictions that have been incorporated into the estimation process are listed.

References

1 Schweppe, F.C., Wildes, J., and Rom, D.B. (1970). Power system static-state estimation, part I: exact model, part II: approximate model, part III: implementation. *IEEE Transactions on Power Apparatus and Systems* PAS-89 (1): 120–135.

2 Monticelli, A. (1999). *State Estimation in Electric Power System. A Generalized Approach*. Kluwer Academic Publishers.

3 Abur, A. and Gomez Exposito, A. (2004). *Power System State Estimation: Theory and Implementation*. Marcel Dekker.

4 Sirisena, H.R. and Brown, E.P.M. (1981). Inclusion of HVDC links in AC power system state estimation. *Proceedings of the Institution of Electrical Engineers* 128 (3): 147–154.

5 Leite Da Silva, A.M., Prada, R.B., and Falcao, D.M. (1985). State estimation for integrated multi-terminal DC/AC systems. *IEEE Transactions on Power Apparatus and Systems* PAS-104: 2349–2355.

6 Jegatheesan, R. and Duraiswamy, K. (1987). AC/multi-terminal DC power system state estimation—a sequential approach. *Electric Machines and Power Systems* 12: 27–42.

7 Sinha, A.K., Roy, L., and Srivastava, H.N. (1994). A decoupled second order state estimator for AC-DC power systems. *IEEE Transactions on Power Systems* 9 (3): 1485–1491.

8 Ding, Q., Chung, T.S., and Zhang, B. (2001). An improved sequential method for AC/MTDC power system state estimation. *IEEE Transactions on Power Systems* 16 (3): 506–512.

9 A. Abur, B. Gou and E. Acha, "State estimation of networks containing power flow control devices", in Proc. 13th PSCC, Trondheim, June–July 1999, pp. 427–433.

10 Xu, B. and Abur, A. (2004). State estimation of systems with UPFCs using the interior point method. *IEEE Transactions on Power Systems* 19 (3): 1635–1641.

11 E.A. Zamora and C.R. Fuerte-Esquivel, "Static state estimation of power containing series and shunt FACTS controllers", in Proc. 15th PSCC, Liege, Belgium, 22–26 August 2005.

12 Zamora, E.A. and Fuerte-Esquivel, C.R. (2011). State estimation of power systems containing facts controllers. *Electric Power Systems Research*, vol. 81,: 995–1002.

13 Qifeng, D., Boming, Z., and Chung, T.S. (2000). State estimation for power systems embedded with FACTS devices and MTDC systems by a sequential solution approach. *Electric Power Systems Research*, 55,: 147–156.

14 Rakpenthai, C., Premrudeepreechacharn, S., and Uatrongjit, S. (2009). Power system with multi-type FACTS devices state estimation based on predictor-corrector interior point algorithm. *Electrical Power and Energy Systems*, 31,: 160–166.

15 de la Villa Jaén, A., Acha, E., and Gómez Expósito, A. (2008). Voltage source converter modeling for power system state estimation: STATCOM and VSC-HVDC. *IEEE Transactions on Power Systems,* 23 (4): 1552–1559.

16 J. Cao, W. Du and H.F. Wang, "The incorporation of generalized VSC MTDC model in AC/DC power system state estimation", International Conference on Sustainable Power Generation and Supply (SUPERGEN 2012), 8–9 September 2012.

17 Thorp, J.S., Phadke, A.G., and Karimi, K.J. (1985). Real-time voltage phasor measurements for static state estimation. *IEEE Transactions on Power Apparatus and Systems* 104 (11): 3098–3107.

18 A. Gómez-Expósito, A. Abur, P. Rousseaux, A. de la Villa Jaén and C. Gómez-Quiles, "On the Use of PMUs in Power System State Estimation", 17th Power System Computation Conference, Stockholm, Sweden, August 22–26, 2011.

19 de la Villa Jaén, A., Beloso Martínez, J., Gómez-Expósito, A., and González Vázquez, F. (2018.). Tuning of measurement weights in state estimation: theoretical analysis and case study. *IEEE Transactions on Power Systems,* 33 (4): 4583–4592,.

20 Acha, E. (2015). *Nodal Analysis & Sparsity.* ADP-London, Kindle Publishing.

21 Stagg, G.W. and El-Abiad, H. (1968). *Computer Methods in Power Systems Analysis.* McGraw-Hill.

22 Mohan, N., Undeland, T.M., and Robbins, W.P. (1995). *Power Electronics. Converters, Applications, and Design.* Wiley.

23 Phadke, A.G. and Thorp, J.S. (2008). *Synchronized Phasor Measurements and Their Applications.* Springer.

24 Zhou, M., Centeno, V.A., Thorp, J.S., and Phadke, A.G. (2006). An alternative for including phasor measurements in state estimation. *IEEE Transactions on Power Systems* 21 (4): 1930–1937.

6

Dynamic Simulations of Power Systems

6.1 Introduction

As elucidated in Chapter 1, today's electric power systems are among the most complex dynamic systems ever built, requiring the intervention of skilful engineers, round the clock. Power systems' engineers ought to be able to assess in advance the various issues that emerge when the power network undergoes drastic changes, such as when new equipment and, particularly, new technologies are added to it, for instance FACTS and VSC-HVDC technologies. This is even more so when one considers the natural dynamic behaviour exhibited by such large systems, which always depend on the whims of societal behaviour and are exposed to the forces of nature.

A great deal of study and research effort is required to master the topic of power system dynamics. At present, comprehensive models of VSC equipment are still required – reliable models with which to evaluate the dynamic impact of these devices in the network, prior to their installation. At the design stage, simulation studies of credible disturbances will reveal how the new equipment will perform and what impact it will have on the power network. Careful consideration needs to be given to the control systems of the VSC to assess the effectiveness, or otherwise, of the voltage and power controls to counteract voltage, power and frequency oscillations in the network – the assessments must be beyond doubt.

Interest in HVDC transmission systems has long been associated with the need to transfer electrical energy over very long distances. However, other applications have emerged in more recent times where the HVDC links are gaining ground, for instance in the tapping of offshore wind energy resources. There is general agreement that the operation of today's power networks is becoming increasingly complex owing to the use of HVDC systems and similar equipment. Moreover, the intermittence of the wind may pose a major risk to the integrity of the power system. In this new scenario, it becomes important that greater attention is given to the dynamics of the power grid than has been the case so far; it is recommended that the HVDC controllers be suitably tuned and that advanced control functions be implemented. The installation of FACTS-type power flow controlling devices seems to offer a realistic solution to ameliorate potential dynamic problems that may arise, as well as to enable the transmission of larger amounts of electrical energy between areas without endangering the stability of the interconnected networks. As discussed in Chapter 1, VSC-HVDC systems are rapidly becoming the most versatile and cost-effective power transmission option in modern power systems networks.

VSC-FACTS-HVDC: Analysis, Modelling and Simulation in Power Grids, First Edition.
Enrique Acha, Pedro Roncero-Sánchez, Antonio de la Villa Jaén, Luis M. Castro and Behzad Kazemtabrizi.

Companion website: www.wiley.com/go/acha_vsc_facts

6.2 Modelling of Conventional Power System Components

The conventional, large synchronous generator continues to be the dominant source of electrical energy in most power systems around the globe. Owing to the wide range of complex dynamic phenomena, to which these synchronous machines are subjected, various models with different ranges of applicability can be found in the open literature. The mechanical equations of a synchronous generator are well established [1–3] and they are only outlined here. Three basic assumptions are made in deriving its equations: (i) the machine rotor speed does not depart greatly from the synchronous speed; (ii) the machine rotational losses due to windage and friction are neglected; and (iii) the mechanical power shaft is smooth, meaning that the power coming from the shaft is constant except for the control action exerted by the speed governor. Therefore, the accelerating power of the generator P_a is the difference between the mechanical input power P_m [p.u.] supplied by the prime mover and the electrical output power P_e [p.u.]. The machine acceleration is expressed as follows:

$$\frac{d^2\delta}{dt^2} = \frac{P_a}{M_g} = \frac{P_m - P_e}{M_g} \tag{6.1}$$

where the second-time derivative of the rotor angle δ [rad] represents the acceleration of the machine and M_g is the angular momentum. The angular momentum may be further defined in terms of the system base frequency f_0 [Hz] and the inertia constant H [MWs/MVA], which does not change much regardless of the size of the machine,

$$M_g = \frac{H}{\pi f_0} \tag{6.2}$$

Eddy currents induced in the rotor iron or in the damping windings produce torques which oppose the motion of the rotor relative to the synchronous speed. To account for these effects, a deceleration power is introduced into the mechanical equation. Additionally, decomposing (6.1) into two first-order differential equations leads to the following expressions:

$$\frac{d\omega}{dt} = \frac{\pi f_0}{H}[P_m - P_e - D(\omega - \omega_0)] \tag{6.3}$$

$$\frac{d\delta}{dt} = \omega - \omega_0 \tag{6.4}$$

where ω [rad s^{-1}] stands for the mechanical angular speed and D [s/rad] is the damping coefficient.

6.2.1 Modelling of Synchronous Generators

In early power system stability studies, the generators were represented by the classical model, which assumes constant flux linkages in both axes and neglects transient saliency [4]. In this simple model, an internal voltage behind the d-axis transient reactance X'_d is determined and its magnitude is assumed constant. To enable a more realistic representation of the power system behaviour, the transient saliency and the varying flux linkages that exist in both axes of the synchronous generator ought to be considered. Figure 6.1 depicts the phasor diagram of the synchronous generator during the transient

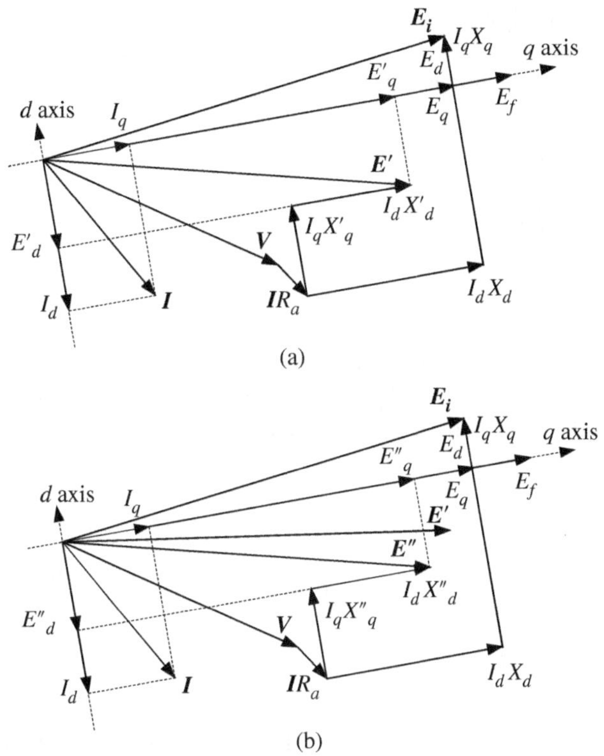

Figure 6.1 Phasor diagram of the synchronous generator: (a) transient state, (b) sub-transient state.

and sub-transient states [5, 6]. Based on these phasor diagrams, the mathematical model of the synchronous generator can be developed.

Taking the case of a synchronous generator undergoing a transient condition, the voltage behind the synchronous reactance is obtained with the internal transient flux voltages, E'_d and E'_q. Then it is necessary to find the voltages which will represent the flux linkages of the rotor windings. The equations that describe the time variation of the voltages representing the rotor flux linkages in the dq reference frame are:

$$T'_{d0}\frac{dE'_q}{dt} = E_f - E'_q + (X_d - X'_d)I_d \tag{6.5}$$

$$T'_{q0}\frac{dE'_d}{dt} = -E'_d - (X_q - X'_q)I_q \tag{6.6}$$

$$E'_d = R_a I_d + X'_q I_q + V_d \tag{6.7}$$

$$E'_q = R_a I_q - X'_d I_d + V_q \tag{6.8}$$

where T'_{d0} and T'_{q0} are the open-circuit transient time constants; E_f is the excitation voltage; V_d and V_q are the generator's terminal voltages; I_d and I_q are the generator's terminal currents; X'_d and X'_q are the transient reactances; X_d and X_q are the synchronous reactances; and R_a is the armature resistance. Notice that except for E_f, all other quantities are given on a per-axis basis: d-axis and q-axis, respectively.

If a more detailed synchronous machine model representation is required, then the sub-transient synchronous generator model must be used. The equations that govern the sub-transient behaviour of the synchronous generator are:

$$T''_{d0}\frac{dE''_q}{dt} = E'_q - E''_q + (X'_d - X''_d)I_d \tag{6.9}$$

$$T''_{q0}\frac{dE''_d}{dt} = E'_d - E''_d - (X'_q - X''_q)I_q \tag{6.10}$$

$$E''_d = R_aI_d + X''_qI_q + V_d \tag{6.11}$$

$$E''_q = R_aI_q - X''_dI_d + V_q \tag{6.12}$$

where T''_{d0} and T''_{q0} are the open-circuit sub-transient time constants, E''_d and E''_q are the internal sub-transient flux voltages, and X''_d and X''_q are the sub-transient reactances. The active and reactive power at the synchronous generator terminal in the *dq* reference frame can be expressed as:

$$P_g = V_dI_d + V_qI_q \tag{6.13}$$

$$Q_g = V_qI_d - V_dI_q \tag{6.14}$$

6.2.2 Synchronous Generator Controllers

Conventional power plants are made up of several components to enable the power conversion from mechanical to electrical in an efficient manner. The main elements that find a representation in most dynamic analyses are the synchronous generators, voltage regulators, speed governors, turbines, boilers and protection equipment. The most relevant models are outlined below.

6.2.2.1 Speed Governors

The speed governor is a frequency controller which plays a crucial role in power system operation. Its main functions are to:

- keep the generator speed as close as possible to its nominal value
- guarantee a fast and automatic participation of the generator following any change in generation or load, aiming at maintaining a balanced generation-load pattern in the system.

To adjust the amount of steam or water input into the turbine, the speed governor senses the difference between the actual rotor speed and the angular frequency and adjusts the turbine valve accordingly. Figures 6.2a,b show the block diagrams used to represent the speed governing system of a steam turbine and hydro turbine, respectively. These are simplified representations of the models in [7].

From Figure 6.2a the following mathematical expression is derived:

$$\frac{dP_{GV}}{dt} = \frac{P_{set} - K\Delta\omega - P_{GV}}{T_{GV}} \tag{6.15}$$

where $\Delta\omega = \omega - \omega_0$, K is the governor gain in per-unit, P_{set} is the power set point in per-unit, T_{GV} is the governor time constant in seconds and P_{GV} is the output power from

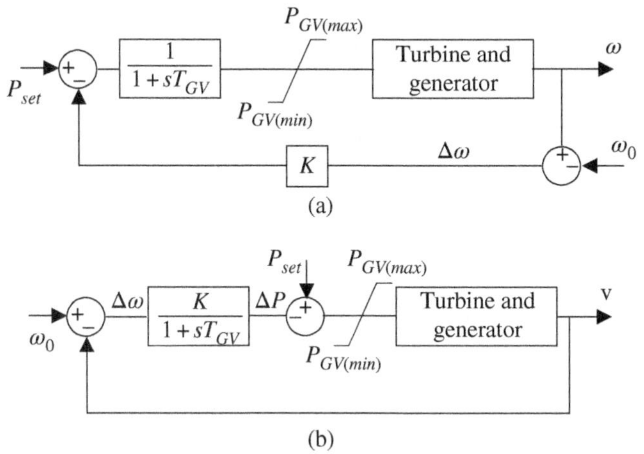

Figure 6.2 Simplified IEEE speed governor model for a (a) steam turbine, (b) hydro turbine. Source: ©IET, 1987.

the control valve in per-unit. For the case of a hydro turbine, the differential equation representing its speed-governing system is given by (6.16), where $P_{GV} = P_{set} - \Delta P$.

$$\frac{d\Delta P}{dt} = \frac{K\Delta\omega - \Delta P}{T_{GV}} \tag{6.16}$$

6.2.2.2 Steam Turbine and Hydro Turbine

A turbine is a fluid machine through which fluid passes continuously, transferring a portion of its energy to an impeller with vanes or blades. The turbine transforms the energy of steam flow or water flow into mechanical energy through a momentum exchange between the working fluid and the impeller, which carries out the energy exchange. Conventional power-generating plants utilize predominantly hydro turbines or steam turbines which drive a synchronous generator to produce electrical energy. The design of each type of turbine differs greatly from the others owing to their vastly different applications.

All compound steam-turbine systems utilize governor-controlled valves at the inlet to the high-pressure or very high-pressure chambers to control the steam flow. The steam chest and inlet piping to the first turbine cylinder and reheaters and crossover piping downstream all introduce delays between valves movements and changes in steam flow. The main objective in modelling the steam system for stability studies is to take account of these delays. For such a purpose, flows in and out of the steam vessels may be related by simple time constants. For steam turbines, a comprehensive model put forward by an IEEE committee report is selected [7] and shown in Figure 6.3.

This block diagram represents a cross-compound, three-stage, single-reheat steam-turbine system. The equations corresponding to this model are:

$$\frac{dP_{HP}}{dt} = \frac{P_{GV} - P_{HP}}{T_{CH}} \tag{6.17}$$

$$\frac{dP_{IP}}{dt} = \frac{P_{HP} - P_{IP}}{T_{RH}} \tag{6.18}$$

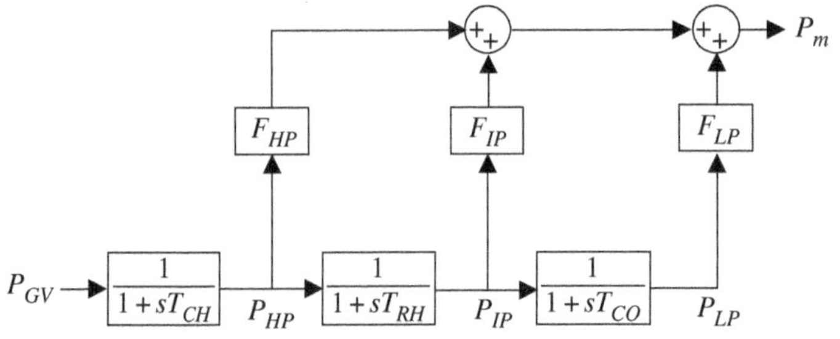

Figure 6.3 Simplified IEEE steam-turbine model. Source: ©IET, 1987.

$$\frac{dP_{LP}}{dt} = \frac{P_{IP} - P_{LP}}{T_{CO}} \tag{6.19}$$

$$P_m = P_{HP}F_{HP} + P_{IP}F_{IP} + P_{LP}F_{LP} \tag{6.20}$$

where T_{CH} is the steam chest time constant in seconds (including the high-pressure *HP* stage of the turbine time constant), T_{RH} is the reheater time constant (including the intermediate-pressure *IP* stage of the turbine time constant), T_{CO} is the steam storage or cross-over time constant (including the low-pressure *LP* stage of the turbine time constant), F_{HP} is the *HP* turbine power fraction, F_{IP} is the *IP* turbine power fraction, F_{LP} is the *LP* turbine power fraction and P_m is the equivalent generator input mechanical power.

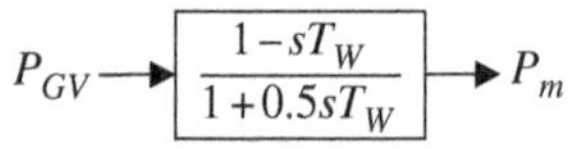

Figure 6.4 IEEE hydro-turbine model. Source: ©IET, 1987.

In a hydro turbine, the motion energy of the water is transformed into rotational motion of a shaft. Typically, this turbine is equipped with an automatic speed regulator, which is essentially a device sensitive to variations in turbine speed. The impeller receives torque from the water pressure, which allows the shaft to obtain a mechanical power. The speed controller is responsible for keeping the turbine at a constant speed in the steady-state regime. Figure 6.4 shows the hydro-turbine model which is selected from the IEEE committee report [7].

The equation arising from the Laplace-based block representing the hydro-turbine model is:

$$\frac{dP_m}{dt} = \frac{P_{GV} - T_W\dot{P}_{GV} - P_m}{0.5T_W} \tag{6.21}$$

where T_W is the water time constant in seconds and P_m is the equivalent generator input mechanical power in per-unit.

6.2.2.3 Automatic Voltage Regulator

The basic function of an automatic voltage regulator (AVR) is to provide the synchronous generator's field winding with a variable direct current, enabling an effective control of the generator terminal voltage and, hence, the enhancement of system stability. In addition, the excitation system can provide a broad number of control and protection functions necessary for the satisfactory performance of the generator and the power system. The protective functions ensure that the capacity limits of both the synchronous machine and the excitation control system will not be exceeded. Several

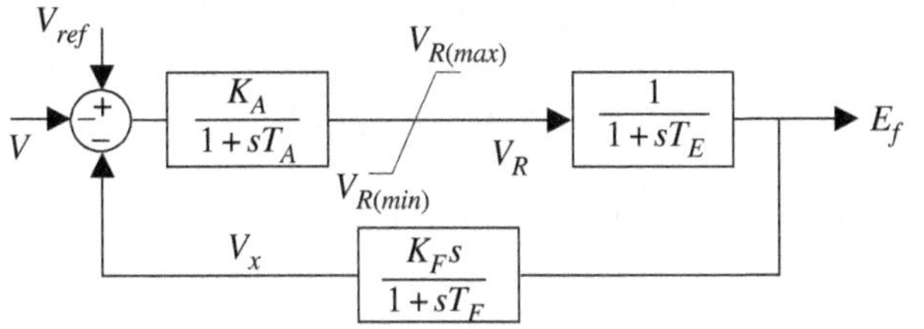

Figure 6.5 Simplified IEEE-type I AVR model.

AVR models for transient stability studies exist in the open literature. The model used in this book is shown in Figure 6.5. It is a simplified version of the STA1 IEEE model [8] with neither saturation function nor transductor function. The latter represents the inherent delay in measuring the voltage at the generator's terminal.

From Figure 6.5 the following expressions are derived for the AVR model:

$$\frac{dV_R}{dt} = \frac{K_A(V_{ref} - V_x - V) - V_R}{T_A} \tag{6.22}$$

$$\frac{dV_x}{dt} = \frac{K_F\dot{E}_f - V_x}{T_F} \tag{6.23}$$

$$\frac{dE_f}{dt} = \frac{V_R - E_f}{T_E} \tag{6.24}$$

where $V_{R(max)}$ is the maximum regulator voltage, $V_{R(min)}$ is the minimum regulator voltage, K_A is the amplifier gain, T_A is the amplifier time constant, sec, K_F is the stabilizer gain, T_F is the stabilizer time constant, sec, and T_E is the field circuit time constant.

6.2.2.4 Transmission Line Model

In fundamental frequency power system studies, such as power flows, state estimation and dynamic studies, it is common to represent a transmission line by its nominal π-circuit, as discussed in (Section 3.2.1). However, during a network disturbance the system frequency may deviate from its nominal value and therefore the series impedance and shunt susceptance of the line may change significantly. Hence, for dynamic studies, both parameters may incorporate a certain degree of frequency dependency, as [9]:

$$Z_s = R_s + jX_s^0\left(\frac{f_k + f_m}{2f_0}\right) \tag{6.25}$$

$$Y_{sh} = G_{sh} + jB_{sh}^0\left(\frac{2f_0}{f_k + f_m}\right) \tag{6.26}$$

where f_k and f_m are the frequency at buses k and m, respectively, X_s^0 and B_{sh}^0 are the series reactance and shunt susceptance obtained at the rated frequency f_0.

6.2.2.5 Load Model

The impact of the load models on the power system dynamics is a matter of great importance. Generally, the load varies with respect not only to voltage magnitude but also to frequency. The case of load dependency with respect to voltage, which is relevant to power flows and state estimation, was discussed in (Section 3.2.7). For a load connected to bus k, the active power P_{Lk} and reactive power Q_{Lk} drawn by the loads may be modelled by a polynomial representation multiplied by a linear factor to account for their voltage and frequency dependence, respectively [10]:

$$P_{Lk} = P_{0k}\left[a_p + b_p\left(\frac{V_k}{V_{0k}}\right) + c_p\left(\frac{V_k}{V_{0k}}\right)^2\right][1 + K_{pf}(f_k - f_0)] \tag{6.27}$$

$$Q_{Lk} = Q_{0k}\left[a_p + b_q\left(\frac{V_k}{V_{0k}}\right) + c_q\left(\frac{V_k}{V_{0k}}\right)^2\right][1 + K_{qf}(f_k - f_0)] \tag{6.28}$$

where P_{0k} and Q_{0k} are the nominal active and reactive powers at rated frequency f_0 and voltage V_{0k}, the parameters a_p, b_p, c_p, a_q, b_q, c_q represent the degree of dependency of the active and reactive powers with respect to voltage magnitude $V_k = \sqrt{(e_k^2 + jf_k^2)}$, and K_{pf} and K_{qf} are the degree of dependency of the active and reactive powers with respect to the frequency f_k of the bus to which the load is connected.

6.3 Time Domain Solution Philosophy

The power flows in a network are calculated very effectively by modelling the power system by a set of nonlinear algebraic equations corresponding to active and reactive power injections at the nodes, valid for a predefined set of system generation and load pattern. Conversely, for dynamic power system assessments, the mathematical model which describes the dynamic behaviour of generators and their controls can be represented by a set of differential equations. By suitable combination of both sets of algebraic and differential equations, the complete power system model aimed at time domain simulations of large-scale power networks may be expressed as an algebraic-differential problem of the form:

$$0 = f(\boldsymbol{x}, \boldsymbol{y}) \tag{6.29}$$

$$\dot{\boldsymbol{y}} = g(\boldsymbol{x}, \boldsymbol{y}, t) \tag{6.30}$$

where f and g are nonlinear vector functions and $\boldsymbol{x}$ and $\boldsymbol{y}$ are vectors of variables that are computed at discrete points in time t. Eq. (6.29) corresponds to the network's power balance equations whereas (6.30) corresponds to the differential equations of all the synchronous generators and controlling devices. Since each generator is coupled only to the electrical system, expression (6.30) is a collection of separate subsets of independent block-like equations. Nevertheless, to efficiently solve Eqs. (6.29) and (6.30) as a function of time, a unified framework is used ([11, 12]. The Newton-Raphson method is selected in this chapter to carry out such a task since it yields reliable numerical solutions and retains the quadratic convergence of the conventional Newton-Raphson power flow, as discussed in Section 3.5.2.

6.3.1 Numerical Solution Technique

On way to solve the algebraic-differential set (6.29) and (6.30) is to transform the differential equations into algebraic equations using the implicit trapezoidal method, a robust technique known for giving reasonably accurate results even when relatively large integration time steps are selected [13, 14]. The implicit trapezoidal method assumes linearity during the integration time step. The transformed algebraic equation of (6.30) for the time interval Δt is:

$$Y_{(t)} = Y_{(t-\Delta t)} + \frac{\Delta t}{2}(\dot{Y}_{(t-\Delta t)} + \dot{Y}_{(t)}) \tag{6.31}$$

Rearranging (6.31) as a mismatch equation yields:

$$F_Y = Y_{(t-\Delta t)} + 0.5\Delta t \dot{Y}_{(t-\Delta t)} - (Y_{(t)} - 0.5\Delta t \dot{Y}_{(t)}) = 0 \tag{6.32}$$

The first step in the application of the trapezoidal method is to express each differential equation in the form of (6.32). Notice that they are expressed in the form of a mismatch equation, in the same fashion as that of the network's active and reactive power mismatch equations – refer to Section 3.5. This enables a suitable combination of both kinds of equations in this unified frame of reference.

The overall dynamic models presented in this chapter are developed in an all-encompassing frame-of-reference where the nonlinear algebraic equations of the transmission network, synchronous generators and their controls are linearized around a base operating point and combined with the discretized differential equations arising from the control devices and synchronous generators, for unified iterative solutions using the Newton-Raphson method [9]. One such iterative solution is valid for a given point in time and its rate of convergence remains quadratic. To this end, the set of differential equations transformed into algebraic equations is appended to the existing set of algebraic equations representing the **Jacobian** matrix of the conventional power flow method, as shown in (6.33):

$$\begin{bmatrix} \Delta P \\ \Delta Q \\ F_Y \end{bmatrix} = \begin{bmatrix} \multicolumn{2}{c}{\textbf{Jacobian}} & \begin{matrix} \dfrac{\partial \Delta P}{\partial Y} \\ \dfrac{\partial \Delta Q}{\partial Y} \end{matrix} \\ \dfrac{\partial F_Y}{\partial e} & \dfrac{\partial F_Y}{\partial f} & \dfrac{\partial F_Y}{\partial Y} \end{bmatrix} \begin{bmatrix} \Delta e \\ \Delta f \\ \Delta Y \end{bmatrix} \tag{6.33}$$

where ΔP and ΔQ are the active and the reactive power mismatch vectors, respectively; F_Y is a vector that contains the discretized differential equations of each generator or controlling device; Δe, Δf and ΔY represent the vectors of incremental changes in the real and imaginary parts of the nodal voltages, as well as the state variables arising from each differential equation. In this unified solution, all the state variables are adjusted simultaneously to compute the new equilibrium point of the power system at every time step. The simultaneous solution of both sets of equations plays a key role in achieving a reliable assessment of the dynamic stability of a power network.

To exemplify the procedure of aggregating the equations arising from any dynamic component into the conventional **Jacobian** matrix, the solution method is applied to the dynamic equations brought about by the synchronous generator transient model represented by Eqs. (6.3)–(6.6):

$$F_\omega = \omega_{(t-\Delta t)} + 0.5\Delta t \dot{\omega}_{(t-\Delta t)} - (\omega_{(t)} - 0.5\Delta t \dot{\omega}_{(t)}) \tag{6.34}$$

$$F_\delta = \delta_{(t-\Delta t)} + 0.5\Delta t \dot{\delta}_{(t-\Delta t)} - (\delta_{(t)} - 0.5\Delta t \dot{\delta}_{(t)}) \tag{6.35}$$

$$F_{E'q} = E'_{q(t-\Delta t)} + 0.5\Delta t \dot{E}'_{q(t-\Delta t)} - (E'_{q(t)} - 0.5\Delta t \dot{E}'_{q(t)}) \tag{6.36}$$

$$F_{E'_d} = E'_{d(t-\Delta t)} + 0.5\Delta t\dot{E}'_{d(t-\Delta t)} - (E'_{d(t)} - 0.5\Delta t\dot{E}'_{d(t)}) \tag{6.37}$$

where,

$$\dot{\omega}_{(t)} = \frac{\pi f_0}{H}[P_{m(t)} - P_{e(t)} - D(\omega_{(t)} - \omega_0)] \tag{6.38}$$

$$\dot{\omega}_{(t-\Delta t)} = \frac{\pi f_0}{H}[P_{m(t-\Delta t)} - P_{e(t-\Delta t)} - D(\omega_{(t-\Delta t)} - \omega_0)] \tag{6.39}$$

$$\dot{\delta}_{(t)} = \omega_{(t)} - \omega_0 \tag{6.40}$$

$$\dot{\delta}_{(t-\Delta t)} = \omega_{(t-\Delta t)} - \omega_0 \tag{6.41}$$

$$\dot{E}'_{q(t)} = T'_{d0}{}^{-1}[E_{f(t)} - E'_{q(t)} + (X_d - X'_d)I_{d(t)}] \tag{6.42}$$

$$\dot{E}'_{q(t-\Delta t)} = T'_{d0}{}^{-1}[E_{f(t-\Delta t)} - E'_{q(t-\Delta t)} + (X_d - X'_d)I_{d(t-\Delta t)}] \tag{6.43}$$

$$\dot{E}'_{d(t)} = T'_{d0}{}^{-1}[-E'_{d(t)} - (X_q - X'_q)I_{q(t)}] \tag{6.44}$$

$$\dot{E}'_{d(t-\Delta t)} = T'_{d0}{}^{-1}[-E'_{d(t-\Delta t)} - (X_q - X'_q)I_{q(t-\Delta t)}] \tag{6.45}$$

The set of Eqs. (6.34)–(6.45) must be solved together with the equations of the whole network when including a synchronous generator dynamic model. Hence, the following linearized equation provides the computing engine with which the time-domain solution is carried out:

$$\begin{bmatrix} \Delta P \\ \Delta Q \\ F_\omega \\ F_\delta \\ F_{E'_q} \\ F_{E'_d} \end{bmatrix} = - \begin{bmatrix} J_{11} & J_{12} \\ J_{21} & J_{22} \end{bmatrix} \begin{bmatrix} \Delta e \\ \Delta f \\ \Delta\omega \\ \Delta\delta \\ \Delta E'_q \\ \Delta E'_d \end{bmatrix} \tag{6.46}$$

where

$$J_{11} = \begin{bmatrix} \frac{\partial \Delta P}{\partial e} & \frac{\partial \Delta P}{\partial f} \\ \frac{\partial \Delta Q}{\partial e} & \frac{\partial \Delta Q}{\partial f} \end{bmatrix}, \quad J_{12} = \begin{bmatrix} \frac{\partial \Delta P}{\partial \omega} & \frac{\partial \Delta P}{\partial \delta} & \frac{\partial \Delta P}{\partial E'_q} & \frac{\partial \Delta P}{\partial E'_d} \\ \frac{\partial \Delta Q}{\partial \omega} & \frac{\partial \Delta Q}{\partial \delta} & \frac{\partial \Delta Q}{\partial E'_q} & \frac{\partial \Delta Q}{\partial E'_d} \end{bmatrix},$$

$$J_{21} = \begin{bmatrix} \frac{\partial F_\omega}{\partial e} & \frac{\partial F_\omega}{\partial f} \\ \frac{\partial F_\delta}{\partial e} & \frac{\partial F_\delta}{\partial f} \\ \frac{\partial F_{E'_q}}{\partial e} & \frac{\partial F_{E'_q}}{\partial f} \\ \frac{\partial F_{E'_d}}{\partial e} & \frac{\partial F_{E'_d}}{\partial f} \end{bmatrix}, \quad J_{22} = \begin{bmatrix} \frac{\partial F_\omega}{\partial \omega} & \frac{\partial F_\omega}{\partial \delta} & \frac{\partial F_\omega}{\partial E'_q} & \frac{\partial F_\omega}{\partial E'_d} \\ \frac{\partial F_\delta}{\partial \omega} & \frac{\partial F_\delta}{\partial \delta} & \frac{\partial F_\delta}{\partial E'_q} & \frac{\partial F_\delta}{\partial E'_d} \\ \frac{\partial F_{E'_q}}{\partial \omega} & \frac{\partial F_{E'_q}}{\partial \delta} & \frac{\partial F_{E'_q}}{\partial E'_q} & \frac{\partial F_{E'_q}}{\partial E'_d} \\ \frac{\partial F_{E'_d}}{\partial \omega} & \frac{\partial F_{E'_d}}{\partial \delta} & \frac{\partial F_{E'_d}}{\partial E'_q} & \frac{\partial F_{E'_d}}{\partial E'_d} \end{bmatrix}.$$

The submatrix J_{11} is the conventional Jacobian matrix which comprises the first-order partial derivatives of the nodal active and reactive power mismatches with respect to the real and imaginary parts of the nodal voltages. Likewise, J_{12} contains the partial derivatives arising from the nodal active and reactive powers with respect to the corresponding variables of the synchronous generator. The matrix J_{21} consists of partial derivatives of the synchronous generator's discretized differential equations with respect to the real and imaginary parts of the nodal voltages. Lastly, J_{22} is a matrix that accommodates the first-order partial derivatives of the synchronous generator's discretized differential equations with respect to its own state variables.

6.3.2 Benchmark Numerical Example

A typical power network is used to illustrate the time domain solution using the Newton-Raphson-based dynamic power flow program. The New England test system [15] shown in Figure 6.6 is a 39-bus network containing 10 synchronous generators,

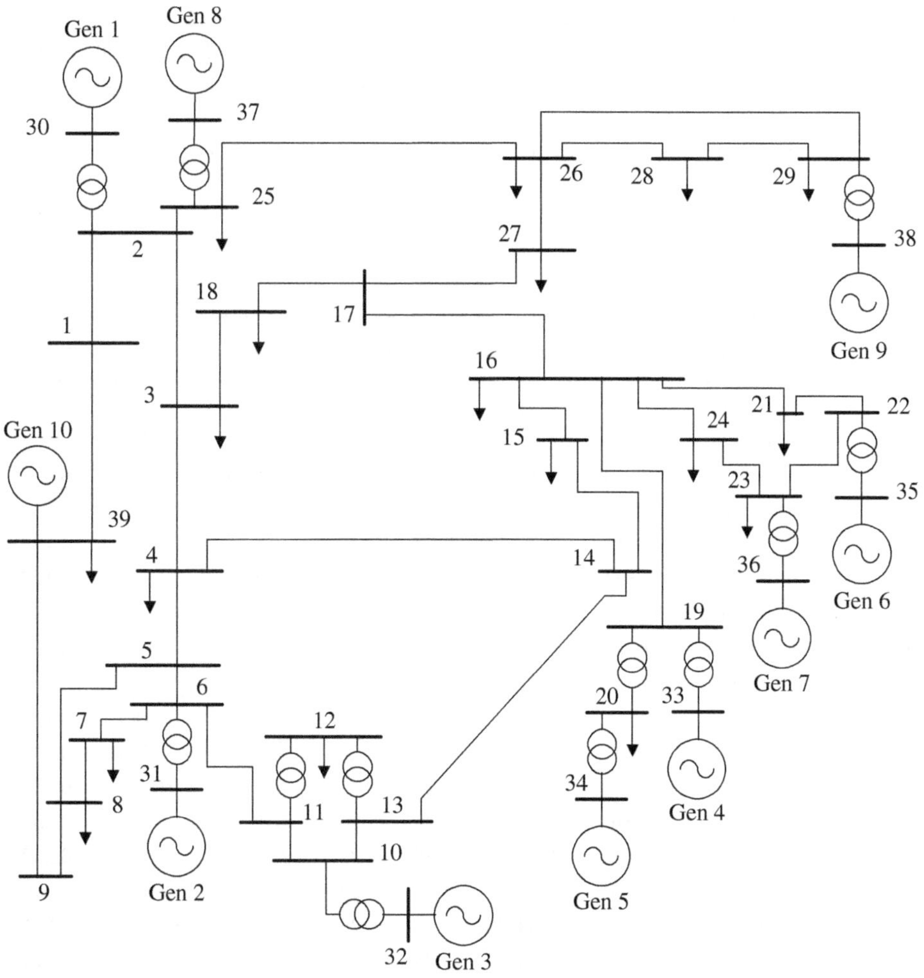

Figure 6.6 New England test system.

34 transmission lines, 12 transformers and 19 loads. For the steady-state solution, the generator number 10 which is connected at node 39 is taken to be the slack generator, while the rest are assumed to be *PV* generators. The data and parameters for the power system are given in per-unit values on a 100 MVA base in Tables 6.1–6.4. The generators

Table 6.1 Synchronous machine parameters.

Bus	P_{nom} (p.u)	R_a (p.u)	X_d (p.u)	X_q (p.u)	X'_d (p.u)	X'_q (p.u)	T'_{do} (s)	T'_{q0} (s)	H (s)	D (s/rad)
30	2.500	0.00	0.1000	0.069	0.0310	0.0080	10.2	0.01	42.0	0.5
31	5.732	0.00	0.2950	0.282	0.0697	0.1700	6.56	1.50	30.3	0.5
32	6.500	0.00	0.2495	0.237	0.0531	0.0876	5.70	1.50	35.80	0.5
33	6.320	0.00	0.2620	0.258	0.0430	0.1660	5.69	1.50	28.60	0.5
34	5.080	0.00	0.6700	0.620	0.1320	0.1660	5.40	0.44	26.0	0.5
35	6.500	0.00	0.2540	0.241	0.0500	0.0814	7.30	0.40	34.80	0.5
36	5.600	0.00	0.2950	0.292	0.0490	0.1860	5.66	1.50	26.40	0.5
37	5.400	0.00	0.2900	0.280	0.0570	0.0911	6.70	0.41	24.30	0.5
38	8.300	0.00	0.2106	0.205	0.0570	0.0587	4.79	1.96	34.50	0.5
39	10.1927	0.00	0.0200	0.019	0.0060	0.0080	7.00	0.70	500	0.5

Table 6.2 Transmission line parameters.

Buses		R (p.u)	X (p.u)	$B/2$ (p.u)	Buses		R (p.u)	X (p.u)	$B/2$ (p.u)
1	2	0.0350	0.0410	0.345	13	14	0.0009	0.0101	0.08625
1	39	0.0100	0.0250	0.375	14	15	0.0018	0.0217	0.183
2	3	0.0130	0.0130	0.125	15	16	0.0009	0.0094	0.0855
2	25	0.0070	0.0086	0.070	16	17	0.0007	0.0089	0.0671
3	4	0.0013	0.0213	0.1107	16	19	0.0016	0.0195	0.152
3	18	0.0011	0.0133	0.1069	16	21	0.0008	0.0135	0.1274
4	5	0.0008	0.0128	0.0671	16	24	0.0003	0.0059	0.034
4	14	0.0008	0.0129	0.0691	17	18	0.0007	0.0082	0.06595
5	6	0.0002	0.0026	0.217	17	27	0.0013	0.0173	0.1608
5	8	0.0008	0.0112	0.0738	21	22	0.0008	0.0140	0.12825
6	7	0.0006	0.0092	0.0565	22	23	0.0006	0.0096	0.09225
6	11	0.0007	0.0082	0.06945	23	24	0.0022	0.0350	0.1805
7	8	0.0004	0.0046	0.039	25	26	0.0032	0.0323	0.2565
8	9	0.0023	0.0363	0.1902	26	27	0.0014	0.0147	0.1198
9	39	0.0010	0.0250	0.600	26	28	0.0043	0.0474	0.3901
10	11	0.0004	0.0043	0.03645	26	29	0.0057	0.0625	0.5145
10	13	0.0004	0.0043	0.03645	28	29	0.0014	0.0151	0.1245

Table 6.3 Transformer parameters.

Buses		R_s (p.u)	X_s (p.u)	*tap*	Buses		R_s (p.u)	X_s (p.u)	*tap*
2	30	0.000	0.0181	1.025	19	33	0.0007	0.0142	1.070
6	31	0.000	0.0250	1.070	20	34	0.0009	0.0180	1.009
10	32	0.000	0.0200	1.070	22	35	0.000	0.0143	1.025
12	11	0.0016	0.0435	1.006	23	36	0.0005	0.0272	1.000
12	13	0.0016	0.0435	1.006	25	37	0.0006	0.0232	1.025
19	20	0.0007	0.0138	1.060	29	38	0.0008	0.0156	1.025

Table 6.4 Load parameters.

Bus	P_L (p.u)	Q_L (p.u)	Bus	P_L (p.u)	Q_L (p.u)	Bus	P_L (p.u)	Q_L (p.u)
3	3.220	0.024	18	1.580	0.300	26	1.390	0.170
4	5.000	1.840	20	6.800	1.030	27	2.810	0.755
7	2.330	0.840	21	2.740	1.150	28	2.060	0.276
8	5.220	1.760	23	2.475	0.846	29	2.835	0.269
12	0.085	0.880	24	3.086	−0.922	31	0.0912	0.046
15	3.200	1.530	25	2.240	0.472	39	11.040	2.500
16	3.294	0.323						

1–5 are driven by a hydro-turbine whilst generators 6–10 are driven by steam-turbines. All synchronous generators are represented by their transient model.

Each generator has an AVR with the following parameters: $T_A = 0.10$ s, $K_A = 75.0$, $T_F = 1.0$ s, $K_F = 0.03$ and $T_E = 0.80$. The parameters for the generators employing hydro-turbine-governing systems are $K = 0.06$, $T_{GV} = 0.30$ s and $T_W = 0.5$ s, whereas for those using steam-turbine-governing systems are $K = 0.06$, $T_{GV} = 0.30$ s, $F_{HP} = 0.3$, $F_{IP} = 0.4$, $F_{LP} = 0.3$, $T_{CH} = 0.3$, $T_{RH} = 7.0$ and $T_{CO} = 0.2$.

The network is subjected to a disconnection of the transmission lines connected between buses 2–25, 2–3 and 3–4, at $t = 0.1$ s (which in steady-state are approximately transmitting 230 MW, 380 MW and 75 MW, respectively). The simulation runs for 5 s with a time step of 1 ms. Because of such a drastic change in the network topology, a rearrangement of power flows takes place in several transmission lines of the system together with changes in the powers drawn by the loads (owing to their voltage dependency). All this brings about variations in power transfers which are accompanied by changes in the angular speed of the synchronous generators as well as in the active and reactive power flows in some of the transmission lines neighbouring the tripped lines, as seen from Figures 6.7–6.9, respectively. Also, such power fluctuations cause significant voltage oscillations in several nodes of the power system, as shown in Figure 6.10. These sample results are representative of what the rest of the nodes in the power network experience following the perturbation.

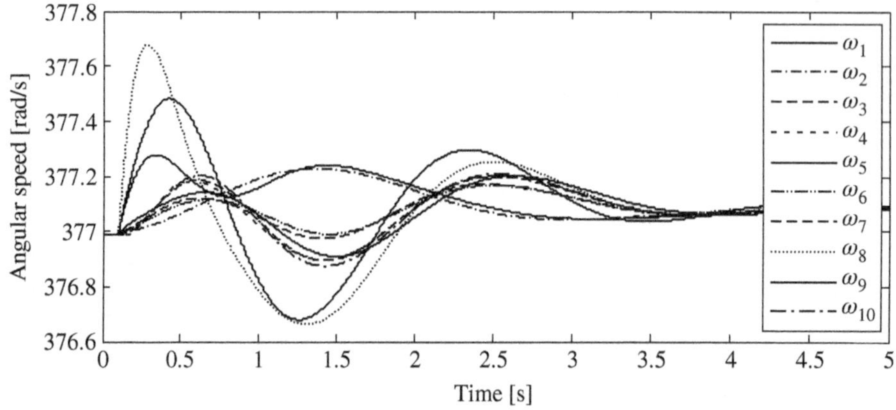

Figure 6.7 Synchronous generators' angular speed.

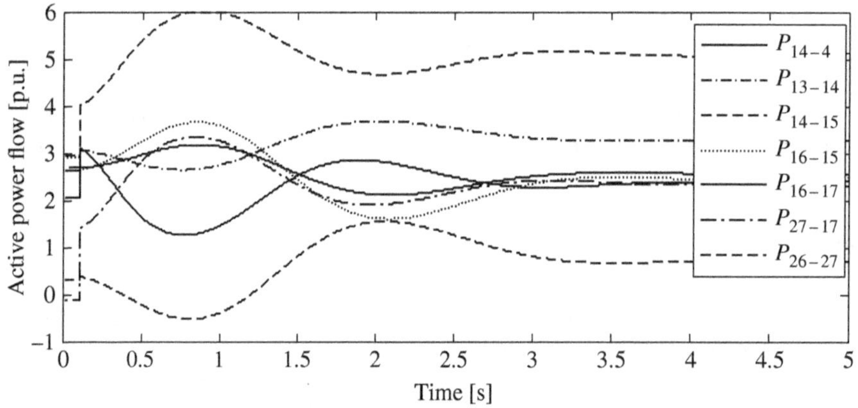

Figure 6.8 Active power flow behaviour in selected transmission lines.

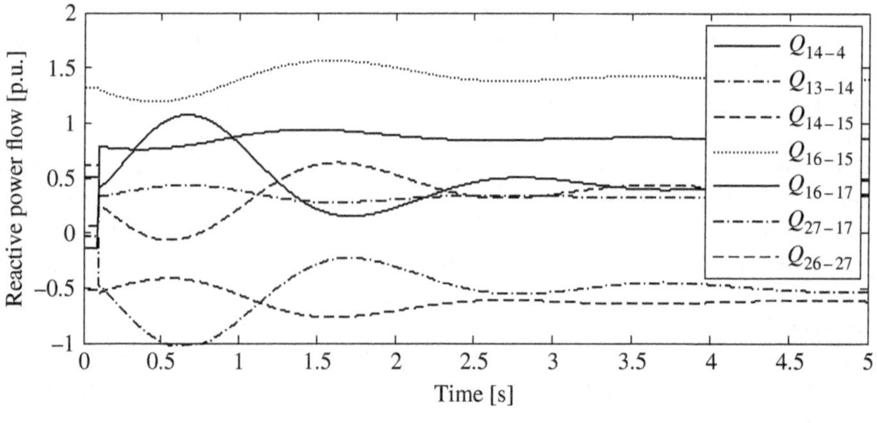

Figure 6.9 Reactive power flow behaviour in selected transmission lines.

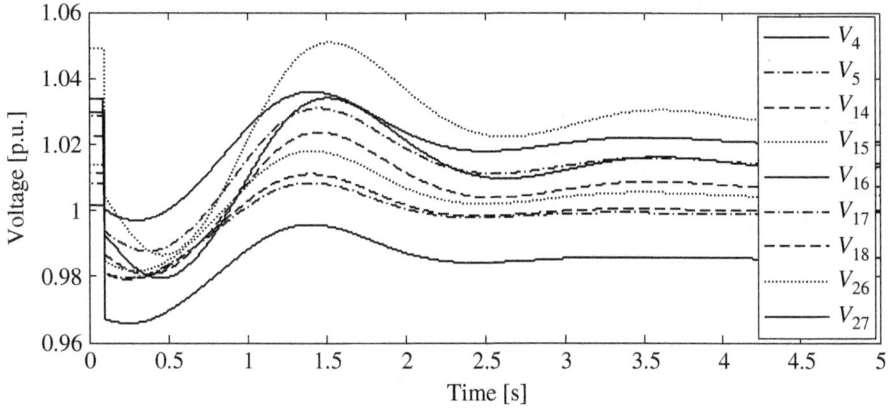

Figure 6.10 Voltage performance at different nodes of the network.

6.4 Modelling of the STATCOM for Dynamic Simulations

The power electronics equipment that emerged from the FACTS initiative [16] and its ramifications has a common purpose: to alleviate one or more operational problems at key locations of the power grid. A case in point are the VSC-based devices such as the STATCOM and VSC-HVDC. The STATCOM was designed to regulate reactive power, at its point of connection with the grid, in response to either fast or slow network voltage variations. Its superior performance, compared with that of the SVC, arises from the fast action of the PWM-driven insulated gate bipolar transistor (IGBT) valves, which enable the VSC to maintain a smooth voltage profile at its connecting node, even in the face of rather severe disturbances in its vicinity.

It was pointed out in Section 1.2.2 of Chapter 1 that the concept of a controllable voltage source behind a coupling impedance has been a popular modelling resource to represent the steady-state, fundamental frequency operation of the VSCs of the STATCOM and VSC-HVDC links [17]. More recently, it has been argued that this simple concept explains well the operation of these devices from the standpoint of the AC network but its usefulness reduces when the requirement involves the assessment of variables relating to their DC buses [18]. The situation is very much the same when looking at the dynamic regime where the standard approach has also been the use of a controllable voltage source. Aiming at alleviating such a key shortcoming, this chapter presents enhanced STATCOM and VSC-HVDC models for assessing the dynamic operating regime of large-scale power grids. Similar to the VSC model developed in Section 3.6 for the purpose of steady-state studies, the VSC model presented below uses a complex tap-changing transformer at its kernel, one where its primary and secondary sides yield, notionally speaking, the AC and DC sides of the VSC, respectively. The AC terminal of the VSC combines easily with the model of the interfacing transformer to make up both the STATCOM and HVDC models, which are elegant and yield great physical insight when employed to carry out time-domain simulations.

Building on the basic concepts and structural characteristics of the STATCOM presented in Chapters 1 and 2, we now turn our attention to develop a comprehensive model for dynamic studies. As shown in Figure 6.11, the VSC station uses a DC capacitor, which

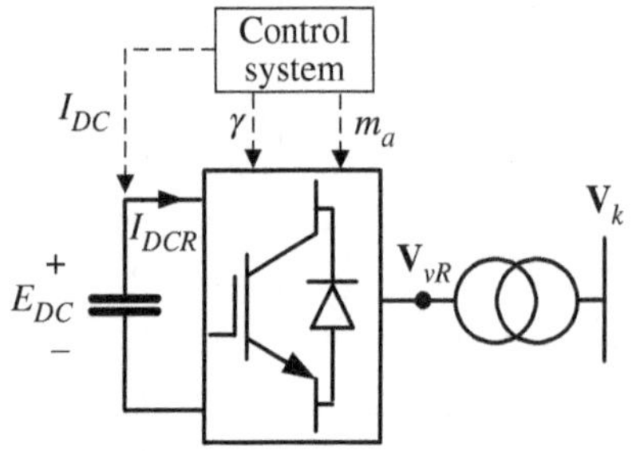

Figure 6.11 A STATCOM and its control variables.

plays a key role in the converter dynamics. It should be recalled that in steady-state the capacitor is assumed to be fully charged, hence its voltage is kept constant at the nominal value E_{DCnom} and the current in the capacitor is zero, $i_c = 0$. However, disturbances on the AC network affect the DC link voltage and current, pushing the capacitor into a charging/discharging state.

The dynamic behaviour of the VSC is adequately captured provided the dynamics of the DC capacitor is suitably represented. For instance, to compute the voltage oscillations at the DC bus, Eqs. (6.47) and (6.48) are used. These yield information of the current I_{DCR}, which is the actual current injected at the converter's DC bus, and the capacitor's current $i_c = -I_{DCR} - I_{DC}$.

$$i_c = C_{DC}\frac{dE_{DC}}{dt} \tag{6.47}$$

$$I_{DCR} = \frac{P_0}{E_{DC}} \tag{6.48}$$

where the power P_0 is the nodal active power injected at the DC bus, which has been previously derived in Section 3.6.3.

Therefore, the following expression serves the purpose of calculating the STATCOM's DC voltage dynamics:

$$\frac{dE_{DC}}{dt} = \frac{-I_{DCR} - I_{DC}}{C_{DC}} \tag{6.49}$$

where I_{DC} represents the DC current control variable acting upon the DC voltage.

This expression enables the calculation of the voltage dynamics in the DC bus. Its capacitor value is estimated as a function of the energy stored in the capacitor: $W_c = \frac{1}{2}C_{DC}E_{DC}^2$. At least from the conceptual point of view, the capacitor's energy storage bears a resemblance to the kinetic energy of the rotating electrical machinery, where this energy can be further related to the inertia constant that impacts the machine's motion equation.

Taking this reasoning further, the electrostatic energy stored in the DC capacitor can be associated with an equivalent inertia constant H_c by means of the following relationship: $W_c = H_c S_{nom}$, where S_{nom} corresponds to the rated apparent power of the VSC. By drawing a parallel between the VSC and a bank of thyristor-switched capacitors of the same rating, this time constant may be taken to be: $H_c \approx \omega_s^{-1}$. However, $H_c \approx 5$ ms has been found to be a more suitable value, to avoid numerical instabilities [19].

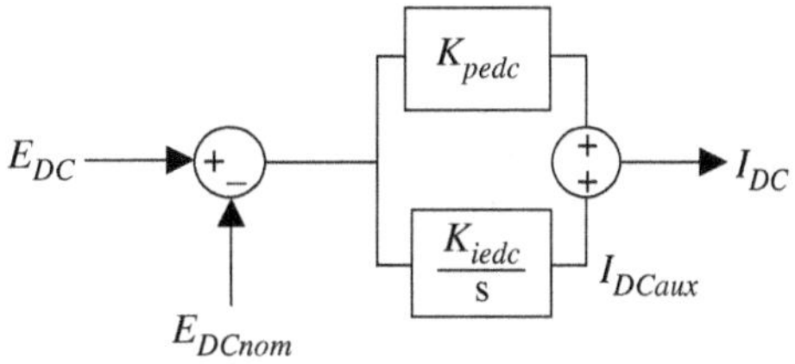

Figure 6.12 DC voltage controller.

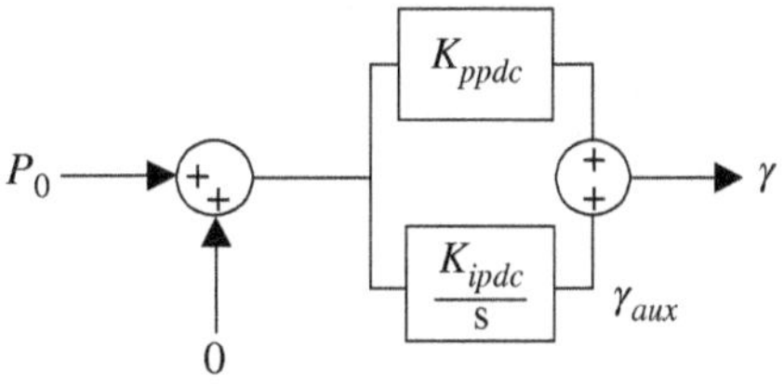

Figure 6.13 DC power controller for the VSC's DC side.

Hence, the per-unit value of the capacitor is:

$$C_{DC} = \frac{2S_{nom}H_c}{E_{DC}^2} \tag{6.50}$$

The VSC controls the voltage at the DC bus, an action carried out by controlling the DC current entering or leaving the converter, I_{DC}. The implementation of the DC voltage controller is shown in Figure 6.12, where the error between the actual voltage E_{DC} and E_{DCnom} is processed through a PI controller to obtain new values of the DC current I_{DC} at every time step. Hence, the DC voltage dynamics is largely determined by the gains K_{pedc} and K_{iedc}.

The equations arising from the block diagram corresponding to the DC voltage controller are:

$$\frac{dI_{DCaux}}{dt} = K_{iedc}(E_{DC} - E_{DCnom}) \tag{6.51}$$

$$I_{DC} = K_{pedc}(E_{DC} - E_{DCnom}) + I_{DCaux} \tag{6.52}$$

The converter keeps the capacitor charged at the required voltage level by making its output voltage lag the AC system voltage by a small angle [16]. This angular difference is computed as $\gamma = \theta_{vR} - \varphi$, where φ is the angle of the phase-shifting transformer and $\theta_{vR} = \tan^{-1}(f_{vR}/e_{vR})$ represents the angle of the VSC's terminal voltage. During the dynamic regime, the power balance on the DC side of the converter, $P_0 = 0$, is achieved by suitable control of the angle γ [20]. The PI controller responsible for this is shown in Figure 6.13.

The differential equation that represents the dynamic behaviour of the power balance controller is:

$$\frac{d\gamma_{aux}}{dt} = K_{ipdc}P_0 \tag{6.53}$$

$$\gamma = K_{ppdc}P_0 + \gamma_{aux} \tag{6.54}$$

The modulation index is responsible for keeping the voltage magnitude $V_{vR} = (e_{vR}{}^2 + f_{vR}{}^2)^{1/2}$ at the desired value; to such an end, the AC-bus voltage controller

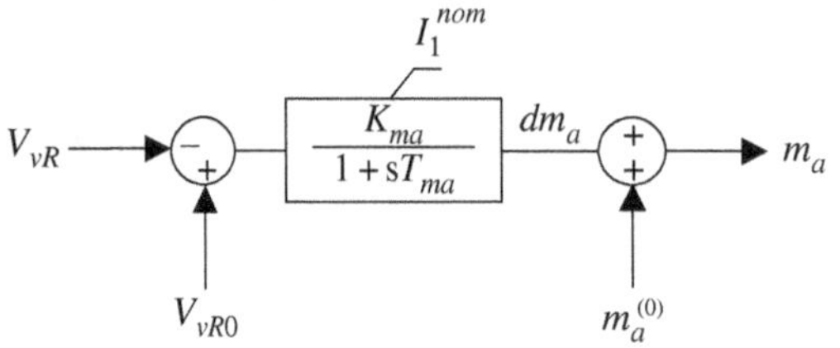

Figure 6.14 AC-bus voltage controller.

depicted in Figure 6.14 is employed. This is a first-order controller which yields small changes in the modulation index dm_a by comparing the actual voltage V_{vR} and the scheduled AC voltage $V_{vR0} = (e_{vR0}{}^2 + f_{vR0}{}^2)^{1/2}$. The modulation index increases or decreases according to the operating requirements [20]. The differential equation for the AC-bus controller is given by (6.55):

$$\frac{d(dm_a)}{dt} = \frac{K_{ma}(V_{vR0} - V_{vR}) - dm_a}{T_{ma}} \tag{6.55}$$

To guarantee the VSC operation within limits, a limit checking of the terminal current must take place, that is, $I_1 \leq I_1^{nom}$. Hence, the overall dynamics of the STATCOM is well captured by means of Eqs. (6.49)–(6.55).

6.4.1 Discretization and Linearization of the STATCOM Differential Equations

The STATCOM differential equations representing the DC voltage dynamics, the DC current controller and the AC voltage controller are discretized as follows:

$$F_{E_{DC}} = E_{DC(t-\Delta t)} + 0.5\Delta t\dot{E}_{DC(t-\Delta t)} - (E_{DC(t)} - 0.5\Delta t\dot{E}_{DC(t)}) \tag{6.56}$$

$$F_{I_{DCaux}} = I_{DCaux(t-\Delta t)} + 0.5\Delta t\dot{I}_{DCaux(t-\Delta t)} - (I_{DCaux(t)} - 0.5\Delta t\dot{I}_{DCaux(t)}) \tag{6.57}$$

$$F_{\gamma_{aux}} = \gamma_{aux(t-\Delta t)} + 0.5\Delta t\dot{\gamma}_{aux(t-\Delta t)} - (\gamma_{aux(t)} - 0.5\Delta t\dot{\gamma}_{aux(t)}) \tag{6.58}$$

$$F_{dm_a} = dm_{a(t-\Delta t)} + 0.5\Delta t\, d\dot{m}_{a(t-\Delta t)} - (dm_{a(t)} - 0.5\Delta t\, d\dot{m}_{a(t)}) \tag{6.59}$$

The discretized differential equations of the VSC are appended to those of the active and reactive power balances of the network, at the AC terminal of the VSC, that is, at node vR. These equations were derived in (Section 3.6.3) but for the sake of fluidity of the text, they are reproduced below:

$$\Delta P_{vR} = -P_{vR} - P_{vR,load} - P_{vR,cal} \tag{6.60}$$

$$\Delta Q_{vR} = -Q_{vR} - Q_{vR,load} - Q_{vR,cal} \tag{6.61}$$

where

$$\begin{aligned} P_{vR} &= G_{vRvR}(e_{vR}^2 + f_{vR}^2) - E_{DC} \\ &\quad \times (e_{vR}[G_{vR0}\cos\phi - B_{vR0}\sin\phi] + f_{vR}[G_{vR0}\sin\phi + B_{vR0}\cos\phi]) \end{aligned} \tag{6.62}$$

$$\begin{aligned} Q_{vR} &= -B_{vRvR}(e_{vR}^2 + f_{vR}^2) - E_{DC} \\ &\quad \times (-e_{vR}[G_{vR0}\sin\phi + B_{vR0}\cos\phi] + f_{vR}[G_{vR0}\cos\phi - B_{vR0}\sin\phi]) \end{aligned} \tag{6.63}$$

Eqs. (6.56)–(6.61) make up the necessary set of equations that must be solved together with the equations of the whole network, including those of the synchronous generators and their controls, to carry out dynamic power flow solutions of a power system containing STATCOM equipment. To solve the nonlinear set of equations, the Newton-Raphson method is employed for reliable dynamic simulations. Hence, the linearized matrix Eq. (6.64) provides the computing framework, with which the time-domain solutions are performed:

$$\begin{bmatrix} \Delta P_{vR} \\ \Delta Q_{vR} \\ F_{E_{DC}} \\ F_{I_{DCaux}} \\ F_{\gamma_{aux}} \\ F_{dm_a} \end{bmatrix} = -\begin{bmatrix} J_{11} & J_{12} \\ J_{21} & J_{22} \end{bmatrix} \begin{bmatrix} \Delta e_{vR} \\ \Delta f_{vR} \\ \Delta E_{DC} \\ \Delta I_{DCaux} \\ \Delta \gamma_{aux} \\ \Delta dm_a \end{bmatrix} \tag{6.64}$$

where J_{11} comprises the first-order partial derivatives of the nodal active and reactive power mismatches with respect to the real and imaginary parts of the AC-side voltage. Likewise, J_{12} contains the partial derivatives arising from the nodal active and reactive powers with respect to the VSC state variables. The matrix J_{21} consists of partial derivatives of the VSC's discretized differential equations with respect to AC voltages. Lastly, J_{22} is a matrix that accommodates the first-order partial derivatives of the VSC discretized differential equations with respect to its own control variables.

$$J_{11} = \begin{bmatrix} \frac{\partial \Delta P_{vR}}{\partial e_{vR}} & \frac{\partial \Delta P_{vR}}{\partial f_{vR}} \\ \frac{\partial \Delta Q_{vR}}{\partial e_{vR}} & \frac{\partial \Delta Q_{vR}}{\partial f_{vR}} \end{bmatrix} \qquad J_{12} = \begin{bmatrix} \frac{\partial \Delta P_{vR}}{\partial E_{DC}} & 0 & 0 & \frac{\partial \Delta P_{vR}}{\partial dm_a} \\ \frac{\partial \Delta Q_{vR}}{\partial E_{DC}} & 0 & 0 & \frac{\partial \Delta Q_{vR}}{\partial dm_a} \end{bmatrix}$$

$$J_{21} = \begin{bmatrix} \frac{\partial F_{E_{DC}}}{\partial e_{vR}} & \frac{\partial F_{E_{DC}}}{\partial f_{vR}} \\ 0 & 0 \\ \frac{\partial F_{\gamma_{aux}}}{\partial e_{vR}} & \frac{\partial F_{\gamma_{aux}}}{\partial f_{vR}} \\ \frac{\partial F_{dm_a}}{\partial e_{vR}} & \frac{\partial F_{dm_a}}{\partial f_{vR}} \end{bmatrix}, \qquad J_{22} = \begin{bmatrix} \frac{\partial F_{E_{DC}}}{\partial E_{DC}} & 0 & 0 & \frac{\partial F_{E_{DC}}}{\partial dm_a} \\ \frac{\partial F_{I_{DCaux}}}{\partial E_{DC}} & \frac{\partial F_{I_{DCaux}}}{\partial I_{DCaux}} & 0 & 0 \\ \frac{\partial F_{\gamma_{aux}}}{\partial E_{DC}} & 0 & \frac{\partial F_{\gamma_{aux}}}{\partial \gamma_{aux}} & \frac{\partial F_{\gamma_{aux}}}{\partial dm_a} \\ 0 & 0 & 0 & \frac{\partial F_{dm_a}}{\partial dm_a} \end{bmatrix} \tag{6.65}$$

where the derivatives of the mismatch powers at terminal vR with respect to the terminal voltage are presented in (Section 3.6.4). The new derivative terms corresponding to VSC's discretized differential equations are given below in explicit form:

$$\frac{\partial F_{E_{DC}}}{\partial e_{vR}} = 0.5\Delta t C_{DC}^{-1} E_{DC}^{-1} \frac{\partial P_0}{\partial e_{vR}}$$

$$\frac{\partial F_{E_{DC}}}{\partial f_{vR}} = 0.5\Delta t C_{DC}^{-1} E_{DC}^{-1} \frac{\partial P_0}{\partial f_{vR}}$$

$$\frac{\partial F_{E_{DC}}}{\partial E_{DC}} = 1.0 + 0.5\Delta t C_{DC}^{-1} \left[E_{DC}^{-1} \frac{\partial P_0}{\partial E_{DC}} - E_{DC}^{-2} P_0 \right]$$

$$\frac{\partial F_{E_{DC}}}{\partial dm_a} = 0.5\Delta t C_{DC}^{-1} E_{DC}^{-1} \frac{\partial P_0}{\partial m_a}$$

$$\frac{\partial F_{I_{DCaux}}}{\partial E_{DC}} = -0.5\Delta t K_{iedc}$$

$$\frac{\partial F_{I_{DCaux}}}{\partial I_{DCaux}} = 1.0$$

$$\frac{\partial F_{\gamma_{aux}}}{\partial e_{vR}} = -0.5\Delta t K_{ipdc} \frac{\partial P_0}{\partial e_{vR}}$$

$$\frac{\partial F_{\gamma_{aux}}}{\partial f_{vR}} = -0.5\Delta t K_{ipdc} \frac{\partial P_0}{\partial f_{vR}}$$

$$\frac{\partial F_{\gamma_{aux}}}{\partial E_{DC}} = -0.5\Delta t K_{ipdc} \frac{\partial P_0}{\partial E_{DC}}$$

$$\frac{\partial F_{\gamma_{aux}}}{\partial \gamma_{aux}} = 1.0$$

$$\frac{\partial F_{\gamma_{aux}}}{\partial dm_a} = -0.5\Delta t K_{ipdc} \frac{\partial P_0}{\partial m_a}$$

$$\frac{\partial F_{dm_a}}{\partial e_{vR}} = 0.5\Delta t K_{ma} T_{ma}^{-1} e_{vR} (e_{vR}^2 + f_{vR}^2)^{-1/2}$$

$$\frac{\partial F_{dm_a}}{\partial f_{vR}} = 0.5\Delta t K_{ma} T_{ma}^{-1} f_{vR} (e_{vR}^2 + f_{vR}^2)^{-1/2}$$

$$\frac{\partial F_{dm_a}}{\partial dm_a} = 1.0 + 0.5\Delta t T_{ma}^{-1}$$

It is worth emphasizing that the dynamic simulation is started from a steady-state equilibrium point, which is obtained a priori using the conventional Newton-Raphson power flow algorithm, detailed in Chapter 3, to ensure a smooth dynamic solution.

6.4.2 Numerical Example with STATCOMs

The New England test system, with suitable modifications, is shown in Figure 6.15. It incorporates STATCOMs at nodes 5 and 27 where large voltage fluctuations are expected to occur following a disturbance at any of the main corridors of the electrical network. The rest of the parameters are the same as those used in the benchmark example of Section 6.3.2. The taps of the load tap changer (LTC) transformers are kept constant during the dynamic simulation. It should be noted that the time response of the servomotor that drives the tap changer is much slower than the time response of the controller that drives the modulation ratio of the VSC. Table 6.5 gives the parameters

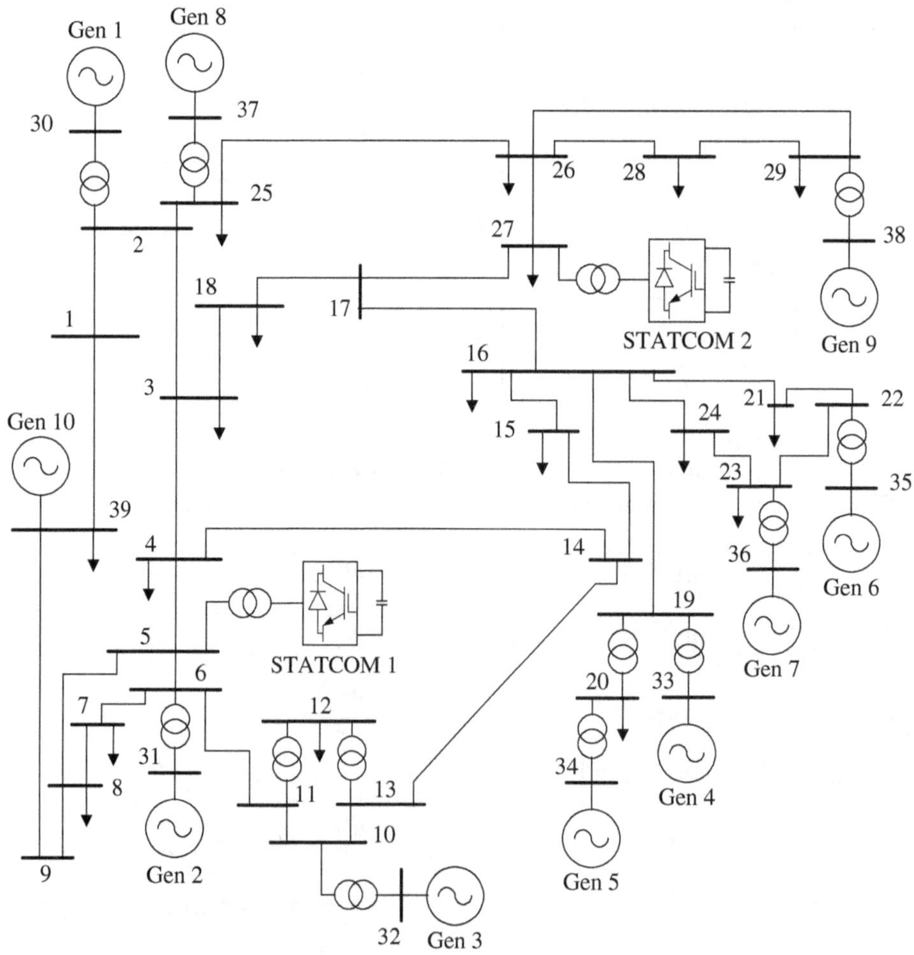

Figure 6.15 Test system used to incorporate two STATCOMs.

Table 6.5 STATCOM parameters.

Bus	S_{nom} (MVA)	E_{DC} (p.u)	G_0 (p.u)	R_1 (p.u)	X_1 (p.u)	K_{pedc}	K_{iedc}	K_{ppdc}	K_{ipdc}	K_{ma}	T_{ma}	X_{ltc} (p.u)
5	100.0	2.00	2e-3	2e-3	0.01	0.20	7.50	0.001	0.15	7.5	0.02	0.05
27	100.0	2.00	2e-3	2e-3	0.01	0.20	7.50	0.001	0.15	7.5	0.02	0.05

Table 6.6 Computed STATCOMs variables by the power flow solution.

	Q_{gen} (MVAR)	m_a	φ (°)	B_{eq} (p.u)	*LTC's tap*	P_{loss} (MW)
STATCOM 1	12.258	0.8311	1.6759	0.1100	1.0060	0.0125
STATCOM 2	35.646	0.8685	2.2568	0.3177	1.0165	0.1142

Table 6.7 Initial values of the STATCOM variables for the dynamic simulation.

	E_{DC} (p.u)	$I_{DC,aux}$ (p.u)	γ_{aux} (rad)	dm_a
STATCOM 1	2.00	0.0	0.2286e-3	0.0
STATCOM 2	2.00	0.0	0.7078e-3	0.0

used in the simulation, on the STATCOM's base, of $S_{nom} = 100$ MVA. Table 6.6 presents a summary of the steady-state power flow results; note that both STATCOMs inject reactive power to the grid in order to uphold their target voltages at 1.01 p.u and 1.04 p.u, respectively.

During steady-state operation, the total power losses incurred by the STATCOMs stand at 0.0125 MW and 0.1142 MW, respectively. The reason for such a difference in these values lies in the quite different currents flowing through the VSCs, which stand at 0.1120 p.u and 0.3380 p.u, respectively. It is also worth recalling that the switching losses are functions of the actual operating conditions, i.e. these losses are scaled by the quadratic ratio of the actual current magnitude to the nominal current of the equipment. In this example, the rated current is 1 p.u for both VSCs.

Once the steady-state power flow solution is determined, the initial values of the control variables with which the dynamic simulation will be carried out are determined. These parameters are shown in Table 6.7. Notice that the following relationships hold for the steady-state regime: $I_{DC} = I_{DC,aux}$, $\gamma = \gamma_{aux}$ and $m_a = m_a^{(0)}$.

The power network undergoes a disturbance and the dynamic response of the STATCOMs is assessed. The transmission lines connecting buses 25–2, 2–3 and 3–4 are tripped at $t = 0.1$ s, with the result that several nodes in the surrounding area experience important voltage drops, as seen from the results shown in Figure 6.10, when no reactive power compensation is available. In contrast, when reactive power support is available in the form of STATCOMs, owing to their very rapid response, the voltage drops are not as severe, as shown by the results presented in Figure 6.16, where the

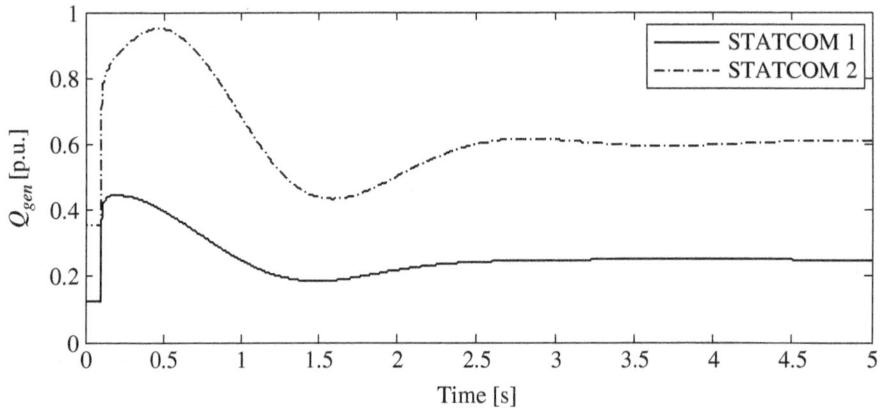

Figure 6.16 Reactive power provided by the STATCOMs.

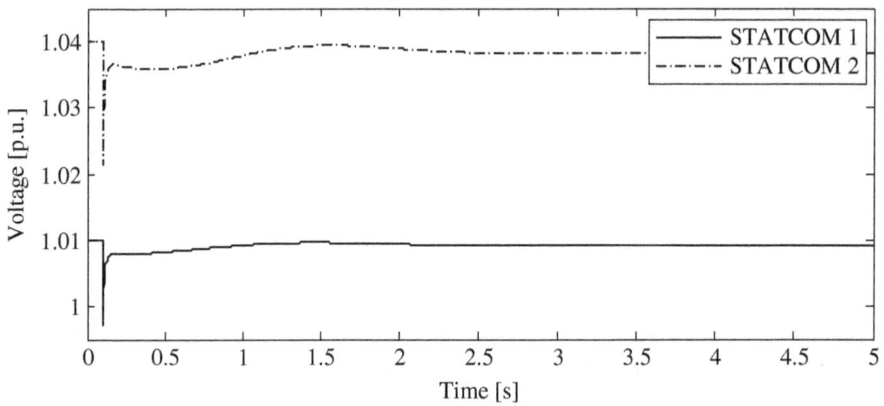

Figure 6.17 Voltage performance at the VSCs' AC network nodes.

reactive power peaks reach values of about 46 MVAR and 98 MVAR, respectively. The nodal voltage at the AC terminal of the converters (low-voltage side of the LTC) changes little owing to the very fast and effective control, as can be seen in Figure 6.17. Undoubtedly, the STATCOMs operation assists not only AC voltage of the converters, but also their neighbouring nodes, throughout the transient period; in the sense that the network voltages are stabilized much faster, as shown in Figure 6.18. This feature makes the STATCOM a very attractive choice to install in weak nodes of the power network.

Following the disturbance, abrupt changes occur in the VSCs' currents, disrupting the power balance on the DC side of the VSCs. This induces variations on the DC voltage, forcing the DC voltage controller to respond by modulating the DC current. This is shown in the results presented in Figures 6.19 and 6.20. It is seen that both control variables stabilize after just a few seconds of the occurrence of the perturbation. It should be noted that the DC current returns to zero once the VSCs and the whole network reach a new equilibrium point. This is an expected behaviour and fully validates the idea that the DC capacitor's current must be zero during steady-state operation.

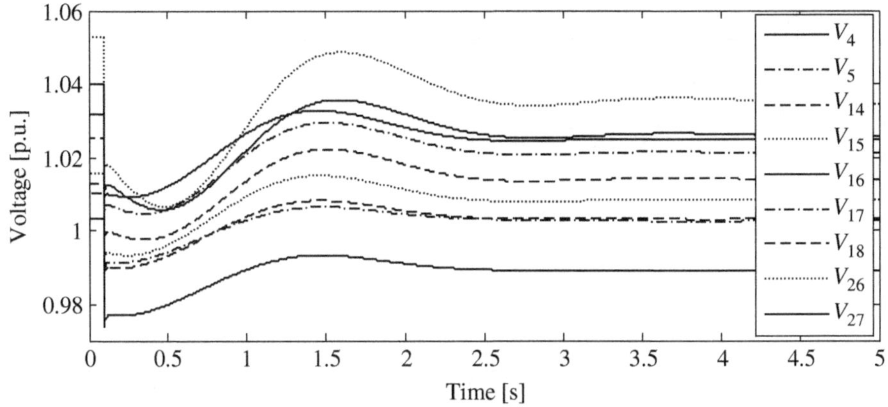

Figure 6.18 Voltage performance at several nodes of the network.

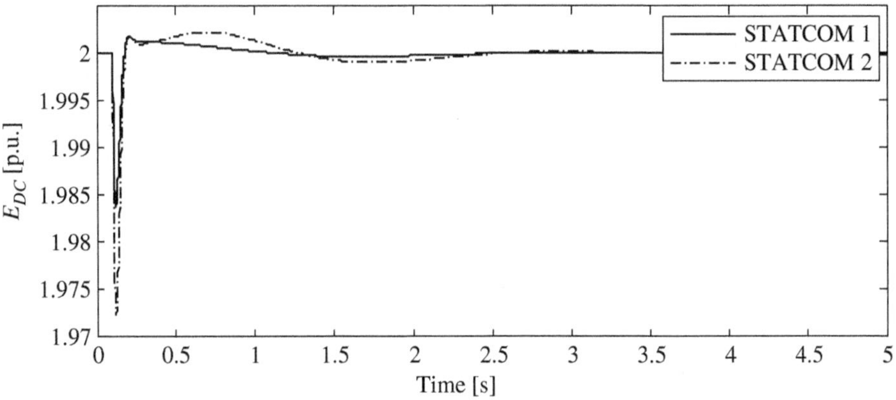

Figure 6.19 STATCOM's DC-bus voltages.

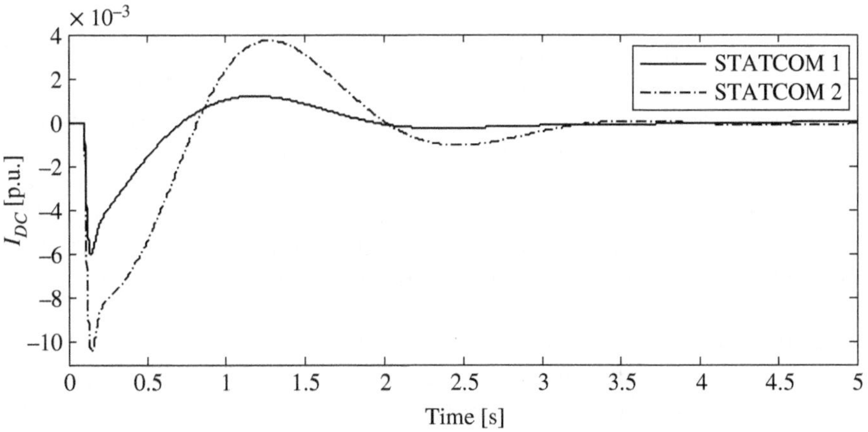

Figure 6.20 STATCOM's DC current.

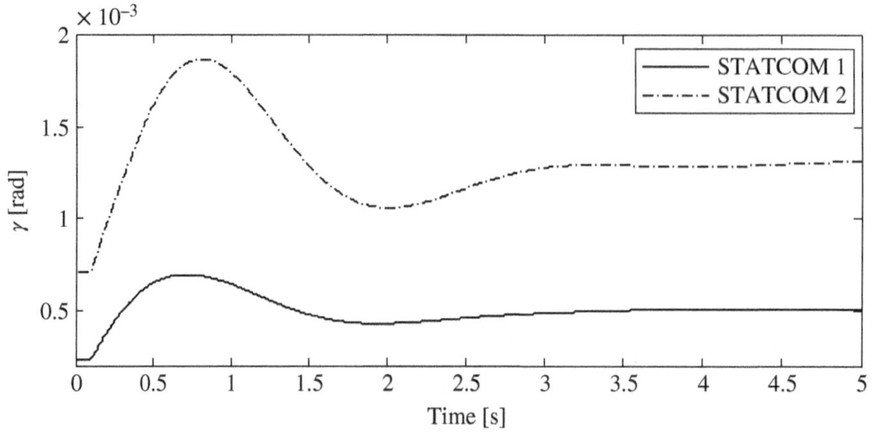

Figure 6.21 STATCOM's angle γ.

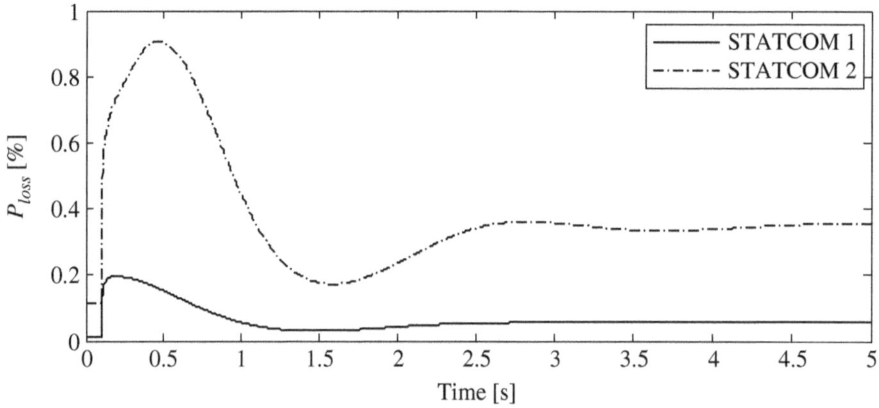

Figure 6.22 Total active power losses incurred by the STATCOMs.

The performance of the angle γ of the DC power controller is shown in Figure 6.21. The reason for the increase in γ is explained by recalling that this angle represents the angular aperture between the phase-shifting angle ϕ and the terminal voltage angle θ_{vR}. As expected, the more current passes through the electronic switches, the more power losses are incurred by the VSCs. The result in Figure 6.22 proves this point rather well. The DC power controller adjusts γ during the transient period aiming at re-establishing the power balance on the VSC's DC side. Notice that the power losses of STATCOM 2 reach almost 1% during the transient period and those of the STATCOM 1 reach approximately 0.2%.

Likewise, the controls of the modulation ratios of the two STATCOMs start exerting voltage regulation just after the disturbance has occurred, as shown in Figure 6.23, whose behaviour is governed by (6.55). From this equation, it is inferred that when the terminal voltage V_{vR} is smaller than V_{vR0}, the derivative of the modulation index with respect to time is positive, this being the reason for the increases in the modulation ratio

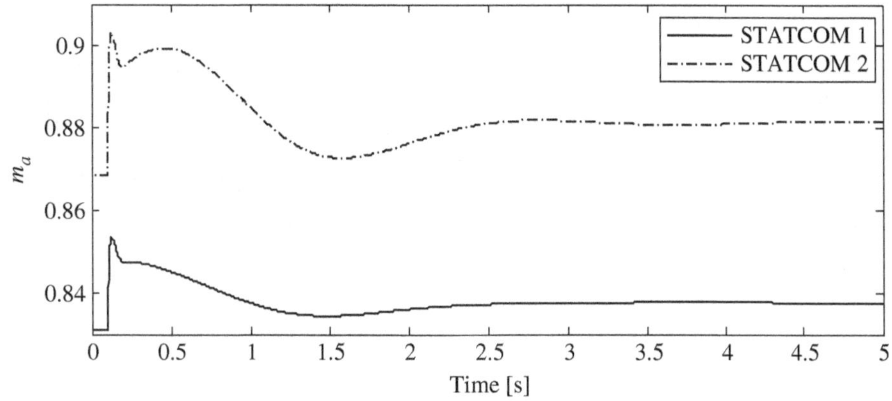

Figure 6.23 Dynamic behaviour of the modulation index of the STATCOMs.

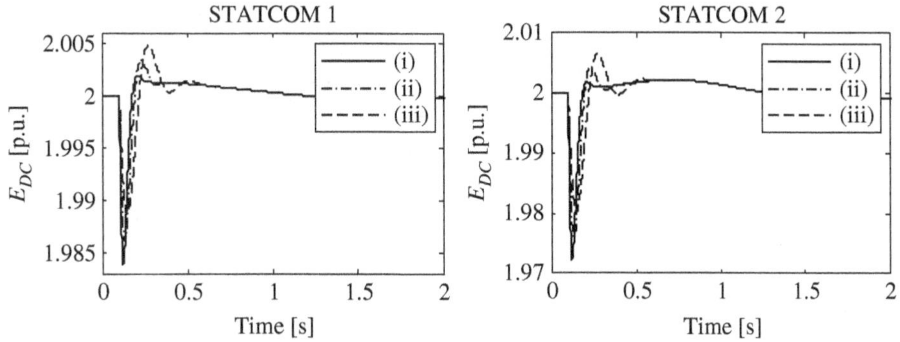

Figure 6.24 DC-bus voltages for different ratings of the capacitors.

of both STATCOMs. Notice that for a very short time, the modulation ratio of STATCOM 2 undergoes a rapid increase, reaching a value of 0.943. Following the transient event, it quickly settles down to a new steady-state value of 0.884.

The internal variables of STATCOM 2 are more sensitive to the disturbance since this device is located closer to the transmission line that connects buses 17 and 18, which becomes the only path available to supply the large loads connected at nodes 3 and 18.

A parametric analysis is carried out to examine more deeply the impact of the size of the capacitors on the dynamics of both the STATCOMs' internal variables along with their reactive power injection into the power grid. It is interesting to reproduce the system's conditions during the disturbance but altering the inertia of the capacitors. The following values are selected: (i) 5 ms, (ii) 10 ms and (iii) 20 ms. The dynamic performance of the DC voltage and modulation ratio is shown in Figures 6.24 and 6.25, respectively. Significant differences are shown with increases in the electrostatic energy stored in the capacitors. It can be seen that the DC voltage dip decreases with increases in the equivalent inertia of the capacitor and that this is accompanied by a DC voltage overshot, which rapidly damps out. The reactive power generated by the STATCOM, for different values of the capacitor's inertia H_c, is presented in Figure 6.26. For this parametric analysis, the gains of the STATCOM control loops have not been altered.

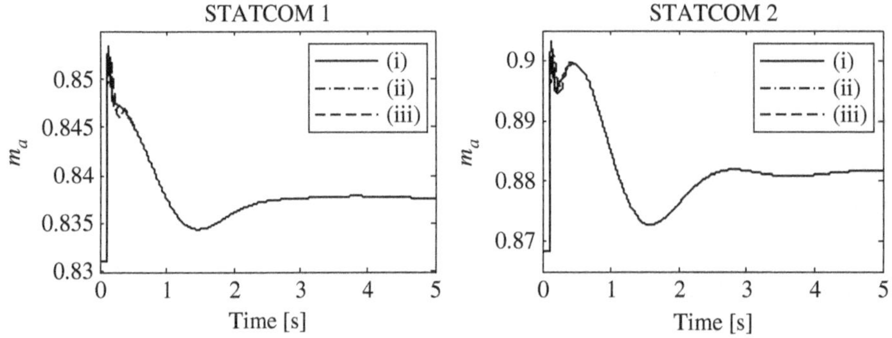

Figure 6.25 Modulation ratios for different ratings of the capacitors.

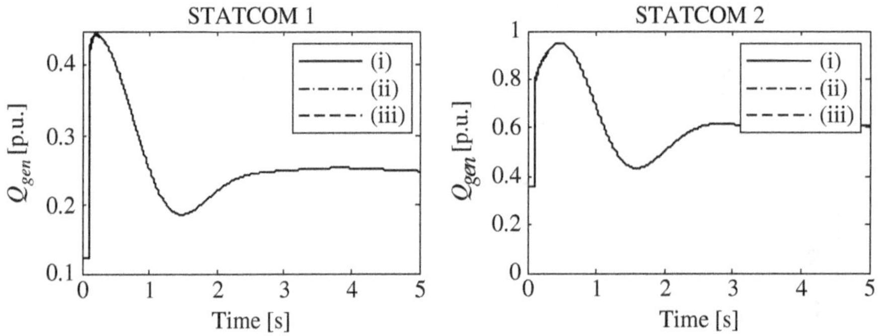

Figure 6.26 Reactive power provided by the VSCs for different ratings of the capacitors.

6.5 Modelling of VSC-HVDC Links for Dynamic Simulations

If two VSC converters are interconnected on their corresponding DC buses, using a cable with a resistance R_{DC}, a VSC-HVDC system forms; this is termed point-to-point configuration. If the cable does not exist, this may be assumed to be a cable of zero resistance and the VSC-HVDC configuration is termed back-to-back. Electric power is taken from one point of the AC network, converted to DC power in the rectifier station, transmitted through the DC cable and then converted back to AC power in the inverter station and injected into the receiving AC network.

In addition to transferring power in DC form, the VSC-HVDC system is capable of supplying/absorbing reactive power at both AC networks, thus providing independent dynamic voltage control. Steady-state and dynamic operating regimes are both fields that need to be covered to the satisfaction of the network's planners and operators, therefore a wide range of power systems-oriented VSC-HVDC models is required, bearing in mind the main operational features of the VSC-HVDC equipment. In large-scale power system studies, the tendency has been to keep the HVDC models as simple as possible in order to keep simulation times and memory requirements at manageable levels. Hence, representing the VSC stations of the VSC-HVDC link as controllable voltage sources has been a popular option. This reduces modelling complexity but sacrifices the acquisition of important VSC parameters since some internal variables may not be readily available.

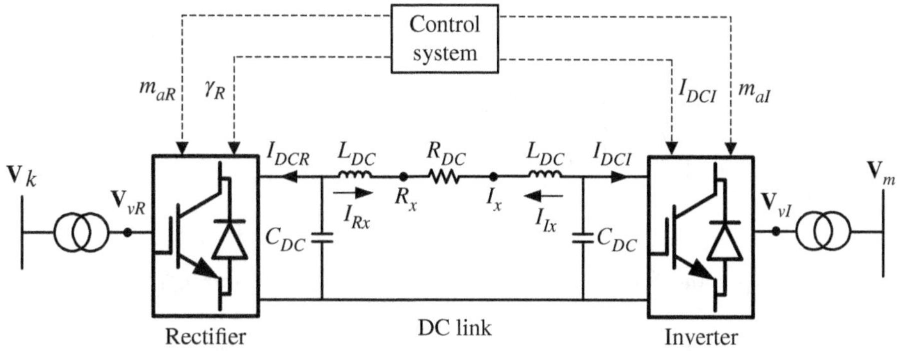

Figure 6.27 Dynamic model of the VSC-HVDC link with control variables.

It has been shown recently that more realistic behaviours of VSC-HVDC systems may be attained by using more advanced models, which are also useful for large-scale power networks. The new models of VSC-HVDC links are made up of basic electric circuit elements, which are suitably combined to represent the main operational features of the VSC-HVDC links during steady-state and dynamic regimes [21]. The most relevant operational parameters of the VSCs become explicitly available and the dynamic performance of the VSC-HVDC model encapsulates very well the behaviour of the AC and DC circuits of the actual equipment. Figure 6.27 depicts a schematic representation of the VSC-HVDC link together with its controls. This model synthesizes rather well the four degrees of freedom found in actual VSC-HVDC installations: simultaneous voltage support on its two AC terminals, DC voltage control at the inverter and DC power regulation at the rectifier end.

The capacitors' and inductors' dynamics play a crucial role in the operation of the HVDC link during voltage and power variations in the external AC networks. In this application, the DC voltage control is exerted to preserve stable operation of the DC link. The converter acting as inverter is usually tasked with this responsibility. The following relationships hold at the capacitors' nodes: $i_{cR} = -I_{DCR} - I_{Rx}$ and $i_{cI} = -I_{DCI} - I_{Ix}$, where i_{cR} and i_{cI} are the capacitor's current, I_{DCR} and I_{DCI} are the currents leaving the rectifier's DC bus and the inverter's DC bus, respectively. Combining these current relationships with those of the capacitor's current $i_c = C_{DC}dE/dt$ and inductor's voltage $E_i = L_{DC}dI/dt$, the ensuing differential equations yields the voltage and current dynamics in the DC link:

$$\frac{dE_{DCR}}{dt} = \frac{-I_{DCR} - I_{Rx}}{C_{DC}}, \frac{dI_{Rx}}{dt} = \frac{E_{DCR} - E_{DCRx}}{L_{DC}} \tag{6.66}$$

$$\frac{dE_{DCI}}{dt} = \frac{-I_{DCI} - I_{Ix}}{C_{DC}}, \frac{dI_{Ix}}{dt} = \frac{E_{DCI} - E_{DCIx}}{L_{DC}} \tag{6.67}$$

where $I_{DCR} = P_{0R}E_{DCR}^{-1}$ and the power P_{0R} has been derived in (Section 3.8.1). The per-unit value of the capacitance is estimated using (6.50) and an equivalent inertia constant H_c representing the electrostatic energy stored in the capacitor. A similar expression may be derived for the per-unit value of the inductor through an inertia constant that accounts for the electromagnetic energy stored in the inductor:

$$L_{DC} = \frac{2S_{nom}H_i}{I_{DC}^2} \tag{6.68}$$

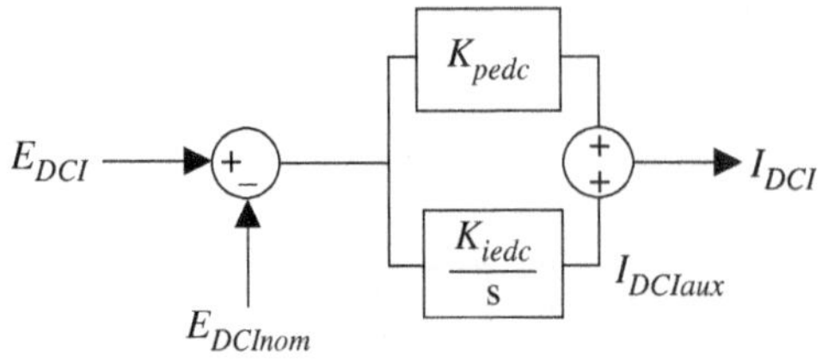

Figure 6.28 VSC-HVDC dynamic controller for the DC voltage. Source: ©Elsevier, 2015.

where H_i and I_{DC} stand for the fictitious inertia constant and nominal current of the inductor.

It may be noticed that the current balance of the back-to-back converter given by (6.66)–(6.67) is akin to the power balance inside the HVDC link for steady-state operation – when the time derivative is zero. When the current/power balance becomes disrupted, voltage variations appear in the DC link and this would need to be brought under quick control. The dynamic control of the DC voltage may be carried out by using the DC current entering the inverter, I_{DCI}, as the control variable, as shown in Figure 6.28. The error between the actual voltage E_{DCI} and the reference E_{DCInom} is passed through a PI controller, with gains K_{pedc} and K_{iedc}, to obtain new values of the DC current I_{DCI}.

The differential and algebraic equations arising from the DC voltage dynamic controller are:

$$\frac{dI_{DCIaux}}{dt} = K_{iedc}(E_{DCI} - E_{DCInom}) \tag{6.69}$$

$$I_{DCI} = K_{pedc}(E_{DCI} - E_{DCInom}) + I_{DCIaux} \tag{6.70}$$

At the same time, the rectifier must ensure that its output active power is kept at the scheduled value P_{sch}. From Figure 6.27, it can be inferred that the power entering the inverter station is the scheduled power P_{sch} minus the power loss incurred in the DC cable R_{DC}. The DC power flowing from the node Rx towards the node Ix may be expressed in terms of DC voltages, as follows:

$$P_{DCR} = (E_{DCRx}^2 - E_{DCRx}E_{DCIx})R_{DC}^{-1} \tag{6.71}$$

Likewise, the power flowing from the inverter towards the rectifier through the DC cable may be expressed as:

$$P_{DCI} = (E_{DCIx}^2 - E_{DCIx}E_{DCRx})R_{DC}^{-1} \tag{6.72}$$

Notice that during steady-state, it is true that $E_{DCR} = E_{DCRx}$ *and* $E_{DCI} = E_{DCIx}$, since the currents taken by both capacitors is zero.

The angular aperture γ_R between the phase-shifting angle of the rectifier φ_R and the voltage angle $\theta_{vR} = tan^{-1}(f_{vR}/e_{vR})$ is related to the power exchange taking place between the network and the rectifier's DC bus at any point in time. Hence, the angular difference $\gamma_R = \theta_{vR} - \varphi_R$ is the parameter that requires regulation with the aim of achieving the scheduled active power transfer P_{sch} in the DC link. The pursued power balance on the DC side is $P_{0R} + P_{sch} = 0$, as shown in Figure 6.29 [21].

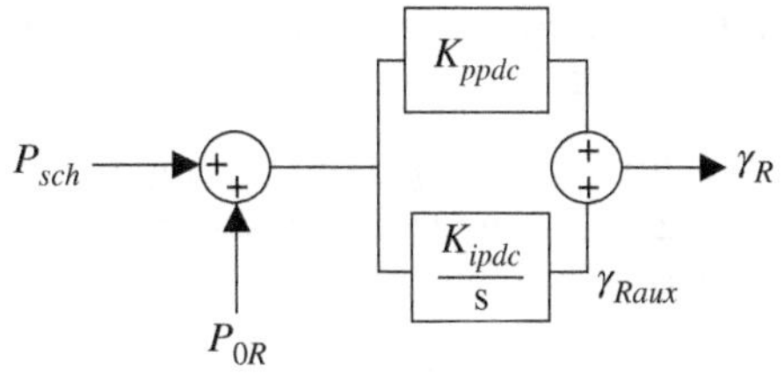

Figure 6.29 DC power transfer controller of the VSC-HVDC link model. Source: ©Elsevier, 2015.

The equations enabling the calculation of the dynamic behaviour of the DC power transmission are:

$$\frac{d\gamma_{Raux}}{dt} = K_{ipdc}(P_{sch} + P_{0R}) \tag{6.73}$$

$$\gamma_R = K_{ppdc}(P_{sch} + P_{0R}) + \gamma_{Raux} \tag{6.74}$$

On the other hand, the AC voltage dynamic control of the VSC-HVDC link requires two control loops. The modulation indices m_{aI} and m_{aR} are responsible for controlling the AC voltage magnitude at the receiving and sending ends of the VSC-HVDC at the scheduled values, V_{vI0} and V_{vR0}. The first-order control blocks shown in Figure 6.30 are employed to provide reactive power support to the network. Each control is designed in such a way that the modulation indices m_{aI} and m_{aR} are readjusted at every time step according to the difference between the scheduled voltage magnitudes and the actual voltage at the nodes where the converters are connected, $V_{vI} = (e_{vI}^2 + f_{vI}^2)^{1/2}$ and $V_{vR} = (e_{vR}^2 + f_{vR}^2)^{1/2}$.

The differential equations for the AC-bus voltage controllers are:

$$\frac{d(dm_{aR})}{dt} = \frac{K_{maR}(V_{vR0} - V_{vR}) - dm_{aR}}{T_{maR}} \tag{6.75}$$

$$\frac{d(dm_{aI})}{dt} = \frac{K_{maI}(V_{vI0} - V_{vI}) - dm_{aI}}{T_{maI}} \tag{6.76}$$

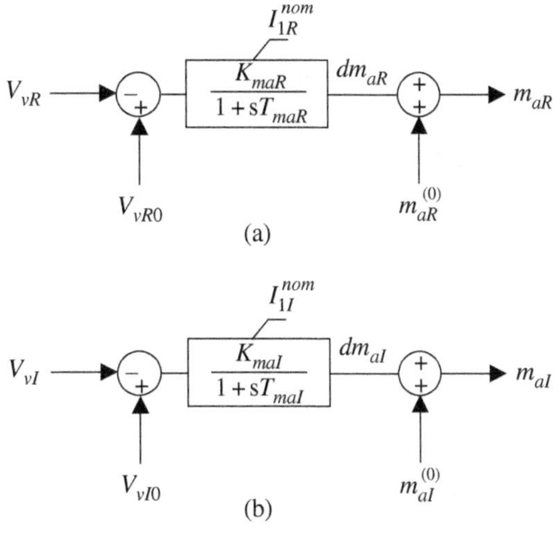

Figure 6.30 AC bus voltage controllers. Source: ©Elsevier, 2015. (a) Rectifier station and (b) inverter station.

6.5.1 Discretization and Linearization of the Differential Equations of the VSC-HVDC

The differential equations arising from the VSC-HVDC controllers are discretized and arranged in a suitable form to facilitate their accommodation into the expanded Jacobian matrix [21]. They are:

$$F_{E_{DCR}} = E_{DCR(t-\Delta t)} + 0.5\Delta t\dot{E}_{DCR(t-\Delta t)} - (E_{DCR(t)} - 0.5\Delta t\dot{E}_{DCR(t)}) \tag{6.77}$$

$$F_{E_{DCI}} = E_{DCI(t-\Delta t)} + 0.5\Delta t\dot{E}_{DCI(t-\Delta t)} - (E_{DCI(t)} - 0.5\Delta t\dot{E}_{DCI(t)}) \tag{6.78}$$

$$F_{I_{Rx}} = I_{Rx(t-\Delta t)} + 0.5\Delta t\,\dot{I}_{Rx(t-\Delta t)} - (I_{Rx(t)} - 0.5\Delta t\,\dot{I}_{Rx(t)}) \tag{6.79}$$

$$F_{I_{Ix}} = I_{Ix(t-\Delta t)} + 0.5\Delta t\,\dot{I}_{Ix(t-\Delta t)} - (I_{Ix(t)} - 0.5\Delta t\,\dot{I}_{Ix(t)}) \tag{6.80}$$

$$F_{I_{DCIaux}} = I_{DCIaux(t-\Delta t)} + 0.5\Delta t\dot{I}_{DCIaux(t-\Delta t)} - (I_{DCIaux(t)} - 0.5\Delta t\dot{I}_{DCIaux(t)}) \tag{6.81}$$

$$F_{\gamma_{Raux}} = \gamma_{Raux(t-\Delta t)} + 0.5\Delta t\dot{\gamma}_{Raux(t-\Delta t)} - (\gamma_{Raux(t)} - 0.5\Delta t\dot{\gamma}_{Raux(t)}) \tag{6.82}$$

$$F_{dm_{aR}} = dm_{aR(t-\Delta t)} + 0.5\Delta t\,d\dot{m}_{aR(t-\Delta t)} - (dm_{aR(t)} - 0.5\Delta t\,d\dot{m}_{aR(t)}) \tag{6.83}$$

$$F_{dm_{aI}} = dm_{aI(t-\Delta t)} + 0.5\Delta t\,d\dot{m}_{aI(t-\Delta t)} - (dm_{aI(t)} - 0.5\Delta t\,d\dot{m}_{aI(t)}) \tag{6.84}$$

The expressions (6.77)–(6.84) establish the dynamic behaviour of the VSC-HVDC model. The first five differential equations capture the DC voltage and current performance of the DC link when the energy balance is disrupted. Likewise, the equation involving the angular aperture γ_R (6.82) deals with the power unbalance present in the DC link. On the other hand, the expressions given in (6.83)–(6.84) enable the computation of the new values of the modulation indices with which the AC voltage set points are achieved.

In order to link the control variables of the VSC-HVDC with the state variables of the network, at nodes vR and vI, the algebraic power mismatch Eqs. (6.85)–(6.88) are used. The power balance equations at the internal nodes Rx, Ix and $0I$, given by the algebraic relationships (6.89)–(6.91), complete the dynamic model:

$$\Delta P_{vR} = -P_{vR} - P_{vR,load} - P_{vR,cal} \tag{6.85}$$

$$\Delta Q_{vR} = -Q_{vR} - Q_{vR,load} - Q_{vR,cal} \tag{6.86}$$

$$\Delta P_{vI} = -P_{vI} - P_{vI,load} - P_{vI,cal} \tag{6.87}$$

$$\Delta Q_{vI} = -Q_{vI} - Q_{vI,load} - Q_{vI,cal} \tag{6.88}$$

$$\Delta P_{Rx} = -E_{DCRx}I_{Rx} + P_{DCR} \tag{6.89}$$

$$\Delta P_{Ix} = -E_{DCIx}I_{Ix} + P_{DCI} \tag{6.90}$$

$$\Delta P_{0I} = E_{DCI}I_{DCI} - P_{0I} \tag{6.91}$$

The active and reactive power injections, derived in (Section 3.8.1), are restated below:

$$P_{vR} = G_{vRvR}(e_{vR}^2 + f_{vR}^2) - E_{DCR}(e_{vR}[G_{vR0}\cos\varphi_R + B_{vR0}\sin\varphi_R] - f_{vR}[G_{vR0}\sin\varphi_R - B_{vR0}\cos\varphi_R]) \tag{6.92}$$

$$Q_{vR} = -B_{vRvR}(e_{vR}^2 + f_{vR}^2) - E_{DCR}(e_{vR}[G_{vR0}\sin\varphi_R - B_{vR0}\cos\varphi_R] + f_{vR}[G_{vR0}\cos\varphi_R + B_{vR0}\sin\varphi_R]) \tag{6.93}$$

$$P_{0R} = (G_{swR} + G_{00R})E_{DCR}^2 - E_{DCR}(e_{vR}[G_{vR0}\cos\varphi_R + B_{vR0}\sin\varphi_R] + f_{vR}[G_{vR0}\sin\varphi_R - B_{vR0}\cos\varphi_R]) \tag{6.94}$$

$$P_{vI} = G_{vIvI}(e_{vI}^2 + f_{vI}^2) - E_{DCI}(e_{vI}[G_{vI0}\cos\varphi_I - B_{vI0}\sin\varphi_I] + f_{vI}[G_{vI0}\sin\varphi_I + B_{vI0}\cos\varphi_I]) \quad (6.95)$$

$$Q_{vI} = -B_{vIvI}(e_{vI}^2 + f_{vI}^2) - E_{DCI}(-e_{vI}[G_{vI0}\sin\varphi_I + B_{vI0}\cos\varphi_I] + f_{vI}[G_{vI0}\cos\varphi_I - B_{vI0}\sin\varphi_I]) \quad (6.96)$$

$$P_{0I} = (G_{swI} + G_{00I})E_{DCI}^2 - E_{DCI}(e_{vI}[G_{vI0}\cos\varphi_I + B_{vI0}\sin\varphi_I] + f_{vI}[G_{vI0}\sin\varphi_I - B_{vI0}\cos\varphi_I]) \quad (6.97)$$

Eqs. (6.77)–(6.91) constitute the set of mismatch equations that must be assembled together with the equations of the whole network, synchronous generators and their corresponding controllers. The linearized form of the VSC-HVDC dynamic model is given by:

$$\Delta F = -\begin{bmatrix} J_{11} & J_{12} \\ J_{21} & J_{22} \end{bmatrix} \Delta z \quad (6.98)$$

$$\Delta F = \begin{bmatrix} \Delta P_{vR} & \Delta Q_{vR} & \Delta P_{vI} & \Delta Q_{vI} & \Delta P_{Rx} & \Delta P_{Ix} & \Delta P_{0I} & \cdots \\ \cdots F_{E_{DCR}} & F_{E_{DCI}} & F_{I_{Rx}} & F_{I_{Ix}} & F_{I_{DCIaux}} & F_{\gamma_{Raux}} & F_{dm_{aR}} & F_{dm_{aI}} \end{bmatrix}^T$$

$$\Delta z = \begin{bmatrix} \Delta e_{vR} & \Delta f_{vR} & \Delta e_{vI} & \Delta f_{vI} & \Delta E_{DCRx} & \Delta E_{DCIx} & \Delta \varphi_I & \cdots \\ \cdots \Delta E_{DCR} & \Delta E_{DCI} & \Delta I_{Rx} & \Delta I_{Ix} & \Delta I_{DCIaux} & \Delta \gamma_{Raux} & \Delta dm_{aR} & \Delta dm_{aI} \end{bmatrix}^T$$

where J_{11} comprises the first-order partial derivatives of the network power mismatch equations and VSC-HVDC mismatch equations with respect to the network's state variables and the state variables of the VSC-HVDC. Likewise, J_{12} contains the first-order partial derivatives arising from the algebraic mismatch equations with respect to the control variables of the VSC-HVDC link. The matrix J_{21} consists of partial derivatives of the VSC-HVDC's discretized differential equations with respect to the AC voltages, the phase-shifting angle φ_I and the DC voltage E_{DCR}. Lastly, J_{22} is a matrix that accommodates the first-order partial derivatives of the VSC-HVDC's discretized differential equations with respect to their own control variables.

$$J_{11} = \begin{bmatrix} \frac{\partial \Delta P_{vR}}{\partial e_{vR}} & \frac{\partial \Delta P_{vR}}{\partial f_{vR}} & 0 & 0 & 0 & 0 & 0 \\ \frac{\partial \Delta Q_{R}}{\partial e_{vR}} & \frac{\partial \Delta Q_{vR}}{\partial f_{vR}} & 0 & 0 & 0 & 0 & 0 \\ 0 & 0 & \frac{\partial \Delta P_{vI}}{\partial e_{vI}} & \frac{\partial \Delta P_{vI}}{\partial f_{vI}} & 0 & 0 & \frac{\partial \Delta P_{vI}}{\partial \varphi_I} \\ 0 & 0 & \frac{\partial \Delta Q_{vI}}{\partial e_{vI}} & \frac{\partial \Delta Q_{vI}}{\partial f_{vI}} & 0 & 0 & \frac{\partial \Delta Q_{vI}}{\partial \varphi_I} \\ 0 & 0 & 0 & 0 & \frac{\partial \Delta P_{Rx}}{\partial E_{DCRx}} & \frac{\partial \Delta P_{Rx}}{\partial E_{DCIx}} & 0 \\ 0 & 0 & 0 & 0 & \frac{\partial \Delta P_{Ix}}{\partial E_{DCRx}} & \frac{\partial \Delta P_{Ix}}{\partial E_{DCIx}} & 0 \\ 0 & 0 & \frac{\partial \Delta P_{0I}}{\partial e_{vI}} & \frac{\partial \Delta P_{0I}}{\partial f_{vI}} & 0 & 0 & \frac{\partial \Delta P_{0I}}{\partial \varphi_I} \end{bmatrix},$$

$$J_{12} = \begin{bmatrix} \frac{\partial \Delta P_{vR}}{\partial E_{DCR}} & 0 & 0 & 0 & 0 & 0 & \frac{\partial \Delta P_{vR}}{\partial dm_{aR}} & 0 \\ \frac{\partial \Delta Q_{vR}}{\partial E_{DCR}} & 0 & 0 & 0 & 0 & 0 & \frac{\partial \Delta Q_{R}}{\partial dm_{aR}} & 0 \\ 0 & \frac{\partial \Delta P_{vI}}{\partial E_{DCI}} & 0 & 0 & 0 & 0 & 0 & \frac{\partial \Delta P_{vI}}{\partial dm_{aI}} \\ 0 & \frac{\partial \Delta Q_{vI}}{\partial E_{DCI}} & 0 & 0 & 0 & 0 & 0 & \frac{\partial \Delta Q_{vI}}{\partial dm_{aI}} \\ 0 & 0 & \frac{\partial \Delta P_{Rx}}{\partial I_{Rx}} & 0 & 0 & 0 & 0 & 0 \\ 0 & 0 & 0 & \frac{\partial \Delta P_{Ix}}{\partial I_{Ix}} & 0 & 0 & 0 & 0 \\ 0 & \frac{\partial \Delta P_{0I}}{\partial E_{DCI}} & 0 & 0 & 0 & 0 & 0 & \frac{\partial \Delta P_{0I}}{\partial dm_{aI}} \end{bmatrix}$$

$$J_{21} = \begin{bmatrix} \frac{\partial F_{E_{DCR}}}{\partial e_{vR}} & \frac{\partial F_{E_{DCR}}}{\partial f_{vR}} & 0 & 0 & 0 & 0 & 0 \\ 0 & 0 & \frac{\partial F_{E_{DCI}}}{\partial e_{vI}} & \frac{\partial F_{E_{DCI}}}{\partial f_{vI}} & 0 & 0 & \frac{\partial F_{E_{DCI}}}{\partial \varphi_I} \\ 0 & 0 & 0 & 0 & \frac{\partial F_{I_{Rx}}}{\partial E_{DCRx}} & 0 & 0 \\ 0 & 0 & 0 & 0 & 0 & \frac{\partial F_{I_{Ix}}}{\partial E_{DCIx}} & 0 \\ 0 & 0 & 0 & 0 & 0 & 0 & 0 \\ \frac{\partial F_{\gamma_{Raux}}}{\partial e_{vR}} & \frac{\partial F_{\gamma_{Raux}}}{\partial f_{vR}} & 0 & 0 & 0 & 0 & 0 \\ \frac{\partial F_{dm_{aR}}}{\partial e_{vR}} & \frac{\partial F_{dm_{aR}}}{\partial f_{vR}} & 0 & 0 & 0 & 0 & 0 \\ 0 & 0 & \frac{\partial F_{dm_{aI}}}{\partial e_{vI}} & \frac{\partial F_{dm_{aI}}}{\partial f_{vI}} & 0 & 0 & 0 \end{bmatrix},$$

$$J_{22} = \begin{bmatrix} \frac{\partial F_{E_{DCR}}}{\partial E_{DCR}} & 0 & \frac{\partial F_{E_{DCR}}}{\partial I_{Rx}} & 0 & 0 & 0 & \frac{\partial F_{E_{DCR}}}{\partial dm_{aR}} & 0 \\ 0 & \frac{\partial F_{E_{DCI}}}{\partial E_{DCI}} & 0 & \frac{\partial F_{E_{DCI}}}{\partial I_{Ix}} & 0 & 0 & 0 & \frac{\partial F_{E_{DCI}}}{\partial dm_{aI}} \\ \frac{\partial F_{I_{Rx}}}{\partial E_{DCR}} & 0 & \frac{\partial F_{I_{Rx}}}{\partial I_{Rx}} & 0 & 0 & 0 & 0 & 0 \\ 0 & \frac{\partial F_{I_{Ix}}}{\partial E_{DCI}} & 0 & \frac{\partial F_{I_{Ix}}}{\partial I_{Ix}} & 0 & 0 & 0 & 0 \\ 0 & \frac{\partial F_{I_{DCIaux}}}{\partial E_{DCI}} & 0 & 0 & \frac{\partial F_{I_{DCIaux}}}{\partial I_{DCIaux}} & 0 & 0 & 0 \\ \frac{\partial F_{\gamma_{Raux}}}{\partial E_{DCR}} & 0 & 0 & 0 & 0 & \frac{\partial F_{\gamma_{Raux}}}{\partial \gamma_{Raux}} & \frac{\partial F_{\gamma_{Raux}}}{\partial dm_{aR}} & 0 \\ 0 & 0 & 0 & 0 & 0 & 0 & \frac{\partial F_{dm_{aR}}}{\partial dm_{aR}} & 0 \\ 0 & 0 & 0 & 0 & 0 & 0 & 0 & \frac{\partial F_{dm_{aI}}}{\partial dm_{aI}} \end{bmatrix} \tag{6.99}$$

It can be seen that the derivative terms corresponding to the two converters working in a coordinated fashion as part of the HVDC link are not different from those corresponding to the VSC operating as a STATCOM. Hence, they can be found by simply using subscripts vR or vI in the equation already available for the STATCOM. However, two new algebraic equations have been added to link the DC voltages and currents between the two converters, one differential equation to capture the inverter's modulation index dynamics and one more relating to the controller responsible for regulating the angle γ_R to suitably guarantee the correct transfer of power.

The additional derivative terms arising from both converters acting as a single VSC-HVDC system are given below:

$$\frac{\partial \Delta P_{Rx}}{\partial E_{DCRx}} = -I_{Rx} + (2E_{DCRx} - E_{DCIx})R_{DC}^{-1}$$

$$\frac{\partial \Delta P_{Rx}}{\partial I_{Rx}} = -E_{DCRx}$$

$$\frac{\partial \Delta P_{Rx}}{\partial E_{DCIx}} = -E_{DCRx}R_{DC}^{-1}$$

$$\frac{\partial F_{I_{Ix}}}{\partial E_{DCI}} = -\frac{\partial F_{I_{Ix}}}{\partial E_{DCIx}} = -L_{DC}^{-1}E_{DCIx}$$

$$\frac{\partial F_{I_{Ix}}}{\partial I_{Ix}} = \frac{\partial F_{I_{Rx}}}{\partial I_{Rx}} = 1.0$$

$$\frac{\partial \Delta P_{0I}}{\partial e_{vI}} = -\frac{\partial P_{0I}}{\partial e_{vI}}$$

$$\frac{\partial \Delta P_{Ix}}{\partial E_{DCIx}} = -I_{Ix} + (2E_{DCIx} - E_{DCRx})R_{DC}^{-1}$$

$$\frac{\partial \Delta P_{Ix}}{\partial I_{Ix}} = -E_{DCIx}$$

$$\frac{\partial \Delta P_{Ix}}{\partial E_{DCRx}} = -E_{DCIx}R_{DC}^{-1}$$

$$\frac{\partial F_{I_{Rx}}}{\partial E_{DCR}} = -\frac{\partial F_{I_{Rx}}}{\partial E_{DCRx}} = -L_{DC}^{-1}E_{DCRx}$$

$$\frac{\partial \Delta P_{0I}}{\partial f_{vI}} = -\frac{\partial P_{0I}}{\partial f_{vI}}$$

$$\frac{\partial \Delta P_{0I}}{\partial \varphi_I} = -\frac{\partial P_{0I}}{\partial \varphi_I}$$

$$\frac{\partial \Delta P_{0I}}{\partial E_{DCI}} = I_{DCI} - \frac{\partial P_{0I}}{\partial E_{DCI}}$$

$$\frac{\partial \Delta P_{0I}}{\partial dm_{aI}} = -\frac{\partial P_{0I}}{\partial m_{aI}}$$

The steady-state condition needed to start the dynamic simulation is calculated using the conventional Newton-Raphson power flow algorithm including the VSC-HVDC link model, derived in Chapter 3. Such a solution provides accurate initial conditions to ensure reliable dynamic simulations, with a smooth start.

6.5.2 Validation of the VSC-HVDC Link Model

The range of validity of the response of the VSC-HVDC link model is established in this section by carrying out a comparison against the widely used electromagnetic transient (EMT)-type simulation software Simulink. Both types of simulation tools enable dynamic assessments of electrical power networks but they take a fundamentally different solution approach. Simulink represents every component of the power grid by means of RLC circuits and their corresponding differential equations require discretization at rather small time steps, in the order of microseconds, to ensure a stable numerical solution. Conversely, the solution of the RMS-type model requires only 'one phase' of the network, i.e. positive sequence network, and uses fundamental frequency phasors of voltages and currents as opposed to the three-phase representation of the network and instantaneous waveforms of voltages and currents used in EMT-type solution approaches such as Simulink.

The VSC-HVDC model comparison is carried out using a simple power system comprising two independent AC networks (2000 MVA, 230 kV, 50 Hz), which are interconnected through a VSC-HVDC link (200 MVA, ±100 kV DC) with a DC cable length of 75 km, as shown in Figure 6.31.

Both converter stations comprise a step-down transformer, AC filter, converter reactor, DC capacitors, DC filters and fixed off-nominal transformers' tap positions. The model of the power system including the VSC-HVDC link together with its parameters can be found in the section of 'demos' in Simulink under the title: VSC-Based HVDC Transmission System (Detailed Model). The parameters of the VSC-HVDC model described in this chapter are shown in Table 6.8. To ensure a reliable numerical solution, the EMT-type simulation package discretizes the power system and the control system

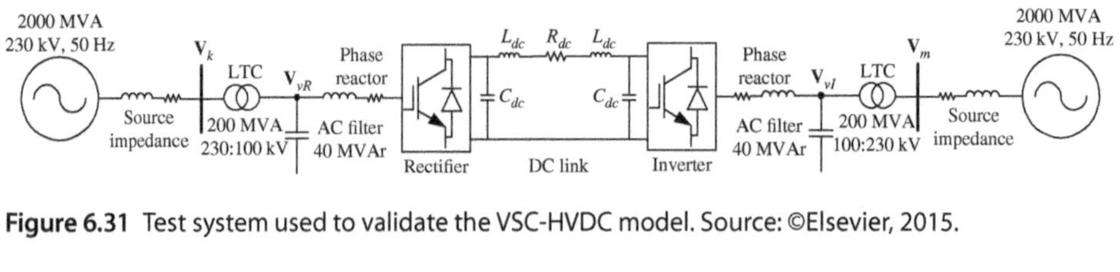

Figure 6.31 Test system used to validate the VSC-HVDC model. Source: ©Elsevier, 2015.

Table 6.8 VSC-HVDC parameters.

S_{nom} (p.u)	P_{sch} (p.u)	R_{DC} (p.u)	E_{DCI} (p.u)	G_{0I}, G_{0R} (p.u)	R_{1I}, R_{1R} (p.u)
2.0	2.0	0.02135	2.00	4e-3	1e-3
X_{1I}, X_{1R} (p.u)	R_{fI}, R_{fR} (p.u)	X_{fI}, X_{fR} (p.u)	B_{fI}, B_{fR} (p.u)	R_{ltc} (p.u)	X_{ltc} (p.u)
0.5e-3	7.5e-4	0.075	0.40	2.5e-3	0.075
H_c, H_i (s)	K_{pedc}	K_{iedc}	K_{ppdc}, K_{ipdc}	K_{maR}, K_{maI}	T_{maR}, T_{maI}
0.014	0.60	35.0	0.0, 5.0	25.0	0.02

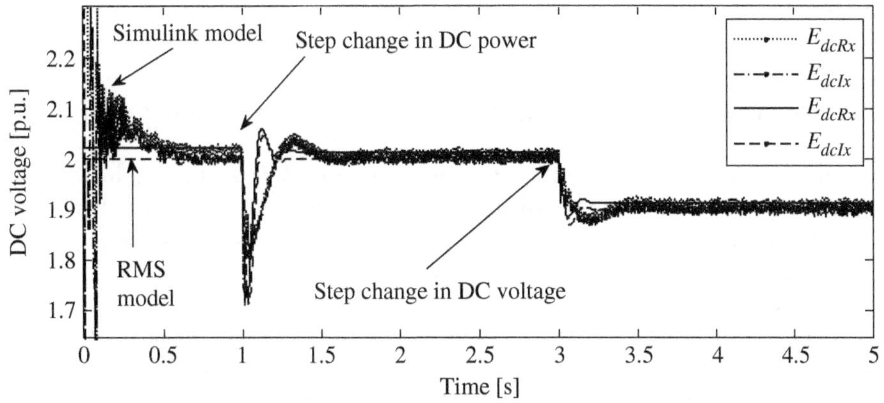

Figure 6.32 DC voltage performance for the RMS-type and Simulink models. Source: ©Elsevier, 2015.

equations with a sample time of 7.406 μs and 74.06 μs, respectively. This contrasts with the RMS-type model, where an integration step of 1 ms is used.

Initially the rectifier station is set to control the active power transmission at P_{sch} = 200 MW (2 p.u.). The inverter is responsible for controlling the DC voltage at E_{dcInom} = 200 kV (2 p.u.). The rectifier and inverter stations are set to comply with a fixed reactive power command of 0 p.u. and −0.1 p.u., respectively. In order to reach the steady-state equilibrium point in Simulink, the simulation is run up to $t = 1$ s. At this point, the active power transmission is reduced from 200 MW to 100 MW, that is, a −50% step is applied to the reference DC power. Furthermore, at $t = 3$ s, a step change of −5% is applied to the reference DC voltage of the inverter, i.e. the DC voltage is decreased from 2 p.u to 1.9 p.u.

The DC voltages corresponding to cases where step changes in the reference DC power and DC voltage are applied are shown in Figure 6.32. As expected, some differences may be seen in these results, owing to the two very different solution techniques. Nevertheless, the dynamic performance of the DC voltages of the RMS-type model follows well the dynamic pattern obtained by the switching-based HVDC model in Simulink.

The difference between the two solution approaches at the start of the simulation (0.5 s of the simulation) is explained by the very different manner in which the two simulations are initialized: the RMS model of the VSC-HVDC system uses accurate starting conditions furnished by a power flow solution whereas the Simulink model starts from

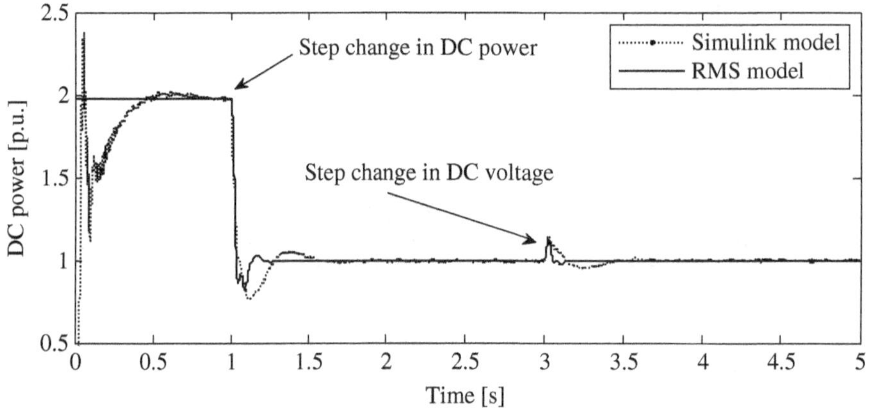

Figure 6.33 DC power performance for the RMS-type and Simulink models. Source: ©Elsevier, 2015.

its customary zero initial condition, i.e. the currents and voltages of the inductors and capacitors, respectively, are set to zero at $t = 0$ s. From the physical vantage, this is akin to assuming that the system was in a de-energized state and that it is energized at a time 0^+. Conversely, the RMS solution assumes that the system and the VSC-HVDC are operating under normal steady-state.

Similar conclusions can be drawn when analyzing the dynamic response of the DC power following the application of the step changes in DC power and DC voltage, as shown in Figure 6.33. As for the change in the DC power reference, it can be seen that the power stabilizes in no more than 0.5 seconds; this shows the rather quick response and robustness afforded by the dynamic controls of the VSC-HVDC link even in the event of a drastic change in the transmitted DC power. On the other hand, the step change in the DC voltage reference causes momentary power flow oscillations in the DC link which are also damped out quite rapidly.

The dynamic behaviour of the modulation indices of the converters is depicted in Figure 6.34. The step change in the DC power reference yields a very abrupt variation in the modulation indices; the dynamic performance of the modulation indices as calculated by both Simulink and the RMS-type VSC-HVDC model follows the same trend although an exact match was not expected. After the first disturbance, a steady-state error of 0.87% and 1.67% is obtained for the modulation indices of the rectifier and the inverter, respectively. Similarly, once the oscillations due to the step change in the reference DC voltage have been damped out, the differences in the modulation indices stand at 0.07% and 1.16%, respectively. These variations in the modulation indices may be explained by the very different modelling and solution approaches employed by the two simulation techniques.

For the sake of comparison, Table 6.9 shows the VSC-HVDC results as obtained by the RMS-type model and the Simulink model at different points in time of simulation. Table 6.9 also shows the computing times required to simulate the test system using both the RMS-type VSC-HVDC model and the EMT-type simulation tool Simulink, with the former model being approximately nine times faster than the EMT simulation. The very significant saving in computing time makes the RMS-type VSC-HVDC link

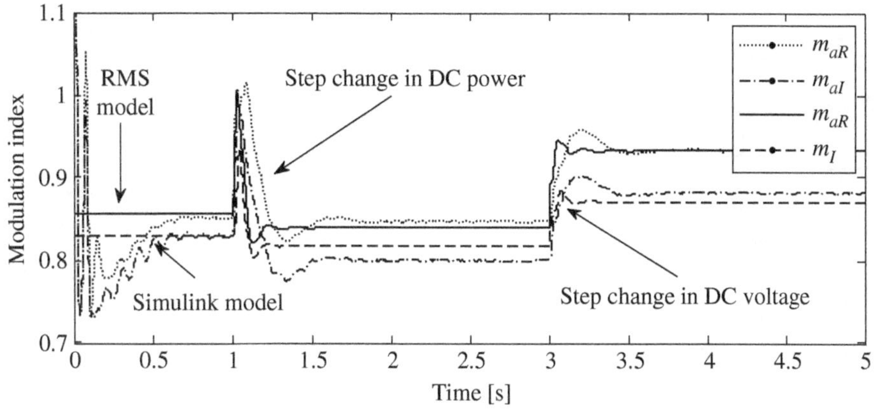

Figure 6.34 Modulation indices performance for the RMS-type and Simulink models. Source: ©Elsevier, 2015.

Table 6.9 VSC-HVDC variables for the RMS-type model and Simulink model.

Time	Developed RMS-type model					Simulink model				
	E_{DCR}	E_{DCI}	m_{aR}	m_{aI}	$-P_{DCI}$	E_{DCR}	E_{DCI}	m_{aR}	m_{aI}	$-P_{DCI}$
$t = 1^-$ s	2.0211	2.000	0.8553	0.8296	1.9791	2.0207	1.9985	0.8499	0.8301	1.9795
$t = 3^-$ s	2.0106	2.000	0.8389	0.8172	0.9947	2.0103	1.9998	0.8476	0.8005	0.9954
$t = 5$ s	1.9112	1.900	0.9329	0.8711	0.9942	1.9113	1.8990	0.9336	0.8827	0.9909
	Computing time: 19.76 seconds					Computing time: 180.39 seconds				

model a recommendable option in cases of large-scale power system simulations, particularly in studies that require long simulation times such as those involving synchronous generators' frequency variations and long-term voltage stability issues.

6.5.3 Numerical Example with an Embedded VSC-HVDC Link

The New England test system is modified, as shown in Figure 6.35, to illustrate the performance of the VSC-HVDC model, with the parameters shown in Table 6.10, when the network is subjected to a disturbance. The rest of the parameters correspond to those of the original case. The transmission line connecting nodes 4 and 14 is replaced by a VSC-HVDC link. The DC cable resistance is assumed to be of 0.08% (100 MVA base), which is the same as that of the replaced transmission line. The rectifier and inverter stations exert voltage control at their respective AC terminals at $V_{vR} = 1.01$ p.u and $V_{vI} = 1.03$ p.u, respectively. For the steady-state condition, the voltage magnitudes of the high-voltage sides of the LTC transformers, which correspond to nodes 4 and 14, are held fixed at the same voltage levels as those of the converters' terminals, V_{vR} and V_{vI}. Under these conditions, the LTCs' taps are computed using the steady-state power

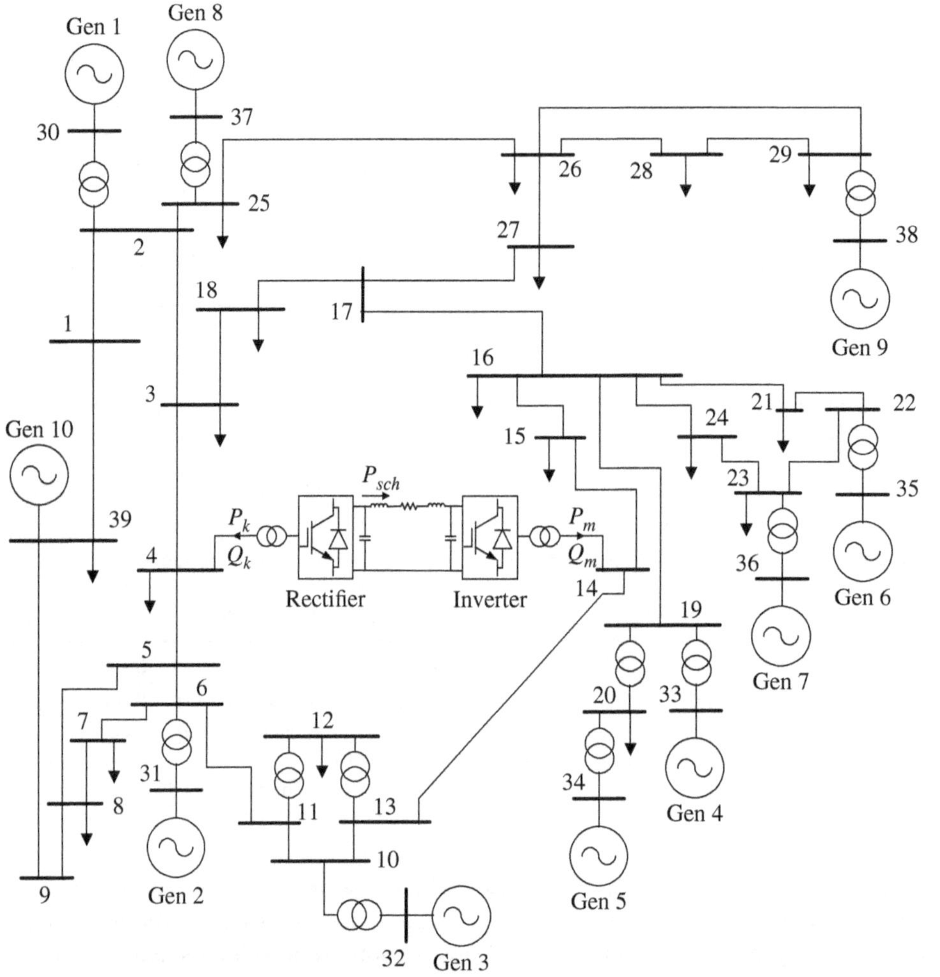

Figure 6.35 Modified New England test system to incorporate a VSC-HVDC link.

Table 6.10 VSC-HVDC parameters.

S_{nom} (p.u)	P_{sch} (p.u)	R_{DC} (p.u)	E_{DCI} (p.u)	G_{0I}, G_{0R} (p.u)	R_{1I}, R_{1R} (p.u)
3.0	1.0	0.0008	2.00	6e-3	6.66e-4
X_{1I}, X_{1R} (p.u)	R_{filtI}, R_{filtR} (p.u)	X_{filtI}, X_{filtR} (p.u)	B_{filtI}, B_{filtR} (p.u)	R_{ltc} (p.u)	X_{ltc} (p.u)
3.33e-3	0.0	0.02	0.30	0.0	0.01667
H_c, H_i (s)	K_{pedc}	K_{iedc}	K_{ppdc}, K_{ipdc}	K_{maI}, K_{maR}	T_{maI}, T_{maR}
15e-3	0.05	1.00	0.002, 0.075	25.0	0.02

flow algorithm, described in Chapter 3, and their values are kept constant during the dynamic solution. In addition to providing reactive power control, the HVDC link performs power regulation at the DC bus of the rectifier station, at $P_{sch} = 100$ MW. This implies that active power is drawn from node 4 and injected into node 14.

During steady-state operation, the rectifier station delivers 153.291 MVAR to the network to uphold its target voltage magnitude with a modulation index of 0.8484, whereas the inverter station operates with a modulation index of 0.8412, injecting 27.3 MVAR into the grid. In the case of the power flowing through the HVDC link, the active power entering the rectifier station stands at 100.869 MW and the active power leaving the inverter station takes a value of 99.667 MW. The difference between these two powers is the total power loss in the HVDC system, including the power loss in the DC cable. Taking as a reference the nominal apparent power of each converter S_{nom}, the total power losses stand at 0.4%, with 0.29% corresponding to the rectifier station and 0.104% to the inverter station, while the power loss produced by Joule's effect in the DC cable stands at 0.006%. It should be recalled that its magnitude is dependent on the length of the DC transmission line. Table 6.11 shows the main VSC-HVDC results, as given by the steady-state power flow solution and used to start the dynamic simulation.

Using the information in Table 6.11, it is a straightforward matter to proceed to the calculation of the initial values of the control variables taking part in the dynamics of the VSC-HVDC link, as shown in Table 6.12. These parameter values are employed to initialize the dynamic simulation where the modified test network is assumed to undergo the same disturbance as the original network, that is, the transmission lines connecting nodes 25-2, 2-3 and 3-4 are tripped at $t = 0.1$ s.

Figure 6.36 shows the voltage magnitudes at key nodes of the network following the disturbance. During the transient period, the target voltage set point is achieved very quickly by the action of the AC-bus voltage controllers that regulate the converters' modulation indices m_{aR} and m_{aI}, as shown in Figure 6.37. The prompt action of both controllers leads to a rapid reactive power injection at both converters' AC terminal, as can be seen in Figure 6.38. This results in a very effective damping of the voltage oscillations, enabling a smooth voltage recovery throughout the grid.

The disturbance in the AC system disrupts the power balance in the DC link and the voltage dips that take place at both converters' AC terminals reduce the power being

Table 6.11 VSC-HVDC results given by the power flow solution.

Converter	P_k, P_m (MW)	Q_k, Q_m (MVAR)	E_{DC} (p.u.)	m_a	φ (°)	B_{eq} (p.u.)	LTC's tap	P_{loss} (MW)
Rectifier	−100.869	153.291	2.0004	0.8484	−3.7380	1.2433	1.0252	0.8697
Inverter	99.667	27.300	2.0000	0.8412	7.9373	−0.0062	1.0044	0.3124

Table 6.12 Initial VSC-HVDC's control variables for the dynamic simulation.

E_{DCI} (p.u)	I_{DCIaux} (p.u)	I_{DCI} (p.u)	γ_{Raux} (rad)	γ_R (rad)	dm_{aR}	dm_{aI}
2.000	0.4999	0.4999	0.0232	0.0232	0.0	0.0

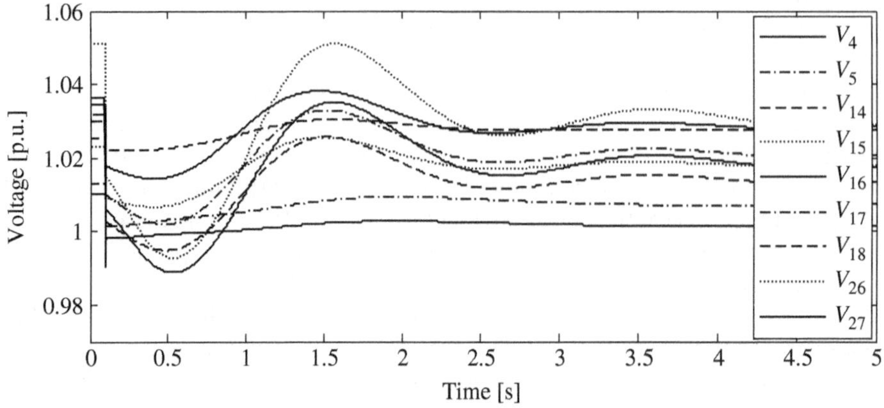

Figure 6.36 Voltage performance at different nodes of the network.

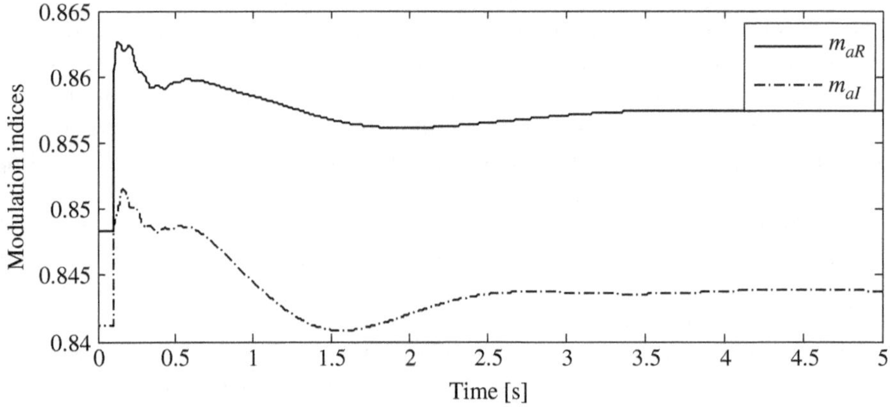

Figure 6.37 Dynamic behaviour of the converters' modulation index.

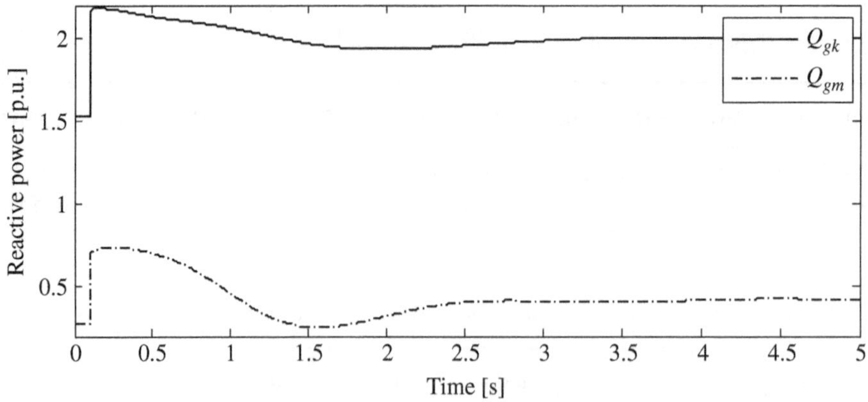

Figure 6.38 Reactive power generated by the HVDC link.

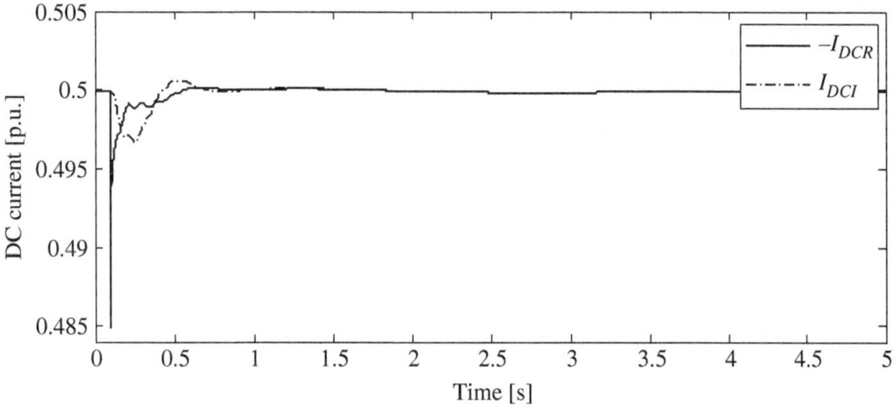

Figure 6.39 DC current behaviour of the rectifier and the inverter.

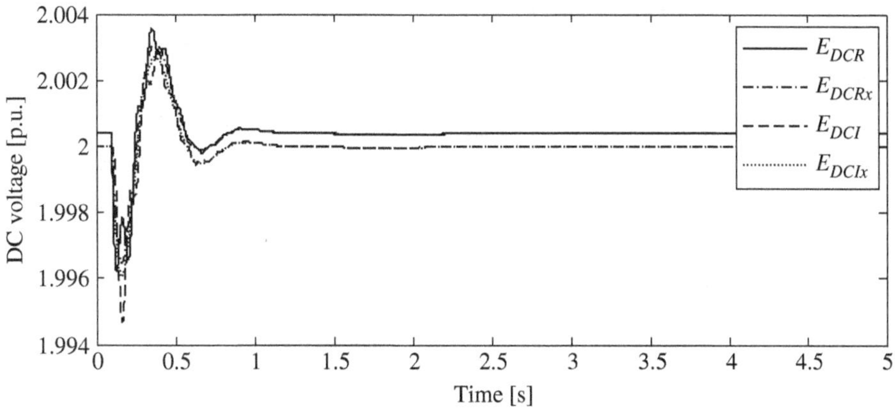

Figure 6.40 Voltage behaviour in the DC link.

transferred through the DC link. Accordingly, the rectifier's current I_{DCR} drops abruptly from 0.4999 p.u. to 0.4849 p.u., as illustrated in Figure 6.39. A momentary mismatch between the DC currents of both converters takes place because this cannot be instantly re-established due to the time constants of the current controller in the inverter station, which have finite values. This leads to DC voltage oscillations, as shown in Figure 6.40. Once the controller starts responding, the current I_{DCI} hunts the rectifier current I_{DCR}, aiming at overcoming the *voltage* drop, as shown in Figure 6.39. This enables a speedy recovery of the DC link voltages. In this case, the cable resistance is relatively small and so is the voltage drop in the DC transmission line. There are minor differences between the voltage plots E_{DCR} and E_{DCI} and the voltages E_{DCRx} and E_{DCIx}, respectively. The oscillations in the voltages E_{DCRx} and E_{DCIx} are smoothened to some extent by the damping effect of the DC inductors.

The simulation results for the active powers and the DC power transfer through the HVDC link, following the disconnection of the transmission lines, are illustrated in Figure 6.41. The power P_{gk} is the active power entering the high-voltage side of the LTC

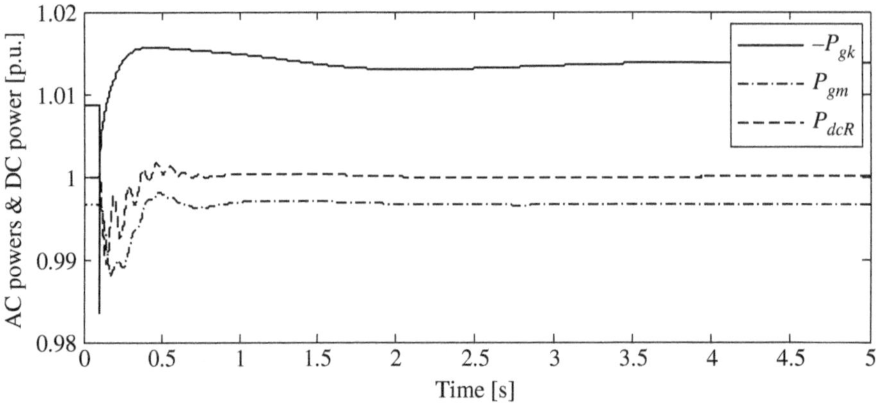

Figure 6.41 VSC-HVDC's AC active powers and DC power transfer behaviour.

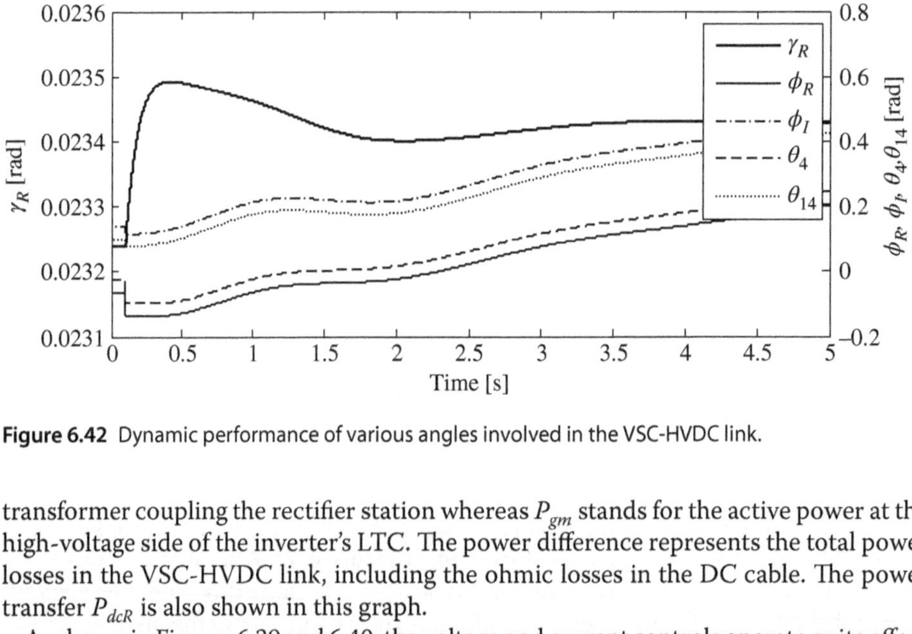

Figure 6.42 Dynamic performance of various angles involved in the VSC-HVDC link.

transformer coupling the rectifier station whereas P_{gm} stands for the active power at the high-voltage side of the inverter's LTC. The power difference represents the total power losses in the VSC-HVDC link, including the ohmic losses in the DC cable. The power transfer P_{dcR} is also shown in this graph.

As shown in Figures 6.39 and 6.40, the voltage and current controls operate quite effectively, leading to a fast power recovery in spite of the rather severe network disturbance. Since the power flow from rectifier to inverter has been brought back to its initial target, $P_{sch} = 1.0$ p.u, the power angle γ_R exhibits only a marginal increase, as observed in Figure 6.42. This small increase is to comply with the new steady-state operating conditions where different currents and, therefore, different power losses are incurred. Figure 6.42 also shows the behaviour of the phase-shifting angles of the rectifier and inverter converters, φ_R and φ_I. It can be seen that these angles follow the same pattern as the AC voltage angles, $\theta = \tan^{-1}(f/e)$, of the nodes where the VSC-HVDC system is connected.

6.5.4 Dynamic Model of the VSC-HVDC Link with Frequency Regulation Capabilities

VSC-HVDC systems find applicability in a wide range of power systems applications, such as interconnecting AC networks of the same or different frequencies, supplying electrical energy to remote islands and offshore oil and gas platforms, in-feeding of high load points in city centres and the evacuation of the electrical energy from offshore wind parks.

Two independent AC networks interconnected by means of an HVDC link exhibit a natural decoupling in terms of both voltage and frequency and it is said that the two AC power systems are interconnected in an asynchronous manner. These kinds of interconnections are primarily aimed at preventing the excursions of oscillations between AC systems, for instance between a strong and a weak electric network. Nevertheless, there are applications where it is desirable to exert the influence of the strong network upon the weak network, through the DC link. Such a situation would arise when power imbalances occur in an AC power network with little or no inertia and fed by a VSC-HVDC link.

In a power network with no inertia, even very small power imbalances would induce rather large frequency rises or drops, depending on the nature of the power unbalance. Furthermore, all AC networks contain a degree of frequency-sensitive loads and VSC converters do not possess the ability to strengthen on their own the inertial response or to aid the primary frequency control of AC networks. This issue is timely because electrical networks with poor inertia are likely to become more common [22]. There has been research progress in resolving this problem, using a range of equipment models and control schemes embedded in software with which simulation studies have been carried out [23–25]. Microgrid structures and control techniques have also been a matter of great research activity, with particular attention being paid to grid-forming, grid-feeding and grid-supporting topologies [26]. These types of grids are likely to become cost-effective solutions for the interconnection of distributed generators in power grids. Hence, power control performance in AC microgrids, with their strong impact on frequency, is receiving research attention [27–29].

Basically, frequency deviations in a network arise from a mismatch between the mechanical power and the electrical power supplied by the synchronous generators. These frequency deviations are larger in the smaller synchronous generators owing to their lighter rotating masses. The generator's inertia is the parameter that determines the size of the frequency deviation following a power imbalance. Besides the inertial response exhibited by synchronous generators, most of them are fitted with speed-governing controls to bring a degree of power controllability into the generating system, which is said to exert primary-frequency control.

In principle at least, a parallel may be drawn between the kinetic energy stored in the rotating mass of a synchronous generator, reflected in its inertia constant, and the electrostatic and electromagnetic energies stored in the converters' DC capacitors and DC inductors of a VSC-HVDC link, which will have an inertia constant $H = H_c + H_i$. Of course, the latter is numerically rather small, but this analogy is useful in the developments presented below. Furthermore, the difference between the DC power entering the inverter $E_{DCI}\, I_{DCI}$ and the power P_{0I} will give rise to a frequency deviation in the network connected to the inverter. Hence, the angular speed ω_I and the angle φ_I of the inverter station are expressed as follows [30]:

$$\frac{d\omega_I}{dt} = K_\omega(E_{DCI}I_{DCI} - P_{0I}) \tag{6.100}$$

$$\frac{d\varphi_I}{dt} = \omega_I - 2\pi f_{Inom} \tag{6.101}$$

where $K_\omega = \pi f_{Inom}/H$ and f_{Inom} represents the nominal electrical frequency [Hz] of the low-inertia AC grid. Its value may be the same as or different from that of the main utility grid connected to the rectifier terminals. It should be remarked that the angle φ_I, as well as the angular speed ω_I, might be taken to be reference signals for the low-inertia grid.

Given that the total fictitious inertia of the inverter H is very small (let us say 10 ms), in the event of relatively small power disturbances in the low-inertia AC grid, large sags/swells in the angular speed will take place. By way of example, take the case of a low-inertia grid operating at 60 Hz, where the value of K_ω would be around 18.85^*10^3. If the power imbalance is in the order of $\Delta P = -1^*10^{-3}$ p.u. MW and it takes one second to exert power control, $\Delta t = 1$ s, then the speed deviation would be $\Delta\omega \approx K_\omega \cdot \Delta P \cdot \Delta t \approx -18.85$ rad s^{-1}, which in terms of frequency is $f_I \approx 57$ Hz, a rather large frequency value which is not operational. This is a critical issue in networks with frequency-sensitive loads.

In cases where the AC grid does not possess frequency-control supporting equipment, e.g. synchronous generators with speed governors or battery-energy storage systems, there is a need to find a solution to mitigate any possible frequency excursion. In this sense, a solution may be to import power from a large, near utility grid through an HVDC link. Notice that this requirement departs from the traditional idea of transmitting a fixed amount of power between two networks interconnected by an HVDC link.

To this end, an auxiliary control loop, such as the one shown in Figure 6.43, is used to enable the HVDC link to import from the strong system the power required to overcome the power mismatches in the low-inertia AC system (passive network). Note that in this application, the inverter station acts as a power electronic source with active and reactive power control capabilities [30].

The control loop in Figure 6.43 measures the actual angular speed of the low-inertia AC network ω_I and compares it with its nominal speed, $\omega_{Inom} = 2\pi f_{Inom}$. The error is processed by a PI controller to adjust the angular aperture between the AC system's voltage phase angle θ_{vR} and the phase-shifting angle of the rectifier φ_R.

It should be remarked that the objective of the controller in Figure 6.29 is to maintain a fixed amount of power transfer in the DC link whereas the objective of the VSC-HVDC controller depicted in Figure 6.43 is to maintain the electrical frequency of the network connected to the inverter within safe limits by varying the power drawn from the grid

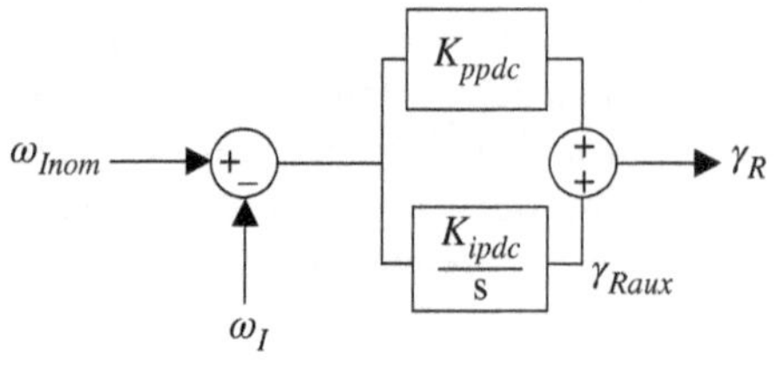

Figure 6.43 Frequency controller of the VSC-HVDC. Source: ©IEEE, 2015.

connected to the rectifier. It follows that the value of the DC power transfer should be sufficient and variable to bring the frequency of the low-inertia AC grid back to its nominal steady-state value.

The equations that enable the assessment of the dynamic performance of the DC power when the inverter station of the HVDC system acts as power electronic source with frequency regulation capabilities are:

$$\frac{d\gamma_{Raux}}{dt} = K_{ipdc}(\omega_{Inom} - \omega_I) \tag{6.102}$$

$$\gamma_R = K_{ppdc}(\omega_{Inom} - \omega_I) + \gamma_{Raux} \tag{6.103}$$

6.5.4.1 Linearization of the Equations of the VSC-HVDC Model with Frequency Regulation Capabilities

The set of power flow equations for the two interconnected AC networks together with the discretized differential equations of the synchronous generators and their controllers, as well as the algebraic and discretized differential equations of the VSC-HVDC link, are solved simultaneously using the trapezoidal method and the Newton-Raphson algorithm.

The mismatch expressions that must be solved together with the mismatch equations of the interconnected networks, at every integration step Δt, are [30]:

$$\Delta P_{vR} = -P_{vR} - P_{vR,load} - P_{vR,cal} \tag{6.104}$$

$$\Delta Q_{vR} = -Q_{vR} - Q_{vR,load} - Q_{vR,cal} \tag{6.105}$$

$$\Delta P_{vI} = -P_{vI} - P_{vI,load} - P_{vI,cal} \tag{6.106}$$

$$\Delta Q_{vI} = -Q_{vI} - Q_{vI,load} - Q_{vI,cal} \tag{6.107}$$

$$\Delta P_{Rx} = -E_{DCRx}I_{Rx} + P_{DCR} \tag{6.108}$$

$$\Delta P_{Ix} = -E_{DCIx}I_{Ix} + P_{DCI} \tag{6.109}$$

$$F_{E_{DCR}} = E_{DCR(t-\Delta t)} + 0.5\Delta t\dot{E}_{DCR(t-\Delta t)} - (E_{DCR(t)} - 0.5\Delta t\dot{E}_{DCR(t)}) \tag{6.110}$$

$$F_{E_{DCI}} = E_{DCI(t-\Delta t)} + 0.5\Delta t\dot{E}_{DCI(t-\Delta t)} - (E_{DCI(t)} - 0.5\Delta t\dot{E}_{DCI(t)}) \tag{6.111}$$

$$F_{I_{Rx}} = I_{Rx(t-\Delta t)} + 0.5\Delta t\,\dot{I}_{Rx(t-\Delta t)} - (I_{Rx(t)} - 0.5\Delta t\,\dot{I}_{Rx(t)}) \tag{6.112}$$

$$F_{I_{Ix}} = I_{Ix(t-\Delta t)} + 0.5\Delta t\,\dot{I}_{Ix(t-\Delta t)} - (I_{Ix(t)} - 0.5\Delta t\,\dot{I}_{Ix(t)}) \tag{6.113}$$

$$F_{I_{DCIaux}} = I_{DCIaux(t-\Delta t)} + 0.5\Delta t\dot{I}_{DCIaux(t-\Delta t)} - (I_{DCIaux(t)} - 0.5\Delta t\dot{I}_{DCIaux(t)}) \tag{6.114}$$

$$F_{\gamma_{Raux}} = \gamma_{Raux(t-\Delta t)} + 0.5\Delta t\,\dot{\gamma}_{Raux(t-\Delta t)} - (\gamma_{Raux(t)} - 0.5\Delta t\,\dot{\gamma}_{Raux(t)}) \tag{6.115}$$

$$F_{dm_{aR}} = dm_{aR(t-\Delta t)} + 0.5\Delta t\,d\dot{m}_{aR(t-\Delta t)} - (dm_{aR(t)} - 0.5\Delta t\,d\dot{m}_{aR(t)}) \tag{6.116}$$

$$F_{dm_{aI}} = dm_{aI(t-\Delta t)} + 0.5\Delta t\,d\dot{m}_{aI(t-\Delta t)} - (dm_{aI(t)} - 0.5\Delta t\,d\dot{m}_{aI(t)}) \tag{6.117}$$

$$F_{\varphi_I} = \varphi_{I(t-\Delta t)} + 0.5\Delta t\,\dot{\varphi}_{I(t-\Delta t)} - (\varphi_{I(t)} - 0.5\Delta t\,\dot{\varphi}_{I(t)}) \tag{6.118}$$

$$F_{\omega_I} = \omega_{I(t-\Delta t)} + 0.5\Delta t\,\dot{\omega}_{I(t-\Delta t)} - (\omega_{I(t)} - 0.5\Delta t\,\dot{\omega}_{I(t)}) \tag{6.119}$$

The set of equations representing the VSC-HVDC dynamic model is reformulated to endow the HVDC link with frequency regulation capabilities. The model incorporates the two differential Eqs. (6.118)–(6.119), which enable the computation of the reference angle and the angular speed of the otherwise independent AC network connected to the inverter. Furthermore, contrary to the customary objective of the HVDC to maintain a fixed amount of power in the DC link, Eq. (6.115) represents the DC power controller that acts upon changes in the frequency of the grid connected to the inverter. This set of equations yields the following linearized form of the VSC-HVDC model:

$$\Delta F = -J_{HVDC}\Delta z \tag{6.120}$$

where

$$\Delta F = \begin{bmatrix} \Delta P_{vR} & \Delta Q_{vR} & \Delta P_{vI} & \Delta Q_{vI} & \Delta P_{Rx} & \Delta P_{Ix} & F_{E_{DCR}} & F_{E_{DCI}} & \cdots \\ \cdots F_{I_{RX}} & F_{I_{Ix}} & F_{I_{DCIaux}} & F_{\gamma_{Raux}} & F_{dm_{ar}} & F_{dm_{aI}} & F_{\varphi_I} & F_{\omega_I} & \end{bmatrix}^T \tag{6.121}$$

$$\Delta z = \begin{bmatrix} \Delta e_{vR} & \Delta f_{vR} & \Delta e_{VI} & \Delta f_{vI} & \Delta E_{DCRx} & \Delta E_{DCIx} & \Delta E_{DCR} & \Delta E_{DCI} & \cdots \\ \cdots & \Delta I_{Rx} & \Delta I_{Ix} & \Delta I_{DCIaux} & \Delta \gamma_{Raux} & \Delta dm_{aR} & \Delta dm_{aI} & \Delta \varphi_I & \Delta \omega_I \end{bmatrix}^T \tag{6.122}$$

The Jacobian matrix J_{HVDC} accommodates the first-order partial derivatives of the algebraic equations and discretized differential equations of the HVDC link model. It should be stressed that Eqs. (6.108)–(6.109) and (6.115) provide the link between the set of equations corresponding to both AC networks. Eq. (6.115) relates the angular aperture of the AC voltage of the rectifier to the actual frequency at the inverter station, for power regulation purposes. Eqs. (6.108)–(6.109) relate the voltages in the DC link through the transmission line resistance.

The VSC-HVDC is used to interconnect two otherwise independent networks and the overall Jacobian matrix $\mathbf{J}$ bears the structure shown in (6.123):

$$\mathbf{J} = \begin{matrix} 1 \\ 2 \\ \vdots \\ k \\ vR \\ \\ \\ \\ \end{matrix} \begin{bmatrix} \mathbf{J}_{ACN1} & & 0 \\ & \ddots & \iddots & \\ & \mathbf{J}_{HVDC} & \\ & \iddots & \ddots & \\ 0 & & \mathbf{J}_{ACN2} \end{bmatrix} \begin{matrix} \\ \\ \\ \\ vI \\ m \\ \vdots \\ 1 \end{matrix} \tag{6.123}$$

where $\mathbf{J}_{ACN1}$ and $\mathbf{J}_{ACN2}$ stand for the Jacobian matrices of the main AC utility grid and the weak AC network upon which frequency support is being exerted, respectively. It should be emphasized that the structure of the Jacobian (6.123) has a similar structure to the nodal admittance matrix of the overall power system. It is also worth stressing that the way the equations of the converters have been assembled enables the unified dynamic solution of both interconnected AC networks together with the VSC-HVDC system, following a straightforward numerical process.

6.5.4.2 Validation of the VSC-HVDC Link Model Providing Frequency Support

The dynamic response of the VSC-HVDC link model with frequency regulation capabilities is compared against a model assembled in the Simulink environment, which uses an EMT-type solution. The comparison is carried out using a rather simple electrical network comprising two independent 50 Hz AC networks interconnected by a VSC-HVDC link. The test system is shown in Figure 6.31. The parameters of the HVDC link are given in Table 6.13. A similar frequency control loop to the one shown in Figure 6.43 is implemented in the Simulink model to endow it with frequency regulation capabilities.

Table 6.13 Parameters of the VSC-HVDC link with frequency regulation capabilities.

S_{nom} (p.u)	P_{sch} (p.u)	R_{DC} (p.u)	E_{DCI} (p.u)	G_{0I}, G_{0R} (p.u)	R_{1I}, R_{1R} (p.u)
2.0	0.2	0.02135	2.00	4e-3	0.0
X_{1I}, X_{1R} (p.u)	R_{filtI}, R_{filtR} (p.u)	X_{filtI}, X_{filtR} (p.u)	B_{filtI}, B_{filtR} (p.u)	R_{ltc} (p.u)	X_{ltc} (p.u)
0.0	7.5e-4	0.075	0.40	2.5e-3	0.075
H_c, H_i (s)	K_{pedc}	K_{iedc}	K_{ppdc}, K_{ipdc}	K_{maI}, K_{maR}	T_{maI}, T_{maR}
0.014	2.50	10.0	0.5e-3, 38e-3	25.0	0.02

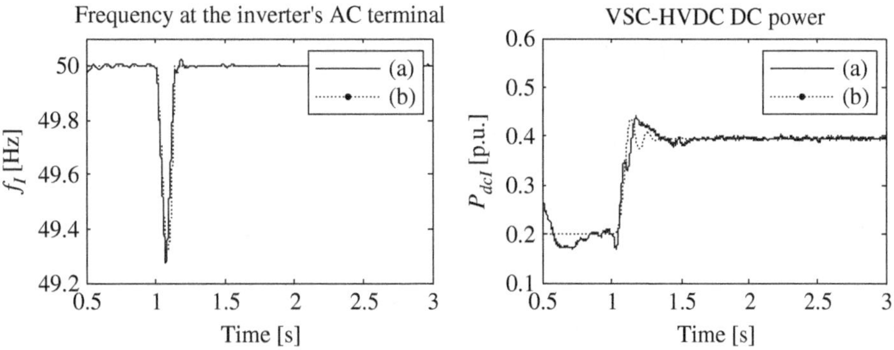

Figure 6.44 Dynamic behaviour of the frequency at the inverter's AC terminal and DC power. Source: ©IEEE, 2015. (a) Simulink model; (b) RMS model.

The initial DC power setpoint of the rectifier station is 20 MW (0.2 p.u.) while the inverter is set to control the DC voltage at $E_{dcInom} = 2$ p.u. The simulation in Simulink is run up to $t = 1.0$ s in order to reach the steady-state. To assess the dynamic response of the HVDC link in the face of rather critical frequency deviations, a sudden frequency drop is imposed by the ideal voltage source connected at the AC terminal of the inverter, as shown in Figure 6.44, something expected to occur in low-inertia networks. Figure 6.44 also depicts the dynamic performance of the DC power transfer for both solution approaches. Some differences exist but it is clear that the dynamic response of the RMS-type HVDC model follows quite well the pattern furnished by the switching-based HVDC model in Simulink.

The abrupt frequency drop seen by the inverter station is rapidly taken care of by the frequency control loop of the HVDC link. This regulator quickly increases the amount of power drawn from the utility grid connected at the rectifier's terminal. The momentary excess of energy in the DC link gives rise to voltage swells in both E_{DCRx} and E_{DCIx}, as seen in Figure 6.45. However the DC voltage controller operates very effectively to damp out the oscillations after only a few milliseconds. The dynamic performance of the modulation indices, as obtained with both HVDC link models with frequency regulation capabilities (i.e. Simulink and RMS-type), is shown in Figure 6.45. Both responses follow the same trend, although an exact match was not expected, owing to the very different modelling approaches used by the two models.

The computing times required to simulate the test system using the RMS-type VSC-HVDC model and the EMT-type model are 16.51 s and 160.42 s, respectively.

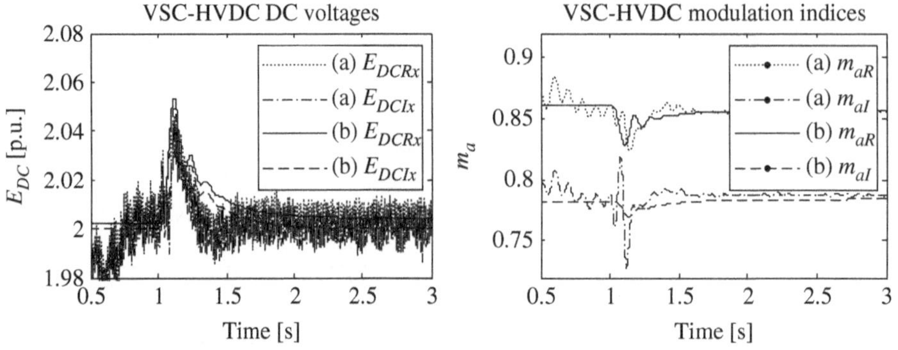

Figure 6.45 Dynamic performance of the DC voltages and modulation indices. Source: ©IEEE, 2015. (a) Simulink model; (b) RMS model.

Both solutions agree quite well with each other but the RMS-type model introduced in this chapter outperforms the switching-based model in Simulink by 10 times, in terms of computing speed. This is mainly due to two reasons: (i) the RMS-type model uses an integration time step of 1 ms whereas a sample time of 7.406 μs is needed for the EMT-type model in order to ensure a reliable numerical solution, and (ii) the RMS-type model requires only the positive sequence representation of the network whereas a three-phase representation is used in the EMT-type model.

6.5.4.3 Numerical Example with a VSC-HVDC Link Model Providing Frequency Support

HVDC Link Feeding into a Low-Inertia AC Network Figure 6.46 shows the schematic representation of a 10-MVA VSC-HVDC link, which interconnects two independent AC networks with very different characteristics. The main utility grid (strong AC network) is represented by the Thevenin equivalent of a large power system with a total demand of 60 GW operating with a lagging power factor of 0.95, which is coupled to a reactive tie-line of value $X_s = 0.06$ p.u.

The low-inertia AC grid (weak network) consists of a 1 MW hydro-generator, a 2 MW wind turbine and a 7 MVA constant load with a lagging power factor of 0.95. The synchronous machine of the hydro-generator is represented by a two-axis model with an inertia constant of 1 ms, resulting in a network with very low inertia. The parameters of the VSC-HVDC and the low-inertia AC network are:

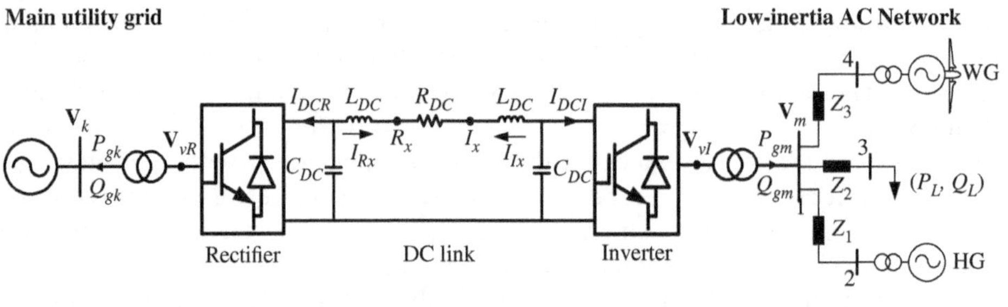

Figure 6.46 VSC-HVDC link feeding into a low-inertia AC network. Source: ©IEEE, 2015.

VSC stations: $S_{nom}=0.1$ p.u, $E_{DCInom}=2.0$ p.u, $R_{DC}=5$ p.u, $G_{0R}=G_{0I}=2e\text{-}4$ p.u, $R_{1R}=R_{1I}=0.02$ p.u, $X_{1R}=X_{1I}=0.7$ p.u., $H_c=7e\text{-}4$ s, $H_i=7e\text{-}3$ s, $K_{pedc}=0.05$, $K_{iedc}=1.0$, $K_{ppdc}=15e\text{-}4$, $K_{ipdc}=30e\text{-}4$, $K_{maR}=2.5$, $T_{maR}=0.02$, $K_{maI}=2.5$, $T_{maI}=0.02$.

Shunt AC filters at nodes vR and vI: $B_{filt}=0.1$ p.u.

LTC transformers: $Z_{LTC}=$ j0.5 p.u.

AC distribution line parameters: $Z_{L1}=Z_{L2}=Z_{L3}=0.1+j0.15$ p.u.

The rectifier and inverter stations are set to exert voltage control at their respective AC nodes at 1 p.u. and 1.025 p.u., respectively. From the results provided by the power flow solution, it is noticed that to meet the power equilibrium at the weak network, an import of 3.9175 MW from the main utility grid is required at the rectifier station of the DC link. The power injected by the inverter station, which stands at 3.7030 MW, takes account of the power losses incurred in the low-inertia AC grid: $P_{vI}+P_g-P_d=52.95$ kW.

The inverter plays the role of a slack generator from the power flow solution standpoint, where the hydro-generator is treated as a PV node with the reference angle provided by the inverter which stands at zero, as shown in Table 6.14. During steady-state operation, the power losses incurred by the rectifier and inverter converters are 16.5 kW and 24.4 kW, respectively, whereas the DC link power loss stands at 173.7 kW.

To show that the VSC-HVDC link can provide dynamic frequency regulation to low-inertia AC networks, the load is increased by 5% at $t=0.5$ s. Two cases are considered below, when the HVDC link is set to provide frequency support and when it is not. Because of the momentary power imbalances, a rearrangement of power flows takes place, causing voltage oscillations and frequency deviations in the AC grid, as shown in Figures 6.47 and 6.48, respectively.

The simulated event leads to temporary frequency deviations of approximately 1 Hz when the VSC-HVDC does not provide frequency support, given the low inertia of the

Table 6.14 VSC-HVDC results given by the power flow solution.

Converter	P_k, P_m (MW)	Q_k, Q_m (MVAR)	E_{DC} (p.u.)	m_a	φ (°)	LTC's tap
Rectifier	3.9175	0.1020	2.0932	0.7803	−2.8295	1.0007
Inverter	3.7030	3.4375	2.0000	0.8570	0	1.0165

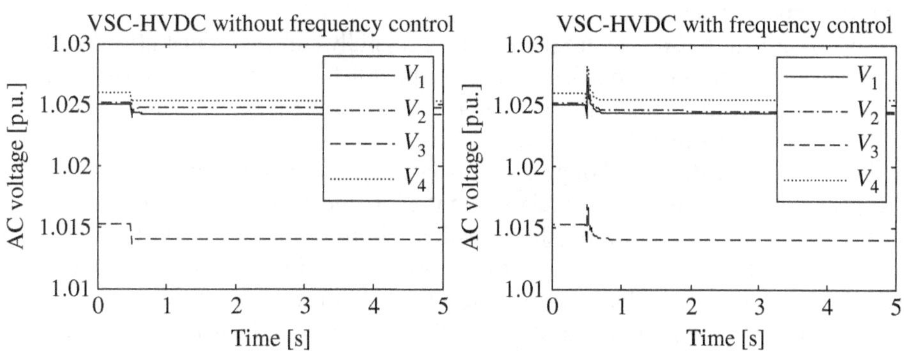

Figure 6.47 Voltage behaviour in the low-inertia AC network. Source: ©IEEE, 2015.

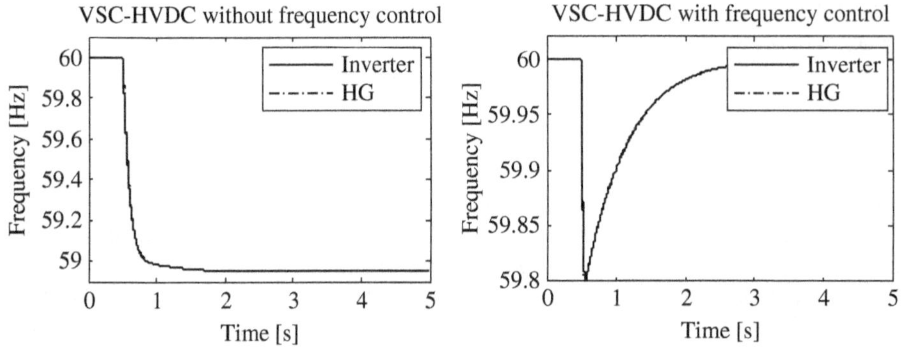

Figure 6.48 Frequency behaviour in the low-inertia AC network. Source: ©IEEE, 2015.

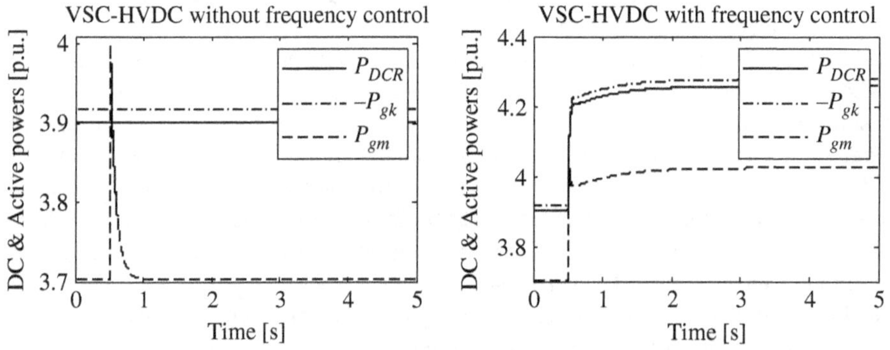

Figure 6.49 Performance of VSC-HVDC link's DC and AC powers. Source: ©IEEE, 2015.

network, as seen in Figure 6.48. However, when frequency control is exerted through the power converters, the frequency drops only to about 59.8 Hz. The power injected to the low-inertia grid, through the inverter, P_{gm}, helps to mitigate the temporary power imbalances brought about by the load increase.

Figure 6.49 shows the dynamic performances of the powers involved in the VSC-HVDC link. Notice that when the converters do not support the frequency, the powers P_{gk} and P_{DCR} are practically constant, with a short duration power peak in P_{gm}. This is the expected behaviour.

As shown in Figure 6.48, the frequency is successfully controlled just after 3 s of the occurrence of the perturbation. This is due to the quick response of both the rectifier and the inverter, which regulate the angular aperture γ_R and the DC current I_{DCI}, respectively, as shown in Figure 6.50.

It should be stressed that although γ_R controls the power being drawn from the main utility grid to provide frequency regulation to the low-inertia network, the current I_{DCI}, which is responsible for controlling the DC voltage, impacts directly on the performance of the DC power P_{DCR}. By examining the performance of the DC voltages shown in Figure 6.51, it is observed that the controller takes less time to bring E_{DCI} back to its nominal value than the time required to lead the frequency in the low-inertia grid back to its rated value, as shown in Figure 6.48.

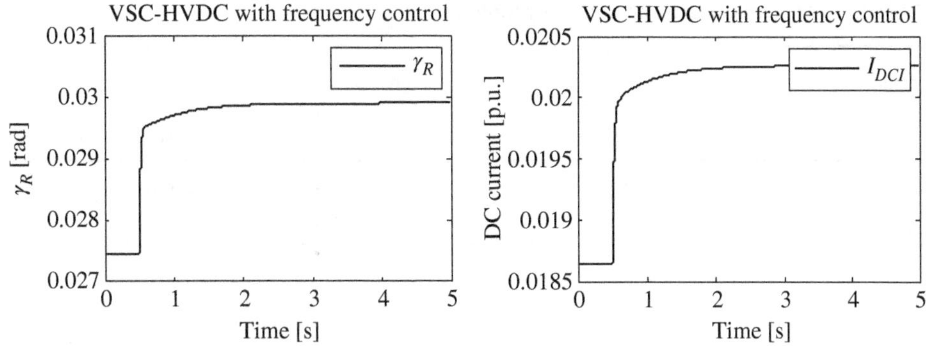

Figure 6.50 Dynamic performances of γ_R and I_{DCI}. Source: ©IEEE, 2015.

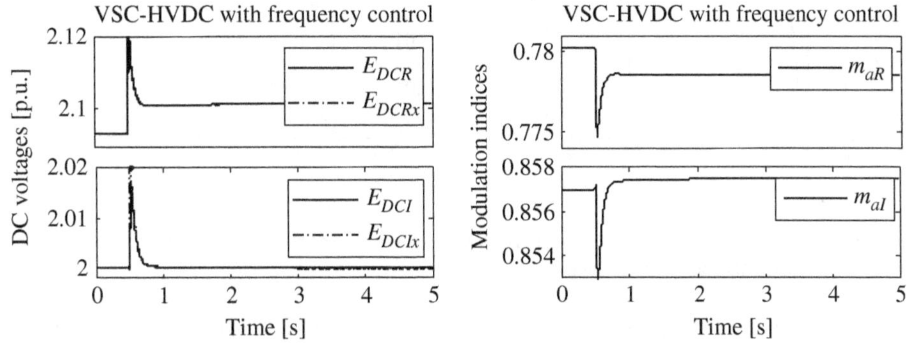

Figure 6.51 Performance of the DC voltages and modulation indices. Source: ©IEEE, 2015.

Small voltage oscillations appear in the low-inertia network because of the power imbalance, as shown in Figure 6.47. During the transient period, the voltage set point is achieved very quickly by the action of the modulation index controllers, as seen from Figure 6.51.

Parametric Analysis of the VSC-HVDC Link Feeding into a Low-Inertia AC Network The VSC-HVDC model provides frequency regulation to networks with near-zero inertia, as is the case of a system fitted with only one small hydro-generator, one wind generator and a fixed load. However, the value of the gains corresponding to the HVDC frequency controller plays a key role in determining the frequency behaviour of the low-inertia grid. To explore this point, a load increase of 5% occurring at $t = 0.5$ s is assessed for different values of gains in the frequency controller, as given in Table 6.15.

Figure 6.52 shows the dynamic performance of the frequency for different gain values in the frequency control loop, namely k_{ppdc} and k_{ipdc}. In cases (i) to (iii), the proportional gain k_{ppdc} is increased while the integral gain k_{ipdc} is kept constant; it stands out that the frequency f_I improves in terms of what is called the inertial response of the network, with the frequency deviation, just after the disturbance, narrowing from 0.56 Hz to 0.12 Hz.

Table 6.15 Different gains for the frequency controller.

Gains	Scenarios					
	(i)	(ii)	(iii)	(iv)	(v)	(vi)
k_{ppdc}	1e-4	15e-4	30e-4	15e-4	15e-4	15e-4
k_{ipdc}	25e-4	25e-4	25e-4	10e-4	30e-4	50e-4

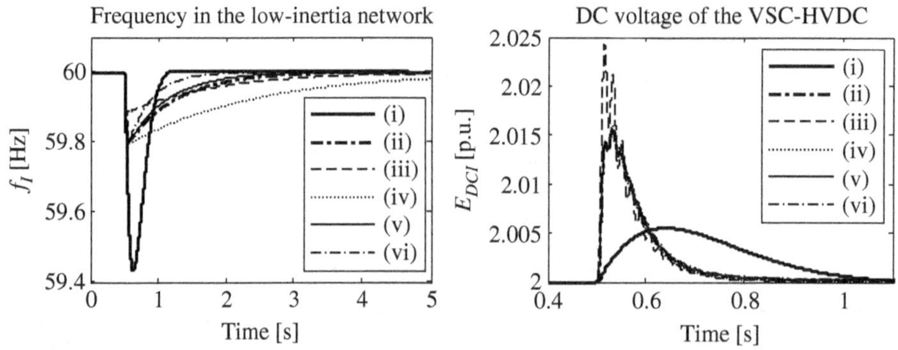

Figure 6.52 Frequency in the low-inertia network and DC voltage of the inverter for different gains of the frequency control loop. Source: ©IEEE, 2015.

For cases (iv) to (vi), k_{ppdc} is kept constant while k_{ipdc} increases gradually. This results in improving what is termed the primary frequency response given that a faster frequency recovery is achieved when increasing k_{ipdc}.

Notice that for the cases being simulated, the smaller the frequency deviations, the bigger the DC voltage offsets, as appreciated in Figure 6.52. The reason is that the power needed to bring the frequency back to its nominal value is injected faster into the network, leading to larger over-voltages in the DC link.

6.6 Modelling of Multi-terminal VSC-HVDC Systems for Dynamic Simulations

The construction and operation of a multi-terminal HVDC system represent a much larger challenge than for a point-to-point HVDC link. At this point, it becomes necessary to develop more versatile and accurate models of multi-terminal HVDC systems to aid power system engineers to confront the challenges relating to the planning, operation and control of such AC-DC systems.

In a multi-terminal VSC-HVDC arrangement, the key objective is to balance the powers of the different VSC stations to ensure a reliable operation of both the whole HVDC system and the various AC networks which connect to it. Power reallocation and control of the voltage profile on the meshed DC network are primary functions pursued in a multi-terminal scheme. The control system of each VSC station enables a rapid response to recover quickly after faults occurring either in the AC system or in the DC system. It is expected that all the benefits gained with the use of two-terminal HVDC systems would be passed onto the multi-terminal schemes. It is also expected that the great resilience exhibited by meshed AC networks, where the power supply to users is maintained even at times when the AC power grid is operating under stress, will be inherited by the multi-terminal VSC-HVDC systems.

At the planning stage, a holistic assessment of the multi-terminal VSC-HVDC system together with the interconnected AC networks will require the availability of suitable VSC models. System-wide dynamic studies of practical networks would benefit from adopting a modular approach to enable efficient simulations with sufficient accuracy of results. Moreover, the availability of essential state and control variables pertaining to the converter stations would be most useful and welcome. It was shown in Chapter 3 that the fundamental VSC station model possesses a high degree of modelling flexibility, where the aggregation of any number of VSC stations is carried out with relative ease to give rise to a generic multi-terminal arrangement [31]. The same modular philosophy is followed for the dynamic operating regime in this chapter.

6.6.1 Three-terminal VSC-HVDC Dynamic Model

Three types of converters were defined in Section 3.9 to conform to specific control strategies and pairing of AC networks: the slack converter controlling its DC voltage (VSC_{Slack}), the converter controlling its DC power (VSC_{Psch}) and the converter feeding into a passive network (VSC_{Pass}). Such a converter classification aimed at steady-state analysis is also suitable for the dynamic modelling of multi-terminal systems addressed in this chapter. To illustrate this point, the basic three-terminal VSC-HVDC system, depicted in Figure 6.53, which uses the three types of converters, is used. Then the three-terminal system is expanded quite naturally to build the model of a generic multi-terminal VSC-HVDC system. The ensuing framework makes provisions for any number of VSC units, which is commensurate with the number of terminals in the HVDC system, suitably accommodated in a unified frame-of-reference suitable for dynamic simulations of power networks using only its positive-sequence representation [32].

The AC-DC transmission system shown in Figure 6.53 comprises three VSC stations. For simplicity of representation, the associated phase reactor and AC filter capacitor of the converters have been omitted from the figure, but they are taken into account. On the DC side of each VSC, the DC bus connects to a network forming a meshed DC power grid with cables of resistance R_{DC4}, R_{DC5} and R_{DC6}. Notice that in a multi-terminal scheme, any converter can play the role of a rectifier or inverter to satisfy the power exchanges between the converters and the DC grid, as well as to satisfy the requirements of the AC network connected at their corresponding AC terminal. Therefore, the subscripts R and I, formerly standing for rectifier and inverter, respectively, have been changed to numbers.

In connection with Figure 6.53, the voltage and current of the capacitors and inductors, respectively, are computed by Eqs. (6.124)–(6.125):

$$\frac{dE_{DC1}}{dt} = \frac{-I_{DC1} - I_{1x}}{C_{DC1}}, \frac{dE_{DC2}}{dt} = \frac{-I_{DC2} - I_{2x}}{C_{DC2}}, \frac{dE_{DC3}}{dt} = \frac{-I_{DC3} - I_{3x}}{C_{DC3}} \tag{6.124}$$

$$\frac{dI_{1x}}{dt} = \frac{E_{DC1} - E_{DC4}}{L_{DC1}}, \frac{dI_{2x}}{dt} = \frac{E_{DC2} - E_{DC5}}{L_{DC2}}, \frac{dI_{3x}}{dt} = \frac{E_{DC3} - E_{DC6}}{L_{DC3}} \tag{6.125}$$

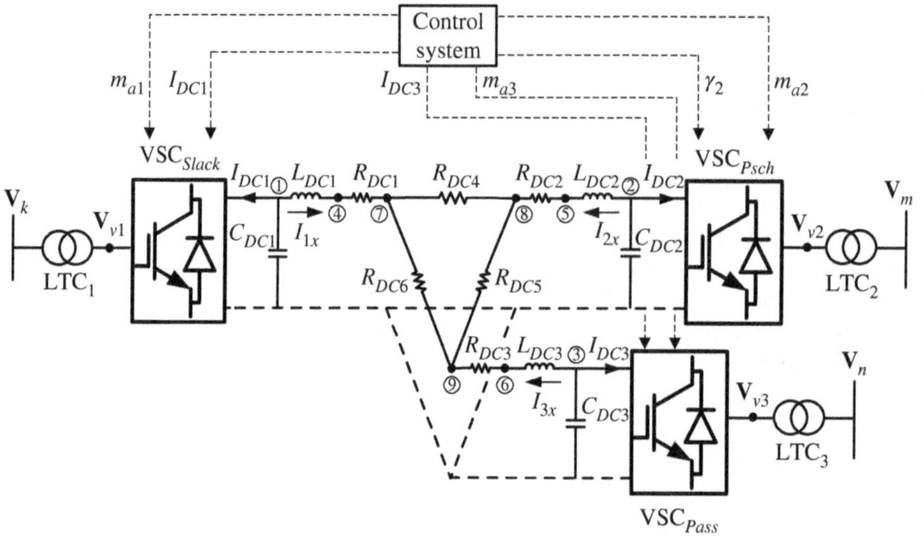

Figure 6.53 Representation of a three-terminal VSC-HVDC link with its control variables. Source: ©IEEE, 2016.

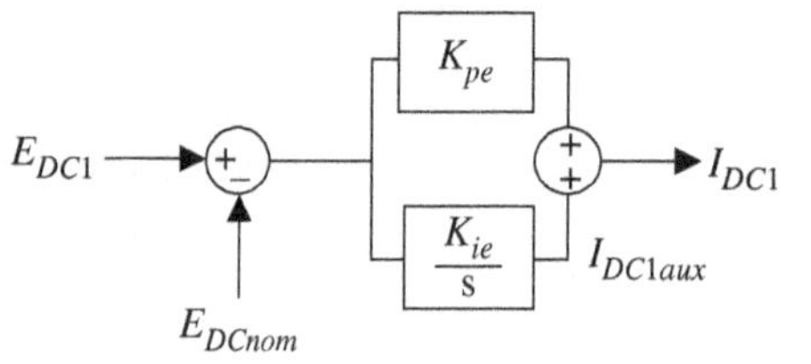

Figure 6.54 Dynamic controller for the DC voltage of the slack converter VSC_{Slack}. Source: ©IEEE, 2016.

Figure 6.54 shows the DC voltage controller for the slack converter VSC_{Slack}. The PI controller processes the error between the actual voltage of the converter's DC bus and its nominal DC voltage, E_{DC1} and E_{DCnom}, respectively. This results in a new value of the DC current I_{DC1}.

Simultaneously, the other two VSC stations are responsible for controlling the power at its corresponding DC bus, but their power control objectives are different from each other. The aim of the converter of type VSC_{Psch} is to achieve a fixed, scheduled DC power transfer. In contrast, the converter of type VSC_{Pass}, whose main goal is to provide frequency support to the network connected at its AC terminal, regulates the power injection into the passive network as a function of the power deviations in the AC grid. Figure 6.55 shows the control loops for these two converter stations.

Notice that the converter of type VSC_{Psch} aims at a power balance at its DC bus by regulating the angular aperture between its phase-shifting angle and its corresponding AC voltage angle. On the other hand, the converter of type VSC_{Pass} acts upon variations on the frequency measured at its AC terminal, increasing or decreasing its DC current depending on the actual operating conditions [32].

In two-terminal HVDC systems, the modulation index of each VSC station is responsible for controlling the voltage magnitudes at their corresponding AC terminal. In the case of the three-terminal VSC-based transmission system, there are three control loops aimed at controlling the voltage magnitudes at each AC terminal. Each control loop

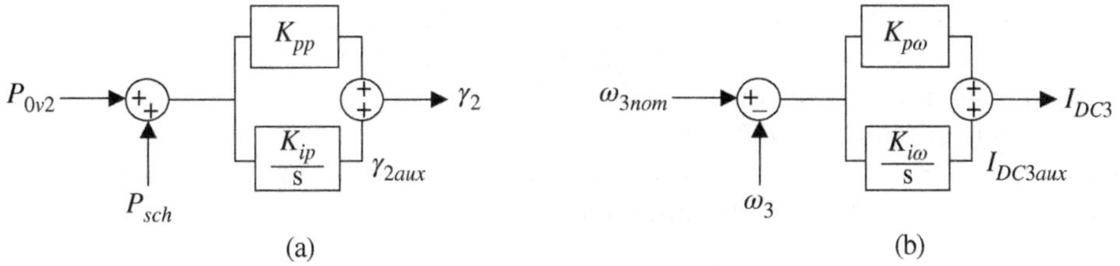

Figure 6.55 Source: ©IEEE, 2016. (a) DC power controller of the station VSC_{Psch}. (b) Frequency controller of the station VSC_{Pass}.

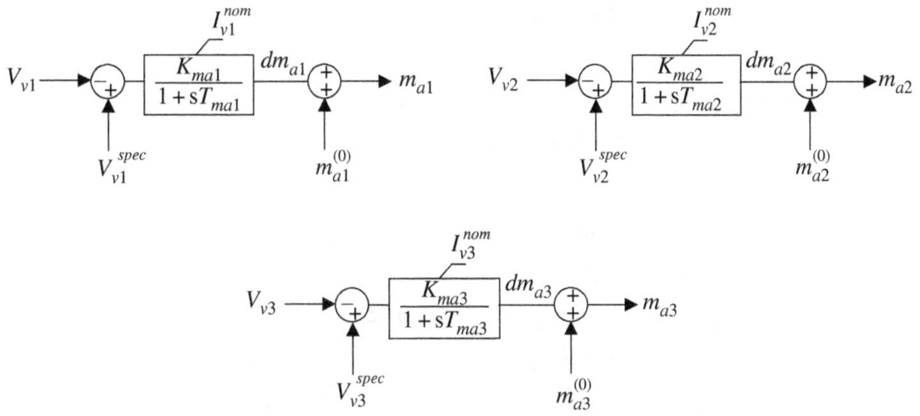

Figure 6.56 Modulation index controllers of the three-terminal HVDC system.

comprises a first-order control block whose objective is to readjust the corresponding modulation index according to the difference between the scheduled voltage magnitudes and the actual voltage at the nodes where the converters are connected, as shown in Figure 6.56.

Notice that each control loop acts autonomously, needing no feedback from the other two modulation index controllers – this implies that no communication is needed between converters in the multi-terminal system, as far as AC voltage control is concerned.

Unlike point-to-point HVDC connections where the resistor of the DC smoothing inductor can be simply added to the DC cable resistor, in a multi-terminal connection, the DC resistance of the DC smoothing inductor produces an additional node in the DC system. However, its inclusion may be carried out with no particular difficulty.

The calculated power injections at each node of the DC system are computed as shown in (6.126). For practical reasons, this expression only includes the computation of the power injections at the DC buses of a purely resistive network, which is formed from bus 4 to bus 9. It should be clear that during steady-state conditions, the voltages at the DC buses of the three converters equal the voltages at buses 4, 5 and 6, respectively.

$$P_{DCj,cal} = E_{DCj}^2 G_{DCjj} + E_{DCj} \sum_{\substack{j=4 \\ m \in j}}^{9} E_{DCm} G_{DCjm} \tag{6.126}$$

The three-terminal HVDC arrangement, shown in Figure 6.53, includes the most representative AC-DC components that significantly affect its dynamics. Reliable dynamic solutions call for the use of an approach similar to the one employed in point-to-point HVDC configurations. The differential equations, shown in (6.127)–(6.140), are derived from the voltage and current dynamics of the capacitors and inductors, as well as those of the controllers of the three-terminal VSC-HVDC model [32].

$$F_{E_{DC1}} = E_{DC1(t-\Delta t)} + 0.5\Delta t \dot{E}_{DC1(t-\Delta t)} - (E_{DC1(t)} - 0.5\Delta t \dot{E}_{DC1(t)}) \tag{6.127}$$

$$F_{E_{DC2}} = E_{DC2(t-\Delta t)} + 0.5\Delta t \dot{E}_{DC2(t-\Delta t)} - (E_{DC2(t)} - 0.5\Delta t \dot{E}_{DC2(t)}) \tag{6.128}$$

$$F_{E_{DC3}} = E_{DC3(t-\Delta t)} + 0.5\Delta t \dot{E}_{DC3(t-\Delta t)} - (E_{DC3(t)} - 0.5\Delta t \dot{E}_{DC3(t)}) \tag{6.129}$$

$$F_{I_{1x}} = I_{1x(t-\Delta t)} + 0.5\Delta t\, \dot{I}_{1x(t-\Delta t)} - (I_{1x(t)} - 0.5\Delta t\, \dot{I}_{1x(t)}) \tag{6.130}$$

$$F_{I_{2x}} = I_{2x(t-\Delta t)} + 0.5\Delta t\, \dot{I}_{2x(t-\Delta t)} - (I_{2x(t)} - 0.5\Delta t\, \dot{I}_{2x(t)}) \tag{6.131}$$

$$F_{I_{3x}} = I_{3x(t-\Delta t)} + 0.5\Delta t\, \dot{I}_{3x(t-\Delta t)} - (I_{3x(t)} - 0.5\Delta t\, \dot{I}_{3x(t)}) \tag{6.132}$$

$$F_{I_{dc1aux}} = I_{DC1aux(t-\Delta t)} + 0.5\Delta t \dot{I}_{DC1aux(t-\Delta t)} - (I_{DC1aux(t)} - 0.5\Delta t \dot{I}_{DC1aux(t)}) \tag{6.133}$$

$$F_{\gamma_{2aux}} = \gamma_{2aux(t-\Delta t)} + 0.5\Delta t \dot{\gamma}_{2aux(t-\Delta t)} - (\gamma_{2aux(t)} - 0.5\Delta t \dot{\gamma}_{2aux(t)}) \tag{6.134}$$

$$F_{I_{dc3aux}} = I_{DC3aux(t-\Delta t)} + 0.5\Delta t \dot{I}_{DC3aux(t-\Delta t)} - (I_{DC3aux(t)} - 0.5\Delta t \dot{I}_{DC3aux(t)}) \tag{6.135}$$

$$F_{\varphi_3} = \varphi_{3(t-\Delta t)} + 0.5\Delta t\, \dot{\varphi}_{3(t-\Delta t)} - (\varphi_{3(t)} - 0.5\Delta t\, \dot{\varphi}_{3(t)}) \tag{6.136}$$

$$F_{\omega_3} = \omega_{3(t-\Delta t)} + 0.5\Delta t\, \dot{\omega}_{3(t-\Delta t)} - (\omega_{3(t)} - 0.5\Delta t\, \dot{\omega}_{3(t)}) \tag{6.137}$$

$$F_{dm_{a1}} = dm_{a1(t-\Delta t)} + 0.5\Delta t\, d\dot{m}_{a1(t-\Delta t)} - (dm_{a1(t)} - 0.5\Delta t\, d\dot{m}_{a1(t)}) \tag{6.138}$$

$$F_{dm_{a2}} = dm_{a2(t-\Delta t)} + 0.5\Delta t\, d\dot{m}_{a2(t-\Delta t)} - (dm_{a2(t)} - 0.5\Delta t\, d\dot{m}_{a2(t)}) \tag{6.139}$$

$$F_{dm_{a3}} = dm_{a3(t-\Delta t)} + 0.5\Delta t\, d\dot{m}_{a3(t-\Delta t)} - (dm_{a3(t)} - 0.5\Delta t\, d\dot{m}_{a3(t)}) \tag{6.140}$$

The above set of expressions determines the whole dynamics in the three-terminal HVDC system. However, the formulation requires the AC and DC power balance equations to be completed. The active and reactive power mismatch equations at the AC terminal of each VSC unit are given by:

$$\Delta P_{v1} = -P_{v1} - P_{v1,load} - P_{v1,cal} \tag{6.141}$$

$$\Delta Q_{v1} = -Q_{v1} - Q_{v1,load} - Q_{v1,cal} \tag{6.142}$$

$$\Delta P_{v2} = -P_{v2} - P_{v2,load} - P_{v2,cal} \tag{6.143}$$

$$\Delta Q_{v2} = -Q_{v2} - Q_{v2,load} - Q_{v2,cal} \tag{6.144}$$

$$\Delta P_{v3} = -P_{v3} - P_{v3,load} - P_{v3,cal} \tag{6.145}$$

$$\Delta Q_{v3} = -Q_{v3} - Q_{v3,load} - Q_{v3,cal} \tag{6.146}$$

With reference to Figure 6.53, the purely resistive network, formed by nodes 7, 8 and 9, produces the mismatch Eqs. (6.150)–(6.152). Furthermore, the coupling between this resistive network and the DC side of the power converters is given by Eqs. (6.147)–(6.149), corresponding to DC nodes 4, 5 and 6.

$$\Delta P_{DC4} = E_{DC4} I_{1x} - P_{DC4,cal} \tag{6.147}$$

$$\Delta P_{DC5} = E_{DC5} I_{2x} - P_{DC5,cal} \tag{6.148}$$

$$\Delta P_{DC6} = E_{DC6} I_{3x} - P_{DC6,cal} \tag{6.149}$$

$$\Delta P_{DC7} = -P_{d7} - P_{DC7,cal} \tag{6.150}$$

$$\Delta P_{DC8} = -P_{d8} - P_{DC8,cal} \tag{6.151}$$

$$\Delta P_{DC9} = -P_{d9} - P_{DC9,cal} \tag{6.152}$$

$$\Delta P_{DC1} = E_{DC1} I_{DC1} - P_{0v1} \tag{6.153}$$

where the powers P_{d7}, P_{d8} and P_{d9} correspond to the powers drawn by the DC loads at the nodes that make up the delta circuit of DC cables. The expressions $P_{DC4,cal}$ to $P_{DC9,cal}$ represent the calculated powers at the corresponding nodes, which are computed using (6.126). In addition to the above expressions, the Eq. (6.153) enables the computation of the power exchange between the AC and DC sides of the slack converter, VSC_{Slack}.

The last equation suggests that the internal power equilibrium of the slack converter is attained 'instantaneously', something that does not occur in converters of type VSC_{Psch} and VSC_{Pass}. In these two types of converters, whose aim is to provide power regulation and frequency support, the internal power equilibrium is reached with a time delay, imposed by the speed of response of their corresponding dynamic controllers, depicted in Figure 6.55. Such a time delay causes a momentary energy mismatch, charging and discharging of the capacitors and inductors of the DC grid. Therefore it can be inferred that the time response of these two types of controllers will directly impact the magnitude of the DC voltage deviations when the AC-DC system undergoes a disturbance.

The overall three-terminal HVDC model for dynamic simulations is defined by Eqs. (6.127)–(6.153), with the linearized form shown in (6.154). The unified frame-of-reference suitably combines the whole set of discretized differential equations and algebraic equations arising from the AC-DC system formed by the three VSC stations [32]:

$$\begin{bmatrix} \mathbf{F}_{\mathbf{Slack}} \\ \mathbf{F}_{\mathbf{Psch}} \\ \mathbf{F}_{\mathbf{Pass}} \\ \mathbf{F}_{\mathbf{DC}} \end{bmatrix}^i = -\begin{bmatrix} \mathbf{J}_{\mathbf{Slack}} & 0 & 0 & \mathbf{J}_{\mathbf{Sdc}} \\ 0 & \mathbf{J}_{\mathbf{Psch}} & 0 & \mathbf{J}_{\mathbf{Psdc}} \\ 0 & 0 & \mathbf{J}_{\mathbf{Pass}} & \mathbf{J}_{\mathbf{Padc}} \\ \mathbf{J}_{\mathbf{dcS}} & \mathbf{J}_{\mathbf{dcPs}} & \mathbf{J}_{\mathbf{dcPa}} & \mathbf{J}_{\mathbf{dc}} \end{bmatrix}^i \begin{bmatrix} \mathbf{\Delta\Phi}_{\mathbf{Slack}} \\ \mathbf{\Delta\Phi}_{\mathbf{Psch}} \\ \mathbf{\Delta\Phi}_{\mathbf{Pass}} \\ \mathbf{\Delta\Phi}_{\mathbf{DC}} \end{bmatrix}^i \tag{6.154}$$

where **0** is a zero-padded matrix of suitable order and the other entries of the Jacobian matrix are:

$$\mathbf{J}_{\mathbf{Slack}} = \begin{bmatrix} \mathbf{J}'_{\mathbf{Slack}} \\ \mathbf{J}''_{\mathbf{Slack}} \end{bmatrix}, \mathbf{J}_{\mathbf{Psch}} = \begin{bmatrix} \mathbf{J}'_{\mathbf{Psch}} \\ \mathbf{J}''_{\mathbf{Psch}} \end{bmatrix}, \mathbf{J}_{\mathbf{Pass}} = \begin{bmatrix} \mathbf{J}'_{\mathbf{Pass}} \\ \mathbf{J}''_{\mathbf{Pass}} \end{bmatrix},$$

$$\mathbf{J}_{\mathbf{Sdc}} = \begin{bmatrix} \mathbf{0} & \mathbf{J}'_{\mathbf{Sdc}} & \mathbf{0} \end{bmatrix}, \mathbf{J}_{\mathbf{Psdc}} = \begin{bmatrix} \mathbf{0} & \mathbf{J}'_{\mathbf{Psdc}} & \mathbf{0} \end{bmatrix}, \mathbf{J}_{\mathbf{Padc}} = \begin{bmatrix} \mathbf{0} & \mathbf{J}'_{\mathbf{Padc}} & \mathbf{0} \end{bmatrix}, \mathbf{J}_{\mathbf{dcS}} = \begin{bmatrix} \mathbf{0} & \mathbf{J}'_{\mathbf{dcS}} & \mathbf{0} \end{bmatrix}^T,$$

$$\mathbf{J}_{\mathbf{dcPs}} = \begin{bmatrix} \mathbf{0} & \mathbf{J}'_{\mathbf{dcPs}} & \mathbf{0} \end{bmatrix}^T, \mathbf{J}_{\mathbf{dcPa}} = \begin{bmatrix} \mathbf{0} & \mathbf{J}'_{\mathbf{dcPa}} & \mathbf{0} \end{bmatrix}^T \tag{6.155}$$

Matrices $\mathbf{J'}_{\mathbf{Slack}}$, $\mathbf{J'}_{\mathbf{Psch}}$ and $\mathbf{J'}_{\mathbf{Pass}}$ accommodate the derivative terms of the AC power mismatches of the three types of converters with respect to their corresponding AC state variables and control variables. Notice that $\mathbf{J'}_{\mathbf{Pass}}$ also makes provision for the derivatives with respect to the electrical reference variables of the passive network, φ_3 and ω_3. On the other hand, the derivative terms of the discretized differential equations arising from the dynamic controllers of the VSC stations with respect to their corresponding AC state variables and control variables are suitably accommodated in matrices $\mathbf{J''}_{\mathbf{Slack}}$, $\mathbf{J''}_{\mathbf{Psch}}$ and $\mathbf{J''}_{\mathbf{Pass}}$.

$$\mathbf{J}'_{\mathbf{Slack}} = \begin{bmatrix} \frac{\partial \Delta P_{v1}}{\partial e_{v1}} & \frac{\partial \Delta P_{v1}}{\partial f_{v1}} & \frac{\partial \Delta P_{v1}}{\partial \phi_1} & 0 & \frac{\partial \Delta P_{v1}}{\partial dm_{a1}} \\ \frac{\partial \Delta Q_{v1}}{\partial e_{v1}} & \frac{\partial \Delta Q_{v1}}{\partial f_{v1}} & \frac{\partial \Delta Q_{v1}}{\partial \phi_1} & 0 & \frac{\partial \Delta Q_{v1}}{\partial dm_{a1}} \\ \frac{\partial \Delta P_{dc1}}{\partial e_{v1}} & \frac{\partial \Delta P_{dc1}}{\partial f_{v1}} & \frac{\partial \Delta P_{dc1}}{\partial \phi_1} & 0 & \frac{\partial \Delta P_{dc1}}{\partial dm_{a1}} \end{bmatrix},$$

$$\mathbf{J}''_{\mathbf{Slack}} = \begin{bmatrix} 0 & 0 & 0 & \frac{\partial F_{I_{DC1aux}}}{\partial I_{DC1aux}} & 0 \\ \frac{\partial F_{dm_{a1}}}{\partial e_{v1}} & \frac{\partial F_{dm_{a1}}}{\partial f_{v1}} & 0 & 0 & \frac{\partial F_{dm_{a1}}}{\partial dm_{a1}} \end{bmatrix} \tag{6.156}$$

$$\mathbf{J}'_{\mathbf{Psch}} = \begin{bmatrix} \frac{\partial \Delta P_{v2}}{\partial e_{v2}} & \frac{\partial \Delta P_{v2}}{\partial f_{v2}} & 0 & \frac{\partial \Delta P_{v2}}{\partial dm_{a2}} \\ \frac{\partial \Delta Q_{v2}}{\partial e_{v2}} & \frac{\partial \Delta Q_{v2}}{\partial f_{v2}} & 0 & \frac{\partial \Delta Q_{v2}}{\partial dm_{a2}} \end{bmatrix},$$

$$\mathbf{J}''_{\mathbf{Psch}} = \begin{bmatrix} \frac{\partial F_{\gamma_{2aux}}}{\partial e_{v2}} & \frac{\partial F_{\gamma_{2aux}}}{\partial f_{v2}} & \frac{\partial F_{\gamma_{2aux}}}{\partial \gamma_{2aux}} & \frac{\partial F_{\gamma_{2aux}}}{\partial dm_{a2}} \\ \frac{\partial F_{dm_{a2}}}{\partial e_{v2}} & \frac{\partial F_{dm_{a2}}}{\partial f_{v2}} & 0 & \frac{\partial F_{dm_{a2}}}{\partial dm_{a2}} \end{bmatrix} \tag{6.157}$$

$$\mathbf{J}'_{\mathbf{Pass}} = \begin{bmatrix} \frac{\partial \Delta P_{v3}}{\partial e_{v3}} & \frac{\partial \Delta P_{v3}}{\partial f_{v3}} & 0 & \frac{\partial \Delta P_{v3}}{\partial dm_{a3}} & 0 & \frac{\partial \Delta P_{v3}}{\partial \varphi_3} \\ \frac{\partial \Delta Q_{v3}}{\partial e_{v3}} & \frac{\partial \Delta Q_{v3}}{\partial f_{v3}} & 0 & \frac{\partial \Delta Q_{v3}}{\partial dm_{a3}} & 0 & \frac{\partial \Delta Q_{v3}}{\partial \varphi_3} \end{bmatrix}$$

$$\mathbf{J}''_{\mathbf{Pass}} = \begin{bmatrix} 0 & 0 & \frac{\partial F_{I_{DC3aux}}}{\partial I_{DC3aux}} & 0 & 0 & 0 \\ \frac{\partial F_{dm_{a3}}}{\partial e_{v3}} & \frac{\partial F_{dm_{a3}}}{\partial f_{v3}} & 0 & \frac{\partial F_{dm_{a3}}}{\partial dm_{a3}} & 0 & 0 \\ \frac{\partial F_{\omega_3}}{\partial e_{v3}} & \frac{\partial F_{\omega_3}}{\partial f_{v3}} & 0 & \frac{\partial F_{\omega_3}}{\partial dm_{a3}} & \frac{\partial F_{\omega_3}}{\partial \omega_3} & \frac{\partial F_{\omega_3}}{\partial \varphi_3} \\ 0 & 0 & 0 & 0 & \frac{\partial F_{\varphi_3}}{\partial \omega_3} & \frac{\partial F_{\varphi_3}}{\partial \varphi_3} \end{bmatrix} \tag{6.158}$$

The terms arising from deriving both the AC power mismatches and the discretized differential equations of the dynamic controllers of the three VSC stations with respect to the capacitor's DC voltage are located in matrices $\mathbf{J}'_{\mathbf{Sdc}}$, $\mathbf{J}'_{\mathbf{Psdc}}$ and $\mathbf{J}'_{\mathbf{Padc}}$. Vectors $\mathbf{J}'_{\mathbf{dcS}}$, $\mathbf{J}'_{\mathbf{dcPs}}$ and $\mathbf{J}'_{\mathbf{dcPa}}$ contain the partial derivatives of the discretized differential equations of the converters' DC voltage dynamics (voltage equations of the capacitors) with respect to their corresponding AC state variables and control variables.

$$\mathbf{J}'_{\mathbf{Sdc}} = \begin{bmatrix} \dfrac{\partial \Delta P_{v1}}{\partial E_{DC1}} \\ \dfrac{\partial \Delta Q_{v1}}{\partial E_{DC1}} \\ \dfrac{\partial \Delta P_{DC1}}{\partial E_{DC1}} \\ \dfrac{\partial F_{I_{DC1aux}}}{\partial E_{DC1}} \\ 0 \end{bmatrix}, \mathbf{J}'_{\mathbf{Psdc}} = \begin{bmatrix} \dfrac{\partial \Delta P_{v2}}{\partial E_{DC2}} \\ \dfrac{\partial \Delta Q_{v2}}{\partial E_{DC2}} \\ \dfrac{\partial F_{\gamma_{2aux}}}{\partial E_{DC2}} \\ 0 \end{bmatrix}, \mathbf{J}'_{\mathbf{Padc}} = \begin{bmatrix} \dfrac{\partial \Delta P_{v3}}{\partial E_{DC3}} \\ \dfrac{\partial \Delta Q_{v3}}{\partial E_{DC3}} \\ 0 \\ 0 \\ \dfrac{\partial F_{\omega_3}}{\partial E_{DC3}} \\ 0 \end{bmatrix}, \mathbf{J}'_{\mathbf{dcS}} = \begin{bmatrix} \dfrac{\partial F_{E_{DC1}}}{\partial e_{v1}} \\ \dfrac{\partial F_{E_{DC1}}}{\partial f_{v1}} \\ \dfrac{\partial F_{E_{DC1}}}{\partial \varphi_1} \\ 0 \\ \dfrac{\partial F_{E_{DC1}}}{\partial dm_{a1}} \end{bmatrix}^T,$$

$$\mathbf{J}'_{\mathbf{dcPs}} = \begin{bmatrix} \dfrac{\partial F_{E_{DC2}}}{\partial e_{v2}} \\ \dfrac{\partial F_{E_{DC2}}}{\partial f_{v2}} \\ \dfrac{\partial F_{E_{DC2}}}{\partial \gamma_{2aux}} \\ \dfrac{\partial F_{E_{DC2}}}{\partial dm_{a2}} \end{bmatrix}^T, \mathbf{J}'_{\mathbf{dcPa}} = \begin{bmatrix} \dfrac{\partial F_{E_{DC3}}}{\partial e_{v3}} \\ \dfrac{\partial F_{E_{DC3}}}{\partial f_{v3}} \\ 0 \\ \dfrac{\partial F_{E_{DC3}}}{\partial dm_{a3}} \\ 0 \\ 0 \end{bmatrix}^T \tag{6.159}$$

Matrix $\mathbf{J_{dc}}$, shown in (6.160), is further subdivided into several matrix blocks to illustrate further the variables involved. This matrix represents the coupling between the DC circuit and the three VSC stations. It contains the derivatives of the discretized differential equations of inductors currents and capacitors voltages together with the nodal power mismatches of the internal DC nodes, 4 to 9, with respect to suitable DC state and control variables.

$$\mathbf{J_{dc}} = \begin{bmatrix} \mathbf{J_{dc1}} & \mathbf{J_{dc2}} & \mathbf{J_{dc5}} \\ \mathbf{J_{dc3}} & \mathbf{J_{dc4}} & 0 \\ \mathbf{J_{dc6}} & 0 & \mathbf{J_{dc7}} \end{bmatrix} \tag{6.160}$$

$$\mathbf{J_{dc1}} = \begin{bmatrix} \dfrac{\partial F_{I_{1x}}}{\partial I_{1x}} & 0 & 0 \\ 0 & \dfrac{\partial F_{I_{2x}}}{\partial I_{2x}} & 0 \\ 0 & 0 & \dfrac{\partial F_{I_{3x}}}{\partial I_{3x}} \end{bmatrix}, \mathbf{J_{dc2}} = \begin{bmatrix} \dfrac{\partial F_{I_{1x}}}{\partial E_{dc1}} & 0 & 0 \\ 0 & \dfrac{\partial F_{I_{2x}}}{\partial E_{dc2}} & 0 \\ 0 & 0 & \dfrac{\partial F_{I_{3x}}}{\partial E_{dc3}} \end{bmatrix},$$

$$\mathbf{J_{dc3}} = \begin{bmatrix} \dfrac{\partial F_{E_{DC1}}}{\partial I_{1x}} & 0 & 0 \\ 0 & \dfrac{\partial F_{E_{DC2}}}{\partial I_{2x}} & 0 \\ 0 & 0 & \dfrac{\partial F_{E_{DC3}}}{\partial I_{3x}} \end{bmatrix} \tag{6.161}$$

$$\mathbf{J_{dc4}} = \begin{bmatrix} \frac{\partial F_{E_{DC1}}}{\partial E_{DC1}} & 0 & 0 \\ 0 & \frac{\partial F_{E_{DC2}}}{\partial E_{DC2}} & 0 \\ 0 & 0 & \frac{\partial F_{E_{DC3}}}{\partial E_{DC3}} \end{bmatrix}, \mathbf{J'_{dc5}} = \begin{bmatrix} \frac{\partial F_{I_{1x}}}{\partial E_{DC4}} & 0 & 0 \\ 0 & \frac{\partial F_{I_{2x}}}{\partial E_{DC5}} & 0 \\ 0 & 0 & \frac{\partial F_{I_{3x}}}{\partial E_{DC6}} \end{bmatrix},$$

$$\mathbf{J'_{dc6}} = \begin{bmatrix} \frac{\partial \Delta P_{DC4}}{\partial I_{1x}} & 0 & 0 \\ 0 & \frac{\partial \Delta P_{DC5}}{\partial I_{2x}} & 0 \\ 0 & 0 & \frac{\partial \Delta P_{DC6}}{\partial I_{3x}} \end{bmatrix} \tag{6.162}$$

$$\mathbf{J_{dc7}} = \begin{bmatrix} \frac{\partial \Delta P_{DC4}}{\partial E_{DC4}} & 0 & 0 & \frac{\partial \Delta P_{DC4}}{\partial E_{DC7}} & 0 & 0 \\ 0 & \frac{\partial \Delta P_{DC5}}{\partial E_{DC5}} & 0 & 0 & \frac{\partial \Delta P_{DC5}}{\partial E_{DC8}} & 0 \\ 0 & 0 & \frac{\partial \Delta P_{DC6}}{\partial E_{DC6}} & 0 & 0 & \frac{\partial \Delta P_{DC6}}{\partial E_{DC9}} \\ \frac{\partial \Delta P_{DC7}}{\partial E_{DC4}} & 0 & 0 & \frac{\partial \Delta P_{DC7}}{\partial E_{DC7}} & \frac{\partial \Delta P_{DC7}}{\partial E_{DC8}} & \frac{\partial \Delta P_{DC7}}{\partial E_{DC9}} \\ 0 & \frac{\partial \Delta P_{DC8}}{\partial E_{DC5}} & 0 & \frac{\partial \Delta P_{DC8}}{\partial E_{DC7}} & \frac{\partial \Delta P_{DC8}}{\partial E_{DC8}} & \frac{\partial \Delta P_{DC8}}{\partial E_{DC9}} \\ 0 & 0 & \frac{\partial \Delta P_{DC9}}{\partial E_{DC6}} & \frac{\partial \Delta P_{DC9}}{\partial E_{DC7}} & \frac{\partial \Delta P_{DC9}}{\partial E_{DC8}} & \frac{\partial \Delta P_{DC9}}{\partial E_{DC9}} \end{bmatrix} \tag{6.163}$$

where $\mathbf{J_{dc5}} = \begin{bmatrix} \mathbf{J'_{dc5}} & \mathbf{0} \end{bmatrix}$ and $\mathbf{J_{dc6}} = \begin{bmatrix} \mathbf{J'_{dc6}} & \mathbf{0} \\ \mathbf{0} & \mathbf{0} \end{bmatrix}$.

Notice that the diagonal matrices shown in (6.161) and (6.162), whose orders correspond to the number of VSC stations, may be expanded with ease to conform to any kind of multi-terminal arrangement of converters.

The mismatch vectors and increments of the state and control variables for the three VSC units and the DC circuit are:

$$\begin{aligned} \mathbf{F_{Slack}} &= \begin{bmatrix} \Delta P_{v1} & \Delta Q_{v1} & \Delta P_{DC1} & F_{I_{DC1aux}} & F_{dm_{a1}} \end{bmatrix}^T \\ \mathbf{F_{Psch}} &= \begin{bmatrix} \Delta P_{v2} & \Delta Q_{v2} & F_{\gamma_{2aux}} & F_{dm_{a2}} \end{bmatrix}^T \\ \mathbf{F_{Pass}} &= \begin{bmatrix} \Delta P_{v3} & \Delta Q_{v3} & F_{I_{DC3aux}} & F_{dm_{a3}} & F_{\omega_3} & F_{\phi_3} \end{bmatrix}^T \\ \mathbf{F_{DC}} &= \begin{bmatrix} F_{I_{1x}} & F_{I_{2x}} & F_{I_{3x}} & F_{E_{DC1}} & F_{E_{DC2}} & F_{E_{DC3}} & \cdots \\ \cdots & \Delta P_{DC4} & \Delta P_{DC5} & \Delta P_{DC6} & \Delta P_{DC7} & \Delta P_{DC8} & \Delta P_{DC9} \end{bmatrix}^T \end{aligned} \tag{6.164}$$

$$\begin{aligned}
\mathbf{\Delta\Phi_{Slack}} &= \left[\Delta e_{v1}\ \Delta f_{v1}\ \Delta\varphi_1\ \Delta I_{DC1aux}\ \Delta dm_{a1}\right]^T \\
\mathbf{\Delta\Phi_{Psch}} &= \left[\Delta e_{v2}\ \Delta f_{v2}\ \Delta\gamma_{2aux}\ \Delta dm_{a2}\right]^T \\
\mathbf{\Delta\Phi_{Pass}} &= \left[\Delta e_{v3}\ \Delta f_{v3}\ \Delta I_{DC3aux}\ \Delta dm_{a3}\ \Delta\omega_3\ \Delta\varphi_3\right]^T \\
\mathbf{\Delta\Phi_{DC}} &= \begin{bmatrix} \Delta I_{1x} & \Delta I_{2x} & \Delta I_{3x} & \Delta E_{DC1} & \Delta E_{DC2} & \Delta E_{DC3} & \dots \\ \dots & \Delta E_{DC4} & \Delta E_{DC5} & \Delta E_{DC6} & \Delta E_{DC7} & \Delta E_{DC8} & \Delta E_{DC9} \end{bmatrix}^T
\end{aligned} \tag{6.165}$$

Notice that vector $\mathbf{\Delta\Phi_{DC}}$ contains the DC voltages as state variables as well as the DC currents of the inductors. The increments of the state variables of the converters and DC circuit, calculated at iteration i, are used to update the state variables as follows:

$$\begin{aligned}
\mathbf{\Phi}^{i+1}_{\mathbf{Slack}} &= \mathbf{\Phi}^{i}_{\mathbf{Slack}} + \mathbf{\Delta\Phi}^{i}_{\mathbf{Slack}} \\
\mathbf{\Phi}^{i+1}_{\mathbf{Psch}} &= \mathbf{\Phi}^{i}_{\mathbf{Psch}} + \mathbf{\Delta\Phi}^{i}_{\mathbf{Psch}} \\
\mathbf{\Phi}^{i+1}_{\mathbf{Pass}} &= \mathbf{\Phi}^{i}_{\mathbf{Pass}} + \mathbf{\Delta\Phi}^{i}_{\mathbf{Pass}} \\
\mathbf{\Phi}^{i+1}_{\mathbf{DC}} &= \mathbf{\Phi}^{i}_{\mathbf{DC}} + \mathbf{\Delta\Phi}^{i}_{\mathbf{DC}}
\end{aligned} \tag{6.166}$$

6.6.2 Validation of the Three-Terminal VSC-HVDC Dynamic Model

The three-terminal VSC-HVDC system, shown in Figure 6.57, is used to carry out a comparison of the RMS-type model against an EMT-type model, which was assembled in the simulation environment afforded by the package SymPowerSystems of Simulink.

With no loss of generality, the AC sub networks connected to VSC_{Slack} and VSC_{Psch} are represented by equivalent networks (2000 MVA, 230 kV, 50 Hz), whereas the passive network fed by VSC_{Pass} is represented by a 50 MW load. Each VSC in this multi-terminal HVDC system is rated at 200 MVA, ±100 kV DC. The resistances of the DC cables, R_{DC4} to R_{DC6}, have value of 1.39×10^{-2} Ω/km, with the cables lengths being 75 km, 100 km and 150 km, respectively. The rest of the parameters, given on a 100 MVA base, are given in Table 6.16.

Initially the slack converter VSC_{Slack} controls its DC voltage at $E_{DCnom} = 2$ p.u., the converter VSC_{Psch} controls its DC power at $P_{sch} = 150$ MW and the converter VSC_{Pass} feeds the 50 MW load. The three converters controlling the reactive power flow to a null value, i.e. $Q_v^{ref} = 0$.

The simulation in Simulink is run up to $t = 1.5$ s to reach steady-state conditions. The following step changes in parameters are applied: (i) a 50 MW reduction of the scheduled power of the converter VSC_{Psch} is applied at $t = 1.5$ s, and (ii) the load fed by the converter VSC_{Pass} is increased by 40% (20 MW) at $t = 3$ s.

Both disturbances cause voltage variations in the DC network, as shown in Figure 6.58. The slack converter controls the voltage at its corresponding DC bus and impacts positively on the DC voltages of the converters VSC_{Psch} and VSC_{Pass}. It can be seen that the pattern of the dynamic response of the RMS-type multi-terminal VSC-HVDC model follows relatively well the pattern of the EMT-type model solution.

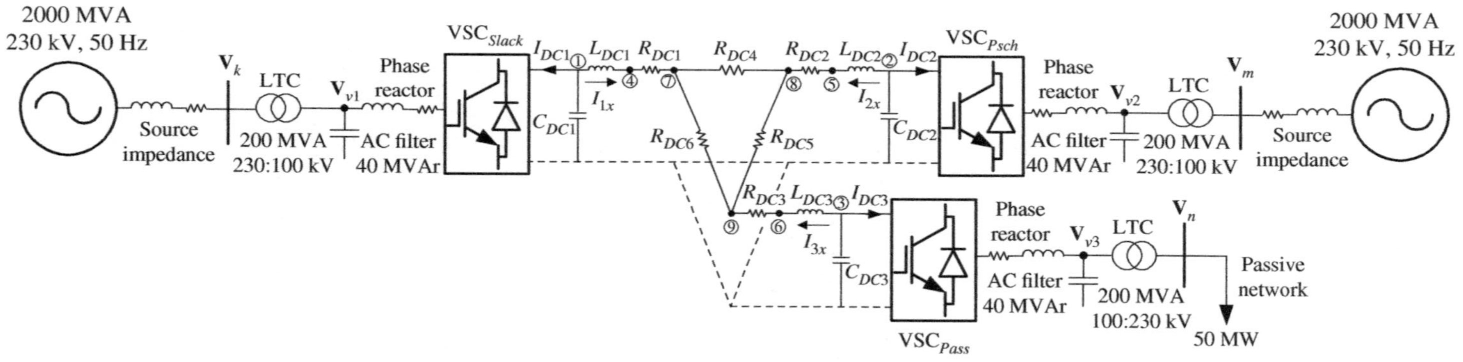

Figure 6.57 Three-terminal VSC-HVDC link used to carry out the validation test.

Table 6.16 Parameters of the three-terminal VSC-HVDC link.

Parameters of the three VSC	S_{nom}(p.u)	G_0 (p.u)	R (p.u)	X (p.u)	R_f (p.u)
	2.0	4e-3	0.0	0.0	7.5e-4
	X_f (p.u)	B_f (p.u)	R_{ltc}(p.u)	X_{ltc}(p.u)	H_c, H_i (s)
	0.075	0.40	2.5e-3	0.075	14e-3, 14e-3
Gains	K_{pe}, K_{ie}	K_{pp}, K_{ip}	$K_{p\omega}, K_{i\omega}$	K_{ma}	T_{ma}
VSC_{Slack}	1.5, 15	—	—	50.0	0.20
VSC_{Psch}	—	0.0, 5.0	—	50.0	0.20
VSC_{Pass}	—	—	0.025, 0.25	50.0	0.20

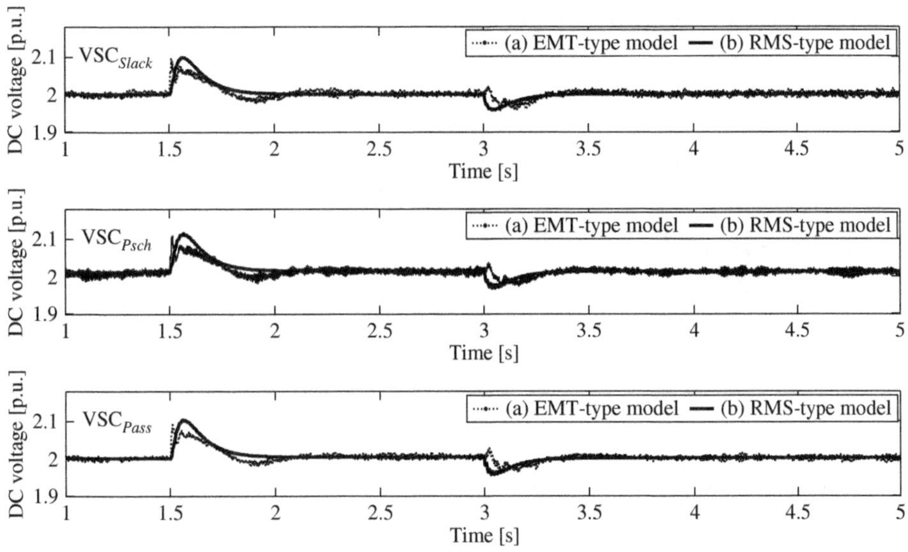

Figure 6.58 DC voltages of the VSC units comprising the three-terminal VSC-HVDC link.

As expected, the power reduction of 50 MW in P_{sch} of the converter VSC_{Psch} produces larger over-voltages than those experienced by the power increase of 20 MW in the load fed by the converter VSC_{Pass}. These DC voltage deviations affect the DC power performance of the three converters, as seen in Figure 6.59. It can be seen that some transient power peaks computed by the Simulink model during the applied step changes are not captured by the RMS-type model. The reason for this is the very different solution approaches taken by models. However, the rest of the DC power responses follow the same trend reasonably well.

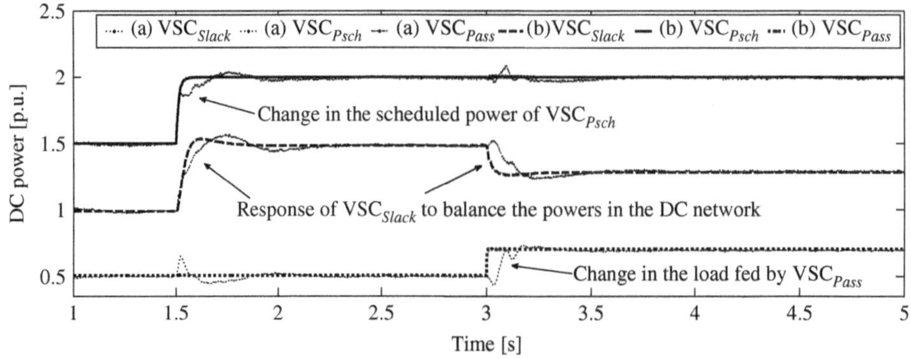

Figure 6.59 DC power behaviour at the DC bus of the three VSC stations. Source: ©IEEE, 2016. (a) EMT-type model. (b) Developed RMS-type model.

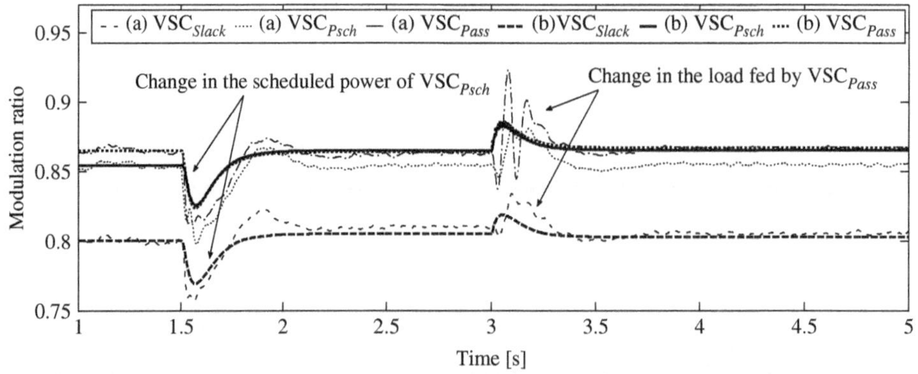

Figure 6.60 Modulation ratio of the three VSC stations. Source: ©IEEE, 2016. (a) EMT-type model. (b) RMS-type model.

The dynamic responses of the modulation indices are shown in Figure 6.60. The most noticeable differences between the two types of responses appear when there is a change in the load fed by the station VSC_{Pass}, i.e. the oscillations of the modulation ratio of VSC_{Pass} are larger than those calculated by the RMS-type model.

In terms of execution times, the RMS-type multi-terminal VSC-HVDC model outperforms by more than eight times the EMT-type model. The former takes 1.04 min to solve the three-terminal VSC-HVDC system whereas the latter takes 8.5 min.

6.6.3 Multi-Terminal VSC-HVDC Dynamic Model

The HVDC dynamic model of the three-terminal system, synthesized by Eq. (6.154), shows the way in which the three types of VSC stations are combined in a unified frame-of-reference, suitable for an iterative solution using the Newton-Raphson method at each time step.

Expanding this representation, a generalized multi-terminal arrangement comprising m converter stations of type VSC_{Slack}, n converter stations of type VSC_{Psch} and r converter stations of type VSC_{Pass} is shown below [32]:

$$\begin{bmatrix} \mathbf{F}_{\mathbf{Slack1}} \\ \vdots \\ \mathbf{F}_{\mathbf{Slack}m} \\ \mathbf{F}_{\mathbf{Psch1}} \\ \vdots \\ \mathbf{F}_{\mathbf{Psch}n} \\ \mathbf{F}_{\mathbf{Pass1}} \\ \vdots \\ \mathbf{F}_{\mathbf{Pass}r} \\ [\mathbf{F}_{\mathbf{DC}}] \end{bmatrix}^i = - \begin{bmatrix} \begin{matrix} \mathbf{J}_{\mathbf{Slack1}} & \cdots & \mathbf{0} \\ \vdots & \ddots & \vdots \\ \mathbf{0} & \cdots & \mathbf{J}_{\mathbf{Slack}m} \end{matrix} & \mathbf{0} & \mathbf{0} & [\mathbf{J}_{\mathbf{Sdc}}] \\ \mathbf{0} & \begin{matrix} \mathbf{J}_{\mathbf{Psch1}} & \cdots & \mathbf{0} \\ \vdots & \ddots & \vdots \\ \mathbf{0} & \cdots & \mathbf{J}_{\mathbf{Psch}n} \end{matrix} & \mathbf{0} & [\mathbf{J}_{\mathbf{Psdc}}] \\ \mathbf{0} & \mathbf{0} & \begin{matrix} \mathbf{J}_{\mathbf{Pass1}} & \cdots & \mathbf{0} \\ \vdots & \ddots & \vdots \\ \mathbf{0} & \cdots & \mathbf{J}_{\mathbf{Pass}r} \end{matrix} & [\mathbf{J}_{\mathbf{Padc}}] \\ [\mathbf{J}_{\mathbf{dcS}}] & [\mathbf{J}_{\mathbf{dcPs}}] & [\mathbf{J}_{\mathbf{dcPa}}] & [\mathbf{J}_{\mathbf{dc}}] \end{bmatrix}^i \times \begin{bmatrix} \boldsymbol{\Delta}_{\mathbf{Slack1}} \\ \vdots \\ \boldsymbol{\Delta}_{\mathbf{Slack}m} \\ \boldsymbol{\Delta}_{\mathbf{Psch1}} \\ \vdots \\ \boldsymbol{\Delta}_{\mathbf{Psch}n} \\ \boldsymbol{\Delta}_{\mathbf{Pass1}} \\ \vdots \\ \boldsymbol{\Delta}_{\mathbf{Pass}r} \\ [\boldsymbol{\Delta}_{\mathbf{DC}}] \end{bmatrix}^i \quad (6.167)$$

In expression (6.167), the mismatch terms $\mathbf{F}$ and the vectors of state variables increments $\boldsymbol{\Delta\Phi}$ take the forms given by (6.164)–(6.165) depending on the type of converter, i.e. VSC_{Slack}, VSC_{Psch} or VSC_{Pass}. Likewise, matrices $\mathbf{J}$ with subscripts **Slack**, **Psch** and **Pass** take the form given by (6.155).

Some caution must be exercised because the non-zero entries of vectors $\mathbf{J}_{\mathbf{Sdc}}$, $\mathbf{J}_{\mathbf{Psdc}}$, $\mathbf{J}_{\mathbf{Padc}}$, $\mathbf{J}_{\mathbf{dcS}}$, $\mathbf{J}_{\mathbf{dcPs}}$ and $\mathbf{J}_{\mathbf{dcPa}}$ need to be suitably staggered, as exemplified in the three-terminal case. Accordingly, the vectors and matrices entries, which have a location corresponding to any of the DC network variables, $\boldsymbol{\Delta\Phi}_{\mathbf{DC}}$, have the following structure:

$$[\mathbf{J}_{\mathbf{Sdc}}] = \begin{bmatrix} \mathbf{0} & \dots & \mathbf{0} & \mathbf{J}'_{\mathbf{Sdc1}} & \cdots & \mathbf{0} & \mathbf{0} & \dots & \mathbf{0} \\ \vdots & \ddots & \vdots & \vdots & \ddots & \vdots & \vdots & \ddots & \vdots \\ \mathbf{0} & \dots & \mathbf{0} & \mathbf{0} & \dots & \mathbf{J}'_{\mathbf{Sdc}m} & \mathbf{0} & \dots & \mathbf{0} \end{bmatrix},$$

$$[\mathbf{J}_{\mathbf{Psdc}}] = \begin{bmatrix} \mathbf{0} & \dots & \mathbf{0} & \mathbf{J}'_{\mathbf{Psdc1}} & \cdots & \mathbf{0} & \mathbf{0} & \dots & \mathbf{0} \\ \vdots & \ddots & \vdots & \vdots & \ddots & \vdots & \vdots & \ddots & \vdots \\ \mathbf{0} & \dots & \mathbf{0} & \mathbf{0} & \dots & \mathbf{J}'_{\mathbf{Psdc}n} & \mathbf{0} & \dots & \mathbf{0} \end{bmatrix},$$

$$[\mathbf{J}_{\mathbf{Padc}}] = \begin{bmatrix} \mathbf{0} & \dots & \mathbf{0} & \mathbf{J}'_{\mathbf{Padc1}} & \cdots & \mathbf{0} & \mathbf{0} & \dots & \mathbf{0} \\ \vdots & \ddots & \vdots & \vdots & \ddots & \vdots & \vdots & \ddots & \vdots \\ \mathbf{0} & \cdots & \mathbf{0} & \mathbf{0} & \cdots & \mathbf{J}'_{\mathbf{Padc}r} & \mathbf{0} & \cdots & \mathbf{0} \end{bmatrix}$$

$$[\mathbf{J}_{\mathbf{dcS}}] = \begin{bmatrix} \mathbf{0} & \dots & \mathbf{0} & \mathbf{J}'_{\mathbf{dcS}1} & \cdots & \mathbf{0} & \mathbf{0} & \dots & \mathbf{0} \\ \vdots & \ddots & \vdots & \vdots & \ddots & \vdots & \vdots & \ddots & \vdots \\ \mathbf{0} & \cdots & \mathbf{0} & \mathbf{0} & \cdots & \mathbf{J}'_{\mathbf{dcS}m} & \mathbf{0} & \cdots & \mathbf{0} \end{bmatrix}^T,$$

$$[\mathbf{J}_{\mathbf{dcPs}}] = \begin{bmatrix} \mathbf{0} & \dots & \mathbf{0} & \mathbf{J}'_{\mathbf{dcPs}1} & \cdots & \mathbf{0} & \mathbf{0} & \dots & \mathbf{0} \\ \vdots & \ddots & \vdots & \vdots & \ddots & \vdots & \vdots & \ddots & \vdots \\ \mathbf{0} & \cdots & \mathbf{0} & \mathbf{0} & \cdots & \mathbf{J}'_{\mathbf{dcPs}n} & \mathbf{0} & \cdots & \mathbf{0} \end{bmatrix}^T,$$

$$[\mathbf{J}_{\mathbf{dcPa}}] = \begin{bmatrix} \mathbf{0} & \dots & \mathbf{0} & \mathbf{J}'_{\mathbf{dcPa}1} & \cdots & \mathbf{0} & \mathbf{0} & \dots & \mathbf{0} \\ \vdots & \ddots & \vdots & \vdots & \ddots & \vdots & \vdots & \ddots & \vdots \\ \mathbf{0} & \cdots & \mathbf{0} & \mathbf{0} & \cdots & \mathbf{J}'_{\mathbf{dcPa}r} & \mathbf{0} & \cdots & \mathbf{0} \end{bmatrix}^T \tag{6.168}$$

The entries in (6.168) are vectors with the generic $j-th$ terms, depending on the number and type of converter, with $j = 1, \ldots, m$, $j = 1, \ldots, n$ and $j = 1, \ldots, r$ for converters VSC_{Slack}, VSC_{Psch} and VSC_{Pass}, respectively. They are given below:

$$\mathbf{J}'_{\mathbf{Sdc}} = \begin{bmatrix} \dfrac{\partial \Delta P_{vj}}{\partial E_{DCj}} \\ \dfrac{\partial \Delta Q_{vj}}{\partial E_{DCj}} \\ \dfrac{\partial \Delta P_{DCj}}{\partial E_{DCj}} \\ \dfrac{\partial F_{I_{DCjaux}}}{\partial E_{DCj}} \\ 0 \end{bmatrix}, \mathbf{J}'_{\mathbf{Psdc}} = \begin{bmatrix} \dfrac{\partial \Delta P_{vj}}{\partial E_{DCj}} \\ \dfrac{\partial \Delta Q_{vj}}{\partial E_{DCj}} \\ \dfrac{\partial E_{DCj}}{\partial F_{\gamma_{jaux}}} \\ \dfrac{\partial F_{\gamma_{jaux}}}{\partial E_{DCj}} \\ 0 \end{bmatrix}, \mathbf{J}'_{\mathbf{Padc}} = \begin{bmatrix} \dfrac{\partial \Delta P_{vj}}{\partial E_{DCj}} \\ \dfrac{\partial \Delta Q_{vj}}{\partial E_{DCj}} \\ 0 \\ 0 \\ \dfrac{\partial F_{\omega_j}}{\partial E_{DCj}} \\ 0 \end{bmatrix},$$

$$\mathbf{J}'_{\mathbf{dcS}} = \begin{bmatrix} \dfrac{\partial F_{E_{DCj}}}{\partial e_{vj}} \\ \dfrac{\partial F_{E_{DCj}}}{\partial f_{vj}} \\ \dfrac{\partial F_{E_{DCj}}}{\partial \varphi_j} \\ 0 \\ \dfrac{\partial F_{E_{DCj}}}{\partial dm_{aj}} \end{bmatrix}^T, \mathbf{J}'_{\mathbf{dcPs}} = \begin{bmatrix} \dfrac{\partial F_{E_{DCj}}}{\partial e_{vj}} \\ \dfrac{\partial F_{E_{DCj}}}{\partial f_{vj}} \\ \dfrac{\partial F_{E_{DCj}}}{\partial \gamma_{jaux}} \\ \dfrac{\partial F_{E_{DCj}}}{\partial dm_{aj}} \end{bmatrix}^T, \mathbf{J}'_{\mathbf{dcPa}} = \begin{bmatrix} \dfrac{\partial F_{E_{DCj}}}{\partial e_{vj}} \\ \dfrac{\partial F_{E_{DCj}}}{\partial f_{vj}} \\ 0 \\ \dfrac{\partial F_{E_{DCj}}}{\partial dm_{aj}} \\ 0 \\ 0 \end{bmatrix}^T \tag{6.169}$$

The Jacobian matrix of the DC network takes the general form of Eq. (6.170):

$$[\mathbf{J}_{\mathbf{dc}}] = \begin{bmatrix} \mathbf{J}_{\mathbf{dc1}} & \mathbf{J}_{\mathbf{dc2}} & [\mathbf{J}_{\mathbf{dc5}}] \\ \mathbf{J}_{\mathbf{dc3}} & \mathbf{J}_{\mathbf{dc4}} & \mathbf{0} \\ [\mathbf{J}_{\mathbf{dc6}}] & \mathbf{0} & \mathbf{J}_{\mathbf{dc7}} \end{bmatrix} \tag{6.170}$$

where $\mathbf{J}_{\mathbf{dc5}} = \begin{bmatrix} \mathbf{J}'_{\mathbf{dc5}} & \mathbf{0} \end{bmatrix}$ and $\mathbf{J}_{\mathbf{dc6}} = \begin{bmatrix} \mathbf{J}'_{\mathbf{dc6}} & \mathbf{0} \\ \mathbf{0} & \mathbf{0} \end{bmatrix}$.

The matrices from $\mathbf{J}_{\mathbf{dc1}}$ to $\mathbf{J}_{\mathbf{dc4}}$, $\mathbf{J}'_{\mathbf{dc5}}$ and $\mathbf{J}'_{\mathbf{dc6}}$ are diagonal matrices with dimensions corresponding to the total number of converters making up the multi-terminal system, i.e. $n_{vsc} = m + n + r$, with $j = 1, \ldots, n_{vsc}$ and $k = n_{vsc} + 1, \ldots, 2n_{vsc}$. The order of matrix $\mathbf{J}_{\mathbf{dc7}}$ corresponds to the number of DC nodes n_{dc} minus the number of converters n_{vsc}, with $\langle j, k\rangle = n_{vsc} + 1, \ldots, n_{dc}$. These matrices are given in explicit form as:

$$\mathbf{J}_{\mathbf{dc1}} = \begin{bmatrix} \frac{\partial F_{I_{jx}}}{\partial I_{jx}} & 0 & 0 \\ 0 & \ddots & 0 \\ 0 & 0 & \frac{\partial F_{I_{jx}}}{\partial I_{jx}} \end{bmatrix}, \mathbf{J}_{\mathbf{dc2}} = \begin{bmatrix} \frac{\partial F_{I_{jx}}}{\partial E_{DCj}} & 0 & 0 \\ 0 & \ddots & 0 \\ 0 & 0 & \frac{\partial F_{I_{jx}}}{\partial E_{DCj}} \end{bmatrix}, \mathbf{J}_{\mathbf{dc3}} = \begin{bmatrix} \frac{\partial F_{E_{DCj}}}{\partial I_{jx}} & 0 & 0 \\ 0 & \ddots & 0 \\ 0 & 0 & \frac{\partial F_{E_{DCj}}}{\partial I_{jx}} \end{bmatrix} \tag{6.171}$$

$$\mathbf{J}_{\mathbf{dc4}} = \begin{bmatrix} \frac{\partial F_{E_{DCj}}}{\partial E_{DCj}} & 0 & 0 \\ 0 & \ddots & 0 \\ 0 & 0 & \frac{\partial F_{E_{DCj}}}{\partial E_{DCj}} \end{bmatrix}, \mathbf{J}'_{\mathbf{dc5}} = \begin{bmatrix} \frac{\partial F_{I_{jx}}}{\partial E_{DCk}} & 0 & 0 \\ 0 & \ddots & 0 \\ 0 & 0 & \frac{\partial F_{I_{jx}}}{\partial E_{DCk}} \end{bmatrix},$$

$$\mathbf{J}'_{\mathbf{dc6}} = \begin{bmatrix} \frac{\partial \Delta P_{DCk}}{\partial I_{jx}} & 0 & 0 \\ 0 & \ddots & 0 \\ 0 & 0 & \frac{\partial \Delta P_{DCk}}{\partial I_{jx}} \end{bmatrix}, \mathbf{J}_{\mathbf{dc7}} = \begin{bmatrix} \frac{\partial \Delta P_{DCj}}{\partial E_{DCk}} & \cdots & \frac{\partial \Delta P_{DCj}}{\partial E_{DCk}} \\ \vdots & \ddots & \vdots \\ \frac{\partial \Delta P_{DCj}}{\partial E_{DCk}} & \cdots & \frac{\partial \Delta P_{DCj}}{\partial E_{DCk}} \end{bmatrix} \tag{6.172}$$

The remaining terms, corresponding to the DC side of the multi-terminal HVDC system, with $j = 1, \ldots, n_{vsc}$ and $k = n_{vsc} + 1, \ldots, n_{dc}$, are:

$$[\mathbf{F}_{\mathbf{DC}}] = \left[F_{I_{jx}} \cdots F_{I_{jx}} \; F_{E_{DCj}} \cdots F_{E_{DCj}} \; \Delta P_{DCk} \cdots \Delta P_{DCk}\right]^T \tag{6.173}$$

$$[\boldsymbol{\Delta\Phi}_{\mathbf{DC}}] = \left[\Delta I_{jx} \cdots \Delta I_{jx} \; \Delta E_{DCj} \cdots \Delta E_{DCj} \; \Delta E_{DCk} \cdots \Delta E_{DCk}\right]^T \tag{6.174}$$

This modelling approach is modular and developing new converter models with different control strategies would not represent major difficulties in terms of their aggregation into the generalized multi-terminal HVDC link model.

During the derivation of the multi-terminal HVDC model, synthetized by (6.167), it has been assumed that each VSC unit connects to a capacitor and a smoothing inductor coupled to a resistor on its DC side. However, any modifications to this arrangement should pose no difficulty.

Unlike the steady-state model of the multi-terminal HVDC system, the DC voltage of the slack converter is part of the variables to be computed through the time-domain solution, where at least one slack converter must exist in the multi-terminal arrangement, to ensure the voltage stability in the DC network.

The calculated power at each node j of the DC network is computed using expression (6.175). The active and reactive power equations that link the multi-terminal HVDC system with the AC networks are given by (6.176)–(6.177), respectively.

$$P_{DCj,cal} = E_{DCj}^2 G_{DCjj} + E_{DCj} \sum_{\substack{j=n_{vsc}+1 \\ m\in j}}^{n_{dc}} E_{DCm} G_{DCjm} \tag{6.175}$$

$$P_{k,cal} = (e_k^2 + f_k^2)G_{kk} + \sum_{m\in k} [e_k(G_{km}e_k - B_{km}f_k) + f_k(B_{km}e_k + G_{km}f_k)] \tag{6.176}$$

$$Q_{k,cal} = -(e_k^2 + f_k^2)B_{kk} + \sum_{m\in k} [f_k(G_{km}e_k - B_{km}f_k) - e_k(B_{km}e_k + G_{km}f_k)] \tag{6.177}$$

where k represents a generic node.

6.6.4 Numerical Example with a Six-Terminal VSC-HVDC Link Forming a DC Ring

The six-terminal VSC-HVDC network, shown in Figure 3.19, which was solved in Chapter 3 for the purpose of a steady-state assessment, is also used here to show the applicability of the multi-terminal VSC-HVDC dynamic model.

It is useful to recall that AC_1, AC_3 and AC_5 stand for the UK Grid, the NORDEL Grid and the UCTE Grid, whose total power demands are 60 GW, 30 GW and 450 GW, respectively. For the sake of simplicity, these AC power networks were represented by their Thevenin equivalents, determined as $R_{th} + jX_{th} = [P + jP\tan(\theta)]^{-1}$, $\theta = \cos^{-1}(\text{pf})$, pf = 0.95, coupled to reactive tie lines. The network AC_2, which represents the in-feed point of the Valhall oil platform, has a power demand of 78 MW. The equivalent power injections of networks AC_4 and AC_6, representing the collector points of the wind parks at the German Bight and Dogger Bank, are rated at 400 MW each. For the purpose of the dynamic simulation, all the AC networks are taken to operate at 50 Hz.

6.6.4.1 Disconnection of a DC Transmission Line

The DC transmission line that connects the nodes d and e, which carries 126 MW during steady-state, is tripped at $t = 0.5$ s. This disturbance causes the DC ring to open, becoming a radial DC grid. In order to redistribute the power flows in the longitudinal DC network, the DC voltages are adjusted, as seen in Figure 6.61. The DC voltage of the slack converter is fittingly controlled and so are the rest of the DC voltages. Accordingly, the power imbalances throughout the DC grid are mitigated promptly, in only a few milliseconds.

The dynamic behaviour of the modulation ratio of the converters is shown in Figure 6.62, where it is observed that only a small readjustment took place following the disconnection of the DC transmission line. The frequencies of the passive networks fed by the converters VSC_b, VSC_d and VSC_f are given in Figure 6.62. It becomes clear from this simulation study that the disturbance in the DC network does not impact adversely the operation of the island AC networks.

6.6.4.2 Three-Phase Fault Applied to AC_3

The behaviour of the six-terminal VSC-HVDC system to AC faults is assessed by simulating a three-phase-to-ground short-circuit fault at node 5. This is applied at

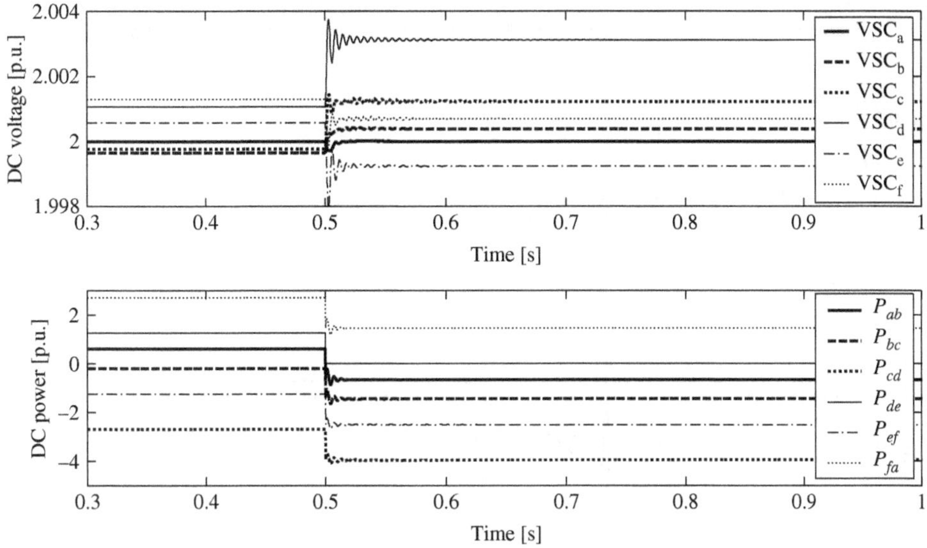

Figure 6.61 DC voltages and power flows in the DC grid.

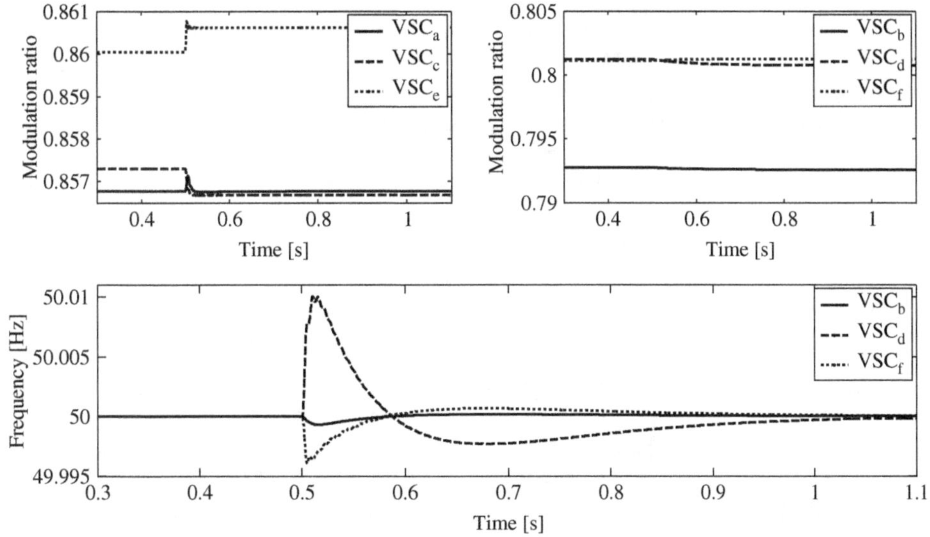

Figure 6.62 Modulation ratio of the VSCs and frequency of the passive grids fed by VSC_b, VSC_d, VSC_f.

$t = 0.5$ s and lasts 120 ms (six cycles of the 50 Hz power system). The point in fault is the high-voltage side of the LTC transformer corresponding to the converter station VSC_c.

The disturbance causes a severe voltage drop at the AC terminal of the converter, as shown in Figure 6.63. As expected, the AC power injected at the fault point, P_{vc}, follows a similar trend to that of the nodal voltage, dropping from 250 MW to 0 MW for the duration of the fault. Since the 250 MW can no longer be delivered to AC_3, this power flows through the DC ring reaching the converter station VSC_a, i.e. the slack converter, as confirmed by the dynamic performance of P_{va} in Figure 6.63.

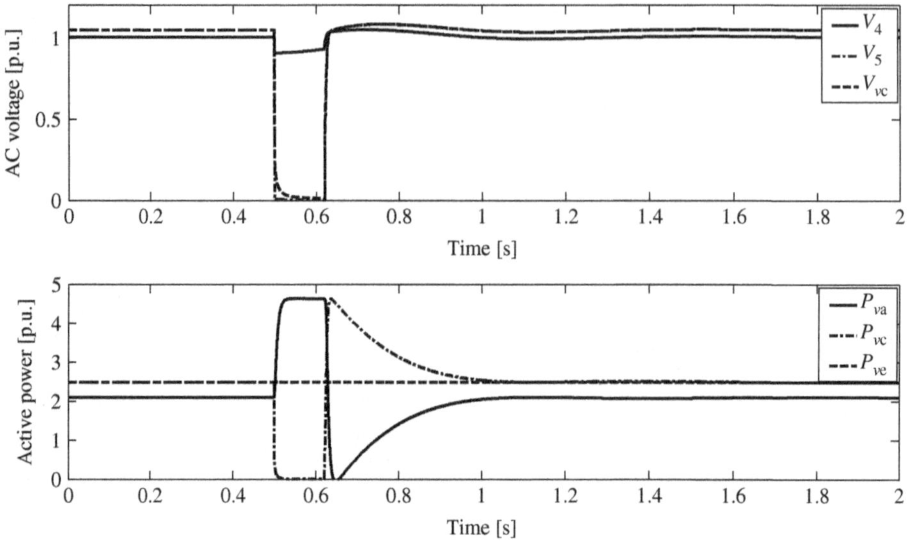

Figure 6.63 AC voltages and AC powers during the three-phase fault.

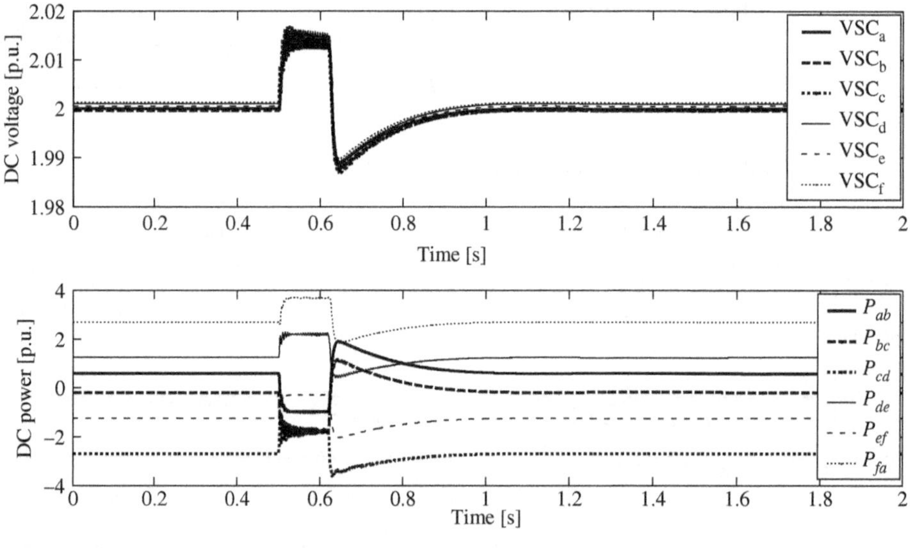

Figure 6.64 Voltages and powers in the DC network during the three-phase fault.

The short-circuit fault also causes transient voltage rises and power fluctuations throughout the DC network, as seen in Figure 6.64.

While the voltage magnitude at node 5 drops to zero for the duration of the three-phase fault, the reactive power injection, Q_{vc}, of the converter station VSC_c suddenly increases, as shown in Figure 6.65. Hence, the modulation index controller, which is exerting a fixed reactive power setpoint, $Q_v^{ref} = 0$, forces the modulation ratio of the converter to decrease rapidly so as to comply with its command.

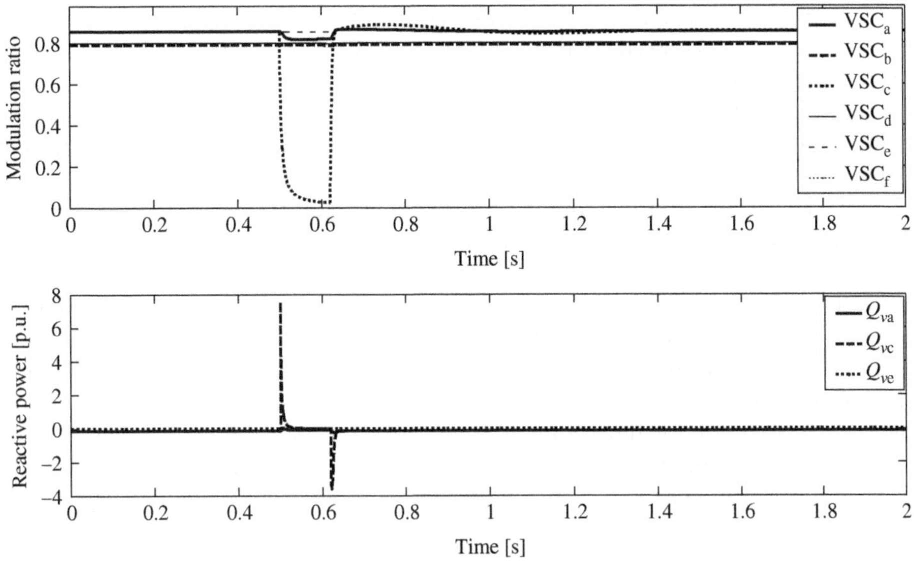

Figure 6.65 Modulation ratio of the converters and reactive power injected by VSC$_c$.

6.7 Conclusion

This chapter has covered the dynamic models of synchronous generators and their controls: the AVR, hydro and steam turbines, and the speed-governing system. More significantly, it has introduced RMS-type dynamic models of the most significant VSC-based equipment: the STATCOM, back-to-back, point-to-point and multi-terminal VSC-HVDC systems. All the key aspects of their modelling implementation within the context of a dynamic power flow algorithm have been elucidated. A salient feature of this modelling approach is that the VSC model does not rely on the customary equivalent voltage source but rather on the use of a complex phase-shifting transformer, a kernel in the modelling of the VSC making up the STATCOM and VSC-HVDC systems of various kinds. The ensuing models yield a reasonably accurate representation of their AC and DC voltage dynamics, where the incorporation of the differential equation that governs both the behaviour of the DC capacitor and DC inductor has turned out to be of paramount importance.

Indeed, the AC bus voltage control and the regulation of active power between the AC networks interconnected through a VSC-HVDC link are promising features of this technology. As for the voltage control, each VSC model is fitted with a first-order regulator that, upon variations in the network's voltage, adjusts the modulation ratio in a dynamic fashion to bring the voltage magnitude back to the setpoint value. The control system responsible for keeping the power balance in the converters, achieved by regulating the angular aperture between the AC terminal voltage angle and the ideal transformers' phase-shifting angle, is a new controller with many potential applications yet to be fully developed.

In the case of a VSC-HVDC model with frequency support capabilities, the steady-state VSC-HVDC model computes the power that the HVDC link must transfer to satisfy the load connected at the inverter's end from the inertial response standpoint. In cases where the low-inertia network contains no synchronous generation of its own, that is, a passive network, the inverter provides the angular reference for this network; it is the angle of the inverter's phase-shifting transformer φ_I. Thus, the AC node of the inverter would act as a slack bus for the low-inertia system. The power flowing through the DC link is not dynamically pegged to a specific constant value; instead, the DC voltages are adjusted to enable a DC power transfer that meets the power requirements of the low-inertia grid fed through the inverter, to mitigate the frequency deviations.

The theory presented in this chapter has shown to be useful for assessing the dynamic model of HVDC multi-terminal systems, with an arbitrary number of VSC units and an arbitrary DC network topology. The classification of VSC stations in three different types of converters – the slack converter, the power-scheduled converter and the passive converter – is key to this development. Their main dynamic features are DC voltage control, DC power and frequency regulation, respectively. In summary, the converters' controllers act independently to cope with specific tasks involving voltage and power transfer control.

In the modelling approach presented in this chapter, the nonlinear equations describing the models of the various VSC-based devices are linearized together with their associated discretized differential equations. Furthermore, the state variables corresponding to the VSC units are combined with the state variables of the synchronous generators and their controls as well as with the network's state variables, all this expressed in a single frame-of-reference for unified iterative solutions. The implicit trapezoidal integration method embedded within a Newton-Raphson iterative technique has been selected to determine, in time domain, the dynamic response of the whole AC-DC network. The sanity of the RMS results, in time domain, has been checked by comparing selected results with comparable time-domain results obtained with Simulink. The overall satisfaction with the closeness of the results is good, considering the very different modelling and solution approaches that the two methods take to solve the same problem. Provided a suitable initialization of the Simulink solution is employed, from the outset, the responses furnished by Simulink are taken to be more accurate than the method put forward in this chapter, since it uses a more detailed representation of the actual equipment; however, this is achieved at the expense of longer computing times. It should be borne in mind that the execution time is an issue that becomes more acute as the size of the AC-DC power system increases.

References

1 Concordia, C. (1951). *Synchronous Machines, Theory and performance.* Wiley.

2 Kimbark, E.W. (1948). *Power System Stability*, vol. 3, Synchronous Machines. IEEE Press.

3 Weedy, B.M. (1967). *Electric Power Systems.* Wiley.

4 Stagg, G.W. and El-Abiad, A.H. (1968). *Computer Methods in Power Systems.* McGraw-Hill.

5 Anderson, P.M. and Fouad, A.A. (1977). *Power System Control and Stability.* IEEE Press.

6 Arrillaga, J. and Arnold, C.P. (1990). *Computer Analysis of Power Systems*. Wiley.

7 IEEE Committee Report (1973). Dynamic models for steam and hydro turbines in power system studies. *IEEE Transactions on Power Apparatus and Systems* 92: 1904–1915.

8 IEEE Committee Report (1968). Computer representation of excitation systems. *IEEE Transactions on Power Apparatus and Systems* 87: 1460–1464.

9 Rafian, M., Sterling, M., and Irving, M. (1987). Real-time power system simulation. *IEE Proceedings* 134 (3): 209–223.

10 IEEE Task Force Report (1993). Load representation for dynamic performance analysis. *IEEE Power Engineering Society* 8: 472–482.

11 Gear, C.W. (1971). Simultaneous numerical solution of differential-algebraic equations. *IEEE Transactions on Circuit Theory* 18: 89–95.

12 Stott, B. (1979). Power systems dynamic response calculations. *IEEE Proceedings* 67: 219–241.

13 Dommel, H.W. and Sato, N. (1972). Fast transient stability solutions. *IEEE Transactions on Power Apparatus and Systems* 91: 1643–1650.

14 Kundur, P. (1994). *Power System Stability and Control*. McGraw-Hill.

15 Chow, J.H. (1982). *Time-Scale Modeling of Dynamic Networks with Applications to Power Systems*. Springer-Verlag.

16 Hingorani, N.G. and Gyugyi, L. (1999). *Understanding FACTS: Concepts and Technology of Flexible AC Transmission Systems*. Wiley-IEEE Press.

17 Acha, E., Fuerte-Esquivel, C.R., Ambriz-Perez, H., and Angeles-Camacho, C. (2005). *FACTS Modeling and Simulation in Power Networks*. Wiley.

18 Acha, E. and Kazemtabrizi, B. (2013). A new STATCOM model for power flows using the Newton-Raphson method. *IEEE Transactions on Power Systems* 28 (3): 2455–2465.

19 De Oliveira M.M. 2000, Power electronics for mitigation of voltage sags and improved control of AC power systems, Doctoral Dissertation, Royal Institute of Technology (KTH), Stockholm.

20 Castro, L.M., Acha, E., and Fuerte-Esquivel, C.R. (2013). A new STATCOM model for dynamic power system simulations. *IEEE Transactions on Power Systems* 28 (3): 3145–3154.

21 Castro, L.M. and Acha, E. (2015). A novel VSC-HVDC link model for dynamic power system simulations. *Electric Power Systems Research* 126: 111–120.

22 Callavik, E.M., Lundber, P., Bahrman, M.P., and Rosenqvist, R.P. (2012). HVDC technologies for the future onshore and offshore grid. *Cigré US National Committee 2012 Grid of the Future Symposium, F-75008* 1–6.

23 Beccuti, G., Papafotiou, G., and Harnefors, L. (2014). Multivariable optimal control of HVDC transmission links with network parameter estimation for weak grids. *IEEE Transactions on Control Systems* 22: 676–689.

24 Guo, C. and Zhao, C. (2010). Supply of an entirely passive AC network through a double-infeed HVDC system. *IEEE Transactions on Power Electronics* 24: 2835–2841.

25 Zhang, L., Harnefors, L., and Nee, H.P. (2011). Modeling and control of VSC-HVDC links connected to island systems. *IEEE Transactions on Power Systems* 26: 783–793.

26 Rocabert, J., Luna, A., Blaabjerg, F., and Rodriguez, P. (2012). Control of power converters in AC microgrids. *IEEE Transactions on Power Electronics* 27: 4734–4749.

27 Chung, I.Y., Liu, W., Cartes, D.A. et al. (2010). Control methods of inverter-interfaced distributed generators in a microgrid system. *IEEE Transactions on Industrial Applications* 46: 1078–1088.
28 Eghtedarpour, N. and Farjah, E. (2014). Power control and management in a hybrid AC/DC microgrid. *IEEE Transactions on Smart Grid* 5: 1494–1505.
29 Sao, C.K. and Lehn, P.W. (2008). Control and power management of converter fed microgrids. *IEEE Transactions on Power Systems* 23: 1088–1098.
30 Castro, L.M. and Acha, E. (2015). On the provision of frequency regulation in low inertia AC grids using HVDC systems. *IEEE Transactions on Smart Grid* 7 (6): 2680–2690.
31 Acha, E. and Castro, L.M. (2016). A generalized frame of reference for the incorporation of multi-terminal VSC-HVDC systems in power flow solutions. *Electric Power Systems Research* 136: 415–424.
32 Castro, L.M. and Acha, E. (2016). A unified modeling approach of multi-terminal VSC-HVDC links for dynamic simulations of large-scale power systems. *IEEE Transactions on Power Systems* 31 (6): 5051–5060.

7

Electromagnetic Transient Studies and Simulation of FACTS-HVDC-VSC Equipment

7.1 Introduction

Electromagnetic transient studies are very useful in the analysis and testing of systems of an electrical nature. They constitute the best solution by which to obtain the system behaviour in steady-state and with regard to its transient response, thus providing valuable information about the dynamic performance of different systems under any operation condition.

Power systems are described by a set of differential equations which can be either linear or nonlinear. These systems are under the influence of certain phenomena such as switching actions, short-circuits and disturbances, among others, which can produce abnormal waveforms, fast changes in the voltage and current or electromechanical transients. Furthermore, modern power systems involve a high degree of complexity owing to, among other things, the increasing use of FACTS and HVDC systems which are based on power electronic converters. In this scenario, a mathematical tool which is able to obtain numerical solutions that accurately describe the behaviour of any power system is therefore necessary.

Various commercial electromagnetic transient simulators exist which not only provide models of basic electrical elements but also power electronic devices and advanced control functions, and are able to achieve an accurate time response for a great variety of power systems with different configurations.

Some of the commercial software packages used for electromagnetic transient analysis are:

- ATP-EMTP – Alternative Transients Program-Electromagnetic Transients Program
- PSCAD/EMTDC – Power System Computer Aided Design/Electromagnetic Transient Direct Current
- SPICE
- Matlab Power System Toolbox.

This chapter deals with the simulation of the transient response of different voltage source-converter based devices. Although any of the above-mentioned packages could be used, the authors have chosen PSCAD/EMTDC owing to the fact that it has been designed to simulate power systems in addition to the implementation of different control functions and power-electronic devices. PSCAD/EMTDC is capable of simulating transient responses in both electric and control systems. Since its first development in

VSC-FACTS-HVDC: Analysis, Modelling and Simulation in Power Grids, First Edition.
Enrique Acha, Pedro Roncero-Sánchez, Antonio de la Villa Jaén, Luis M. Castro and Behzad Kazemtabrizi.

Companion website: www.wiley.com/go/acha_vsc_facts

1975, PSCAD/EMTDC has evolved up to the current version, PSCAD X4 (release of version 4.6.2 in June 2017). Detailed information on the features of PSCAD/EMTDC can be found in [1].

The electromagnetic transient responses of four different FACTS and HVDC test systems are studied throughout this chapter. Section 7.2 analyzes the behaviour of a STATCOM based on a conventional two-level voltage source converter connected to a distribution grid. An extension of the STATCOM using a three-level flying capacitor converter is developed in Section 7.3. The case of a point-to-point HVDC system of two terminals based on use of multilevel voltage source converters is studied in Section 7.4. Finally, a multi-terminal HVDC system, built with multilevel voltage source converters, is simulated in Section 7.5.

The design of the various schemes used to control the different power systems is comprehensively explained in the chapter by means of the root-locus technique, and all four examples have been simulated using the educational licence of the latest available version of PSCAD X4.

7.2 The STATCOM Case

Reactive-power compensation was traditionally carried out by means of SVCs which have the ability to inject either inductive or capacitive current into the electrical grid. However, these devices may cause resonance problems with other elements of the grid. The modern alternative is the STATCOM, which has the same ability as the SVC but with added advantages: its transient response is faster than that achieved with the SVC, and its enhanced controllability is greater since it is based on the VSC principle, which makes it easier to regulate voltage magnitude and reactive power support.

The STATCOM may be used both in transmission systems as a reactive power compensator to enable voltage regulation and in distribution systems to improve power quality at the point of common coupling (PCC) – in the latter case it is termed distribution STATCOM or DSTATCOM [2].

The most basic structure of a three-phase STATCOM comprises a two-level VSC with a DC energy storage device which is connected to the AC mains by means of a coupling transformer. The VSC can be operated by a control system. Figure 7.1 shows the schematic diagram of the DSTATCOM in which a capacitor is used as the DC energy storage device and two different loads are connected at the PCC. Figure 7.2 depicts the one-line diagram of the STATCOM connected to the distribution system at the PCC, where the VSC is represented by an ideal voltage source u, v_s is the grid voltage, v is the voltage at the PCC and i is the current injected into the grid by the STATCOM. The parameters R and L are the resistance and inductance of the coupling transformer, respectively.

The state-variable model of the system shown in Figure 7.2 can be expressed using Park's transformation in the synchronous reference frame (SRF) together with the PCC voltage. Moreover, the three-phase system may be assumed to be balanced and the following equation is then obtained [3]:

$$\frac{d}{dt}\begin{bmatrix} i_d \\ i_q \end{bmatrix} = \begin{bmatrix} -\frac{R}{L} & \omega \\ -\omega & -\frac{R}{L} \end{bmatrix}\begin{bmatrix} i_d \\ i_q \end{bmatrix} + \frac{1}{L}\begin{bmatrix} 1 & 0 \\ 0 & 1 \end{bmatrix}\begin{bmatrix} u_d - v_d \\ u_q \end{bmatrix} \tag{7.1}$$

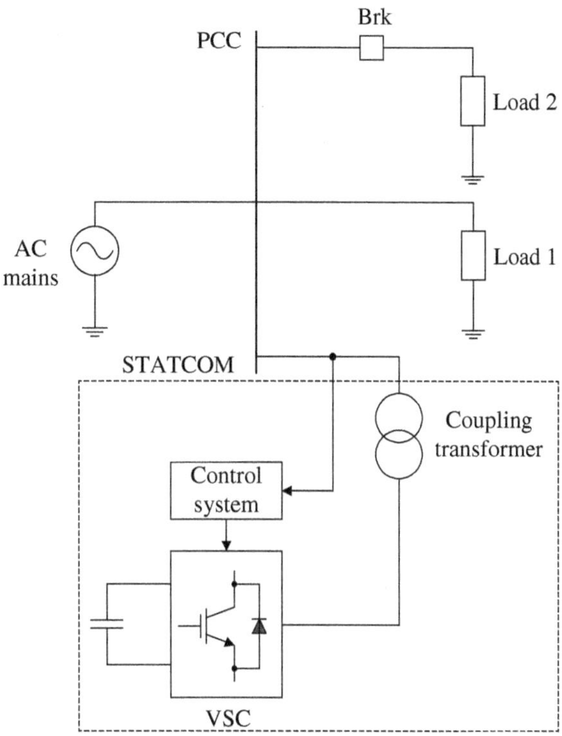

Figure 7.1 Basic scheme of a STATCOM connected to a distribution system. Source: ©MDPI, 2014.

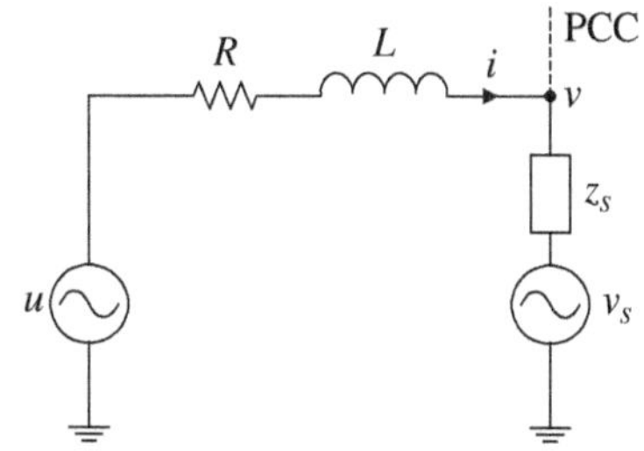

Figure 7.2 One-line equivalent circuit of the STATCOM connected at PCC. Source: ©MDPI, 2014.

where the state variables are the d and q components of the current, namely i_d and i_q, the variables u_d and u_q are the components of the VSC (control inputs), and ω is the angular speed of the SRF, i.e. the angular frequency of the grid voltage. This implies that the PCC voltage has only the d component, v_d, which is a signal that can be measured. In these conditions, all the sinusoidal variables of frequency ω become DC-like magnitudes and the instantaneous active power, p, and the instantaneous reactive power, q, injected into the grid by the converter are [4]:

$$p = v_d i_d \tag{7.2}$$

$$q = -v_d i_q \tag{7.3}$$

p and q are therefore DC quantities that can be controlled by the i_d and i_q components, respectively. Unfortunately, model (7.1) is coupled, and changes in the current i_d will therefore produce variations in i_q and vice versa. This issue may be circumvented by obtaining an equivalent decoupled model of the following form:

$$\frac{d}{dt}\begin{bmatrix} i_d \\ i_q \end{bmatrix} = \begin{bmatrix} -\frac{R}{L} & 0 \\ 0 & -\frac{R}{L} \end{bmatrix}\begin{bmatrix} i_d \\ i_q \end{bmatrix} + \frac{1}{L}\begin{bmatrix} 1 & 0 \\ 0 & 1 \end{bmatrix}\begin{bmatrix} z_d \\ z_q \end{bmatrix} \tag{7.4}$$

where z_d and z_q are the yet-to-be-defined control inputs. Once variables z_d and z_q have been calculated using a specific control law, the variables u_d and u_q will be obtained as:

$$\begin{bmatrix} u_d \\ u_q \end{bmatrix} = L\begin{bmatrix} 0 & -\omega \\ \omega & 0 \end{bmatrix}\begin{bmatrix} i_d \\ i_q \end{bmatrix} + \begin{bmatrix} z_d \\ z_q \end{bmatrix} + \begin{bmatrix} v_d \\ 0 \end{bmatrix} \tag{7.5}$$

Furthermore, the power p_c extracted from the capacitor of the DC energy storage by the VSC can be written as:

$$\frac{dv_c}{dt} = \frac{1}{Cv_c}p_c = -\frac{1}{Cv_c}p_{vsc} \tag{7.6}$$

where v_c is the voltage of the DC capacitor C and p_{vsc} is the power of the lossless VSC:

$$p_{vsc} = u_d i_d + u_q i_q \tag{7.7}$$

The average voltage of the DC capacitor must be controlled to reduce the energy extracted from the DC capacitor and to guarantee proper control of the reactive power injected into the grid. Moreover, a fast controller will make it possible to reduce the capacitor value.

Equation (7.6) is nonlinear. In order to obtain a simpler capacitor-voltage equation, (7.6) can be written as:

$$2v_c\frac{dv_c}{dt} = \frac{d(v_c^2)}{dt} = -\frac{2}{C}p_{vsc} \tag{7.8}$$

where v_c^2 is the new state variable.

Since the VSC is assumed to be lossless, the power extracted from the capacitor equals the power injected into the grid plus the losses in the coupling transformer:

$$p_{vsc} = p + p_T = u_d i_d + u_q i_q \tag{7.9}$$

The substitution of (7.1) into (7.9) yields:

$$p_{vsc} = \overbrace{\underbrace{Ri_d^2 + Ri_q^2}_{p_R} + \underbrace{\frac{L}{2}\frac{di_d^2}{dt} + \frac{L}{2}\frac{di_q^2}{dt}}_{p_L}}^{p_T} + \overbrace{v_d i_d}^{p} \tag{7.10}$$

where the term p_R represents the losses in the transformer resistance, which are normally quite small ($R \approx 0$). Although the term p_L (losses in the inductance) will always be zero in steady-state, it will not be zero during transients owing to changes in the current. A current controller should be designed for maximum speed of response in order to reach the steady-state as quickly as possible. A simple approach aimed at fulfilling this premise is to assume that $p_{vsc} \approx p$ [5].

The block diagram of the control scheme for a DSTATCOM is depicted in Figure 7.3. It shows two inner control loops with two identical PI regulators for the d and q components of the current injected into the grid (i.e. the active and reactive powers).

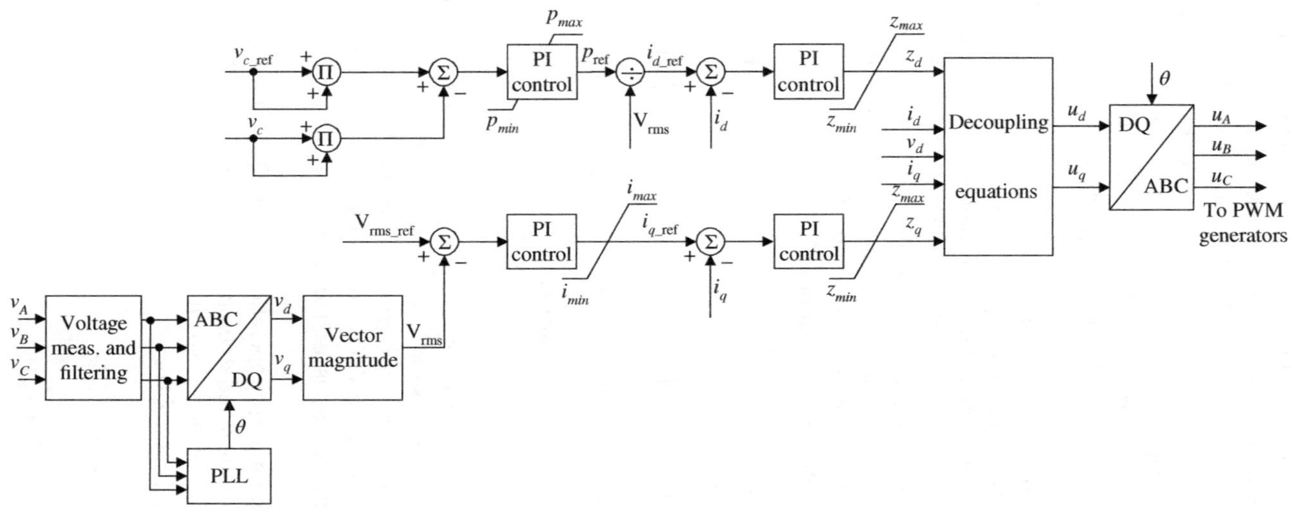

Figure 7.3 Control system scheme for the DSTATCOM case. Source: ©MDPI, 2014.

The outputs of these controllers are the compound variables z_d and z_q. The voltage of the DC capacitor is controlled by means of an outer PI controller that is slower than the current controllers and which includes an anti-windup action. It should be noted that this PI regulator controls the state variable v_c^2. The output of this controller is the reference of the active power that is necessary to maintain the capacitor voltage at the desired value. This reference is divided by the value of the PCC voltage in order to obtain the reference of the d axis current. Another outer PI regulator is also tailored in order to maintain the voltage at the PCC: the output of this controller is the reference of the q component of the current (i.e. the DSTATCOM injects or absorbs reactive power) and, as in the case of the DC voltage regulator, it is slower than the current PI controllers.

Since the overall control system is designed in an SRF, a phase locked loop (PLL) is used to obtain the angle required to carry out such a transformation. The inputs of the PLL are the filtered measurements of the grid voltages. This transformation is a power invariant $dq0$ transformation [6] and will be used throughout the chapter. The chosen transformation has the additional advantage that for balanced sinusoidal conditions, the RMS value of the line-to-neutral voltages and the line currents of the three-phase system equal the magnitude of the resulting vectors in the SRF.

Equation (7.11) shows this transformation:

$$\begin{bmatrix} x_0 \\ x_d \\ x_q \end{bmatrix} = \sqrt{\frac{2}{3}} \begin{bmatrix} \frac{1}{\sqrt{2}} & \frac{1}{\sqrt{2}} & \frac{1}{\sqrt{2}} \\ \cos\theta & \cos\left(\theta - \frac{2\pi}{3}\right) & \cos\left(\theta - \frac{4\pi}{3}\right) \\ -\sin\theta & -\sin\left(\theta - \frac{2\pi}{3}\right) & -\sin\left(\theta - \frac{4\pi}{3}\right) \end{bmatrix} \begin{bmatrix} x_A \\ x_B \\ x_C \end{bmatrix} \tag{7.11}$$

where x_A, x_B and x_C are the electrical magnitudes of the three-phase system, and x_0, x_d and x_q are the transformed variables in the SRF, which rotates at an angular speed ω such as $\theta = \int \omega dt$.

The decoupled Eq. (7.5) is used to obtain the outputs u_d and u_q, which are transformed into variables of a three-phase system by means of the inverse matrix of the $dq0$ transformation. These control variables are used to drive the PWM process and to generate the necessary firing signals of the VSC using a sinusoidal PWM scheme [7]. A conventional two-level, three-phase VSC is used to illustrate the control laws/theory described above.

It should be noted that for proper operation of the DSTATCOM, the inner control loop (i.e. the PI controllers for the d and q components of the current) should respond much faster than the outer control loop. Two different alternatives can be used to implement the proportional-integral control law:

$$z_x = k_{ix} \int (i_{x\_ref} - i_x)dt - k_{px} i_x \tag{7.12}$$

$$z_x = k_{ix} \int (i_{x\_ref} - i_x)dt + k_{px}(i_{x\_ref} - i_x) \tag{7.13}$$

where k_{ix} and k_{px} are the controller gains and the subscript x represents either the d axis or the q axis. For equal gain values, the associated closed-loop system will have the same poles for both equations. However, Eq. (7.13) yields zeros in the transfer function of the closed-loop system, which results in a more adverse transient response than that obtained with control law (7.12). Hence, Eq. (7.12) is normally used in the design of current regulators.

The PI controller for the voltage of the DC capacitor is designed according to the following equation:

$$p_{\text{ref}} = k_{ic} \int (v^2_{c\_\text{ref}} - v^2_c)dt + k_{pc}(v^2_{c\_\text{ref}} - v^2_c) \tag{7.14}$$

where k_{ic} and k_{pc} are the controller gains. This PI regulator also includes an anti-windup action with its respective upper and lower limits, which resets the integral term of the controller when the limits are reached. This action avoids the integral windup effect [6].

Finally, the voltage regulator at the PCC can be designed only by integrating the error between the reference and the measured voltage:

$$i_{q\_\text{ref}} = k_{iv} \int (V_{\text{rms\_ref}} - V_{\text{rms}})dt \tag{7.15}$$

where k_{iv} is the integral gain, $V_{\text{rms\_ref}}$ is the PCC voltage reference and V_{rms} is the filtered value of the measured voltage. Since the Thevenin equivalent is seen from PCC changes with load variations, these parameters should be avoided in the voltage control design. Moreover, the dynamics of the filter used in the measurement of the voltage must be taken into account. The parameter k_{iv} must therefore be chosen with care.

A three-phase distribution system such as the one described in Figure 7.1 has been simulated using PSCAD/EMTDC. The RMS line-to-line voltage of the grid is 13.8 kV, with rated frequency equal to 50 Hz. Two different loads are connected to the PCC with the following features:

- *Load 1*: Active power 20 MW and reactive power 15 MVAR.
- *Load 2*: Active power 60 MW and reactive power 50 MVAR.

Load 1 is permanently connected and a circuit breaker is used to control the connection of Load 2 to the system. A three-phase transformer with a winding ratio 20 kV/62.5 kV is used to connect the VSC to the system. The transformer's primary and secondary windings are star and delta connected, respectively. The DSTATCOM is connected to the secondary side of the transformer. The respective resistance and leakage inductance of the transformer referred to the 20 kV, primary side are 5 mΩ and 4 mH. A 660 μF capacitor is used in the DC side of the STATCOM as the DC energy storage system.

The parameters of the different controllers have been obtained using the root-locus technique, which has been successfully applied in [8]: the desired poles for the inner control loop have been placed at $s_1 = s_2 = -1000$ rad s^{-1}, as shown in Figure 7.4, and the corresponding control gains of Eq. (7.13) are $k_{ix} = 4000$ and $k_{px} = 7.995$.

The time response for a step input in the reference is plotted in Figure 7.5, which shows that no overshoot takes place and the steady-state is reached in approximately 10 ms.

The PI regulator for the DC capacitor voltage has been designed to obtain a time response that is much slower than that obtained with the current controllers. In this situation, the dynamics of the inner control loop can be neglected in comparison with the dynamics of the DC voltage control loop. In this example, the chosen poles have been placed at $s_1 = -20$ rad/s and $s_2 = -100$ rad/s and the resulting parameters for control law (7.14) are $k_{ic} = -0.66$ and $k_{pc} = -0.0396$. Figure 7.6 shows the location of the closed-loop system poles, whereas Figure 7.7 exhibits the time response for a step

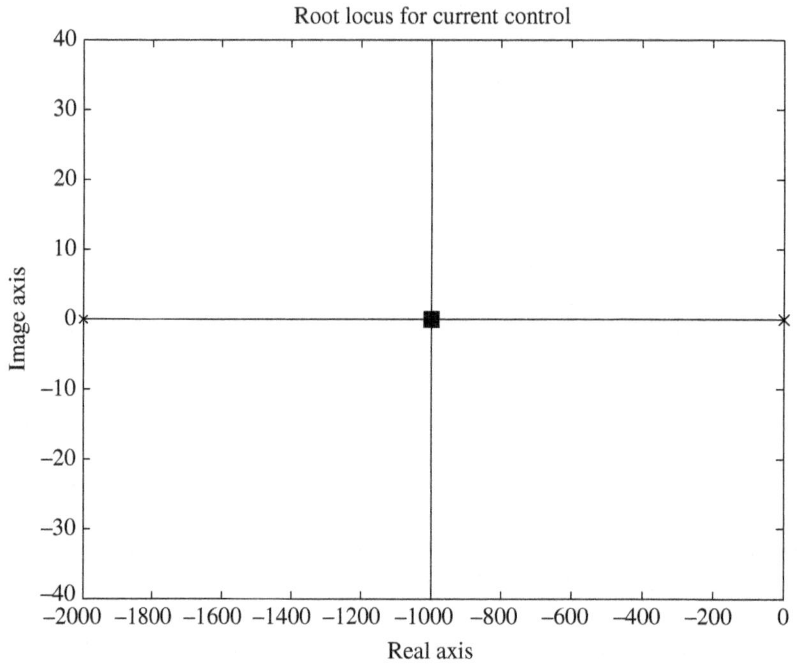

Figure 7.4 Pole location for the inner control loop. Source: ©MDPI, 2014.

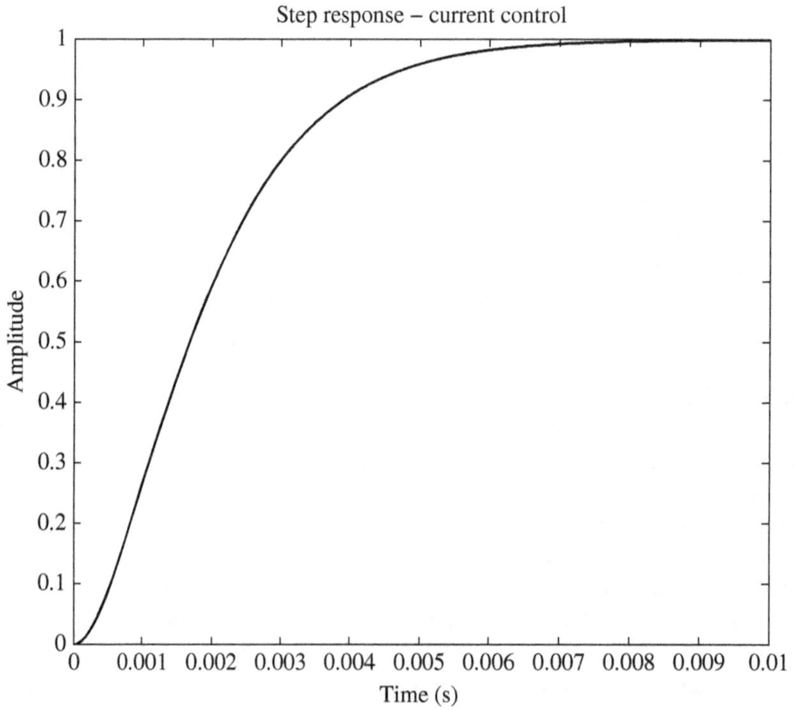

Figure 7.5 Time response of the inner loop obtained for a step reference. Source: ©MDPI, 2014.

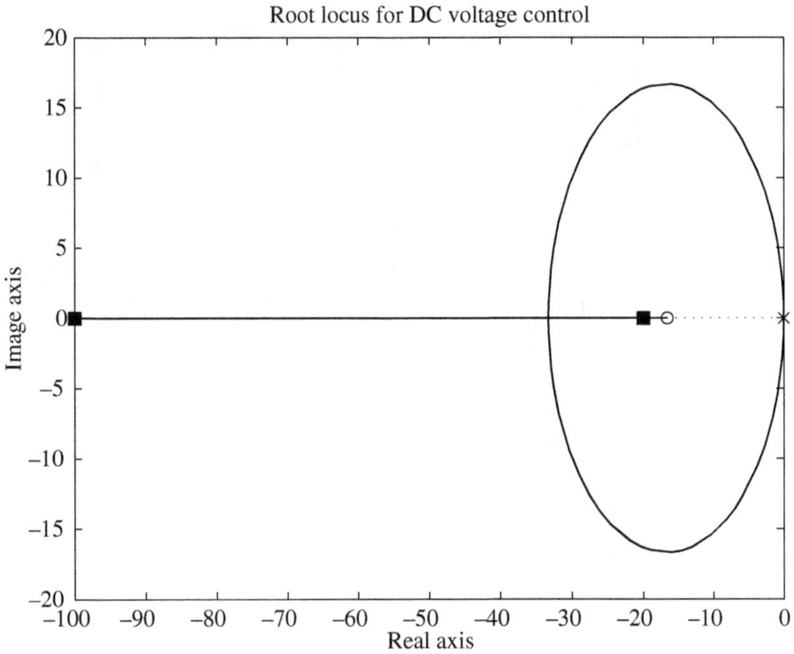

Figure 7.6 Location of the poles for DC voltage control loop. Source: ©MDPI, 2014.

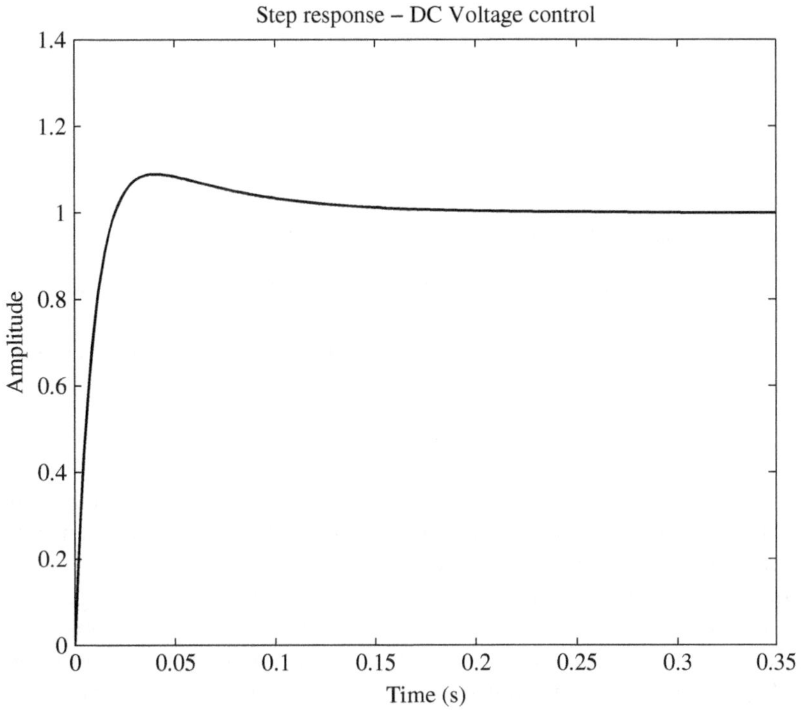

Figure 7.7 Time response of the DC voltage control loop for a step reference. Source: ©MDPI, 2014.

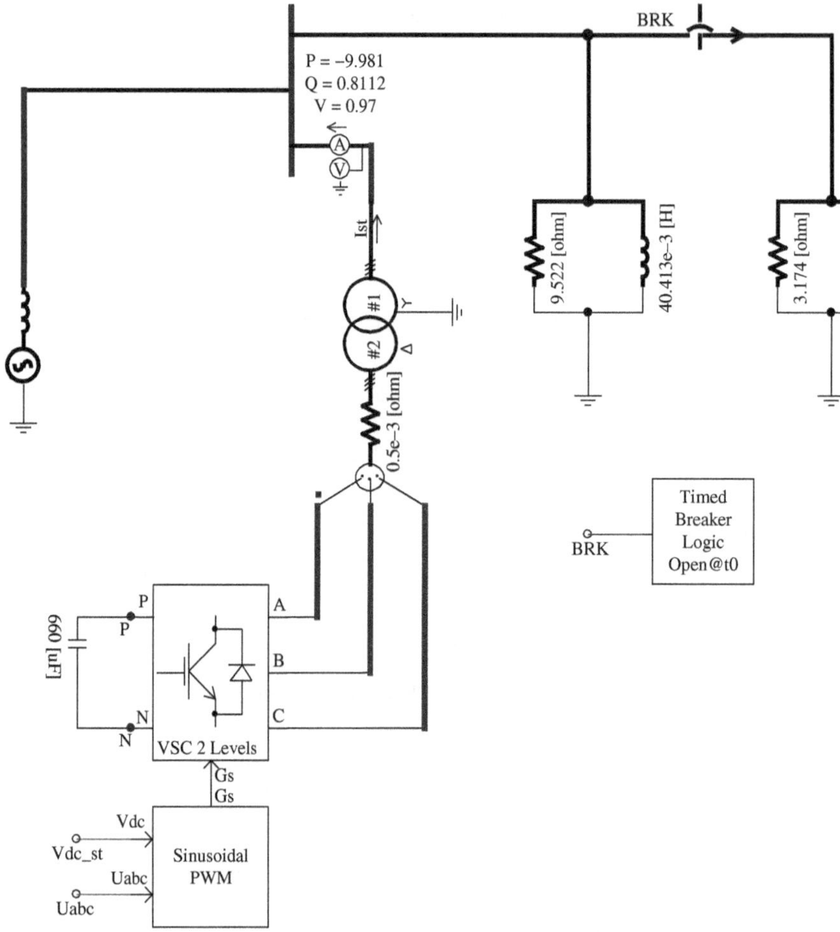

Figure 7.8 Circuit of a VSC-based STATCOM implemented in PSCAD/EMTDC.

input in the reference: no significant overshoot is present (less than 10%) and the time response settles in approximately 0.15 s.

With regard to the PCC voltage control, the measurement filter of the voltage has the following first-order differential equation:

$$T_f \frac{dv_f}{dt} + v_f = v_m \tag{7.16}$$

where v_m is the measured voltage, v_f is the filtered voltage and T_f is the smoothing time constant. For this example, $T_f = 10$ ms, and the parameter k_{iv} has been chosen to be equal to 100 after an adjusting process involving different simulations.

The implementation of the STATCOM in the distribution system of Figure 7.1 is shown in Figure 7.8, while the two-level, three-phase VSC implemented in PSCAD/EMTDC is plotted in Figure 7.9. Note that the VSC was previously modelled as an ideal voltage source for solely control design purposes, but the implementation of the VSC in PSCAD/EMTDC is carried out using elements available in the master library, such as IGBTs and diodes. Furthermore, the three different control schemes are

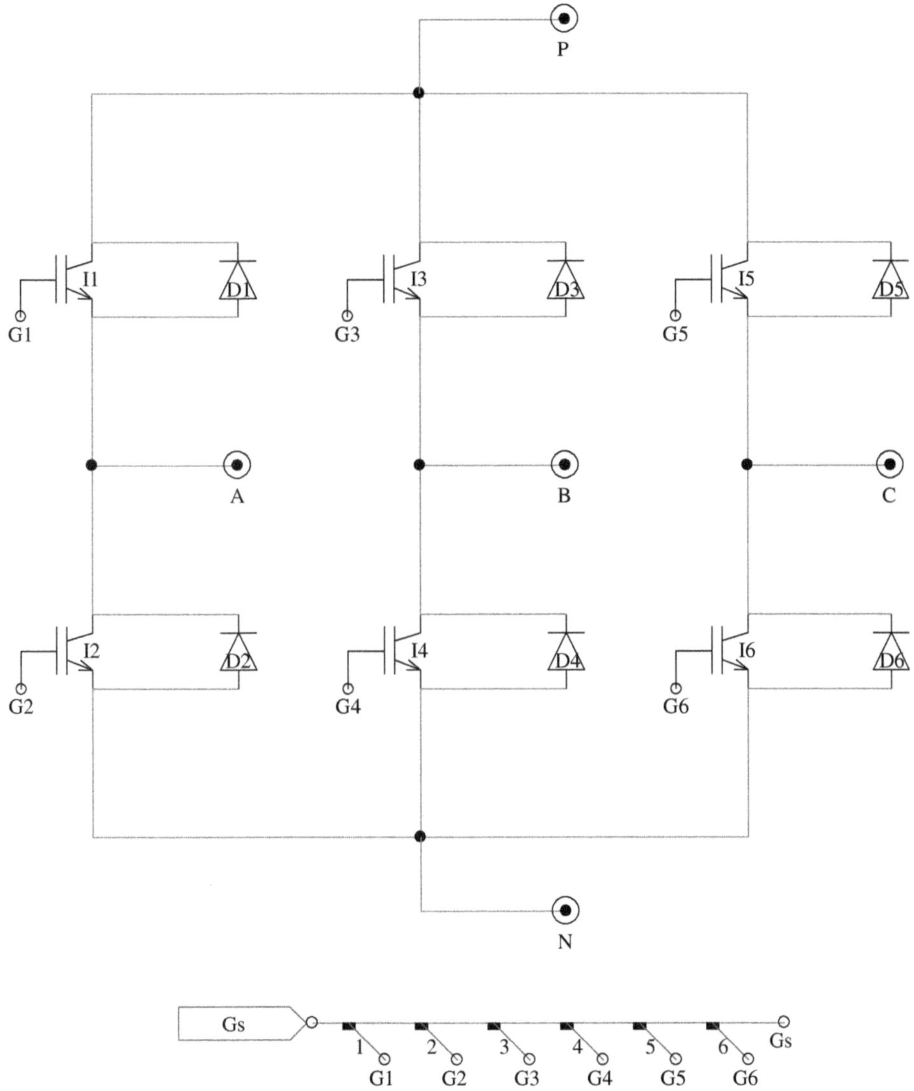

Figure 7.9 Implementation of the two-level VSC using IGBTs and diodes from the PSCAD/EMTDC master library.

depicted in Figures 7.10–7.12. Finally, Figure 7.13 shows the sinusoidal PWM scheme implemented in PSCAD/EMTDC with a switching frequency of 3150 Hz.

The experiment is carried out as follows: Load 1 and the STATCOM are initially connected to the grid; the DC capacitor is discharged and, following a transient period, the DC voltage regulator controls the level of charge of the capacitor until it reaches 100 kV. At $t = 0.3$ s, Load 2 is connected, producing a balanced voltage sag of approximately 15%. In order to compensate this voltage sag, the PCC voltage control system is connected at $t = 0.5$ s and the STATCOM injects reactive power to increase the voltage at the PCC. The simulation ends at $t = 1$ s.

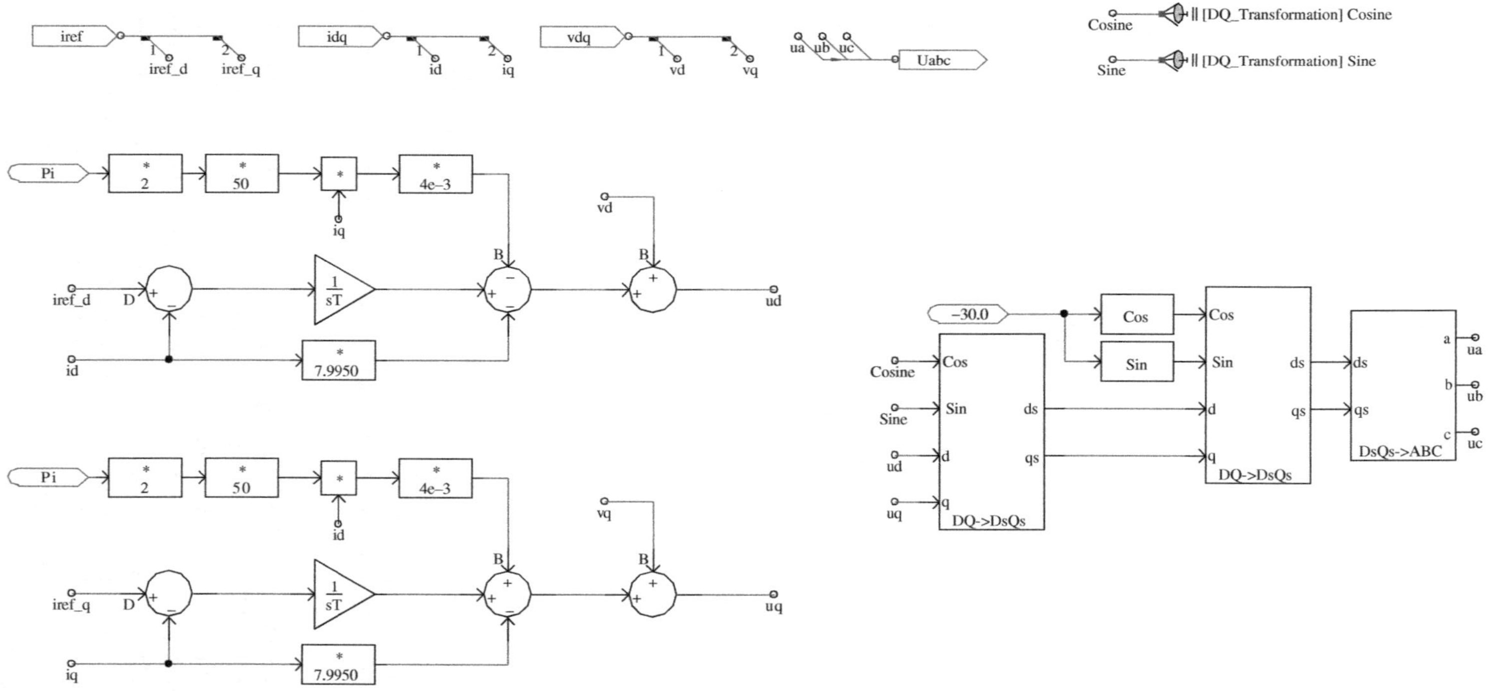

Figure 7.10 Inner control loop (current controllers) for the STATCOM implemented in PSCAD/EMTDC.

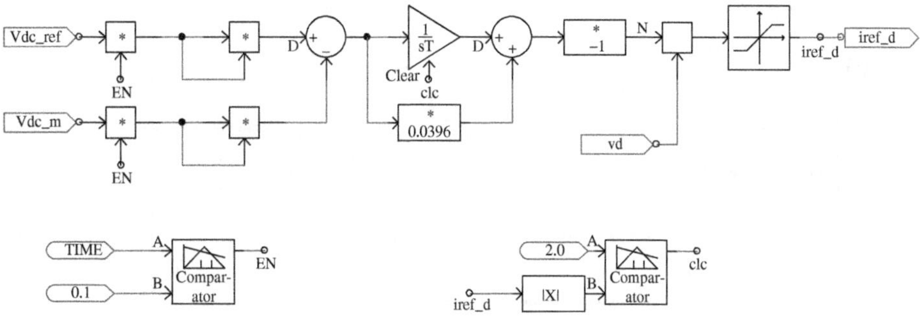

Figure 7.11 Control system of the voltage of the DC capacitor implemented in PSCAD/EMTDC.

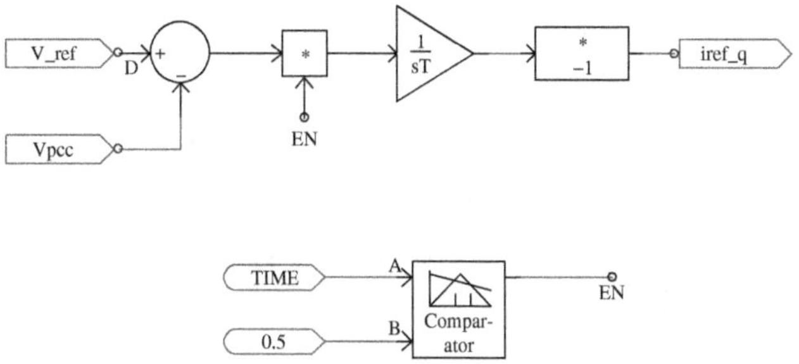

Figure 7.12 Control system of the PCC voltage implemented in PSCAD/EMTDC.

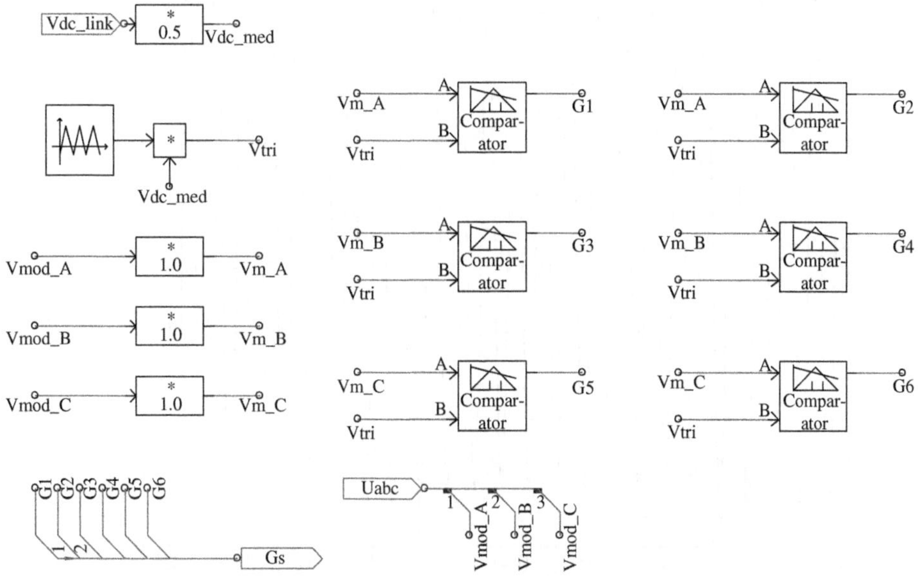

Figure 7.13 Sinusoidal PWM modulation scheme for two-level VSC of the STATCOM implemented in PSCAD/EMTDC.

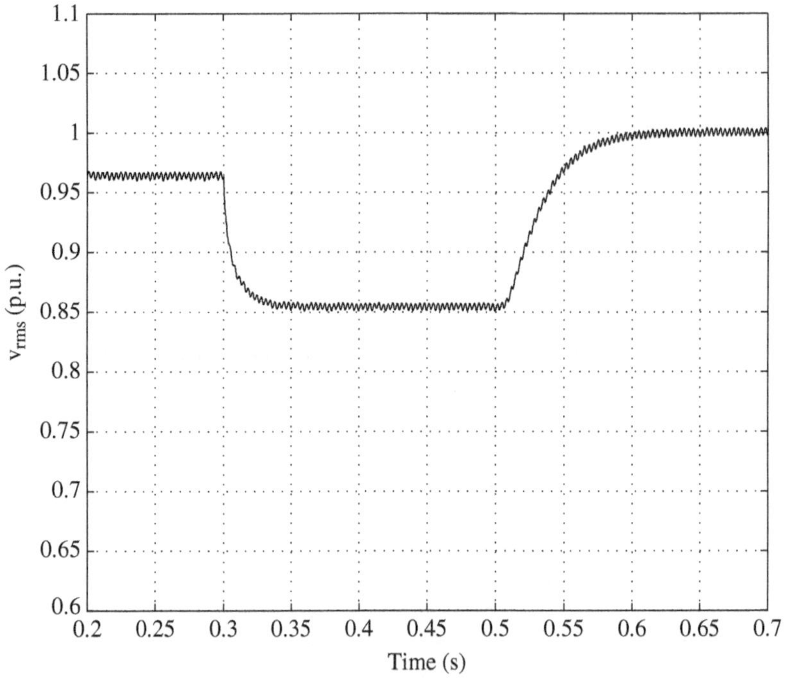

Figure 7.14 RMS voltage at the PCC in p.u.

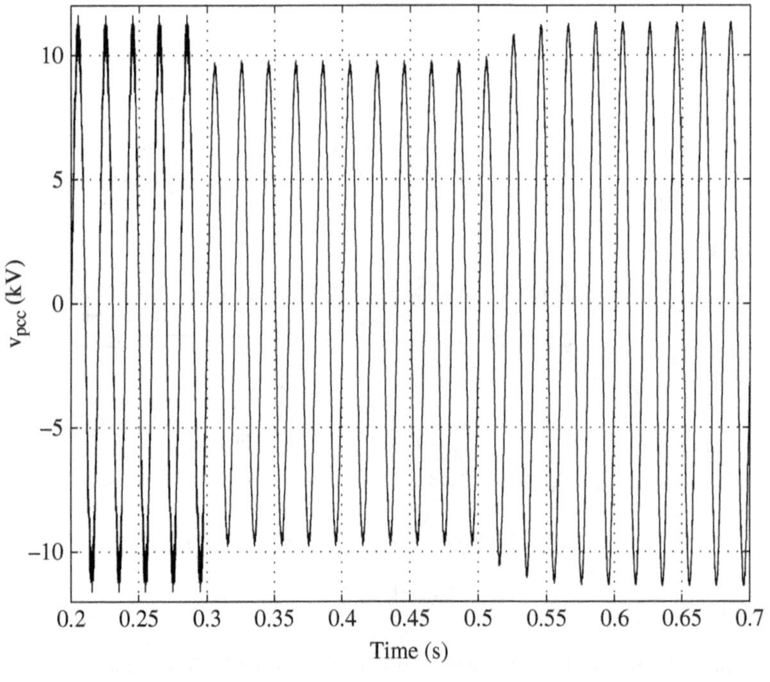

Figure 7.15 Line-to-neutral voltage at the PCC in kV.

Figure 7.14 shows the RMS voltage at the PCC produced by PSCAD/EMTDC. When the STATCOM starts operating as a voltage regulator, the voltage sag is cancelled out and the voltage at the PCC reaches its target value of 1 p.u. The transient response of the line-to-neutral voltage at the PCC is plotted in Figure 7.15 for phase A: it will be observed that the voltage amplitude is kept constant when the STATCOM compensates the voltage sag and the voltage waveform has a low distortion.

The d- and q-axis currents are shown in Figure 7.16a, while the active and reactive powers that the STATCOM exchanges with the grid are plotted in Figure 7.16b. It will be observed that the active power is proportional to i_d, the reactive power is proportional to $-i_q$ and the transient responses are decoupled. The STATCOM provides the necessary reactive power to overcome the voltage sag, while the active power absorbed from the grid remains close to zero.

Figure 7.17 shows the DC capacitor voltage, which is kept constant and equal to 100 kV during steady-state operation. Notice that there is a small transient disturbance in the

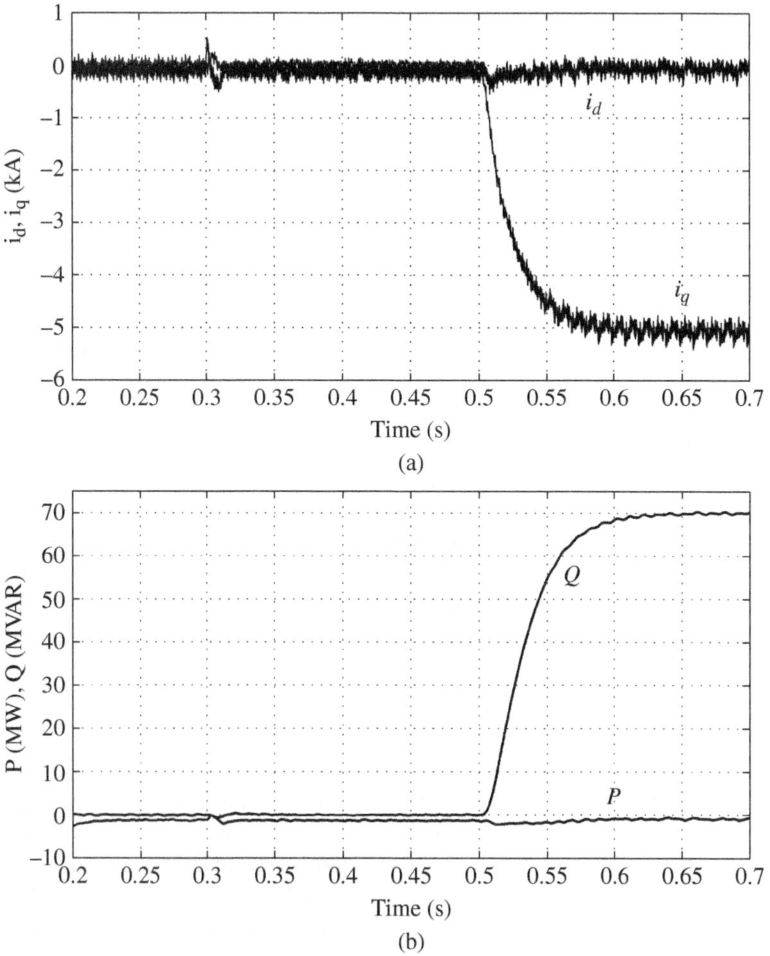

Figure 7.16 STATCOM waveforms: (a) d and q components of the current injected into the grid; (b) active and reactive powers injected into the grid.

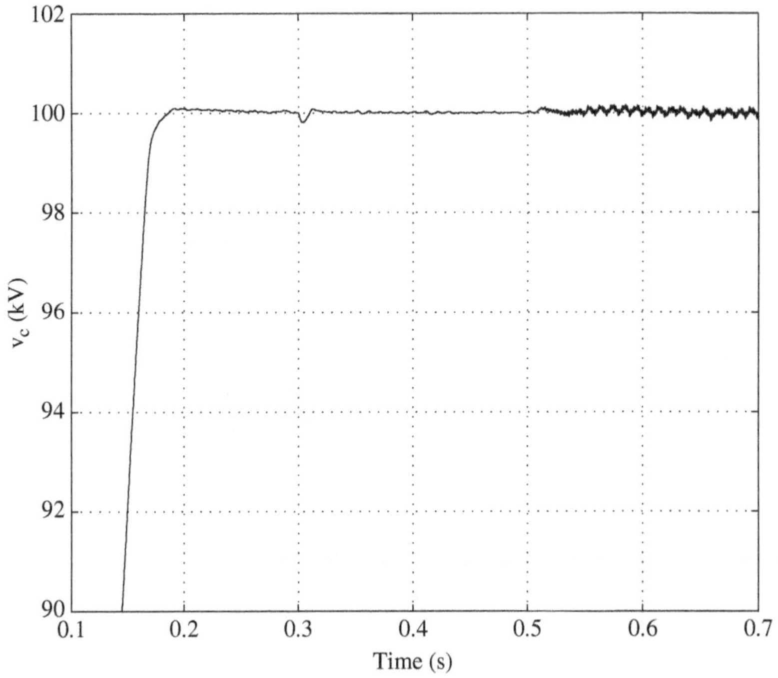

Figure 7.17 DC capacitor voltage in kV.

response when Load 2 is connected and when the STATCOM begins to operate in voltage control mode. These disturbances are quickly corrected by the control system. Since no switching losses are included and the transformer losses are very low in this study case, the active power that the STATCOM needs to absorb from the grid in order to maintain the capacitor voltage level at 100 kV is very small, as shown in Figure 7.16b. It should be noted that the regulator is designed to attain control of v_c^2, and this action implies that the control maintains the DC capacitor voltage at its reference value, as shown in Figure 7.17.

It is quite clear from these results that the injection of reactive power into the grid can be used to regulate voltage at the PCC and to improve on the poor voltage quality caused by the voltage sags. Moreover, the only exchange of active power between the grid and the STATCOM is owing to the ohmic losses in the coupling transformer.

7.3 STATCOM Based on Multilevel VSC

The previous example illustrates how a simple STATCOM can be used to compensate the necessary reactive power in an electrical grid in order to maintain the voltage at the PCC and how it is implemented in PSCAD/EMTDC. Two-level converters are normally used for low-voltage and low-power applications. Nonetheless, as the rating of STATCOMs continues to increase in the realm of reactive-power compensation, the power electronic converters are beginning to be higher-voltage points of the grid, and the use of two-level VSCs is more difficult to justify owing to the high voltages that the switches must block. Multilevel converter topologies have thus been advanced as a means to reduce the voltage stress on the switching devices [9].

A number of multilevel converter topologies have been put forward, although the most popular are neutral-point-clamped converters (NPC), flying capacitor converters (FC) and H-bridge converters [10]. All of them have benefits and drawbacks, and various PWM techniques can be used to draw on the best control characteristics of these converters [11].

This case study addresses the control capabilities of a multilevel STATCOM based on the FC topology, which has been shown to be more flexible than NPC topology when it comes to increasing the number of levels [12].

Figure 7.18 shows a three-phase, three-level FC converter where the capacitor C_1 is the flying capacitor, one per leg, which is charged at $^{2v_c}/_2 = v_c$. Hence, each switch device must block only half of the voltage in comparison to the voltage that the switch of a conventional two-level converter would withstand, thus enabling the use of devices of a lower rating, albeit with a higher number of switches.

The study case developed in Section 7.2 is again used, but this time the two-level VSC is replaced with a three-level FC converter, as shown in Figure 7.18. No parameter change takes place but the phase-shifted pulse width modulation method is used to control the output voltage of the FC converter: for a converter with n levels, n-1 triangular carriers with frequency f_{sw} must be compared with a common sinusoidal modulating waveform per phase with frequency f_m, and $f_{sw} \gg f_m$. The switching instants are obtained at the intersection between the modulating signal and the various carrier signals. A shifting phase of $^{2\pi}/_{n-1}$ is introduced at each carrier signal, thus obtaining an effective switching frequency of $(n-1)f_{sw}$ and resulting in a significant improvement in the total harmonic distortion of the output voltage [13].

A control scheme with which to balance the FC voltages is also needed to cancel out voltage imbalances in the flying capacitors. These voltage imbalances may be caused by

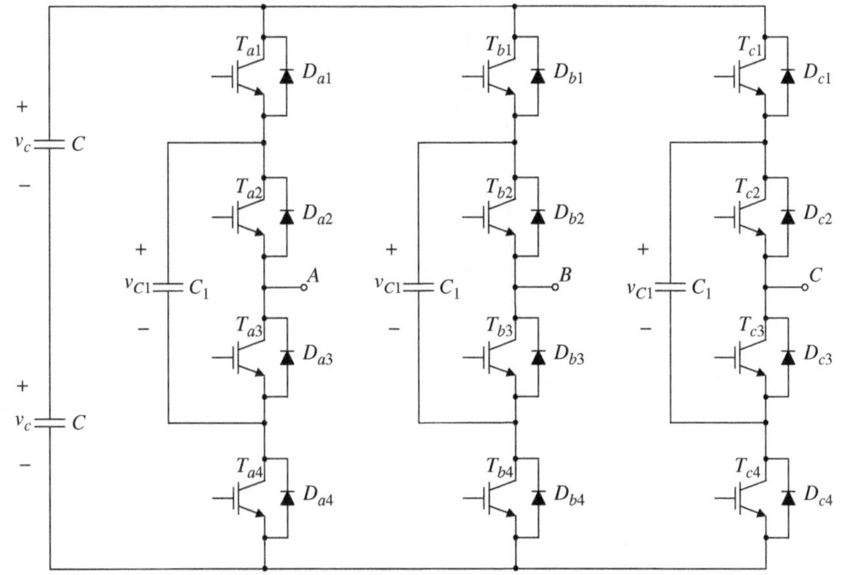

Figure 7.18 Topology of a three-phase, three-level flying capacitor converter.

asymmetrical conditions in the circuit parameters in the bridge, or by differences in the switching of the valves. A closed-loop control system modifies the modulating signal by adding a square waveform to increase or reduce the voltage in the flying capacitors [14]. The control method put forward in this publication for the case of n number of converter levels and m switches connected to the positive terminals of the various flying capacitors, where $m = 1, \ldots, n-1$, makes use of a square waveform v_{sq} which is added to the modulating signal:

$$v_{sq} = A \cdot D \cdot \text{sign}(i) \tag{7.17}$$

where A is the amplitude of the waveform v_{sq}; D is a function that indicates that the duty cycle decreases ($D = -1$), increases ($D = 1$) or remains unchanged ($D = 0$); and variable i stands for the phase current while 'sign' is the sign function.

The FC voltage is compared with the reference voltage and the result of this comparison is used to calculate the value of D by means of a logic function.

Figure 7.19 shows the test system implemented in PSCAD/EMTDC with the three-level FC converter and the control blocks necessary for its correct operation,

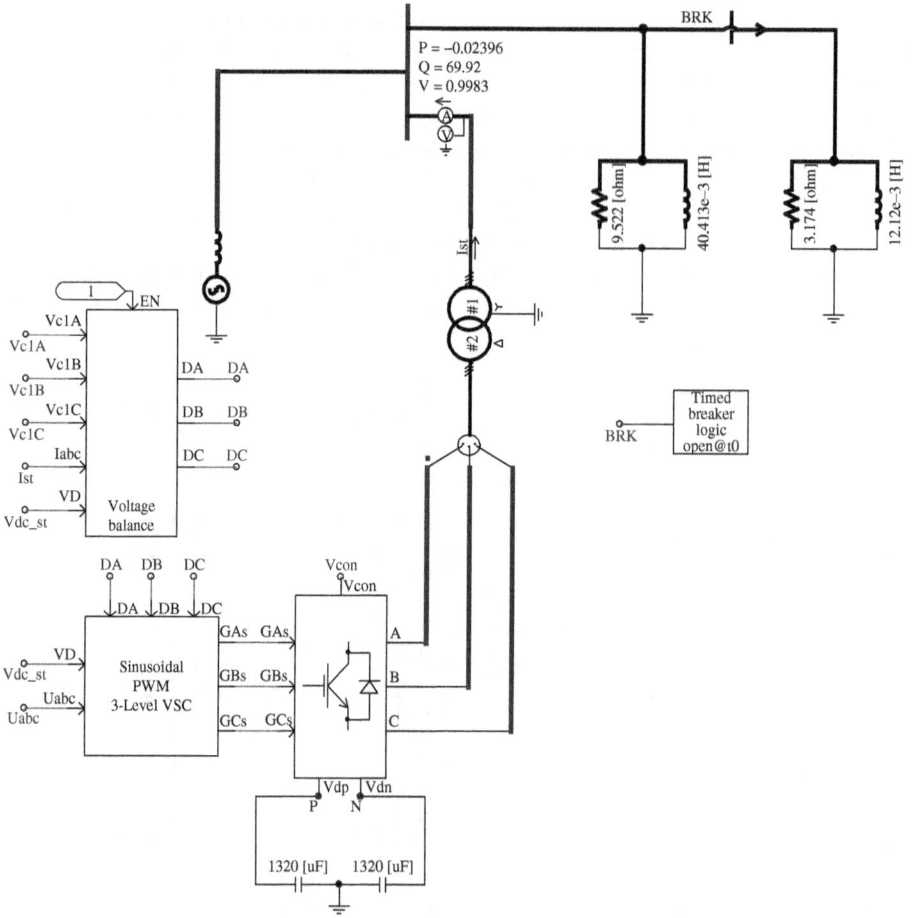

Figure 7.19 STATCOM based on the three-level FC converter implemented in PSCAD/EMTDC.

namely the PWM control process and the method used to balance the flying capacitor voltages.

The parameters of the two loads and the grid remain the same as in the example shown in Section 7.2. The DC capacitor is made of two identical capacitors of 1320 μF each, connected in series and with a midpoint grounded. Experience has shown that by splitting the overall value of capacitor in this manner, the simulation process speeds up.

The PSCAD/EMTDC scheme of the FC converter is shown in Figure 7.20. For this example, the value of the flying capacitors has been set at 110 μF in order to maintain a trade-off between capacitor size and voltage ripple.

The balancing control scheme of the flying capacitor voltages is depicted in Figure 7.21. The amplitude A of the square waveform is initially set at 20 kV in order to quickly charge the flying capacitors to the reference voltage (i.e. half of the voltage of the DC capacitor). This amplitude value is quite high and yields very significant harmonic components in the output voltage of the FC converter. In order to reduce this problem to manageable levels, the amplitude value is reduced to 1 kV after 200 ms, which represents a good compromise to maintain the balance of the flying capacitor voltages and to reduce the harmonic content in the output voltage.

Figure 7.22 shows the implementation of the phase-shifted PWM method to control the FC converter: the two different carrier signals are displaced from one to another by π radians and compared with the modulating signal of each phase in order to obtain the switching instants. The modulating signals are the result of the waveforms obtained by

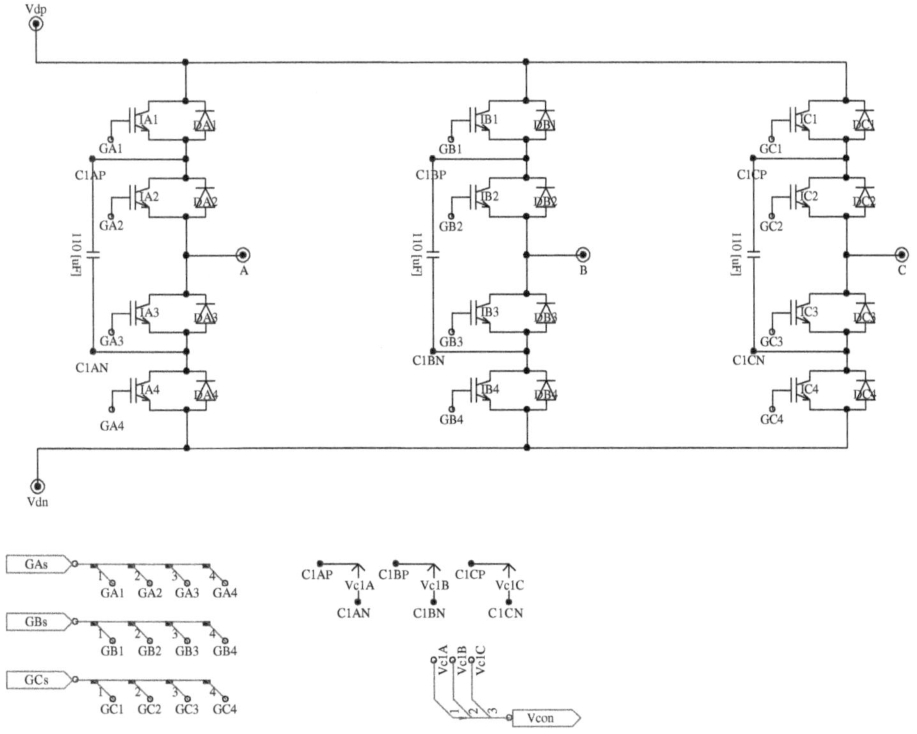

Figure 7.20 Three-level flying capacitor converter implemented in PSCAD/EMTDC.

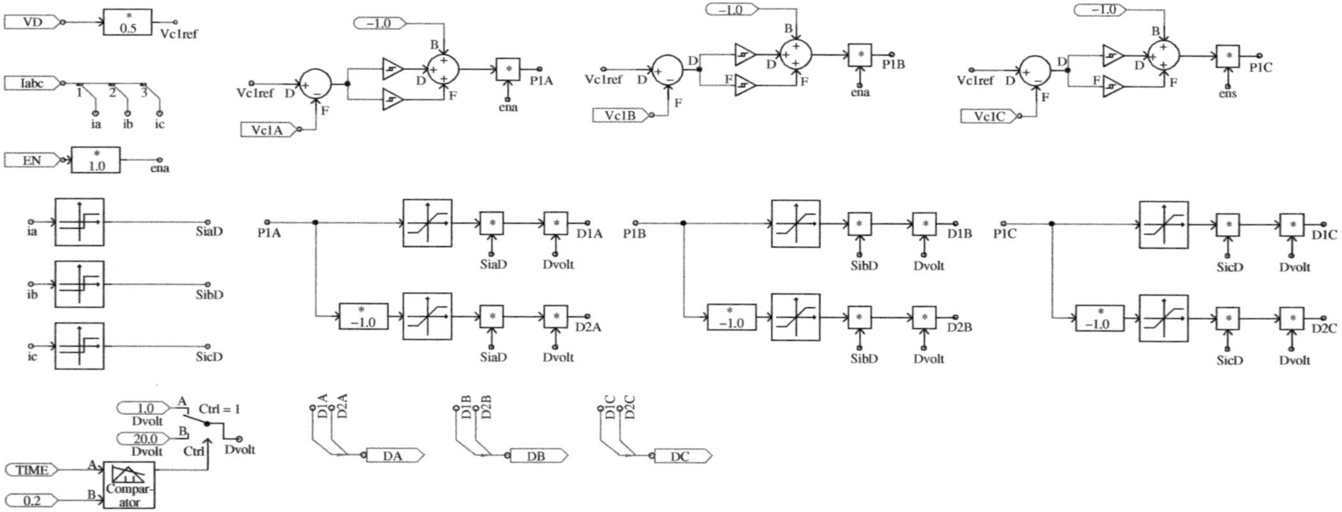

Figure 7.21 Voltage-balance method for the flying capacitors implemented in PSCAD/EMTDC.

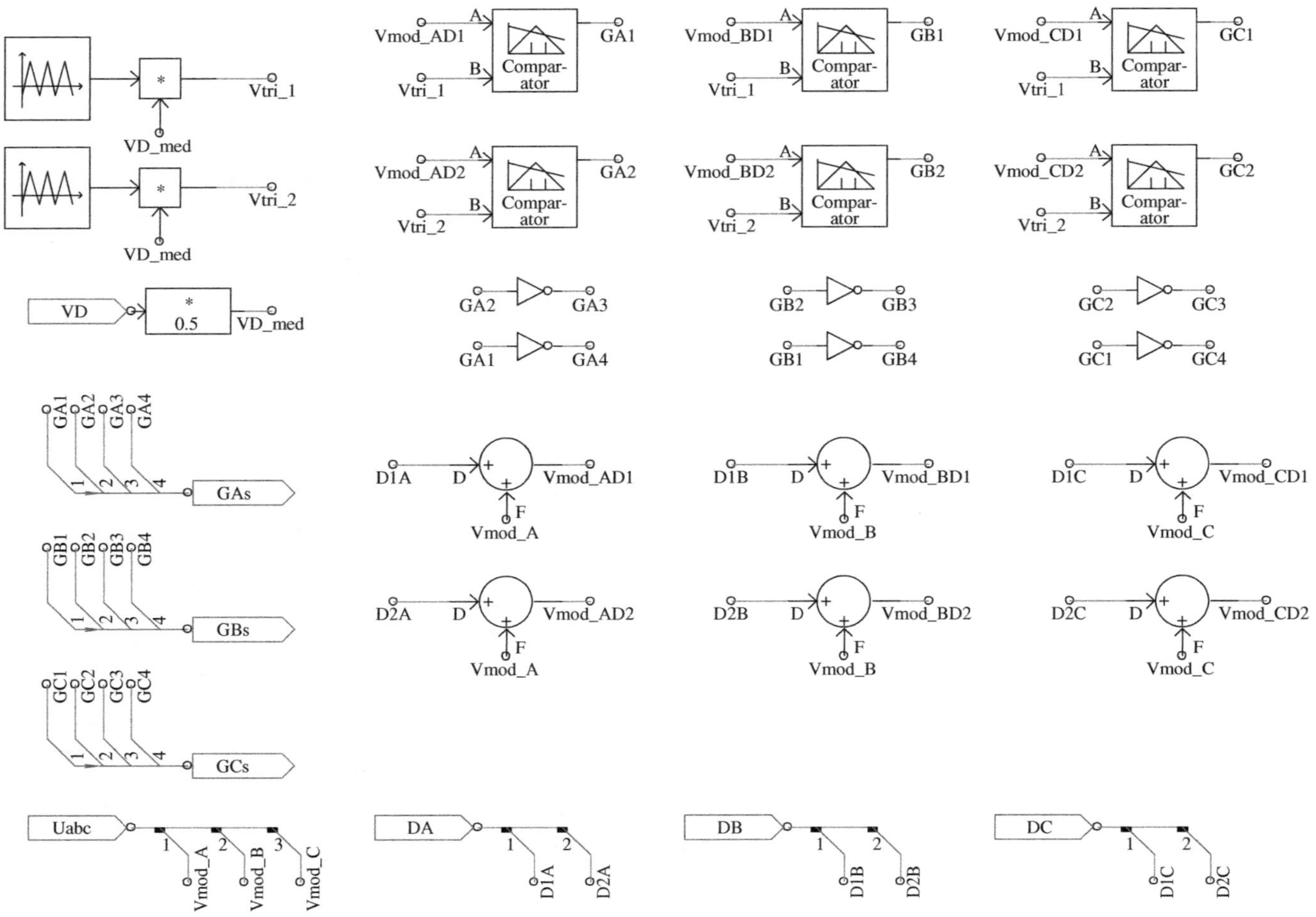

Figure 7.22 Phase-shifted PWM modulation scheme for the three-level FC converter implemented in PSCAD/EMTDC.

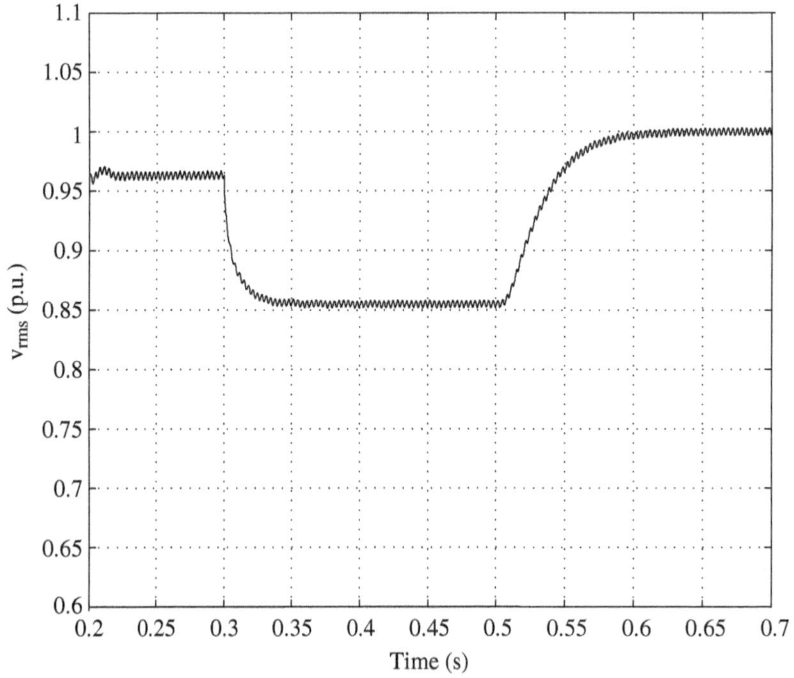

Figure 7.23 RMS voltage at the PCC obtained with the three-level FC converter.

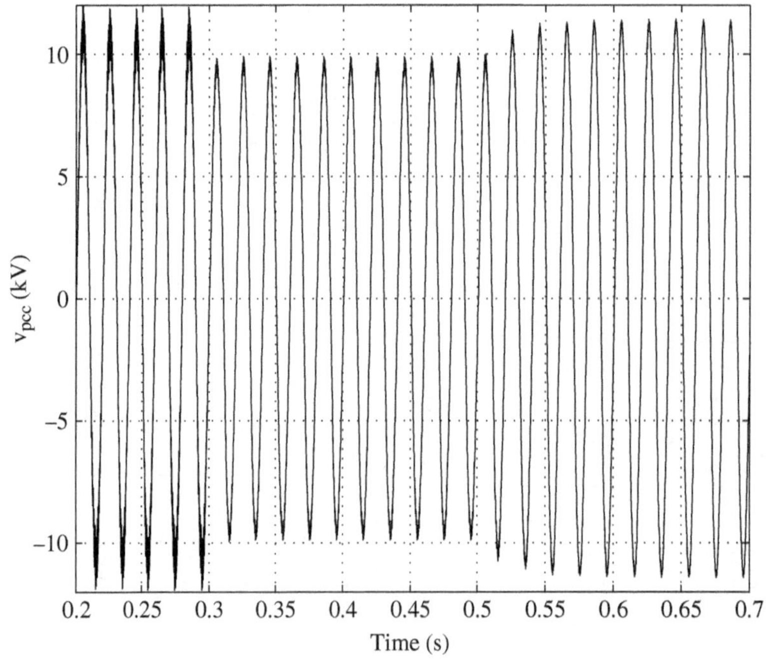

Figure 7.24 Instantaneous line-to-neutral voltage at the PCC.

the current controller plus the square signals generated to maintain the voltage balance of the flying capacitors. The frequency of the carrier signals (i.e. the switching frequency) is set at 1950 Hz.

Figure 7.23 shows the RMS voltage at the PCC: the voltage sag caused by the connection of Load 2 is compensated when the multilevel STATCOM injects reactive power at time $t = 0.5$ s. The corresponding instantaneous line-to-neutral voltage at PCC is shown in Figure 7.24: it should be noted that the waveform is very similar to that obtained with the two-level VSC. A thorough analysis of the harmonic content is shown in Figure 7.25: the harmonic spectrum of the voltage obtained with the two-level VSC is higher than that produced with the three-level FC converter. The voltage total harmonic distortions (THDs) are 2.5% and 1.4%, respectively. Furthermore, the harmonics caused by the switching frequency are located at around 3150 Hz for the two-level case and at around

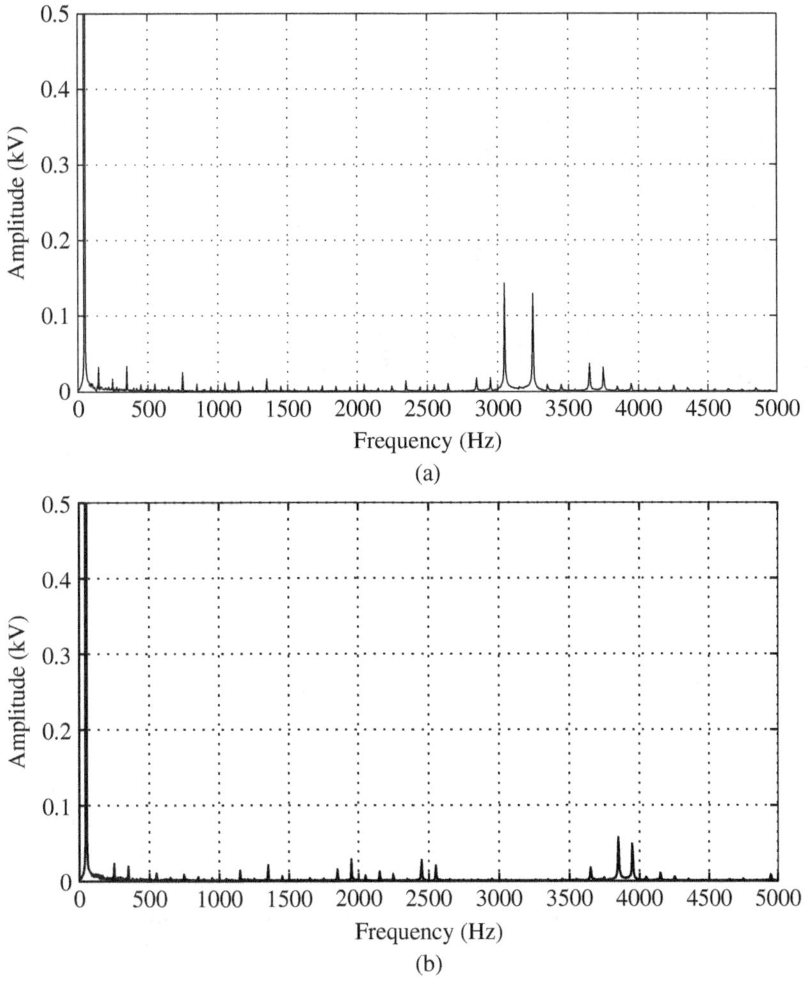

Figure 7.25 Detail of the spectrum of the line-to-neutral voltage at the PCC with: (a) conventional two-level VSC, and (b) three-level FC converter.

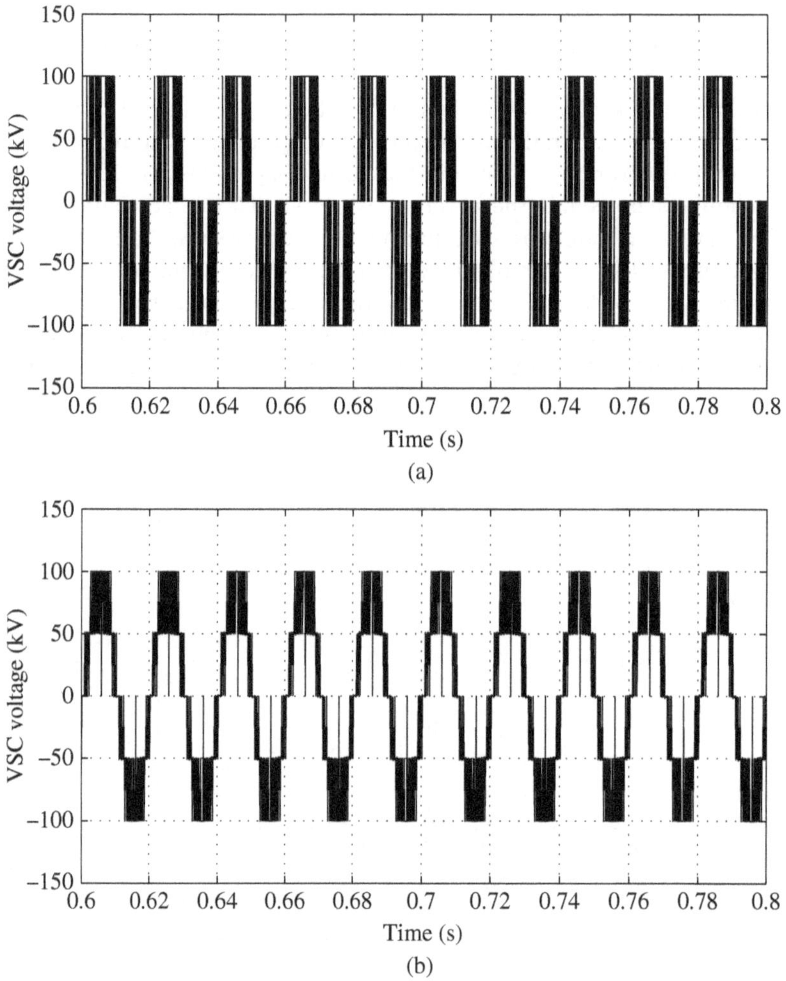

Figure 7.26 Line-to-line output voltage of the VSC obtained with: (a) conventional two-level VSC, and (b) three-level FC converter.

3900 Hz for the three-level case – it should be recalled that the switching frequency was set at 1950 Hz for the FC converter. A detailed view of the converter-output voltage for both types of converters is plotted in Figure 7.26, in which the three levels in the output voltage of the FC converter (Figure 7.26b) can be clearly appreciated. This represents a clear advantage over the waveform of the two-level converter (Figure 7.26a). It could be argued that the former follows a sinusoidal waveform more closely than the latter.

Figure 7.27 shows the d and q current components and the active and reactive powers exchanged with the grid. The time responses of the two current components are again decoupled; the active power exchanged with the grid is proportional to i_d and is very close to zero: the small value is necessary to compensate the losses of the transformer and the switching internal losses of the valves. In this test case, no switching losses are considered. The reactive power is proportional to $-i_q$, and the injected amount of

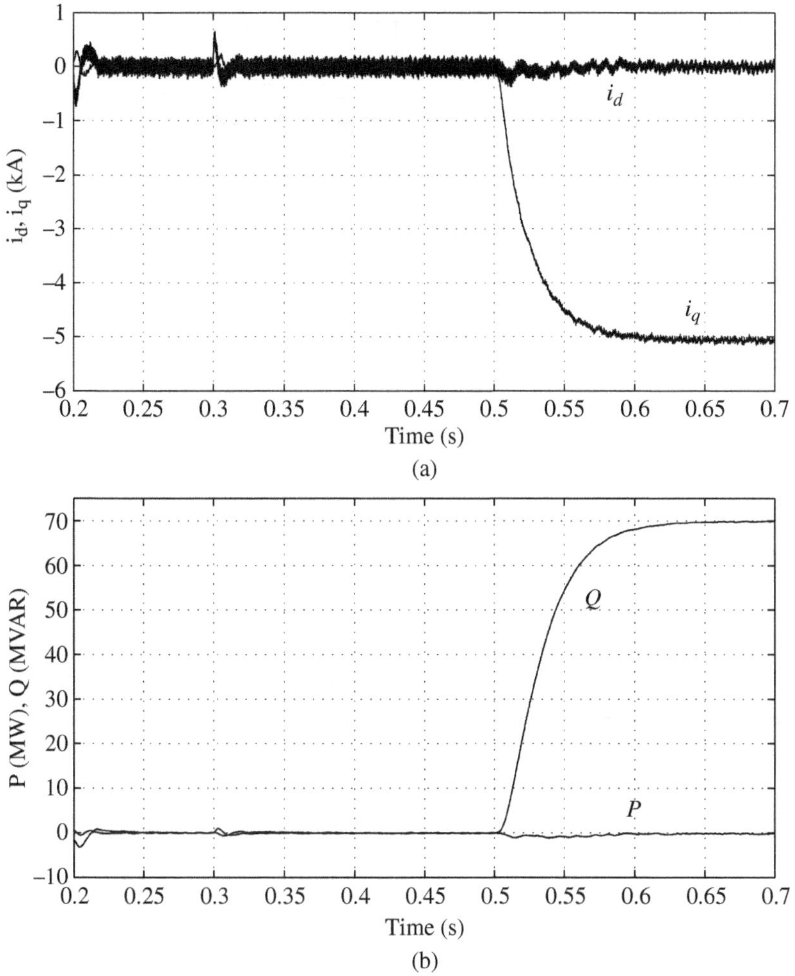

Figure 7.27 Waveforms of the STATCOM obtained with the three-level FC converter: (a) *d* and *q* components of the current injected into the grid; and (b) active and reactive powers injected into the grid.

reactive power increases to compensate the voltage sag and to maintain the RMS voltage at around 1 p.u., as shown in Figure 7.23.

Figure 7.28 shows the time response of the DC-capacitor voltage: the steady-state value is kept constant at around 100 kV. There are three disturbances at different time instants that the control system of the DC-capacitor voltage quickly corrects, as can be appreciated in the figure.

The voltages of the flying capacitors are plotted in Figure 7.29. The amplitude of the square waveform was initially set at 20 kV in order to charge the capacitors to the reference voltage of 50 kV and, after 200 ms, the amplitude decreases to 1 kV. Figure 7.29 shows that in the time interval $t = 0.2$ s and $t = 0.5$ s, the control system of the voltage balance is not able to maintain the voltages at their reference value, showing small deviations. This is owing to the fact that the value of the current during this interval is almost

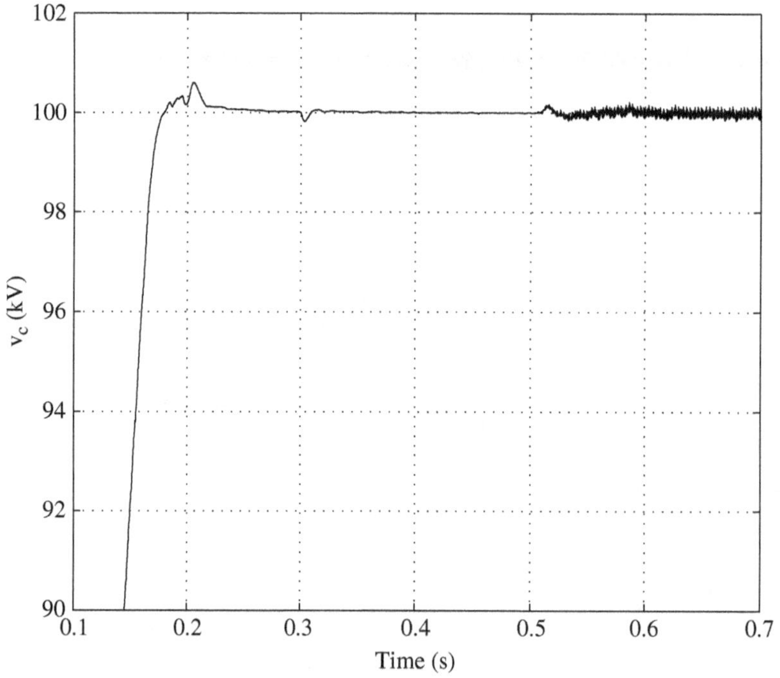

Figure 7.28 Voltage of the DC capacitor in kV.

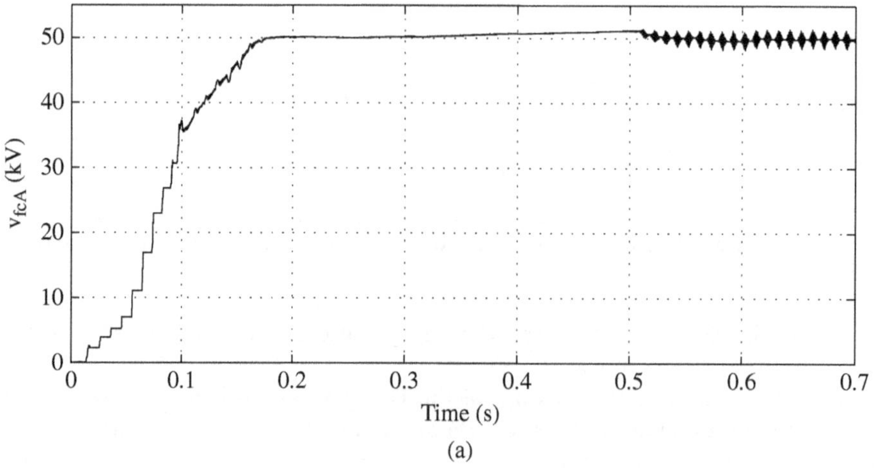

Figure 7.29 Voltages of the three flying capacitors: (a) phase *A*, (b) phase *B*, and (c) phase *C*.

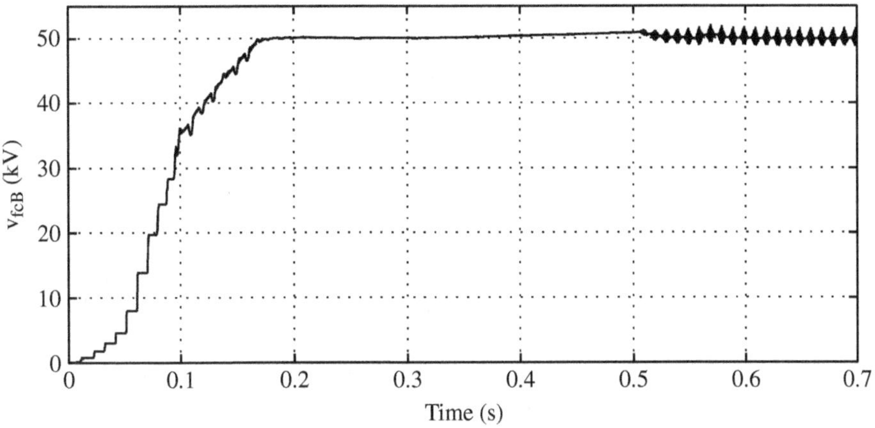

(b)

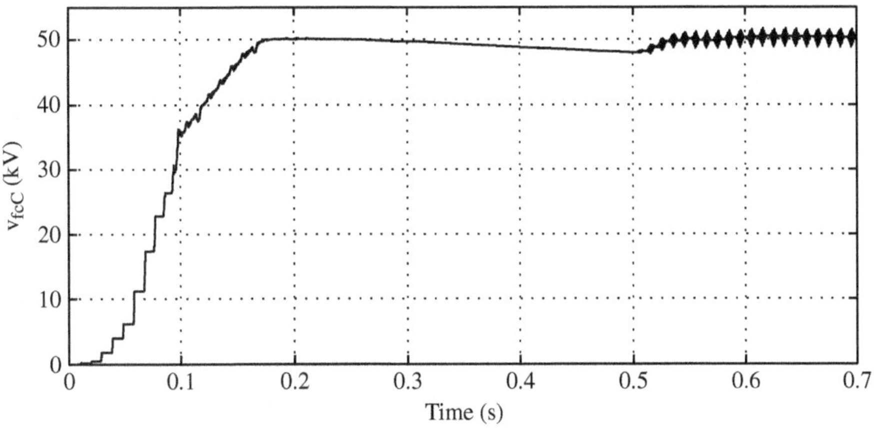

(c)

Figure 7.29 (*Continued*)

zero, as shown in Figure 7.27a. This issue is resolved at $t = 0.5$ s when the q component of the current changes drastically and the voltages of the flying capacitors are effectively balanced.

7.4 Example of HVDC based on Multilevel FC Converter

The following example illustrates the implementation in PSCAD/EMTDC of a monopolar, point-to-point HVDC system with two terminals which employs two three-level FC converters. Figure 7.30 shows the schematic diagram of the HVDC link: this connects two different electrical grids, one to the AC side of the rectifier VSC 1 and the other to the VSC 2. In each case, the connection is carried out by using coupling transformers. Each converter has a capacitor on its DC side as an energy-storage device, and both converters are connected through a DC line. The DC line is modelled as an inductor in series

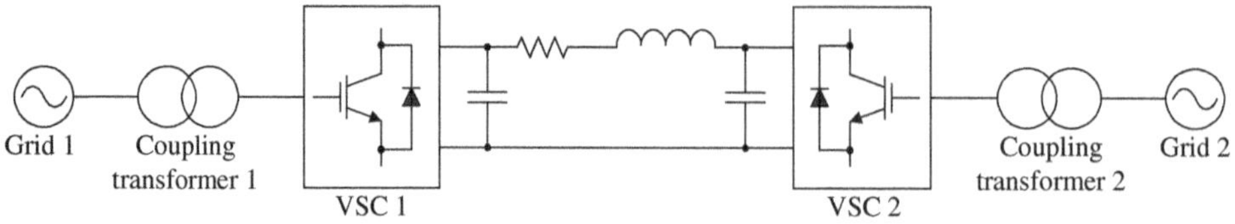

Figure 7.30 Basic scheme of an HVDC system.

with a resistance; these parameters reduce with the length of the line and become zero in the back-to-back HVDC schemes. In these cases, the resistance and the inductance are not taken into account and both converters share one equivalent capacitor.

In the scheme shown in Figure 7.30, the active power flow is bidirectional since the two converters can be operated either in inverter mode or in rectifier mode. Moreover, both converters are able to provide ancillary services such as VAR support, since each VSC controls its own reactive power. One of the converters controls the DC voltage in order to maintain the power balance, e.g. VSC 1, and the other converter controls the active power setpoint, e.g. VSC 2, [11]. The DC voltage control system should be designed to counteract disturbances, such as power imbalances and losses, and even possible modelling errors caused by inaccuracies in the data set and random variations.

In Figure 7.30, the state-space equation of the capacitor C_1 associated with terminal 1 is:

$$2v_{c1}\frac{dv_{c1}}{dt} = \frac{d(v_{c1}^2)}{dt} = \frac{2}{C_1}(p_2 - p_{loss} - p_1) \tag{7.18}$$

where v_{c1} is the voltage of the VSC 1 capacitor, p_2 is the power of terminal 2, p_{loss} is the term owing to losses and p_1 is the power of terminal 1.

From the discussion above, it follows that the control system of terminal 1 coincides with that of the STATCOM case and this is plotted in Figure 7.31 for completeness. The control scheme of terminal 2, meanwhile, is shown in Figure 7.32, where it will be observed that the converter VSC 2 does not control the DC voltage but rather follows an active-power reference. It should be noted that the value of this reference can be either positive or negative since the active-power flow is bidirectional.

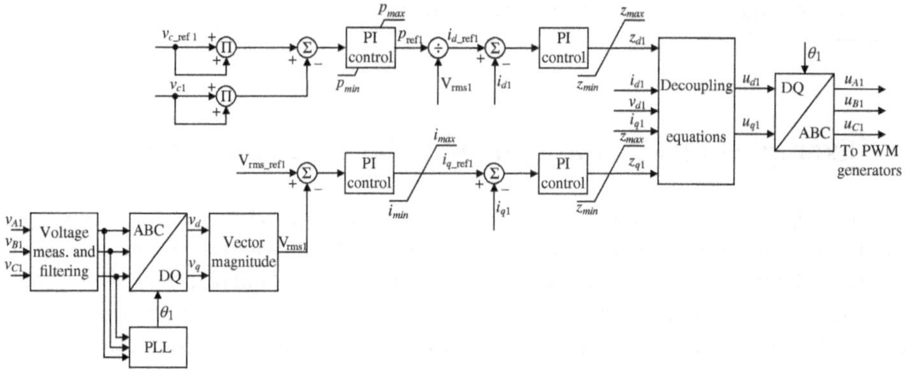

Figure 7.31 Control system for terminal 1 of the HVDC. Source: ©MDPI, 2014.

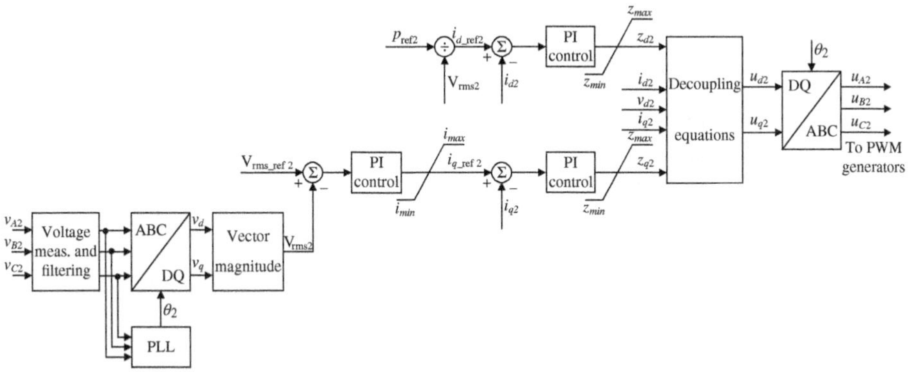

Figure 7.32 Control system for terminal 2 of the HVDC.

The method used to design the various PI regulators is the same as that explained in Section 7.2.

The two-terminal HVDC system implementation in PSCAD/EMTDC is shown in Figure 7.33. The two electrical grids with different network parameters and various kinds of loads connected to their respective points of common coupling have the following characteristics: the line-to-line voltage and frequency of the grid connected to terminal 1 are 13.8 kV and 50 Hz, while the rated voltage and frequency of the grid connected to terminal 2 are 13.8 kV and 60 Hz. The loads connected to both power systems are used to illustrate the ancillary service capabilities that the HVDC system should be able to provide to the two electrical grids, e.g. voltage support. The load characteristics are as follows:

- Grid 1:
  - *Load 1.* Constant active power 30 MW.
  - *Load 2.* Active power 4.76 MW and inductive reactive power 12.12 MVAR. Load 2 is connected by means of a circuit breaker at time instant $t = 0.25$ s.
- Grid 2:
  - *Load 1.* Constant active power 80 MW and constant inductive reactive power 15 MVAR.
  - *Load 2.* Constant capacitive reactive power 40 MVAR. Load 2 is connected by means of a circuit breaker at time instant $t = 1.2$ s.

The coupling transformers have winding ratios of 20 kV/62.5 kV with a star/delta connection and the same parameters as those used in the transformers in the example shown in Section 7.2. Each terminal uses a 660 μF capacitor as DC energy storage: these capacitors are made up of two capacitors of 1320 μF in series with their midpoints grounded in order to speed up the simulation process, as explained in the example in Section 7.3. The DC line is modelled as a resistance of 2.5 Ω in series with an inductance of 2 mH. The three-level FC converters use 47 μF flying capacitors. The reference for the DC voltage is 100 kV and it is controlled by terminal 1.

The switching frequencies of the converters are 3150 Hz and 3780 Hz for terminal 1 and terminal 2, respectively. In both cases, the frequency modulation index is $m_f = 63$.

The parameters of the various controllers have been obtained by using the root-locus technique and they have the same values as those obtained in Section 7.2, since

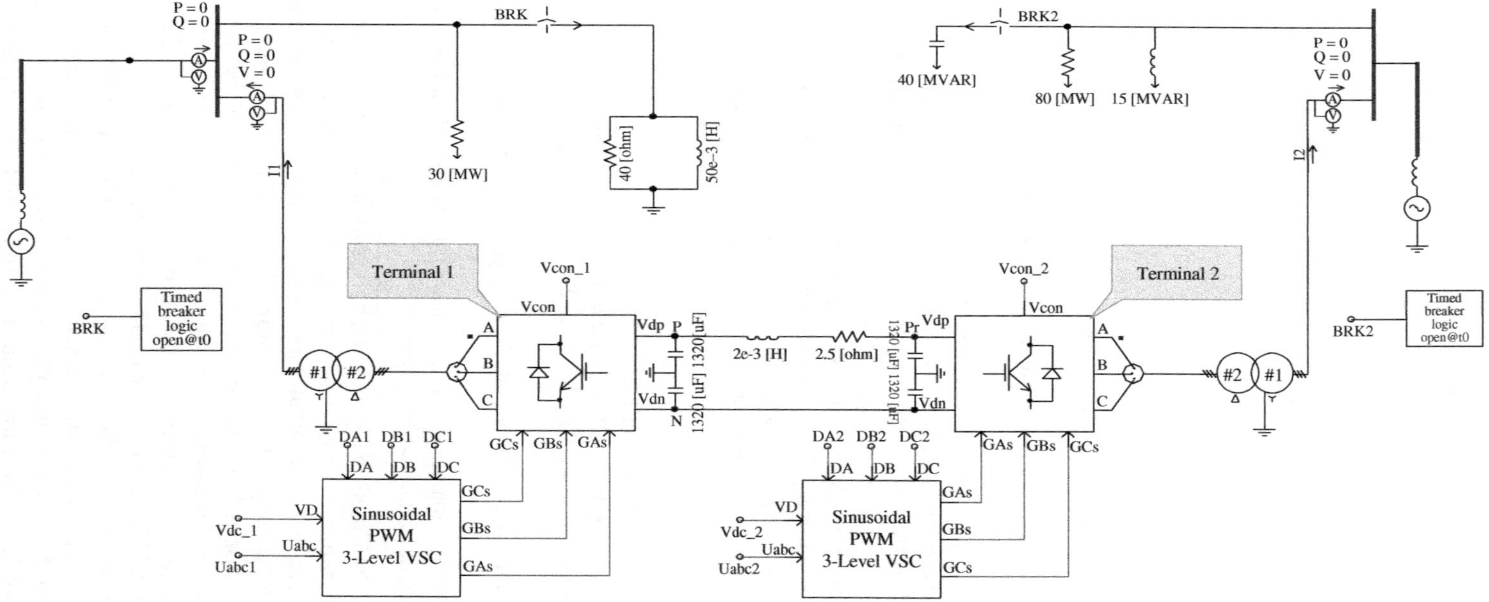

Figure 7.33 Two-terminal HVDC system implemented in PSCAD/EMTDC.

the parameters of the coupling transformers and the DC capacitors have remained unchanged.

The control scheme used to balance the flying capacitor voltages in Section 7.3 has been used again. The amplitude of the square waveform is initially set at 12 kV in order to charge the flying capacitors very quickly, and this amplitude is then reduced to 600 V at time instant $t = 0.2$ s to maintain the voltages of the flying capacitors at around their reference values.

The event sequence is the following: the current controllers, the DC voltage regulator at terminal 1 and the controllers used to balance the flying capacitor voltages in both terminals are, initially, the only control systems enabled. At this time interval, the reference of active power, i.e. the reference of i_d, in terminal 2 is set at zero. After 250 ms, the DC voltage reaches its reference value of 100 kV and the control systems of the PCC voltages are enabled at both terminals. This action releases the full STATCOM capabilities of the two converters to their respective electrical grids. At time instant $t = 0.5$ s, the active-power reference in terminal 2 is set at 40 MW, i.e. terminal 2 injects active power into grid 2, and at $t = 0.9$ s this reference value changes to -10 MW, i.e. terminal 2 absorbs active power from grid 2. The simulation ends at $t = 1.5$ s.

Figure 7.34 shows the most relevant parameters of grid 1, while those of grid 2 are plotted in Figure 7.35. The HVDC system is able to compensate in steady-state the variations of voltage at their respective points of common coupling, supplying or absorbing the necessary reactive power to maintain these voltage values at 1 p.u., as shown in Figures 7.34a and 7.35a, respectively. A salient point of this compensation is when Load 2 of grid 2 is connected at $t = 1.2$ s: the converter at terminal 2 of the HVDC link injects reactive power into the grid and, at this instant, Load 2 provides a very significant amount of additional reactive power which is quickly absorbed by the converter at terminal 2 in order to maintain the PCC voltage equal to 1 p.u., as can be seen in Figure 7.35a and, more importantly, in Figure 7.35b. The HVDC transmits power from one terminal to another: when the active-power reference at terminal 2 is set at 40 MW, the converter at terminal 1 absorbs active power from grid 1 and transmits it in DC form through the cable to reach the converter at terminal 2 and into grid 2, excepting the losses. When the power reference changes from 40 to -10 MW, the converter at terminal 2 absorbs active power from grid 2 and this energy, excepting the losses, is injected into the converter at terminal 1 and into grid 1. These results are illustrated in Figures 7.34b and 7.35b.

As in the previous examples, the power is proportional to the d component of the current, whereas the reactive power is proportional to the q component of the current, as can be seen by comparing Figures 7.34b,c for the case of grid 1 and Figures 7.35b,c for the case of grid 2.

Figure 7.36 shows a detailed view of the voltages at the points of common coupling for both grids. The time interval is $0.6\ \text{s} < t < 0.8$ s, where the converter at terminal 1 absorbs 40.7 MW from grid 1 and injects, approximately, 19 MVAR into grid 1. The converter at terminal 2 injects 40 MW and 18 MVAR into grid 2. The multilevel-converter configuration achieves its operation with low-distortion output voltages, signifying that the harmonic content of the PCC voltages is very much reduced.

The voltages of the DC capacitors are plotted in Figure 7.37 for both converters. A detailed view of these voltages is shown in Figure 7.37c for the capacitor voltage of the converter at terminal 1 and in Figure 7.37d for the capacitor voltage of the converter at terminal 2. Note that the capacitor voltage of the converter at terminal 1 has a value of 100 kV in steady-state and that the difference between the two capacitor voltages

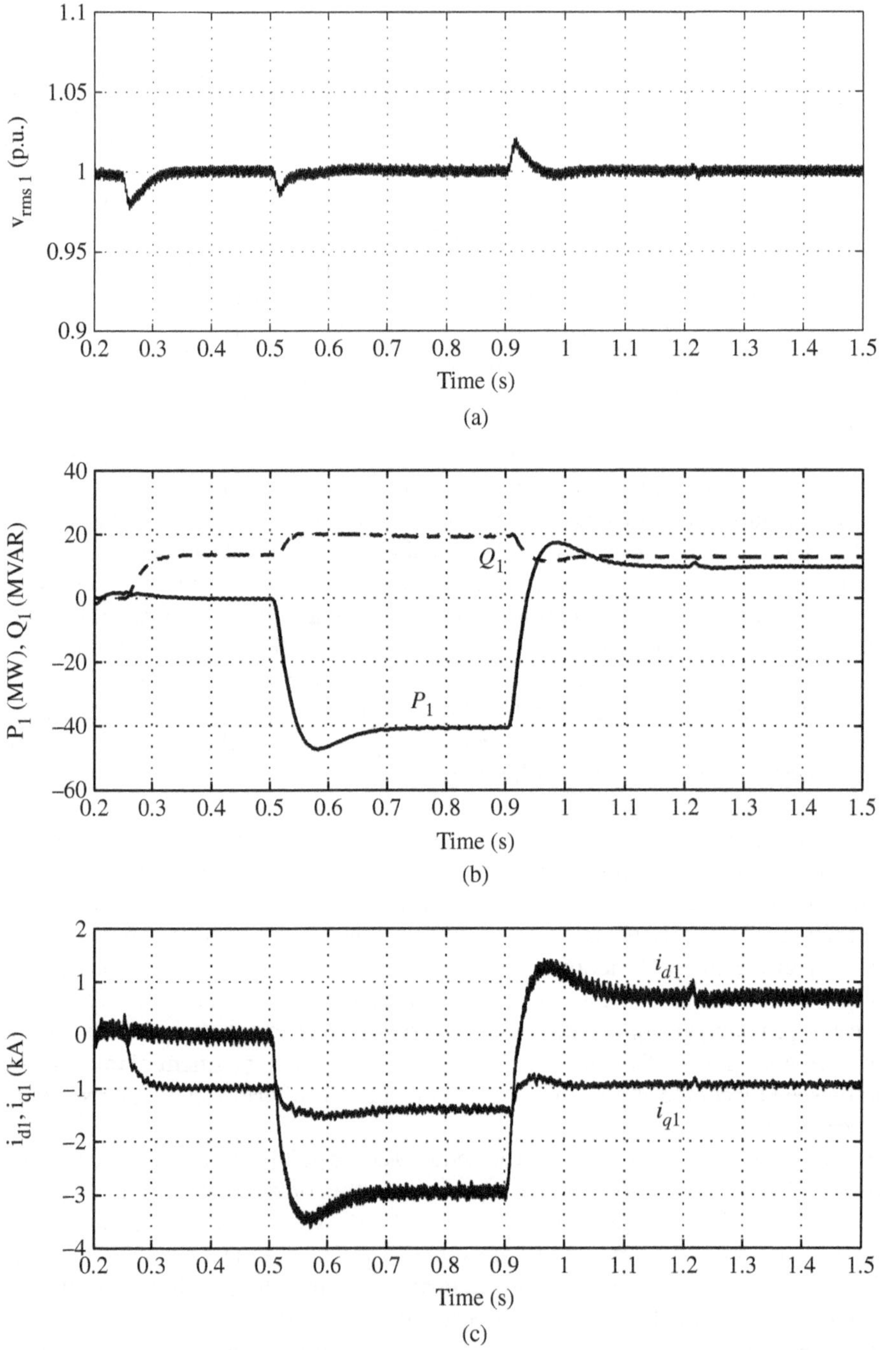

Figure 7.34 Main electrical magnitudes of grid 1: (a) RMS voltage at PCC, (b) active and reactive powers, and (c) *d* and *q* components of the current injected into the grid by the VSC 1.

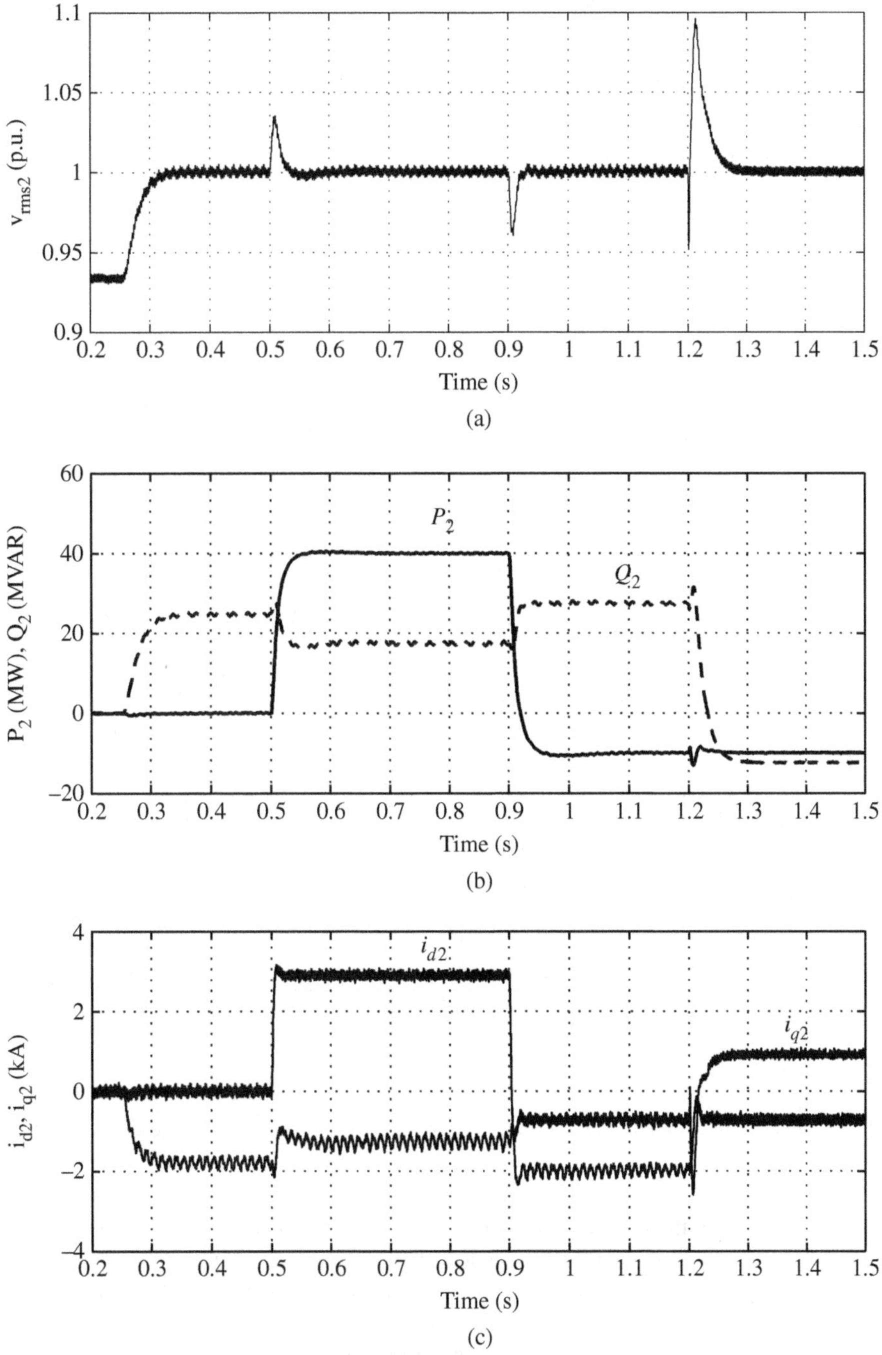

Figure 7.35 Main electrical magnitudes of grid 2: (a) RMS voltage at PCC, (b) active and reactive powers, and (c) *d* and *q* components of the current injected into the grid by the VSC 2.

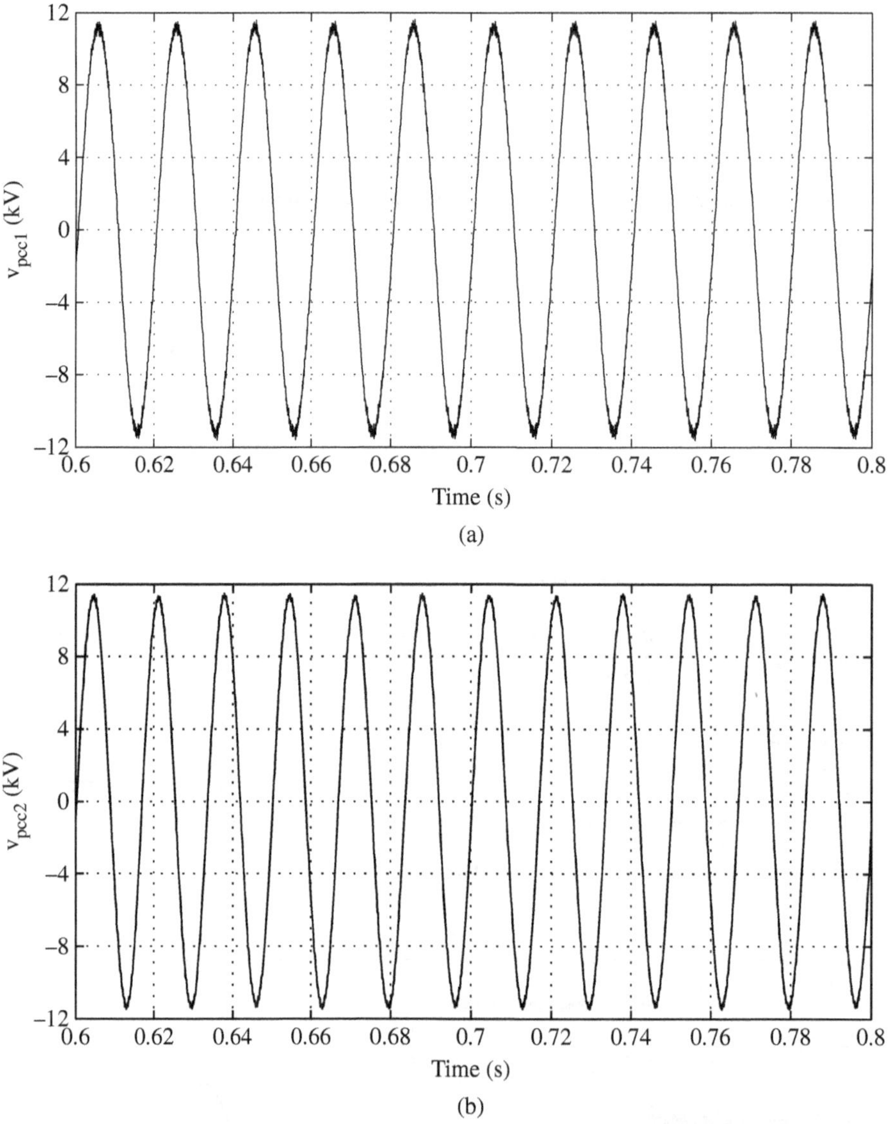

Figure 7.36 Detail of the PCC voltages for the time interval 0.6 s < t < 0.8 s: (a) voltage at PCC 1, and (b) voltage at PCC 2.

is owing to the losses in the DC line. When the converter at terminal 2 injects active power into grid 2, its capacitor voltage is lower than the capacitor voltage of the converter at terminal 1, i.e. 100 kV in steady-state. This can be observed in the time interval 0.5 s < t < 0.9 s. Conversely, when the converter at terminal 2 absorbs active power from grid 2, i.e. t > 0.9 s, the capacitor voltage of the converter at terminal 2 is higher than the capacitor voltage of the converter at terminal 1.

Figures 7.38 and 7.39 show the flying capacitor voltages of the three-level converters associated with terminal 1 and terminal 2, respectively. The transient response quickly

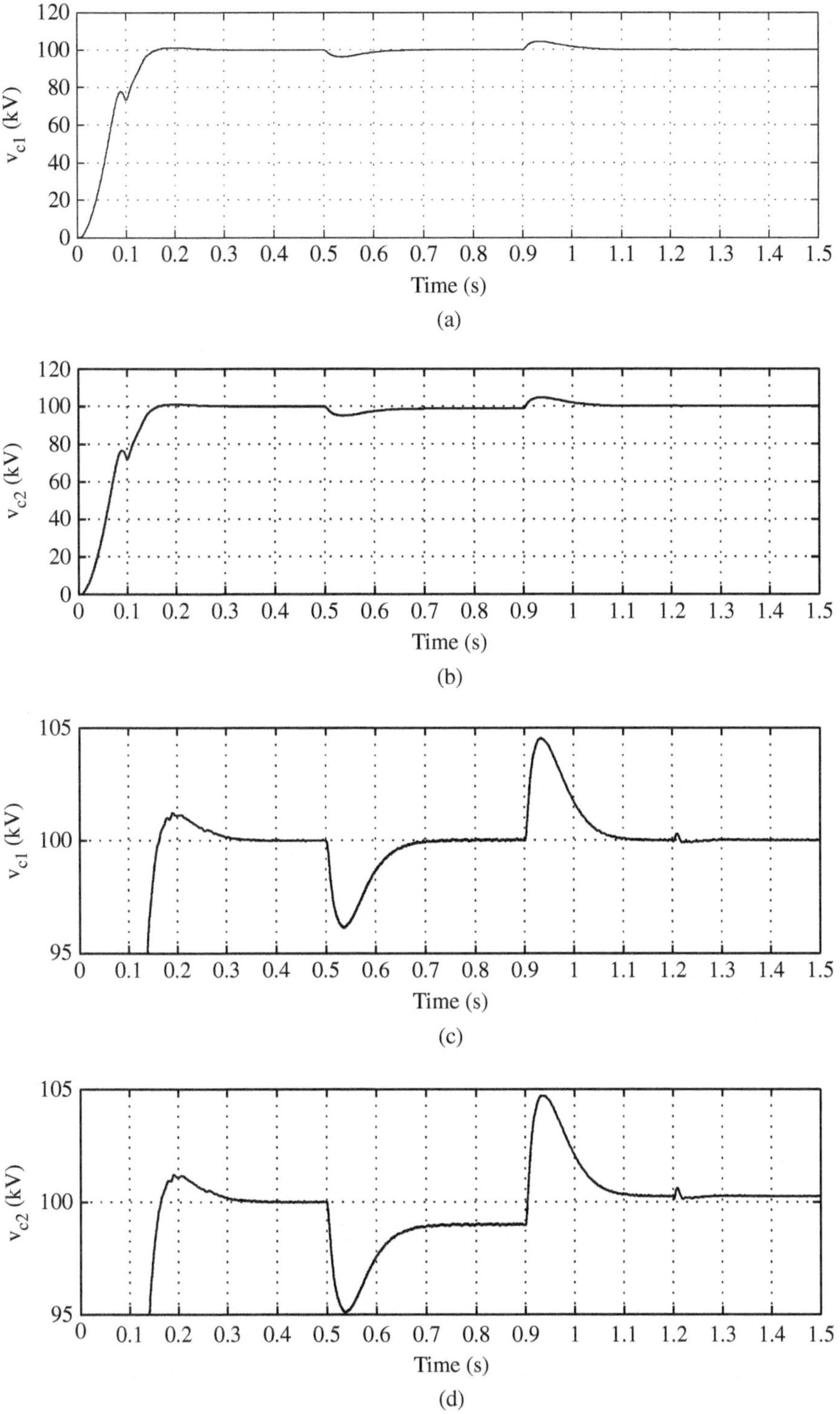

Figure 7.37 Detail of the PCC voltages for the time interval 0.6 s < t < 0.8 s: (a) voltage at PCC 1, and (b) voltage at PCC 2.

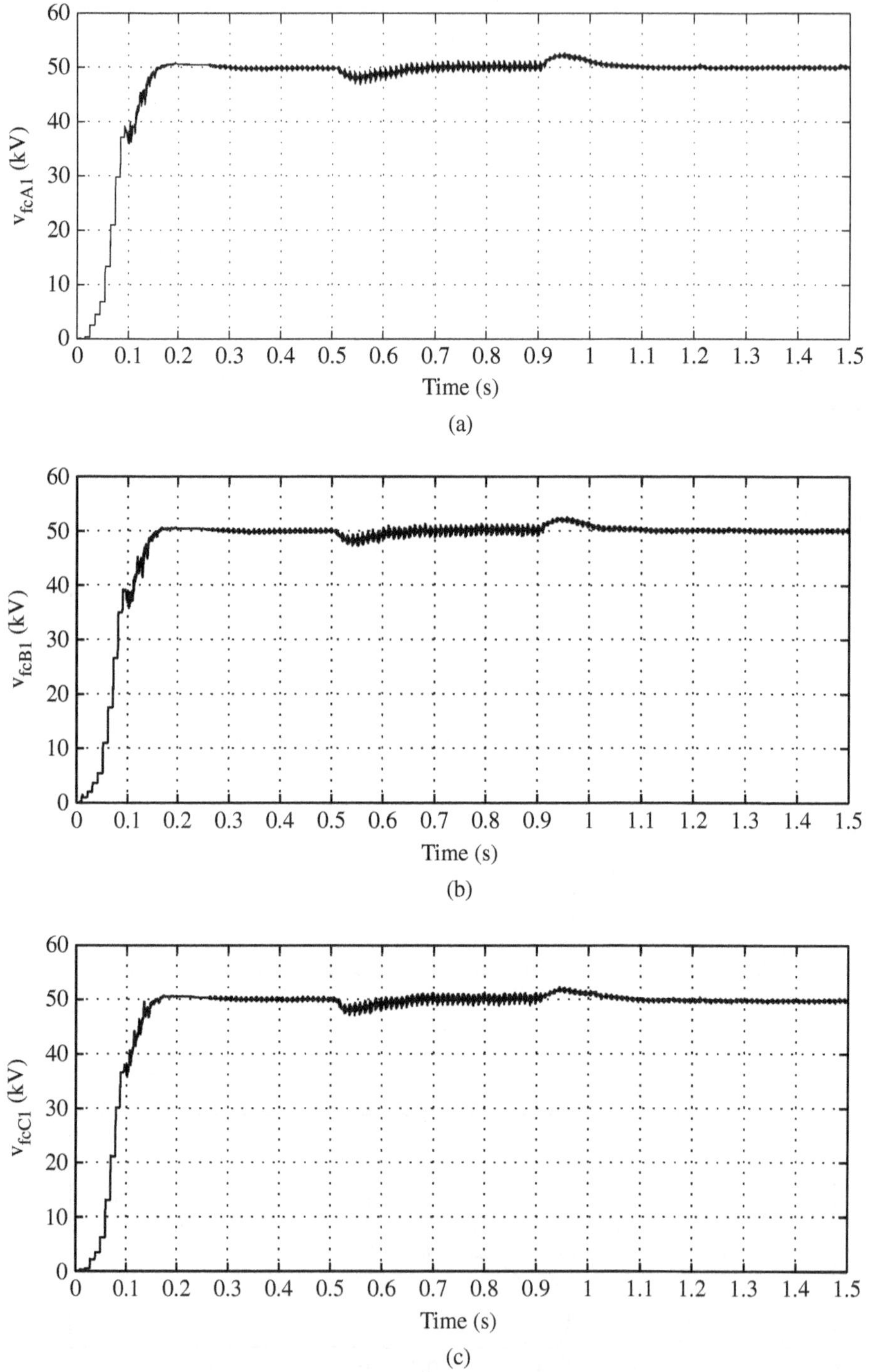

Figure 7.38 Voltages of the three flying capacitors of terminal 1: (a) phase *A*, (b) phase *B*, and (c) phase *C*.

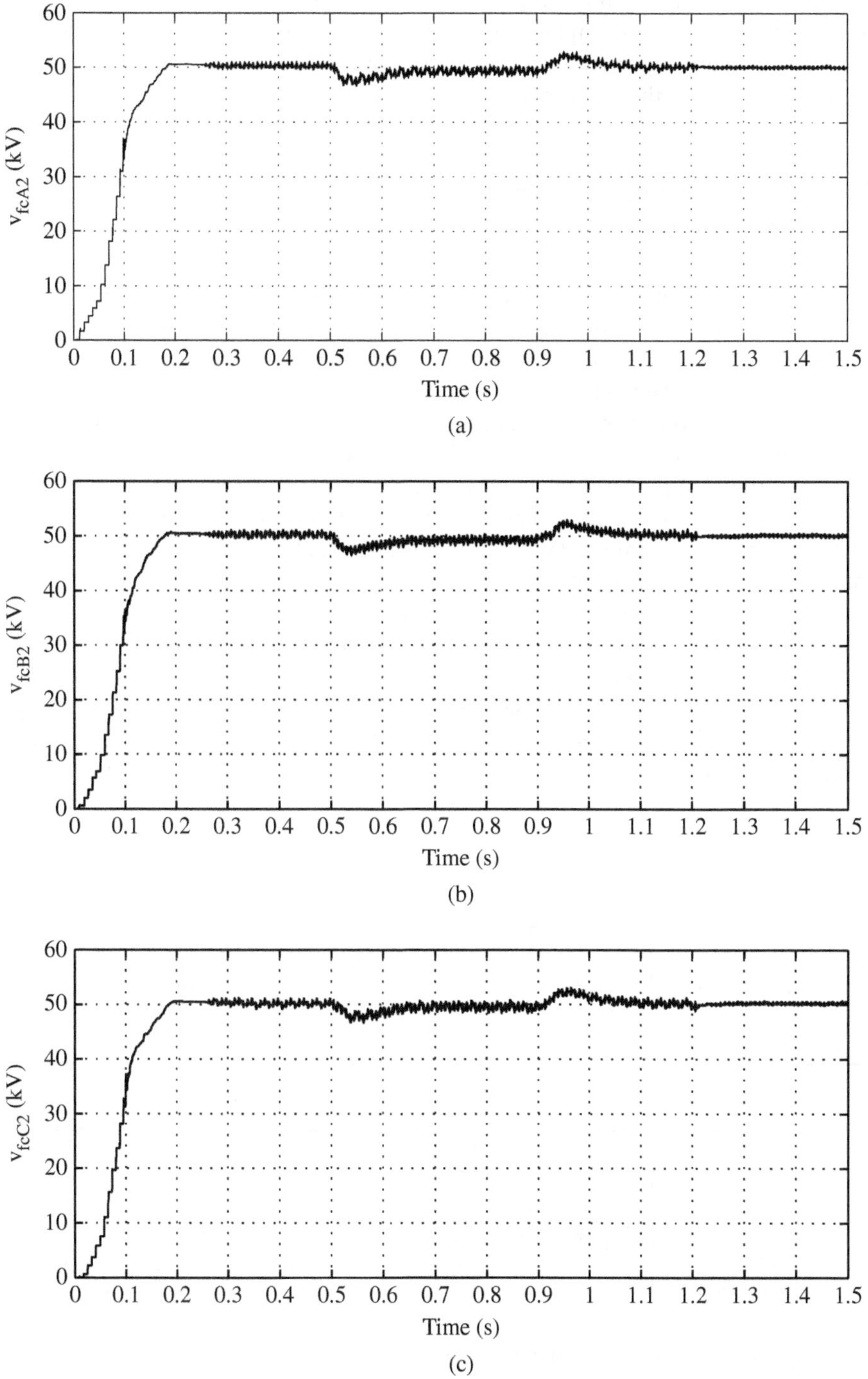

Figure 7.39 Voltages of the three flying capacitors of terminal 2: (a) phase *A*, (b) phase *B*, and (c) phase *C*.

reaches 50 kV owing to the high amplitude of the square waveform (i.e. 12 kV) used at the initial time interval. The amplitude of the square waveform is then reduced to 600 V, which is a sufficiently high value to balance the voltages of the three capacitors in both converters throughout the rest of the simulation experiment.

7.5 Example of a Multi-Terminal HVDC System Using Multilevel FC Converters

There is a great deal of interest in exploring the feasibility of multi-terminal HVDC systems as technical solutions that may yield more competitive solutions than a conventional AC transmission system from the technical and economic vantage points. For instance, the perceived wisdom is that HVDC systems require less investment and smaller footprints for equal power-transmission capabilities. Furthermore, multi-terminal HVDC systems exhibit additional advantages such as the ability to connect a number of power grids in an asynchronous manner, unlike point-to-point HVDC systems which can only exchange active power between two terminals [15].

A number of multi-terminal HVDC system topologies have been put forward in the literature, such as those reported in [15] and in [16] relating to the offshore wind farm installations in the North Sea under current consideration.

A generic multi-terminal HVDC system is shown in Figure 7.40 in which four terminals are connected to a common DC bus. This configuration allows the connection of four different electrical grids, and it has been suggested that this would establish a typical configuration of DC distribution systems for city centres [17], although many other topologies are possible.

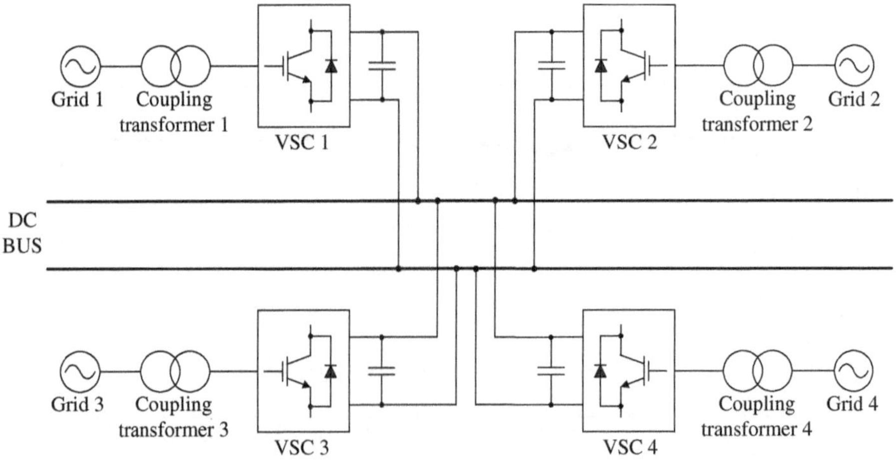

Figure 7.40 Scheme of a multi-terminal HVDC system.

In multi-terminal HVDC systems, one of the terminals is dedicated to the control of the DC bus voltage and in many ways acts as the voltage reference, whereas the other terminals operate in active-power control mode. All the terminals are initially disconnected from the DC bus and they work in DC voltage control mode. This feature allows the converter capacitors at each terminal to be independently energized from the AC side in order to attain the same voltage value. At this time, all the converters are connected to the DC bus and one and only one of the converters takes charge of controlling the DC bus voltage while the others can be used to control the exchange of active power with their respective AC sides.

The block diagram for the terminal which controls the voltage at the DC bus is depicted in Figure 7.3, which coincides with that of DSTATCOM. Figure 7.41 shows the general control scheme for the remaining terminals of the HVDC system: the subscript n applies to any of the non-voltage control type converters; instead, they have a setpoint for active power. The terminal VSCs initially charge their capacitors at the desired voltage levels. When this voltage level is reached, the terminal converters are connected to their respective DC buses and their control systems switch to the active-power control mode. The rest of the control functions used in the multi-terminal HVDC system are identical to those described for the terminal that controls the DC bus voltage.

The PSCAD/EMTDC implementation of a four-terminal HVDC system is shown in Figure 7.42. Each terminal comprises a three-level FC converter with a DC capacitor of 330 μF which plays the role of DC energy storage system. Terminal 1 controls the DC bus voltage while the other three terminals operate in their active-power control mode. A total equivalent capacitor of 1320 μF is obtained when the four terminals are connected to the DC bus. The main parameters of the coupling transformers associated with each terminal are summarized in Table 7.1, while the parameters of the different AC grids are shown in Table 7.2.

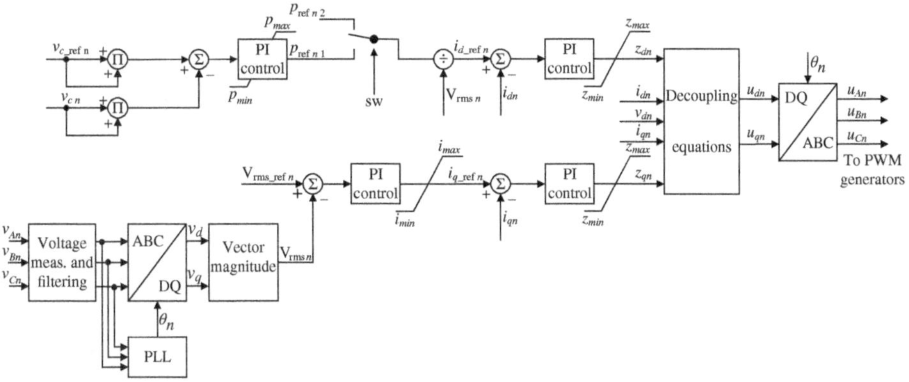

Figure 7.41 Control scheme of a terminal with active-power reference.

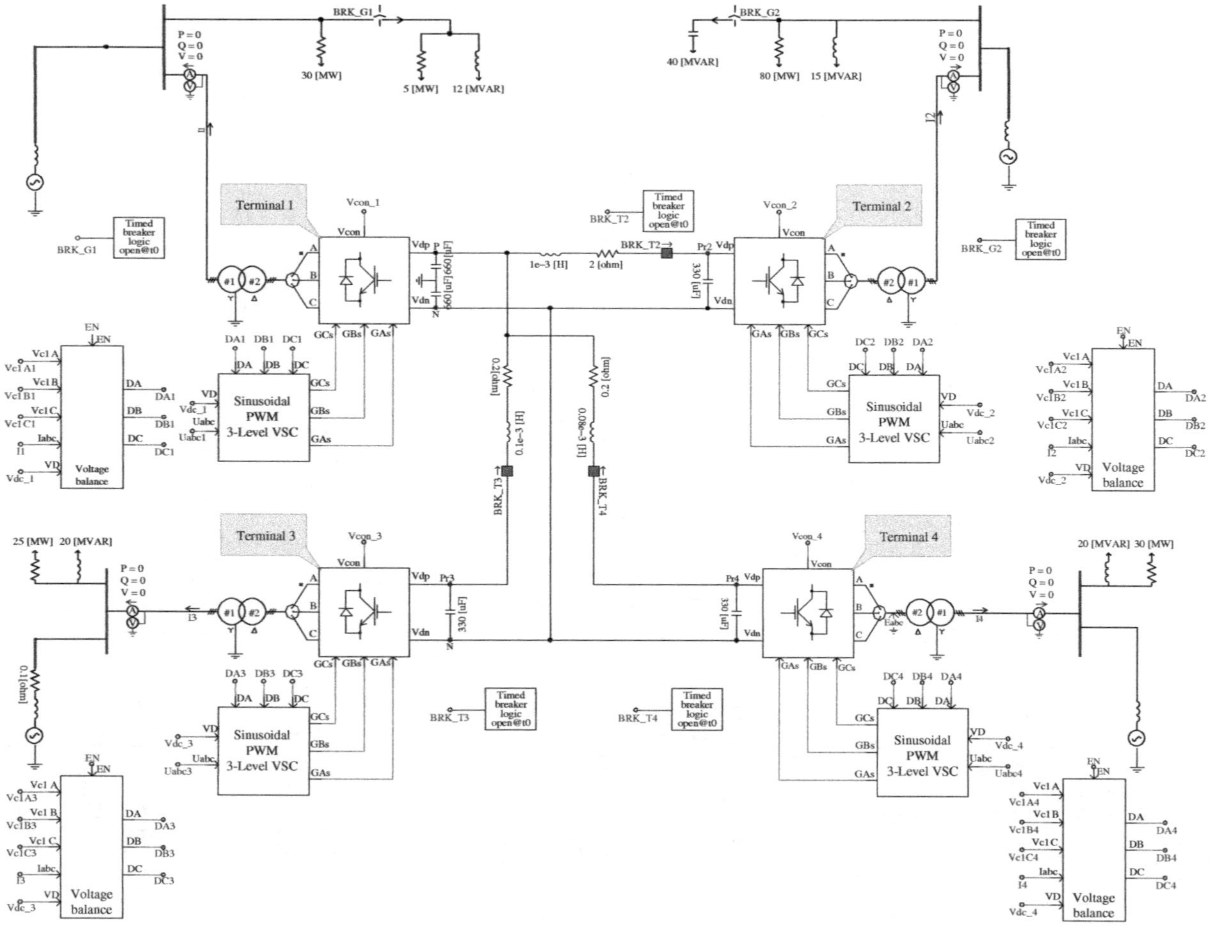

Figure 7.42 Four-terminal HVDC system implemented in PSCAD/EMTDC.

Table 7.1 Main parameters of the coupling transformers.

Coupling transformers	Winding voltages	Leakage inductance (mH)	Copper losses (mΩ)	Operation frequency (Hz)
Transformer 1	20 kV/62.5 kV (Y/Δ)	4	5	50
Transformer 2	20 kV/62.5 kV (Y/Δ)	4	5	60
Transformer 3	15 kV/62.5 kV (Y/Δ)	3	5	50
Transformer 4	24.5 kV/62.5 kV (Y/Δ)	5	8	50

Table 7.2 Electric parameters of the different grids.

Grids	Rated voltage (kV)	Grid inductance (mH)	Grid resistance	Operation frequency (Hz)
Grid 1	13.8	1.5	–	50
Grid 2	13.8	1.8	–	60
Grid 3	11	1.3	100 mΩ	50
Grid 4	20	1	–	50

The various loads connected at the PCC of each grid have the following characteristics:

- Grid 1:
  - *Load 1.* Constant active power of 30 MW. The load is permanently connected.
  - *Load 2.* Constant active power of 5 MW, constant inductive reactive power of 12 MVAR. This load is connected by means of a circuit breaker at the time instant $t = 0.25$ s.
- Grid 2:
  - *Load 1.* Constant active power of 80 MW, constant inductive reactive power of 15 MVAR. The load is permanently connected.
  - *Load 2.* Constant capacitive reactive power of 40 MVAR. The load is connected by means of a circuit breaker at the time instant $t = 1.1$ s.
- Grid 3:
  - *Load.* Constant active power of 25 MW, constant inductive reactive power of 20 MVAR. The load is permanently connected.
- Grid 4:
  - *Load.* Constant active power of 30 MW, constant inductive reactive power of 20 MVAR. The load is permanently connected.

The state-variable model of each of the terminals is the same as that described in Section 7.2 and the locations of the closed-loop poles of the inner control loop are therefore chosen to be the same as those used in the STATCOM test case, i.e. $s_1 = s_2 = -1000$ rad s^{-1}. However, the pole locations of the closed-loop system for the

Table 7.3 Parameters of the inner control loop for each terminal.

Terminal	k_{ix} Gain	k_{px} Gain
Terminal 1	$k_{ix} = 4000$	$k_{px} = 7.995$
Terminal 2	$k_{ix} = 4000$	$k_{px} = 7.995$
Terminal 3	$k_{ix} = 3000$	$k_{px} = 5.995$
Terminal 4	$k_{ix} = 5000$	$k_{px} = 9.995$

Table 7.4 Parameters of the DC voltage control loop for each terminal.

Terminal	k_{ic} Gain	k_{pc} Gain
Terminal 1	$k_{ic} = 0.66$	$k_{pc} = 0.0462$
Terminal 2	$k_{ic} = 0.33$	$k_{pc} = 0.0198$
Terminal 3	$k_{ic} = 0.33$	$k_{pc} = 0.0198$
Terminal 4	$k_{ic} = 0.33$	$k_{pc} = 0.0198$

DC voltage control loop have been placed at $s_1 = -20$ rad s^{-1} and $s_2 = -50$ rad s^{-1} for terminal 1, which is responsible for controlling the voltage at the DC bus, and $s_1 = -20$ rad s^{-1} and $s_2 = -100$ rad s^{-1} for the remaining three terminals. It should be noted that the last three converters control the voltage of their respective DC capacitors only until they have reached the reference voltage and they are then switched to the active-power control mode.

The parameters of the inner control loop and the DC voltage control loop are summarized in Table 7.3 and in Table 7.4, respectively.

The parameter of the PCC voltage control loop has the same setting as that used in the example in Section 7.2, i.e. $k_{iv} = 100$.

The three-level FC converters use the control scheme described in Section 7.3 to balance the flying capacitor voltages of the converters at the four terminals: the initial amplitude of the square waveform has been set at 12 kV and is reduced to 600 V when the flying capacitor voltages have been charged to their reference voltages, i.e. at instant $t = 0.2$ s. The modulation process of all the FC converters uses a frequency modulation index $m_f = 63$, and the switching frequency for terminals 1, 3 and 4 is therefore 3150 Hz. The switching frequency for terminal 2 is 3780 Hz. The reference value for the voltage at the DC bus is 100 kV.

The sequence of events from this experiment is as follows. All the terminals are initially disconnected from the DC bus and work in DC voltage control mode. The loads of the different grids are connected to the corresponding point of common coupling by following the load characteristics previously stated for each grid. All the terminals provide VAR support at $t = 0.25$ s, thus compensating possible voltage sags/swells caused by the connection/disconnection of loads. The DC side of the converter at terminal 3 is connected to the DC side of terminal 1 at $t = 0.4$ s, working in active-power control mode with a constant reference value of −20 MW, i.e. the order is that terminal 3 absorbs active power from grid 3. At the time instant $t = 0.5$ s, the DC side of the converter at terminal 2 is connected to the DC bus with an active-power reference value of 40 MW, which is changed from 40 to −10 MW at $t = 0.9$ s. Finally, the DC side of the converter at terminal 4 is connected to the DC bus at $t = 1.3$ s with a constant setpoint for active power equal to 40 MW. The total simulation time is 1.6 s.

Figure 7.43 shows the active powers injected by the four terminals in their respective grids – if the measured active power is negative, the terminal absorbs active power from the grid. It should be noted that all measured values equal their respective references in steady-state, i.e. zero tracking error. Furthermore, it can be seen that the four terminals maintain the power balance in steady-state owing to the control of the voltage of the DC bus. Figure 7.44 plots the voltages of the DC capacitors at each converter, showing that the converter at terminal 1 maintains the voltage at 100 kV in steady-state and the differences between this value and those measured on the DC sides of the rest of terminals are caused solely by the losses in the various DC lines. The results shown in Figures 7.43 and 7.44 demonstrate that the multi-terminal HVDC system can correctly control the active-power flow between the four different grids.

The various electrical magnitudes of the four grids are plotted in Figures 7.45–7.48. In particular, the RMS voltage at PCC, the reactive powers injected by the FC converters and a detailed view of the time responses of the line-to-neutral voltage at PCC are shown. In all cases, the RMS voltages at PCC are kept equal to 1 p.u. in steady-state owing to the reactive power compensation carried out by the FC converters. A point of interest is the situation shown in Figures 7.46a,b where it can be appreciated that the converter at terminal 2 injects/absorbs reactive power in order to cancel out the voltage sags/swells produced by the connection of different loads while maintaining the RMS value of the PCC voltage at 1 p.u. The details of the voltages at PCC have also been plotted to show how close to a sinusoidal waveform these voltages really are.

The currents injected by the four FC converters into their respective grids are shown in Figure 7.49. Similarly to the previous examples, the active power is proportional to the d-axis current and the reactive power is proportional to the q-axis current, a fact that can be appreciated by a careful comparison of Figure 7.43 with Figure 7.49, and Figures 7.45–7.48 with Figure 7.49.

The three flying capacitor voltages of all four FC converters are plotted in Figures 7.50–7.53. As shown in these figures, once the flying capacitor voltages reach their reference values, i.e. 50 kV, the control scheme of the FC voltages is able to compensate the voltage variations and therefore guarantees the voltage balance of the flying capacitors.

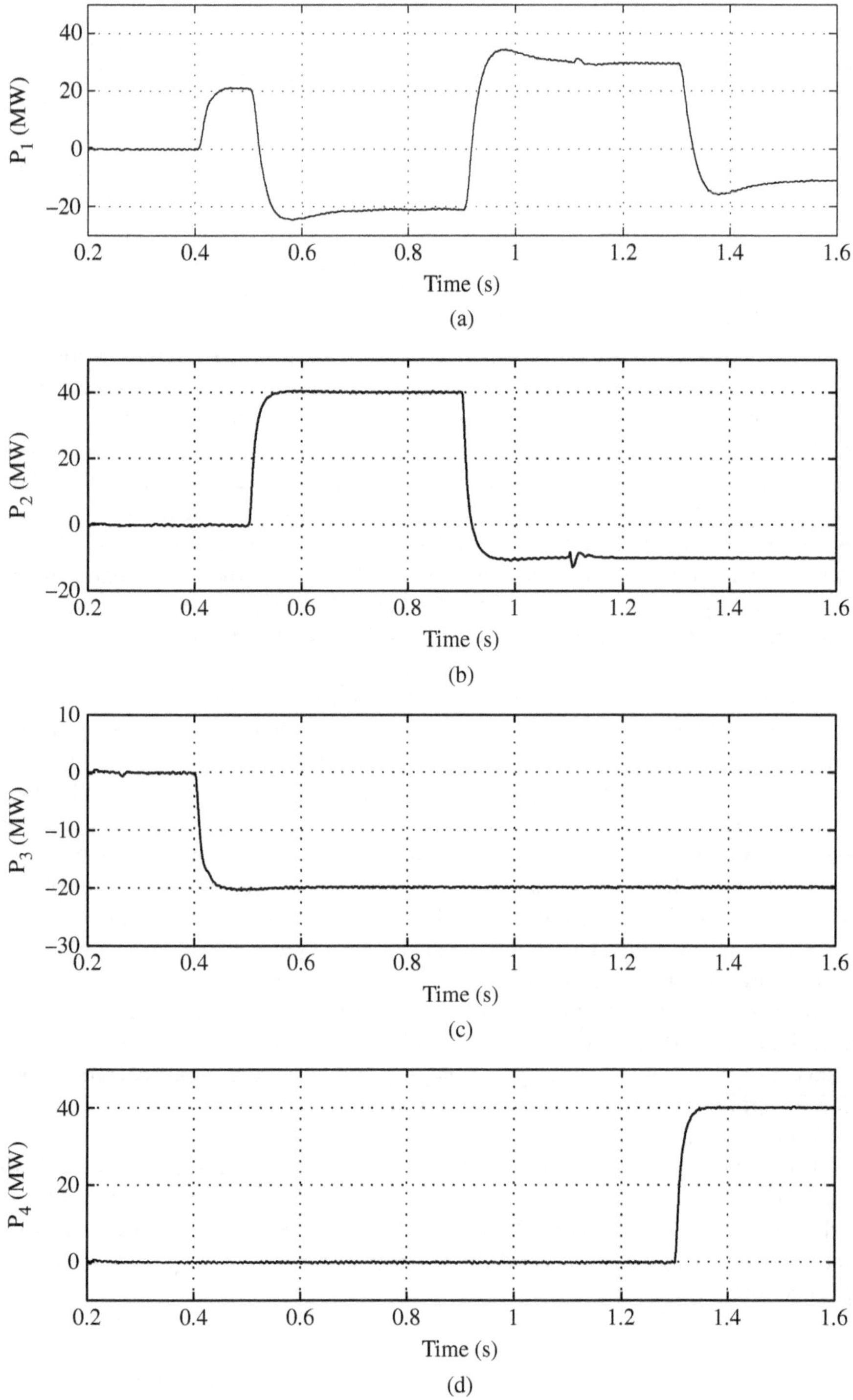

Figure 7.43 Measured active powers of the four terminals: (a) active power of terminal 1, (b) active power of terminal 2, (c) active power of terminal 3, and (d) active power of terminal 4.

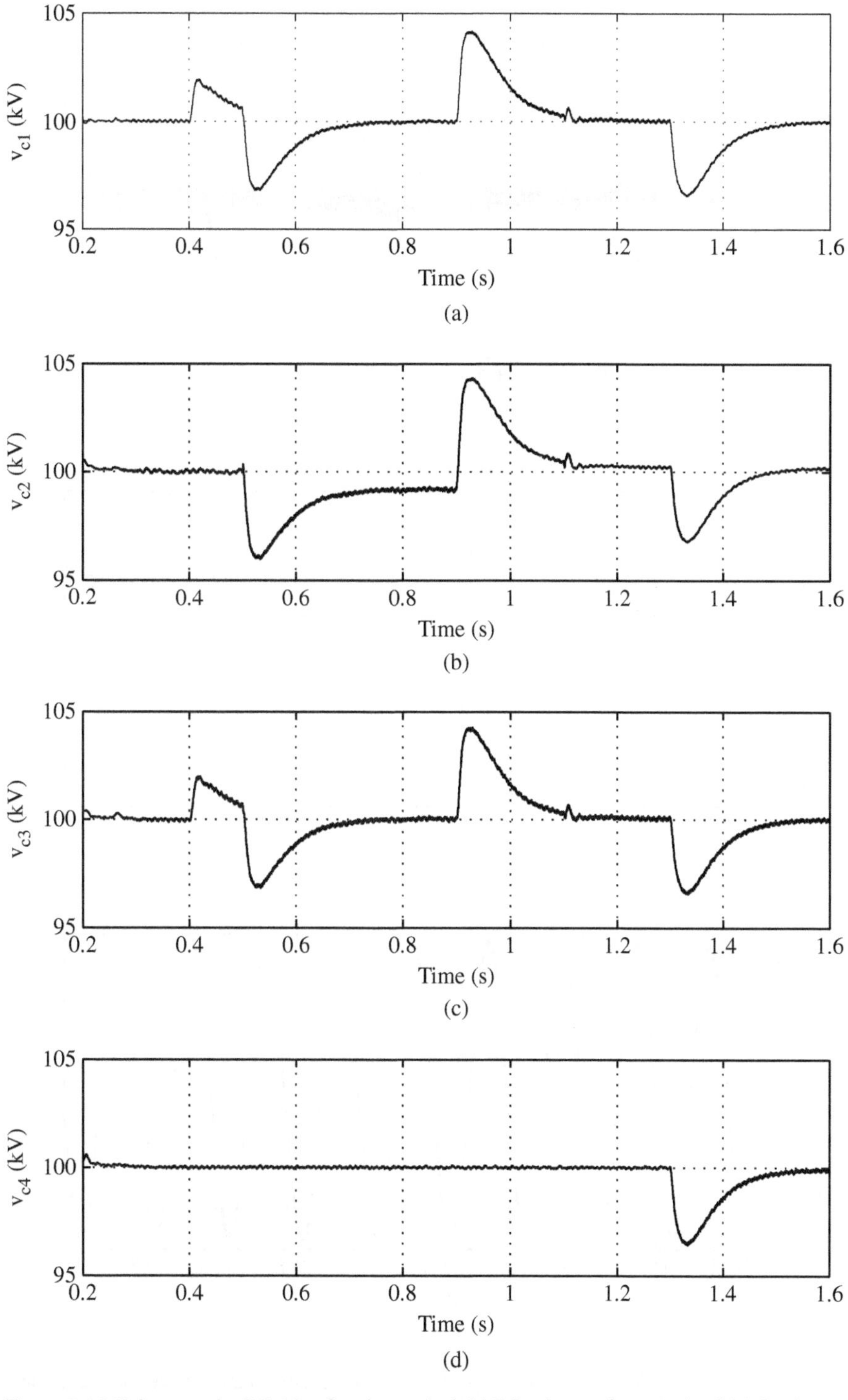

Figure 7.44 Voltage at the DC side of each terminal: (a) DC voltage of terminal 1, (b) DC voltage of terminal 2, (c) DC voltage of terminal 3, and (d) DC voltage of terminal 4.

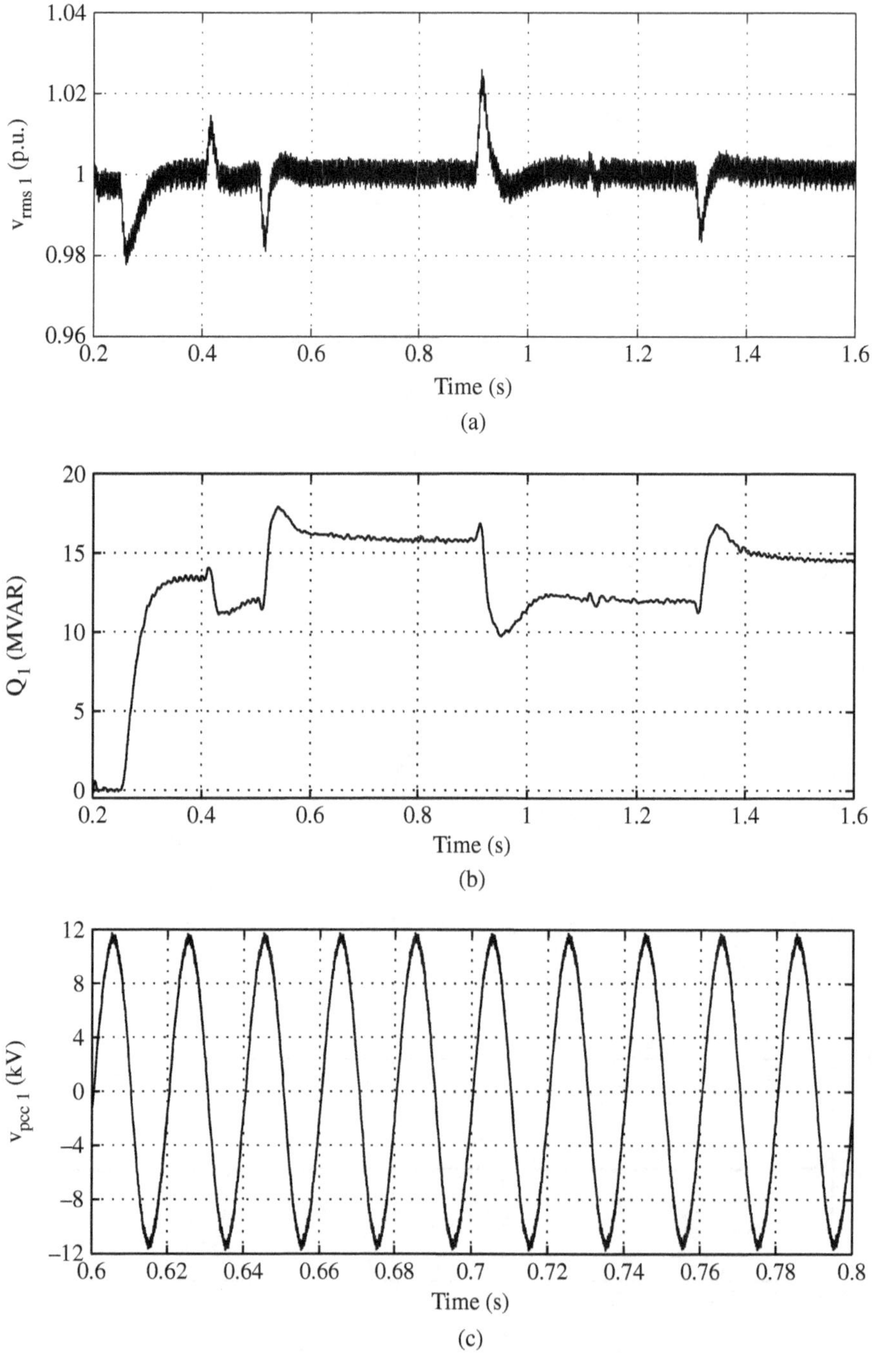

Figure 7.45 Electrical magnitudes of grid 1: (a) RMS voltage at PCC, (b) reactive power, and (c) detail of the line-to-neutral voltage at PCC.

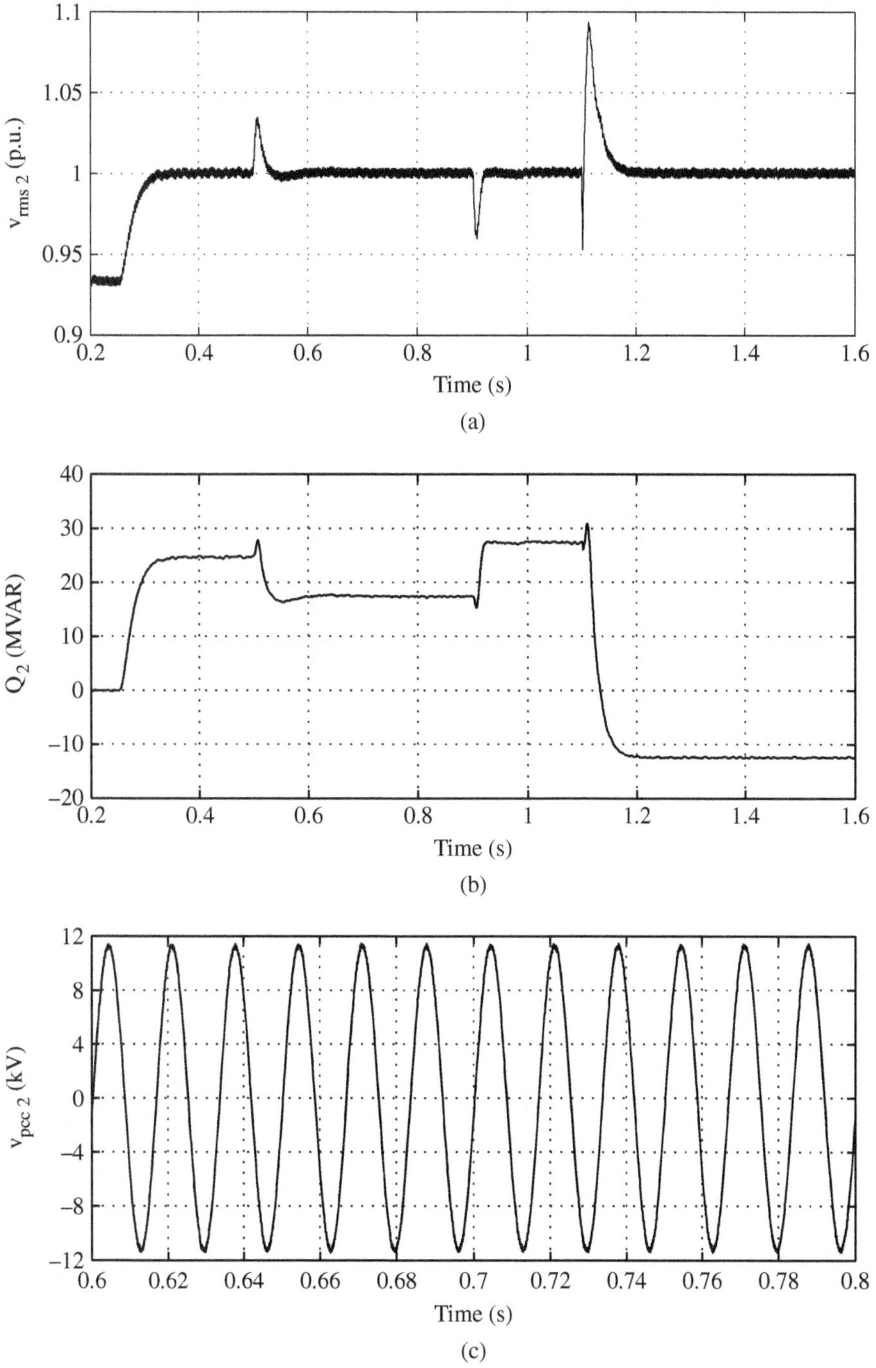

Figure 7.46 Electrical magnitudes of grid 2: (a) RMS voltage at PCC, (b) reactive power, and (c) detail of the line-to-neutral voltage at PCC.

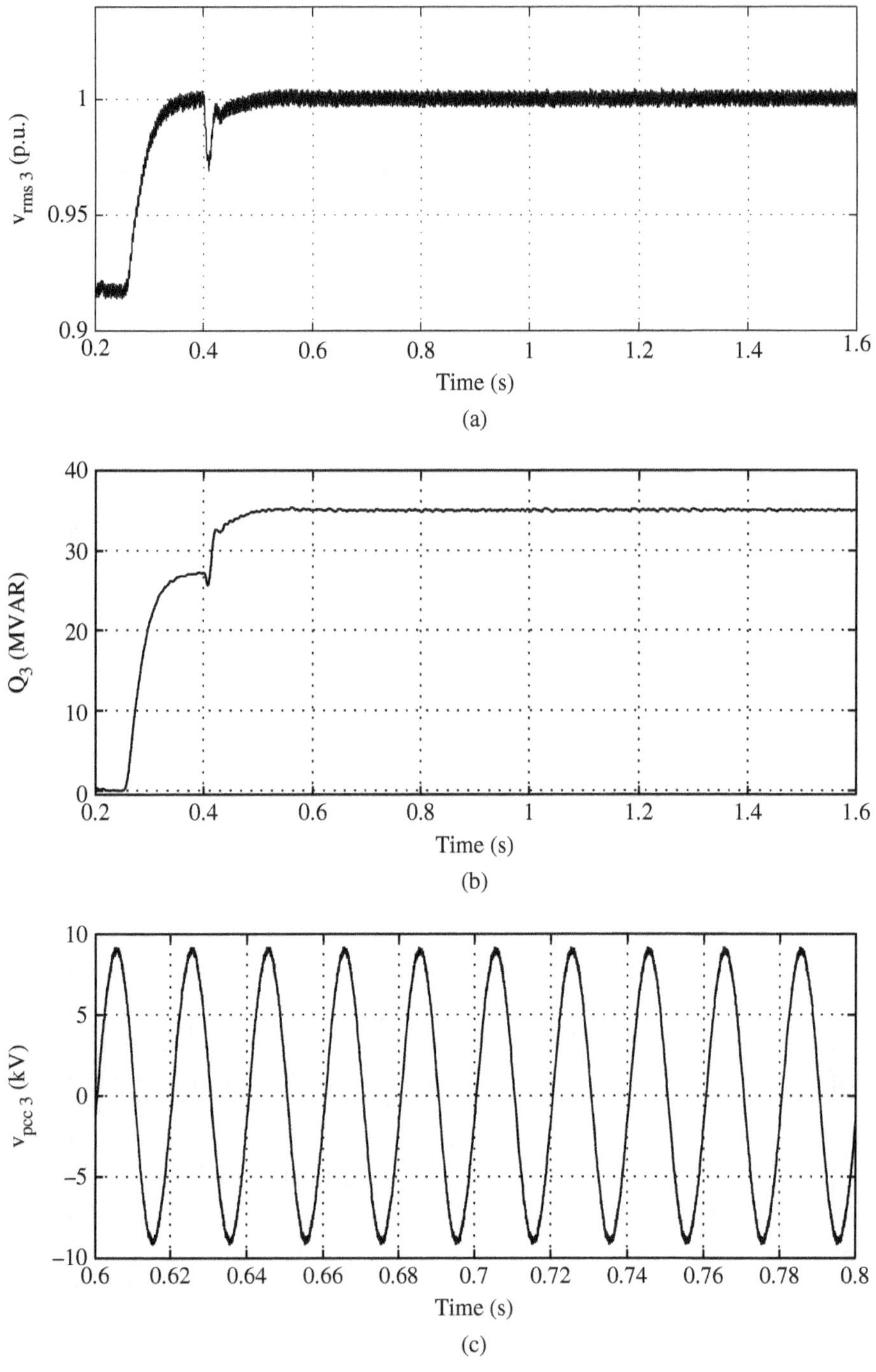

Figure 7.47 Electrical magnitudes of grid 3: (a) RMS voltage at PCC, (b) reactive power, and (c) detail of the line-to-neutral voltage at PCC.

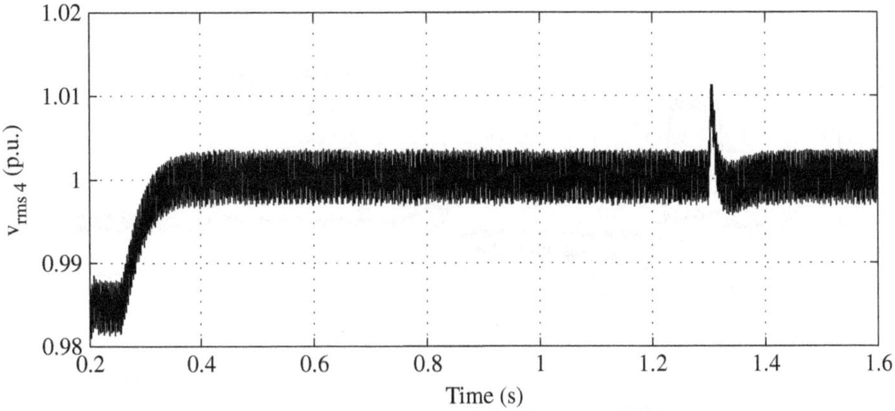

(a)

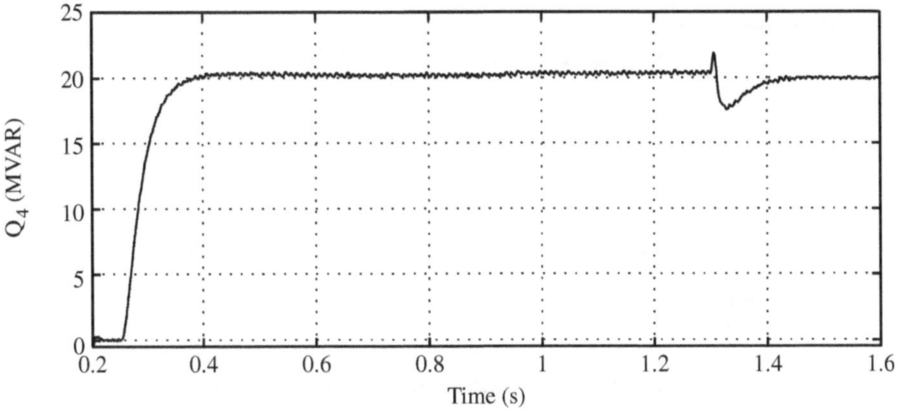

(b)

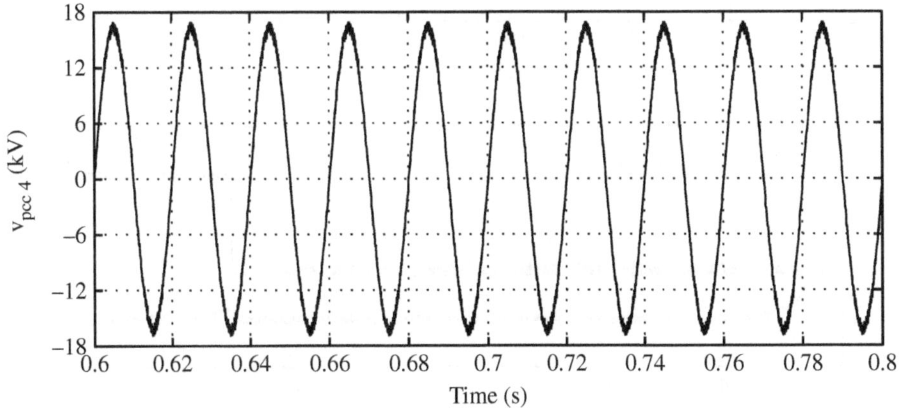

(c)

Figure 7.48 Electrical magnitudes of grid 4: (a) RMS voltage at PCC, (b) reactive power, and (c) detail of the line-to-neutral voltage at PCC.

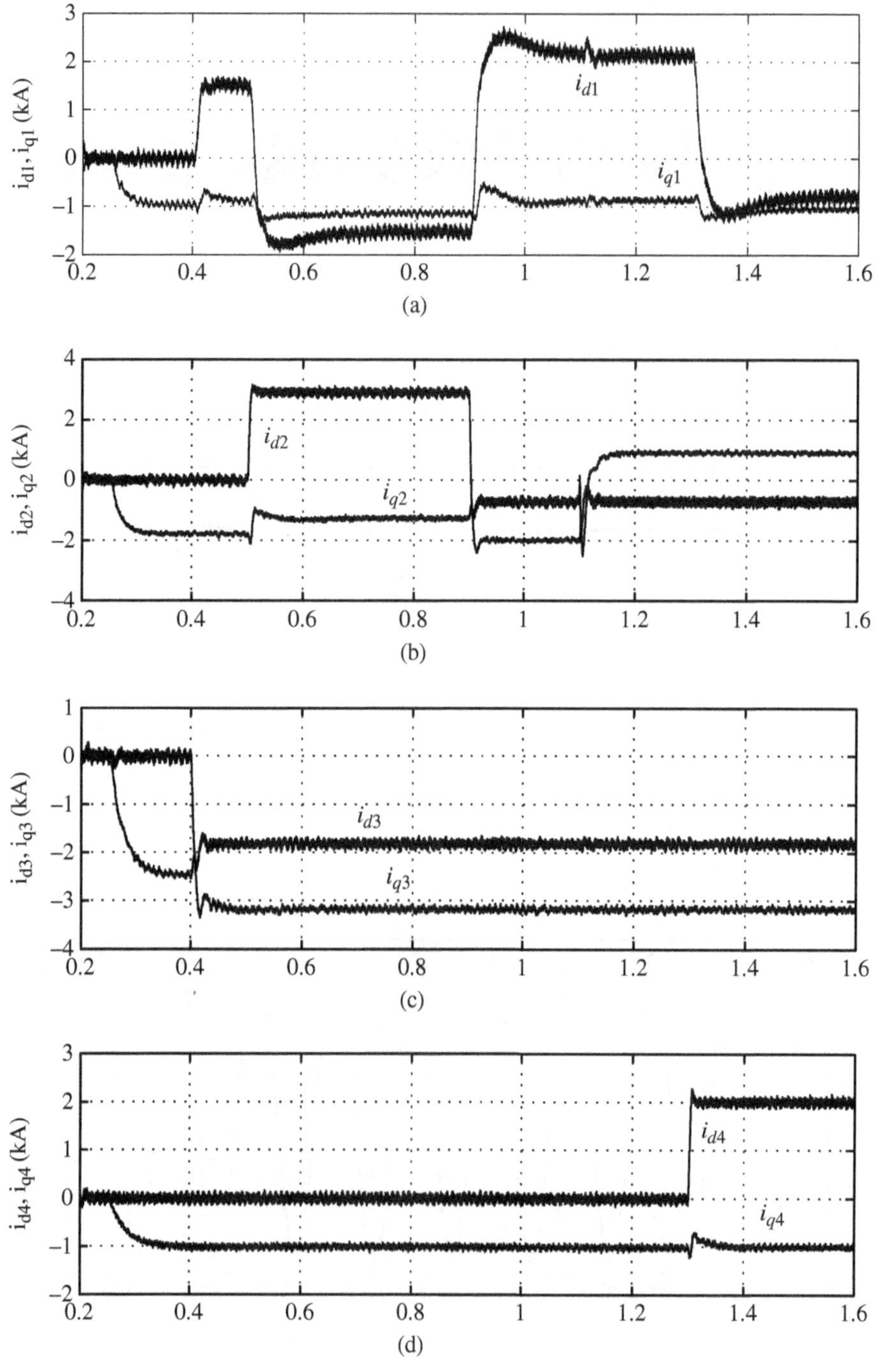

Figure 7.49 *d* and *q* components of the currents injected by: (a) terminal 1, (b) terminal 2, (c) terminal 3, and (d) terminal 4.

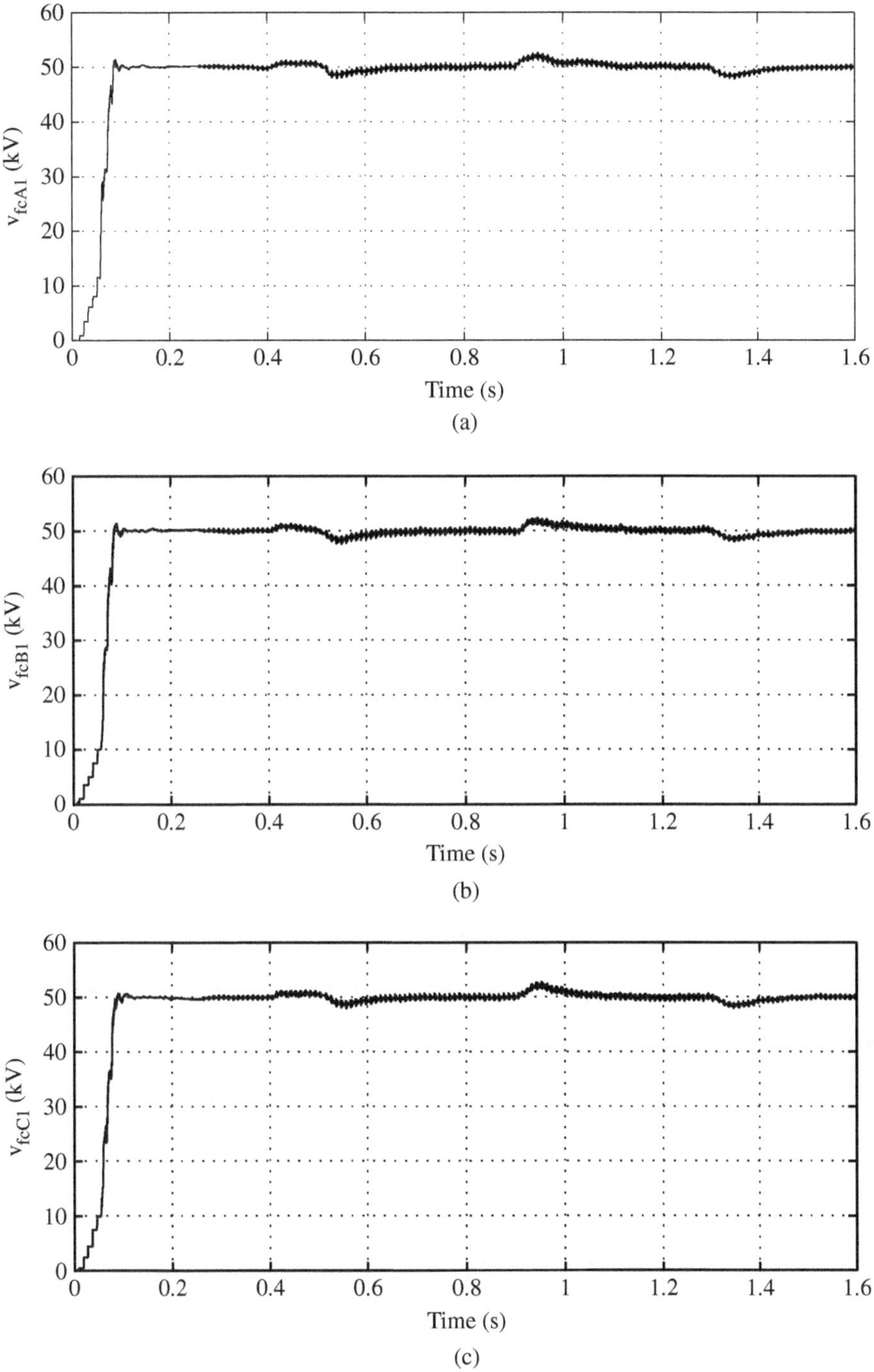

Figure 7.50 Voltages of the three flying capacitors of terminal 1: (a) phase *A*, (b) phase *B*, and (c) phase *C*.

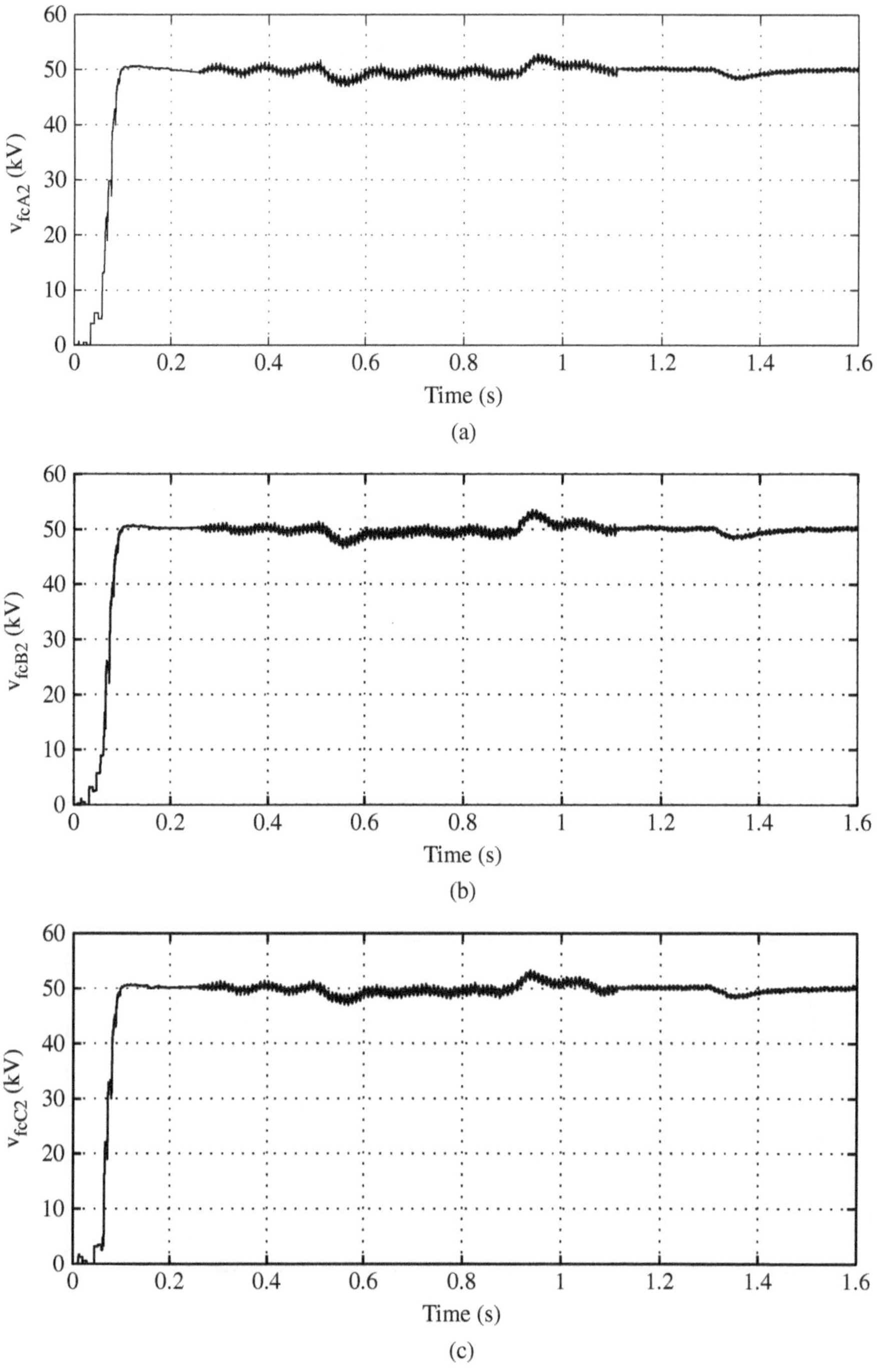

Figure 7.51 Voltages of the three flying capacitors of terminal 2: (a) phase *A*, (b) phase *B*, and (c) phase *C*.

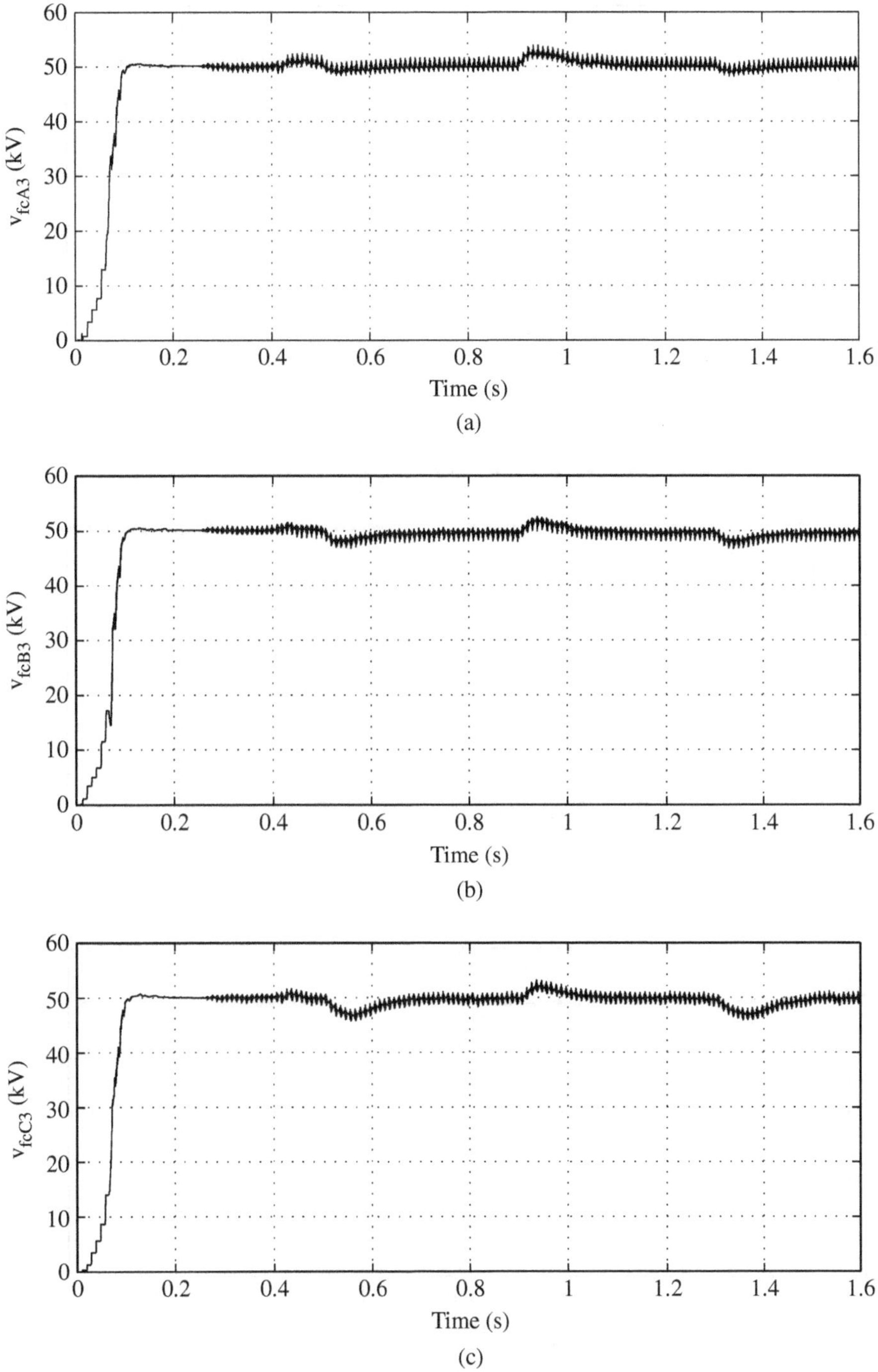

Figure 7.52 Voltages of the three flying capacitors of terminal 3: (a) phase *A*, (b) phase *B*, and (c) phase *C*.

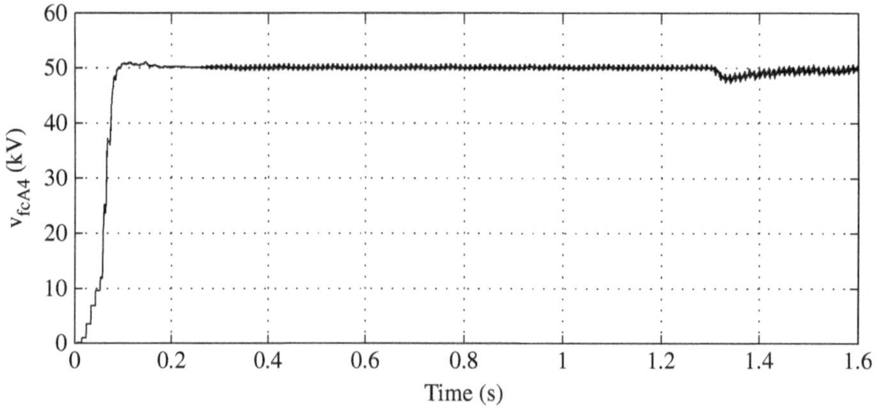

(a)

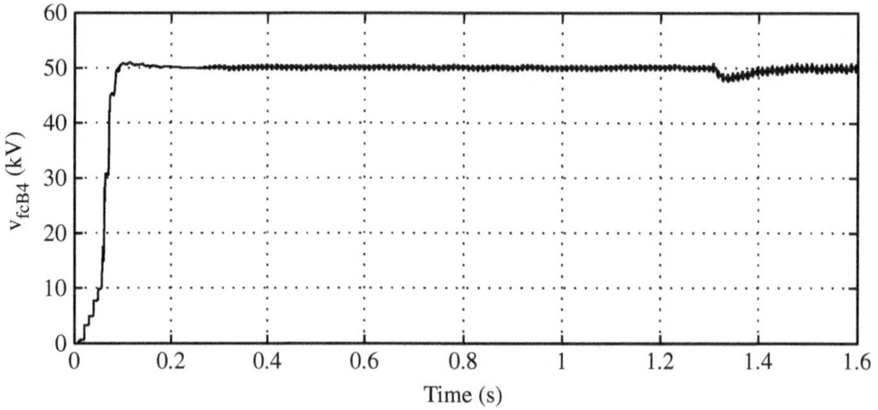

(b)

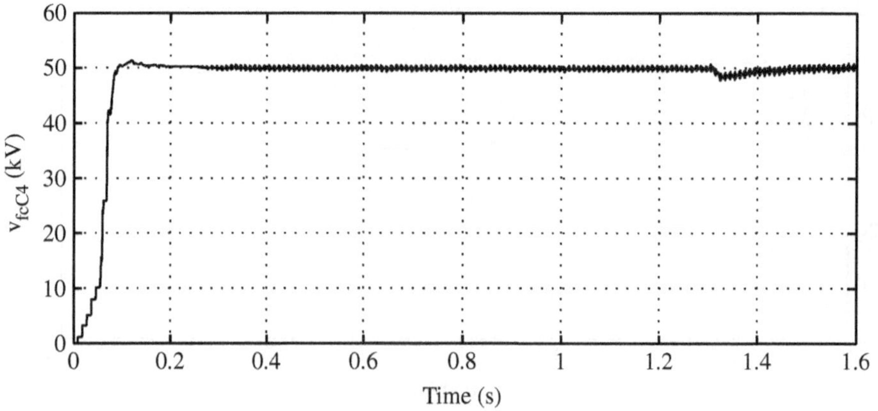

(c)

Figure 7.53 Voltages of the three flying capacitors of terminal 4: (a) phase *A*, (b) phase *B*, and (c) phase *C*.

7.6 Conclusions

This chapter has presented time-domain simulations of several examples of FACTS devices, including HVDC systems. The simulation software PSCAD/EMTDC has been chosen owing to its easy use and control implementation. The reader can of course implement these examples with other simulation software packages.

The control systems designed for the different cases are very simple and easy to implement, but they are useful to show how to obtain decoupled control of the active and reactive powers exchanged between FACTS and the grid. These control schemes have demonstrated an efficient behaviour with which to provide the necessary reactive power and therefore maintain the PCC voltage at a desired value, and they have allowed the active-power flow between several electrical grids. Moreover, these controllers are independent of the topology of the power-electronic converter, up to certain limits, and they can either be used with a conventional two-level VSC or implemented in three-level FC converters.

Multilevel voltage source converters, operated by an efficient control strategy, have proved to be very useful in reducing the switching frequency and the voltage stress of the switching devices in HVDC system applications.

Finally, the reader can add more complexity to these examples by increasing the number of levels of the multilevel FC converters, using more advanced controllers or with different grid topologies.

References

1 PSCAD (2016). *User's Guide*. Manitoba: HVDC Research Centre.

2 Mishra, M.K., Ghosh, A., and Joshi, A. (2003). Operation of a DSTATCOM in voltage control mode. *IEEE Transactions on Power Delivery* 18 (1): 258–264.

3 Krause, P.C. (1986). *Analysis of Electric Machinery*. McGraw-Hill Inc.

4 Akagi, H., Kanazawa, Y., and Nabae, A. (1984). Instantaneous reactive power compensators comprising switching devices without energy storage components. *IEEE Transactions on Industry Applications* IA–20 (3): 625–630.

5 García-González, P. and García-Cerrada, A. (2000). Control system for a PWM-based STATCOM. *IEEE Transaction on Power Delivery* 15 (4): 1252–1257.

6 Kundur, P. (1994). *Power System Stability and Control*. McGraw-Hill, Inc.

7 Mohan, N., Undeland, T.M., and Robbins, W.P. (2002). *Power Electronics: Converters, Applications, and Design*, 3e. Wiley.

8 Roncero-Sánchez, P. and Acha, E. (2014). Design of a control scheme for distribution static synchronous compensators with power-quality improvement capability. *Energies* 7 (4): 2476–2497.

9 Gupta, R., Ghosh, A., and Joshi, A. (2008). Switching characterization of cascaded multilevel-inverter-controlled systems. *IEEE Transactions on Industrial Electronics* 55 (3): 1047–1058.

10 Rodríguez, J., Lai, J.-S., and Peng, F.Z. (2002). Multilevel inverters: a survey of topologies, controls, and applications. *IEEE Transactions on Industrial Electronics* 49 (4): 724–738.

11 Arrillaga, J., Liu, Y.H., and Watson, N.R. (2007). *Flexible Power Transmission. The HVDC Options*. Chichester: Wiley.

12 Feng, C., Liang, J., and Agelidis, V.G. (2007). Modified phase-shifted pwm control for flying capacitor multilevel converters. *IEEE Transactions on Power Electronics* 22 (1): 178–185.

13 Liang, Y. and Nwankpa, C.O. (1999). A new type of STATCOM based on cascading voltage-source inverters with phase-shifted unipolar SPWM. *IEEE Transactions on Industry Applications* 35 (5): 1118–1123.

14 Xu, L. and Agelidis, V.G. (2004). Active capacitor voltage control of flying capacitor multilevel converters. *IEE Proceedings on Electric Power Applications* 151 (3): 313–320.

15 Häusler M.; Schaltanlagen C.E., "Multiterminal HVDC for high power transmission in Europe," in *Proceedings of the Central European Power Exhibition and Conference (CEPEX99)*, Poznan, Poland, 1999.

16 Haileselassie, T.M. (2008). *Control of Multi-terminal VSC-HVDC, Master of Science in Energy and Environment*. Norwegian University of Science and Technology.

17 Sood, V.K. (2004). *HVDC and FACTS Controllers. Applications of Static Converters in Power Systems*. Boston: Kluwer Academic Publishers.

Index

VSC-FACTS-HVDC: Analysis, Modelling and Simulation in Power Grids, First Edition.
Enrique Acha, Pedro Roncero-Sánchez, Antonio de la Villa Jaén, Luis M. Castro and Behzad Kazemtabrizi.

Companion website: www.wiley.com/go/acha_vsc_facts

e

f

o

p